D0152361

Biomimetic Chemistry

Biomimetic Chemistry

David Dolphin, EDITOR
University of British Columbia

Charles McKenna, EDITOR
University of Southern California

Yukito Murakami, EDITOR
Kyushu University

Iwao Tabushi, EDITOR
Kyoto University

Based on a symposium cosponsored
by the ACS Divisions of Organic
Chemistry and Inorganic Chemistry
at the ACS/CSJ Chemical Congress
(177th ACS National Meeting),
Honolulu, Hawaii, April 2–5, 1979.

ADVANCES IN CHEMISTRY SERIES **191**

AMERICAN CHEMICAL SOCIETY
WASHINGTON, D. C.   1980

574.192
B6155

Library of Congress CIP Data

Biomimetic chemistry.
  (Advances in chemistry series; 191 ISSN 0065–2393)

  Includes bibliographies and index.
  "Papers . . . presented as invited talks at the Symposium on Biomimetic Chemistry."

  1. Biological chemistry—Congresses.
  I. Dolphin, David. II. Symposium on Biomimetic Chemistry, Honolulu, 1979. III. American Chemical Society. Division of Organic Chemistry. IV. American Chemical Society. Division of Inorganic Chemistry. V. Series.

QD1.A355 no. 191 [QP501] 540s [574.19′2] 80-22864
ISBN 0–8412–0514–0 ADCSAJ 191 1–437 1980

Copyright © 1980

American Chemical Society

All Rights Reserved. The appearance of the code at the bottom of the first page of each article in this volume indicates the copyright owner's consent that reprographic copies of the article may be made for personal or internal use or for the personal or internal use of specific clients. This consent is given on the condition, however, that the copier pay the stated per copy fee through the Copyright Clearance Center, Inc. for copying beyond that permitted by Sections 107 or 108 of the U.S. Copyright Law. This consent does not extend to copying or transmission by any means—graphic or electronic—for any other purpose, such as for general distribution, for advertising or promotional purposes, for creating new collective works, for resale, or for information storage and retrieval systems.

The citation of trade names and/or names of manufacturers in this publication is not to be construed as an endorsement or as approval by ACS of the commercial products or services referenced herein; nor should the mere reference herein to any drawing, specification, chemical process, or other data be regarded as a license or as a conveyance of any right or permission, to the holder, reader, or any other person or corporation, to manufacture, reproduce, use, or sell any patented invention or copyrighted work that may in any way be related thereto.

PRINTED IN THE UNITED STATES OF AMERICA

Advances in Chemistry Series

M. Joan Comstock, *Series Editor*

Advisory Board

David L. Allara

Kenneth B. Bischoff

Donald G. Crosby

Donald D. Dollberg

Robert E. Feeney

Jack Halpern

Brian M. Harney

Robert A. Hofstader

W. Jeffrey Howe

James D. Idol, Jr.

James P. Lodge

Leon Petrakis

F. Sherwood Rowland

Alan C. Sartorelli

Raymond B. Seymour

Gunter Zweig

MAY 29 '81

UNIVERSITY LIBRARIES
CARNEGIE-MELLON UNIVERSITY
PITTSBURGH, PENNSYLVANIA 15213

FOREWORD

ADVANCES IN CHEMISTRY SERIES was founded in 1949 by the American Chemical Society as an outlet for symposia and collections of data in special areas of topical interest that could not be accommodated in the Society's journals. It provides a medium for symposia that would otherwise be fragmented, their papers distributed among several journals or not published at all. Papers are reviewed critically according to ACS editorial standards and receive the careful attention and processing characteristic of ACS publications. Volumes in the ADVANCES IN CHEMISTRY SERIES maintain the integrity of the symposia on which they are based; however, verbatim reproductions of previously published papers are not accepted. Papers may include reports of research as well as reviews since symposia may embrace both types of presentation.

CONTENTS

PREFACE

The interdisciplinary areas between the traditional inorganic and organic divisions of chemistry and those of biochemistry have seen a remarkable growth in the past few years. This increase in interest and effort has been marked by the establishment of new journals in these areas.

Just as in the past when the artificial divisions between chemistry and biochemistry became indistinct, so more recently have the boundaries between bioorganic and bioinorganic chemistry. Indeed, investigators in these fields need, of necessity, to be well versed in all of these disciplines, and it is appropriate that Breslow coined the term "biomimetic" to describe those endeavours that include all areas relating to our understanding of the processes, enzymatic and otherwise, controlled by nature.

Of major importance to biomimetic studies have been the development of model systems that mimic the natural ones. Indeed, the basis for our understanding of coenzyme functions stems primarily from such studies with model systems. We can anticipate for the future, and the reader will see in the pages of this monograph, promise of the ability to understand and parallel in vitro the specificity and catalytic acceleration of enzymes themselves.

Areas of biomimetic chemistry relating to enzyme systems that function both with and without the benefit of coenzymes are included. Special emphasis has been placed on the following subjects: vitamin B_{12} and flavins; oxygen binding and activation; bioorganic mechanisms; and nitrogen and small molecule fixation.

Biomimetic chemistry not only provides means of elucidating enzyme and coenzyme functions through manipulation of model systems, but also opens a way to the development of novel polyfunctional catalysts and materials that may or may not exist in nature. On the basis of the advancement of research described in this book, we now stand on the edge of an interdisciplinary valley so that the jump beyond the mimetic chemistry becomes possible in the future.

DAVID DOLPHIN
CHARLES McKENNA
YUKITO MURAKAMI
August 8, 1980.   IWAO TABUSHI

Adjusting the Lock and Adjusting the Key in Cyclodextrin Chemistry

An Introduction

RONALD BRESLOW

Department of Chemistry, Columbia University, New York, NY 10027

Processes in which cyclodextrins (cycloamyloses) act as reagents or catalysts toward substrates bound in the cyclodextrin cavity can be improved markedly by careful attention to the geometry. As one approach, cyclodextrin was modified with groups that cap and partly fill the cavity. The resulting new geometry of binding leads to an increased reaction rate for m-t-butylphenyl acetate. As another approach, the shape of the substrate was modified, by examining derivatives of ferrocene and of cinnamic acid, so as to bring the geometry of bound substrate close to that for reaction; large rate increases resulted. When cyclodextrin is converted to a bis-imidazole derivative, the resulting catalyst cleaves catechol cyclic phosphates. The examination of several such substrates, and of cyclodextrin monoimidazole and cyclodextrin heptaimidazole as catalysts, help show that the mechanism is similar to that of the enzyme ribonuclease.

Since this is a symposium devoted to biomimetic chemistry, it might be worthwhile to review briefly the history of this term. For a number of years we have been concerned with trying to imitate enzymes in simple organic systems. Part of this work has been concerned with trying to imitate some selective enzymatic transformations by using geometric control with reagents and templates. Norman Deno has been interested in achieving selective functionalization reactions by a different approach, using charges in the substrate to direct a functionalization reaction away from the charged region. As we were discussing this field one day, Deno remarked that he tends to think of

0-8412-0514-0/80/33-191-001$05.00/0

© 1980 American Chemical Society

this as an area which he called enzyme-mimetic chemistry, since we are trying to mimic the results of enzymatic reactions.

As I thought about this later, I concluded that while enzyme-mimetic might well capture the spirit of the field, it was not an euphonius phrase. I thus coined the word biomimetic (1) to describe all such efforts to imitate either the result or the style of biochemical reactions, which in particular means reactions catalyzed by enzymes. This word has caught on to the point where it now seems to be part of the language. It describes a range of activities—from attempts to build artificial enzymes on the one hand to synthetic reactions following biogenetic pathways on the other hand. Since biomimetic seems here to stay, I think it is desirable here to call attention to the contribution of Norman Deno in his stimulating alternative phrase.

Of course, the attempts to imitate enzymes in organic chemistry long predate the coinage of the word biomimetic. A number of systems have been examined as possible enzyme mimics; I will not review them here. Suffice it to say that one of the most attractive systems is the class of cyclodextrins, cyclic polyglucosides, with six (α-cyclodextrin, cyclohexaamylose) or seven (β-cyclodextrin, cycloheptaamylose) glucose residues in a toroidal ring. The interest in this molecule is that it has a cavity large enough to include and surround completely various normal-sized organic compounds. The earliest work on these molecules was reviewed in the excellent book by Cramer (2), and a more recent review of this field has been published by Bender (3). It is an active area, with contributions from many laboratories around the world.

The chemistry of interest when cyclodextrin or its derivatives are used as enzyme mimics involves two features. First of all, the substrate binds into the cavity of the cyclodextrin as the result of hydrophobic or lyophobic (4) forces. Then the bound substrate undergoes a reaction, which may involve the cyclodextrin as a reagent or as a catalyst. The speed of this reaction is promoted generally by the proximity induced by binding, and in addition the reactions are often selective because of geometric constraints in the transition state. This selectivity may involve the selective reaction of one potential substrate relative to another, selective production of one regiochemical isomer compared with another, or selective production of one stereoisomer relative to another. This last area, selective stereochemistry and asymmetric synthesis, is still one of the most neglected areas of cyclodextrin chemistry.

Linus Pauling pointed out many years ago (5) that enzymes are catalysts because they can bind strongly the transition states of chemical reactions. Typical accelerations by cyclodextrin or its derivatives are in the range between 10- and 300-fold; this very modest acceleration has led many people to conclude that cyclodextrins do not have the capability of producing strong acceleration in a chemical reaction or catalysis. However, it seemed to us that the best reactions had not been examined yet. That is, the problem of binding the transition state of a reaction is related to the problem of having a key that fits a lock very well. Examination of molecular models convinced us that the substrates that had been examined with cyclodextrin, the "keys", could bind to the cavity of the cyclodextrin "lock" well enough, but that this was not equally true for the transition states of the reactions being examined.

A good example of this is the classic work by Bender (6) on the reaction of *m-t*-butylphenyl acetate. This substrate binds well into the cavity, and the substrate then undergoes an acetyl transfer reaction in which a cyclodextrin hydroxyl group is acetylated. The reaction can be compared with the first step in the action of a serine esterase, or a serine protease acting on an ester substrate. However, the acceleration of this acetyl transfer, compared with simple hydrolysis by the medium, was only 250-fold.

When we examined molecular models of this system we discovered that the tetrahedral intermediate for acetyl transfer has a geometry such that the *t*-butylphenyl group is pulled somewhat out of the cavity. That is, the substrate is bound nicely but the transition state is bound more poorly, at least judged from models. This is the opposite of the situation needed to achieve maximum acceleration. Thus, we set out to improve this kind of reaction in two directions. The goal here was to discover if we could indeed produce accelerations that were of enzymatic magnitude by careful attention to geometrical relationships between the lock and the key.

It is important to consider explicitly the goal of such studies to make it clear how to proceed. Work in this field at the present time is not aimed necessarily at the acceleration of some particular reaction of current practical importance. Given a choice, everyone would prefer to achieve a useful rather than a simply interesting result, but at the current time this does not seem to me to be the most important goal. Instead, it is important to understand just how much acceleration one can achieve by careful attention to details, and to understand as well how important each of the details is. At a later stage of this field, when this understanding is available, it will become easier and more worthwhile to try to design specific catalysts for reactions of practical interest, building on the knowledge achieved from the first kind of study. Thus there is no reason to ask if anyone really wants to accelerate the rate of an acetyl transfer from *m-t*-butylphenyl acetate. The answer is that one indeed wants to do so if one can understand what factors have led to very large accelerations in such a reaction, with the full confidence that this understanding will then be translatable to other catalysts and other reactions.

Our earliest work on this was taken up by Jack Emert as part of his Ph.D. thesis (7). Emert converted the well known β-cyclodextrin heptatosylate (8) to the heptamethylamine and the heptaethylamine. These charged derivatives of cyclodextrin are of some interest in their own right, but Emert converted them to the N-formyl derivatives to neutralize charge and retain water solubility. It was expected that in water solution the methyl groups and ethyl groups would tend to cluster, producing a floor on the cyclodextrin cavity. We described this as a flexible floor (9). Tabushi has described subsequently a compound in which a rigid floor is attached to the cyclodextrin molecule (10); we will return to this later.

Although the Emert compounds were expected to put a floor on the cyclodextrin cavity, this floor should also, from models, intrude part way into the cavity and make it shallower. It was this feature that was of particular interest with respect to our ideas about lock and key chemistry. That is, one of the difficulties with the reaction of cyclodextrin itself with m-t-butylphenyl acetate is that the substrate can bind deeply into the cavity, while the transition state must have shallower binding. If we were right about the conformations of the Emert compounds, then they should bind the substrate and the transition state in a shallower cavity. In this way, they should accelerate the reaction within the complex, because it would not have to climb out of a deep potential well on going from substrate to transition state.

First, it was important to get some evidence on the existence of this floor; this proved relatively easy. It was known (6) that adamantanecarboxylic acid binds to cyclodextrin. Molecular models suggested that this binding could not involve complete inclusion of the adamantane nucleus, but instead must involve occlusion in which the adamantane penetrates only partly into the cavity. Striking confirmation of this idea was found in Emert's observation (9) that the complexing involves two adamantanecarboxylates per cyclodextrin molecule. One of them occludes onto a cyclodextrin, only partially filling the cavity, and the resulting molecule has an even greater affinity for a second adamantanecarboxylate because the resulting floor on the cavity produces a much larger hydrophobic surface. The second binding was so much stronger than the first that we were not able to get a really good estimate for the first binding constant. However, when the N-methyl- or N-ethylformamide cyclodextrins were examined, it was found that they bound only a single adamantanecarboxylate, and with a binding constant about twenty times stronger than that estimated for the one-to-one complex of adamantanecarboxylate with cyclodextrin itself. Thus, the N-methyl and N-ethyl groups occupy one face of the cyclodextrin, preventing binding of a second adamantanecarboxylate.

Furthermore, they are providing something of a floor, which leads to increased binding of the first adamantanecarboxylate.

Interesting results came from the study of the reaction rates of these flexibly capped cyclodextrins with *m-t*-butylphenyl acetate and *m*-nitrophenyl acetate (9). The *m-t*-butylphenyl acetate was bound more weakly to the N-methyl derivative by a factor of 2.3, even with a new hydrophobic floor. This was reasonable if the geometry of the system were such that the substrate could not penetrate now as deeply into the cavity as it normally does. The result is in contrast with the situation of adamantanecarboxylate, which does not penetrate completely the cavity in the first place. On the other hand, the reaction rate for acetyl transfer within the complex was increased by a factor of 9. This was also expected, since now the new, more shallow, binding geometry of the substrate was closer to the geometry required in the transition state for acetyl transfer, so there was a smaller potential well from which the substrate had to climb.

Related observations were made with *m*-nitrophenyl acetate. The binding of the substrate to the N-methyl compound was approximately as strong as it was with cyclodextrin, although with the N-ethyl compound the binding was 5-fold weaker. However, with the N-methyl compound there was also a 10-fold increase in the rate of acetyl transfer, and this was a 20-fold increase when we used the N-ethyl compound. Thus, with ethyl groups on the nitrogen the cavity is even shallower than with methyl groups. The result is somewhat weaker binding, but better acceleration since the transition state is closer in geometry to the geometry of the bound substrate. This series of modifications of the lock thus confirmed our idea that the geometry of these transition states was not ideal for simple cyclodextrin reactions. We thus took up a search for better substrates, that is, better keys for the cyclodextrin lock.

Before we could take up this study in general, we had to solve one of the more bothersome aspects of cyclodextrin chemistry. It was believed strongly that cyclodextrin would bind substrates only in pure water solution, and this was a serious defect. First of all, it severely restricted the range of substrates that could be examined, since many interesting molecules have low water solubility. As a second point, it made it difficult to examine another feature of enzyme-catalyzed reactions. One of the roles that can be ascribed to the large protein mass, which contains the functional groups of an enzyme, is the function of water exclusion. That is, enzyme reactions can be considered to be operating in a nonaqueous or only partially aqueous medium.

Chemists have long recognized that the easiest way to evaluate this situation is to examine model reactions in organic or mixed aqueous phases. We have described a model system that mimicked an enzyme in such a mixed solvent but not in water (*11*). Thus, it was important to examine the possibility that cyclodextrin chemistry could be performed in polar organic solvents or mixed solvents, to solve both of these problems. This study was undertaken by Brock Siegel, who found that a variety of hydrocarbon molecules could be bound into the β-cyclodextrin cavity even in pure dimethylsulfoxide solution (4). The binding was stronger in mixed H_2O–DMSO solutions, and in such solutions it was easy to examine a variety of organic substrates that were insoluble in pure water. One of the most interesting substrates he examined was ferrocene, a material with a cylindrical geometry that is almost ideal to fit the β-cyclodextrin cavity. As we will see later, this work laid the foundation for an investigation of substrates based on the ferrocene nucleus.

The other point that was discovered was that some reaction rates were accelerated by operating in a mixed solvent rather than in pure water. The one that was examined most carefully was the acetyl transfer from bound m-t-butylphenyl acetate to β-cyclodextrin with buffers that in water give a pH of 9.5. It was observed that the reaction was almost 50-fold accelerated in a 60% DMSO–H_2O solvent compared with the reaction rate in pure water. Part of this acceleration came from an increase in the apparent basicity of the medium, since relative pK's are solvent dependent; part of it was also a solvent effect on the reaction rate of the cyclodextrin anion with the substrate. Thus, in 60% DMSO–H_2O the β-cyclodextrin reaction with this substrate was 13,000-fold faster than was the rate of hydrolysis of the substrate in an aqueous buffer of the same composition. Of this approximately 50-fold acceleration over cyclodextrin in water, about 10-fold was caused by changes in the pK's in the system and about 5-fold was caused by a change in the reaction rate of the cyclodextrin.

Molecular model building convinced us that an entirely new kind of substrate could be much better, in the sense that its transition state could retain the full binding of the starting material. This substrate was designed by constructing a model of the tetrahedral intermediate for an acyl transfer to cyclodextrin, and then considering what shape of substrate might be attached to the acyl group so that it could still fit the cyclodextrin cavity. The surprising answer was that a derivative of ferrocene looked particularly attractive. That is, a chain attached to the ferrocene nucleus comes out in a line directly perpendicular to the axis of the cyclodextrin cavity; if this chain carried a carbonyl group at the end, then the tetrahedral intermediate would fit without requiring any upward movement of the ferrocene nucleus out of the cavity.

One structure that seemed attractive was ferrocinnamic acid, a ferrocene derivative carrying an acrylic acid chain. The substrate examined was the p-nitrophenyl ester of this acid (I) (12). It was clear from the known binding constants of ferrocene (4) and of p-nitrophenyl acetate that the ferrocene nucleus should be the one bound in the cyclodextrin cavity. Our kinetic studies (12) demonstrated that, indeed, this molecule forms only a one-to-one complex with cyclodextrin, as revealed by the Eadie plot for the reaction of this substrate with cyclodextrin. One must be careful about such plots, since they could also be linear if a two-to-one complex were formed but only one of the cyclodextrins were dissociating over the concentration region examined. In our case, we could exclude this possibility, since undissociated binding of a cyclodextrin to the substrate throughout the concentration range examined would have resulted in a different concentration of free cyclodextrin, because of the binding of one molecule, and this difference would have led to a curvature in our plot well outside the experimental error.

The reason for interest in the details is that the acceleration with **I** was very large compared with that which had been observed previously with cyclodextrin reactions. In fact, in a system with 60% DMSO–40% H_2O and hydroxide supplied by a buffer that in H_2O would have a pH of 6.8, the V_{max} was 0.18 sec^{-1}. This represented an acceleration 750,000-fold compared with hydrolysis by the hydroxide ion alone in this same buffered medium in the absence of cyclodextrin. The product of this reaction was the ferrocinnamate ester of cyclodextrin, which then hydrolyzed slowly in a second step to the salt of the free acid.

The reaction was first order in hydroxide in this neutral region, indicating that it did involve the attack of a cyclodextrin anion on the bound ester, and the rate of the process near neutrality was comparable with the rate of reaction of the enzyme chymotrypsin with *p*-nitrophenyl acetate. This was especially remarkable since our substrate is approximately 20 times more slowly hydrolyzed than is *p*-nitrophenyl acetate by aqueous buffer alone. The overall reaction rate for acylation of this cyclodextrin in our mixed DMSO–H_2O medium is 18 million times as fast as is the hydrolysis of the substrate under the same conditions in a simple aqueous buffer. Thus, combining the well-defined geometry of our complex with the medium effect that an enzyme is expected to contribute, we have achieved an acceleration that is good enough to bring this into the rate range for chymotrypsin with a similar substrate and that is close to that needed to mimic other enzymes with optimal substrates. The actual target acceleration is about 100 times larger to achieve a kinetic mimic of the best enzyme cases.

M. F. Czarniecki (13) also has examined the ferrocene ester (**II**) carrying a propiolic acid sidechain. This also fits very well, since the somewhat shorter distance involved with an acetylene makes up for the zig-zag arrangement of an olefin. He finds that the acceleration of reaction of this acetylenic *p*-nitrophenyl ester on reaction with cyclodextrin under our conditions is 100,000-fold, close to that for the

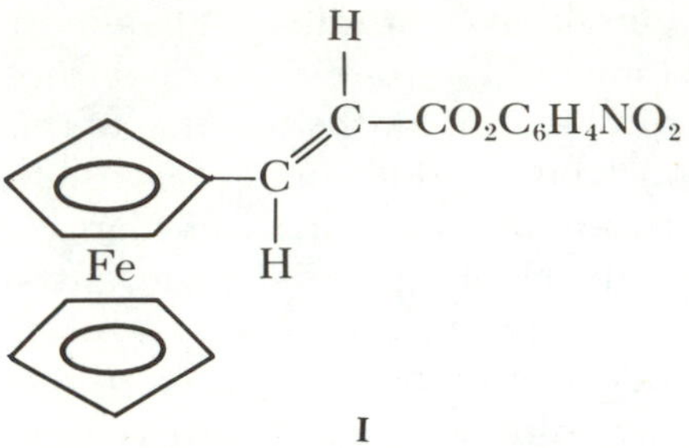

I

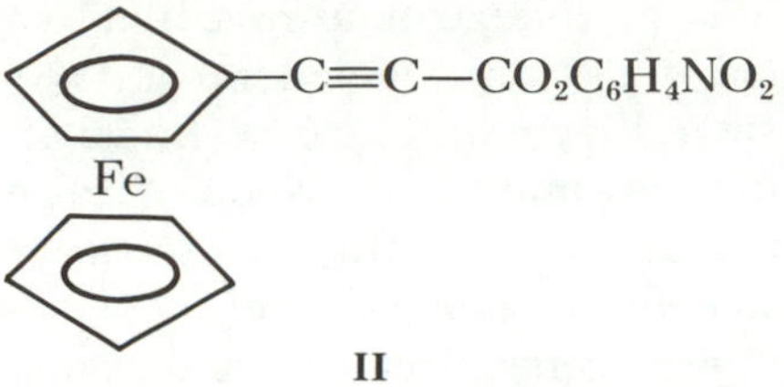

II

olefinic compound. This finding is consistent with our models if the major factor is the geometry of binding, and it helps to exclude various other effects that might have contributed to the previous rate, such as deconjugation of the olefin in the ferrocinnamic acid derivative.

Czarniecki also has examined another cinnamic acid derivative (12), in an attempt to explore the relevance of our ideas about binding geometry. Since the t-butylphenyl group binds to β-cyclodextrin in these mixed solvents about as well as a ferrocene nucleus does (4), Czarniecki examined substrate **III**, the p-nitrophenyl ester of cinnamic acid carrying a m-t-butyl group. The methoxyl group is present on the substrate for synthetic simplicity. This substrate binds nicely to β-cyclodextrin in our medium, and again an Eadie plot shows that a one-to-one complex is formed. However, the acceleration in this case is only 1200-fold compared with hydrolysis of the substrate by the same buffer, so this is approximately 600 times less effective as a sub-

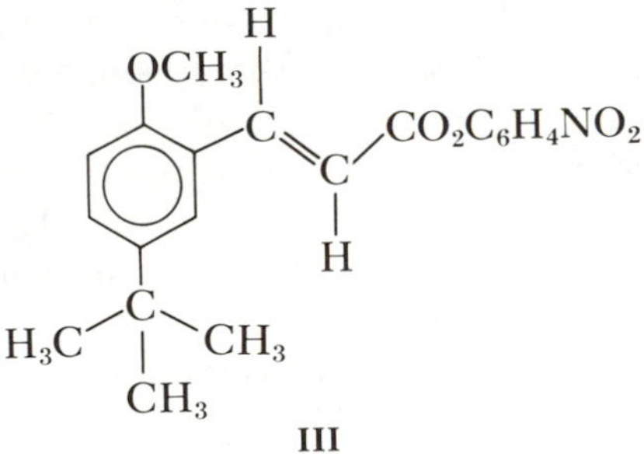

III

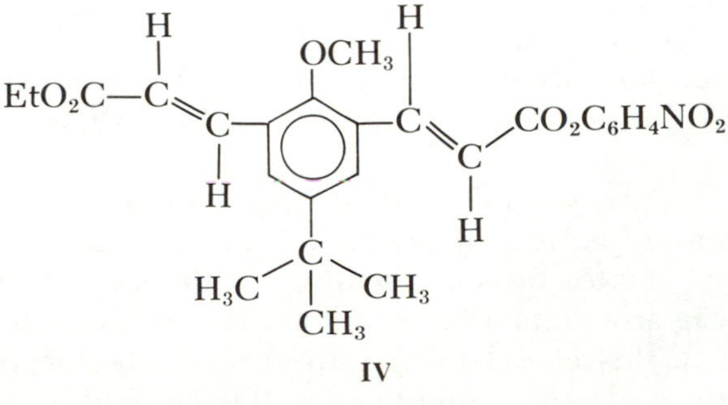

IV

strate then was the ferrocinnamate case. Molecular models show that in this system the *m-t*-butyl phenyl group must be pulled up somewhat from the cavity when the substrate goes to the transition state. Thus, it seems clear that this is a fatal flaw that must be avoided in the design of effective substrates.

Interestingly, this problem could be overcome somewhat by attaching an additional small projection to the phenyl ring to catch on the cyclodextrin lip and prevent the substrate from binding too deeply into the cavity. That is, Czarniecki prepared substrate **IV**, in which an additional small chain is placed so as to prevent deep binding of the *t*-butylphenyl ring. This substrate also undergoes an acyl transfer reaction with β-cyclodextrin in our medium, with one-to-one stoichiometry, but in **IV** the binding is 1.5-fold weaker than that for substrate **III** without such a projection. More striking, the rate of acylation with this substrate is now 4900-fold compared with its rate of hydrolysis, so it is indeed a better substrate for the catalyst.

Thus, this mode of adjusting the binding geometry between substrate and catalyst also works, and obviously is related in spirit to the adjustment we made by using the Emert flexibly capped systems. The better approach, however, is to produce a substrate that can take full advantage of the binding cavity of the cyclodextrins but permit the binding to be retained in the transition state for the reaction, or preferably even increased. Only time will tell if this leads to even greater accelerations in rates, up to the 100,000,000-fold or so accelerations that we want to imitate for the best enzymes.

Cyclodextrin derivatives can act as catalysts, not just as reagents. We are focussing on an attempt to develop a mimic for the enzyme ribonuclease A that incorporates the functional groups of the enzyme, binds an appropriate substrate, and then catalyzes the hydrolysis of such a substrate by a mechanism used by the enzyme itself. Although we want to imitate the mechanism, the selectivity, and the rate of the enzyme, our systems do quite well only with the first two points. They are still quite slow compared with the real enzyme.

The starting point for these systems was a compound (**V**) reported by Tabushi et al. (*10*) in which β-cyclodextrin had reacted with a rigid disulfonyl chloride. We have examined this substance in some detail. Although by many criteria it appears to be pure, high-pressure liquid chromatography revealed that it was a mixture of two closely related compounds. Since molecular models indicated that the reagent used by Tabushi et al. should be able to reach between glucose residues A and C and also between glucose residues A and D, we assigned (*14*) the structures of the 6A,6C, and 6A,6D isomers to this mixture. Some of our earliest work (*14*) was done with this substance (**V**), which we called the *C*-capped compound. More recently, we have used the disulfonyl chloride derived from diphenyl ether. From what is known of the geometry of diphenyl ethers, this reagent should be somewhat longer than that with a carbon between the phenyls. In line with this, we find that this reagent caps β-cyclodextrin to give a material (VI), which by HPLC criteria seems to be a single isomer. We assign it the 6A,6D structure, although more structural work will be needed before these assignments can be considered to be final. We call VI the O-capped β-cyclodextrin.

Either the C-capped or the O-capped material can be heated with an excess of imidazole to produce β-cyclodextrin bisimidazole (VII). On the basis of the above discussion, we believe that the material produced from the C-capped compound is a 6A,6C and 6A,6D isomeric mixture, while that produced from the O-capped compound is largely the 6A,6D isomer. As it turns out, we have not detected yet significant differences between these two materials as catalysts in our kinetic and product studies.

With the existence of this new cyclodextrin lock, it was again important to select a key to fit it and to serve as substrate. For this we wanted a cyclic phosphate ester that this cyclodextrin bisimidazole could hydrolyze. The enzyme ribonuclease hydrolyzes cyclic phosphates as the second step in the hydrolysis of RNA, and cyclic phosphates are used as assay substrates for the enzyme. The advantage to us of such a substrate was that the geometry of the transition state would be relatively well-defined, so that it should be possible to design congruence between the catalyst and the transition state. Molecular model building indicated that a possible substrate was the cyclic phosphate derived from 4-t-butylcatechol (VIII). Indeed, the cyclodextrin bisimidazole (VII) is a catalyst for the cleavage of cyclic phosphate (VIII) (14).

We found (14) that the pH–rate profile for the cleavage of (VIII) by (VII) shows a bell-shaped curve, which can be duplicated by the theoretical curve if catalysis involves two groups, one protonated and one

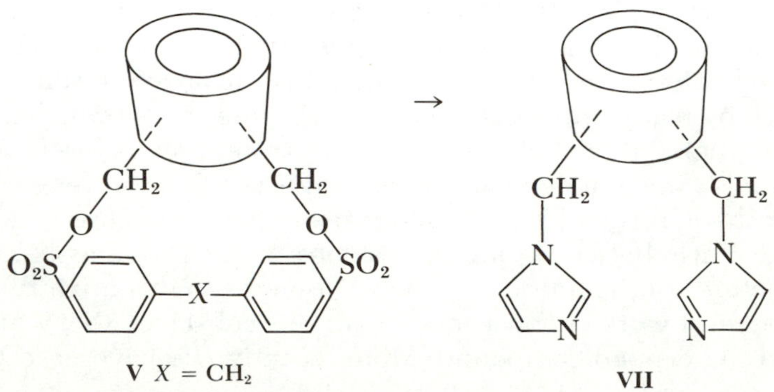

V X = CH₂
VI X = O

VII

unprotonated, each with a pK_a of 6.2. The kinetic pK's are not displaced detectably from the titration pK's of the imidazoles. By contrast, catalysis by β-cyclodextrin 6-monoimidazole showed weak catalysis by the protonated catalyst and somewhat stronger catalysis by the unprotonated imidazole, but no such bell-shaped curve. On this basis, one can conclude that this catalyst (**VII**) may be attacking by the same mechanism usually invoked for the enzyme ribonuclease, in which an imidazole ring acts as a general base to deliver a water to the phosphate while the imidazolium ring acts as a general acid to protonate the leaving group. Of course, as with all such kinetic evidence, other kinetically equivalent mechanisms also must be considered until they are excluded.

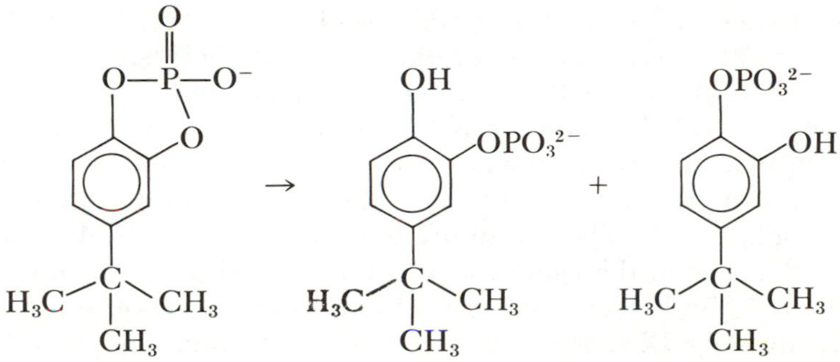

Perhaps the most striking observation was that the catalyzed cleavage of cyclic phosphate (**VIII**) was selective (*14*). That is, the product from the reaction was, within experimental error, only **IX**, not **X**. This contrasts with the simple hydrolysis of **VIII** by base, in which a mixture of **IX** and **X** is produced with **X** actually being the dominant isomer. Molecular model building accounts for this result. The transition state for cleavage of the phosphate ester involves a 180° relationship between the attacking water and the departing oxygen in a trigonal bipyramid at phosphorus. (This assumes that pseudorotation is not involved; it has been shown not to be involved with the enzyme ribonuclease.) For the attacking water to be delivered by an imidazole from the catalyst, it is necessary that the path of approach be more or less perpendicular to the axis of the cyclodextrin cavity; this is only possible if oxygen-1, which lies on that axis, is the leaving group. A transition state in which oxygen-2 was the leaving group would require attack by water from a line that is further out, and it would make that water out of reach of the imidazole that is to deliver it. Thus, this system is selective, and opens the cyclic phosphate only in the direction expected from our molecular models of the transition state. However, the system certainly is not optimized yet: the rate acceleration for this hydrolysis compared with simple hydrolysis with the same buffer is only 17-fold. An improvement of this acceleration will require a better definition of the geometry in the transition state, and this is currently under study. The catalytic groups will have to be immobilized so that they are held correctly to bind the atoms in the transition state, rather than held so freely that the catalytic conformation is only one of many that are available.

The geometric selectivity is of interest with respect to the chemistry catalyzed by cyclodextrin monoimidazole, a poorer catalyst but still a real one. On the basic side, in which the free imidazole is the catalyst, the product from the catalyzed reaction is again **IX**, as was expected from these arguments. The geometric control invoked involved the imidazole acting as base to the attacking water molecule, while models suggest that the substrate should be able to adjust itself to permit either oxygen to be protonated by the imidazolium ion (assuming that this is actually the catalytic function of this acidic group). However, on this basis one might have expected that this geometric selectivity would be lost when the catalyst is a cyclodextrin imidazolium ion, that is, when the cyclodextrin monoimidazole is examined at acidic pH's. The extent of the catalysis here is quite small, so the products from the catalyzed reaction are mixed with those from simple hydrolysis. However, again the catalytic reaction seems to be producing only **IX** along with the random products produced by a concurrent uncatalyzed hydrolysis. This suggests that for cyclodextrin monoimidazole, the mechanism on the acidic side is not simply a general-acid reaction. Instead, it involves the kinetically equivalent use of the unprotonated imidazole as a general base with the proton serving another function, protonating either the leaving group oxygen or one of the negatively charged phosphate anion oxygens. Again, the geometric prediction that general-base delivery of a water should occur along a line perpendicular to the axis of the cavity would explain the product distribution.

Another kinetically equivalent possibility must be considered: the free imidazole in these catalysts may be acting as a nucleophile at phosphorus, not as a base. This is not yet excluded experimentally, but it seems unlikely. No intermediate phosphoimidazole is detectable, but of course it might be hydrolyzing rapidly. However, molecular models for such a reaction of **VII** with **VIII** are very strained, and essentially impossible with the naphthalene substrate to be described below. Only with the extra water molecule of the general-base mechanism do the models fit well.

One might wonder if the catalytic groups of **VII** are located on the wrong face of the cyclodextrin reagent. That is, there has been a suggestion (15) that substrates tend to bind so as to prefer interactions with catalysts on the secondary face of cyclodextrin, although our investigations (16) with specifically placed phosphate catalytic groups on either of these faces do not support this. However, we also have examined this question here by mounting an imidazole ring on the secondary face of cyclodextrin to evaluate its catalytic abilities compared with cyclodextrin monoimidazole bound to the primary face.

Iwakura et al. have reported (15) that the secondary hydroxyls of cyclodextrin can be sulfonated selectively by carrying out reactions in aqueous solution so as to take advantage of the binding of toluenesulfonyl chloride into the cavity, but we cannot confirm this for at least β-cyclodextrin. Reaction of β-cyclodextrin with toluenesulfonyl chloride under the conditions described (14) leads to a very bad mixture of products, in which one of the components is certainly the simple primary tosylate. Perhaps this situation is different for α-cyclodextrin. Accordingly, we have adopted a more mundane approach.

We find that β-cyclodextrin reacts smoothly with t-butyldimethylsilylchloride to afford the heptasilyl derivative in which all of the primary hydroxyls are blocked. This then can be sulfonated on the secondary face; for ease of separation, our preferred reagent is triisopropylbenzene sulfonyl chloride. The resulting monotrypsilated cyclodextrin is, from NMR evidence, sulfonated on hydroxyl-2. This is not easily displaced, but on treatment with base it is converted to the 2,3-epoxide, and this is relatively reactive towards nucleophiles.

Using this system, Nakasuji (17) has prepared successfully two different cyclodextrin derivatives carrying imidazole rings on the secondary face. The t-butyldimethylsilyl groups were removed at the end of the sequence by the normal procedure using fluoride anion. These compounds were inferior to the primary cyclodextrin monoimidazole as catalysts with **VIII**. Furthermore, they have so far shown diminished catalytic ability toward ferrocene esters compared with simple β-cyclodextrin itself. Thus, at the present time there is no strong argument for the idea that our ribonuclease models would be even better if the groups were on the secondary carbons.

One possible improvement in the cyclodextrin bisimidazole catalyst would be to add further catalytic groups. That is, the enzyme ribonuclease uses not only two imidazole rings, but also a lysine cation that probably assists in the binding of the phosphate anion and in the activation of that phosphate by hydrogen bonding to it. All of this might be available from further protonated imidazole groups. Consequently, we have taken cyclodextrin heptatosylate and have displaced it with an excess of imidazole to produce cyclodextrin heptaimidazole, in which all of the rings are located on the primary face. At pH 6.2 this compound is actually a twofold better catalyst for the cleavage of t-butylcatechol cyclic phosphate (**VIII**). The pH–rate profile for the system shows that the catalyst maintains activity to much lower pH's, as several protonated imidazoles can be present in the transition state

UNIVERSITY LIBRARIES
CARNEGIE-MELLON UNIVERSITY
PITTSBURGH, PENNSYLVANIA 15213

and the residual unprotonated one can still serve the basic catalytic function. Thus, there is a slight improvement in the catalytic abilities of the system with these extra groups, but obviously much more than this will be required to bring this whole system up to the catalytic rate of the enzyme itself.

We have done a little lock and key chemistry with cyclodextrin imidazoles. For instance, Bovy (18) has prepared the cyclic phosphate derived from naphthalenediol (XI) and from a tetralindiol (XII). Both of these are hydrolyzed by our 6A,6D cyclodextrin bisimidazole (VII) but these substrates are hydrolyzed less effectively than is the t-butylcatechol cyclic phosphate (VIII). In XI and XII, the phosphorus atom will lie on the axis of the cavity, rather than displaced to one side as with t-butylcatechol, and in particular the attacking water molecules must come from a direction further out than the perpendicular line to the cavity axis. Thus, we would expect them to be less well delivered by imidazole. We are now examining other substrates that might be better keys for this particular cyclodextrin lock.

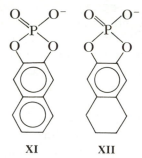

XI   XII

Finally, we might mention briefly some of our experiments directly relating the cyclodextrin bisimidazole (VII) or heptaimidazole to the enzyme ribonuclease. First of all, we ask if our catalysts also can cleave some of the normal enzyme substrates. The answer so far is a little ambiguous. Thus, cyclodextrin heptaimidazole gives a small acceleration in the rate of hydrolysis of the cyclic 2,3-phosphate of adenosine and also of the cyclic 2,3-phosphate of cytidine, but the effect is so small that it may be simply the result of a polyimidazole system with no special contribution from the binding cavity. This is supported by the fact that adenine is known to bind reasonably well to the cyclodextrin cavity but that cytosine binds less well. It looks as if our catalyst is not suited particularly well to binding and cleaving the substrates of the natural enzymes. However, the obverse is also true. Ribonuclease is completely unable to catalyze the cleavage of our t-butylcatechol cyclic phosphate, and indeed this substrate for our pseudoenzyme VII is an inhibitor for the natural enzyme, binding to it but not being cleaved by it.

It will be of some interest to learn how to build catalysts to handle the particular substrates that natural enzymes cleave, at a rate comparable to the rates of those enzymatic reactions. However, one of the aims of biomimetic chemistry is to extend the kinds of rates and selectivities of enzymatic reactions into reactions for which natural enzymes have not been optimized and to substrates that are neither recognized nor handled by normal enzymes. It is clear that we already have achieved this, even though our ribonuclease model system has some distance to go before it can approach the kinds of rates we have observed in the cyclodextrin ferrocinnamate ester reaction, for instance. In lock and key chemistry, the keys that fit artificial enzymes best are not the same as the keys that open the natural enzyme locks.

Acknowledgment

I would like to acknowledge the support of our work over the years by the NIH and the experimental contributions of my coworkers, most of whom are named in the references.

Literature Cited

1. Breslow, R. *Chem. Soc. Rev.* **1972**, *1*, 553.
2. Cramer, F. "Einschlussverbindungen"; Springer–Verlag: Berlin, 1954.
3. Bender, M. L.; Komiyama, M. "Cyclodextrin Chemistry"; Springer–Verlag: Berlin, 1978.
4. Siegel, B.; Breslow, R. *J. Am. Chem. Soc.* **1975**, *97*, 6869.
5. Pauling, L.; *Chem. Eng. News* **1946**, *24*, 1375.
6. van Etten, R. L.; Sebastian, J. F.; Clowes, G. A.; Bender, M. L. *J. Am. Chem. Soc.* **1967**, *89*, 3242.
7. Emert, J. Ph.D., Thesis, Columbia University, New York, NY, 1974.
8. Laustch, W.; Wieckert, R.; Lehmann, H. *Kolloid-Z.* **1954**, *135*, 134.
9. Emert, J.; Breslow, R. *J. Am. Chem. Soc.* **1975**, *97*, 670.
10. Tabushi, I.; Shimokawa, K.; Shimizu, N.; Shirakata, H.; Fujita, K. *J. Am. Chem. Soc.* **1976**, *98*, 7855.
11. Breslow, R.; McClure, D. E. *J. Am. Chem. Soc.* **1976**, *98*, 258.
12. Czarniecki, M. F.; Breslow, R. *J. Am. Chem. Soc.* **1978**, *100*, 7771.
13. Czarniecki, M. F., personal communication.
14. Breslow, R.; Doherty, J.; Guillot, G.; Lipsey, C. *J. Am. Chem. Soc.* **1978**, *100*, 3227.
15. Iwakura, Y.; Uno, K.; Toda, F.; Onozuka, S.; Hattori, K.; Bender, M. L. *J. Am. Chem. Soc.* **1975**, *97*, 4432.
16. Siegel, B.; Pinter, A.; Breslow, R. *J. Am. Chem. Soc.* **1977**, *99*, 2309.
17. Nakasuji, H.; personal communication.
18. Bovy, P.; personal communication.

RECEIVED May 14, 1979.

Models for Enzymic Systems
An Introduction

FRANK H. WESTHEIMER

Department of Chemistry, Harvard University, Cambridge, MA 02138

During the last two decades, the mechanisms of many enzymic processes have been established, and model systems have been developed that effectively mimic their action. In particular, the roles of thiamin, NAD, pyridoxal, flavins, B_{12}, ferridoxin, and metals in many enzymic processes now are understood. Model systems have been developed to imitate the action of decarboxylases and esterases, to imitate the action of enzymes in binding their substrates, and to approach the stereospecificity of enzymes. Our laboratory recently has found phosphorylating agents that release monomeric methyl metaphosphate, which in turn carries out phosphorylation reactions, including some at carbonyl oxygen atoms, that suggest the actions of ATP. The ideas of biomimetic chemistry are illustrated briefly in terms of the processes mentioned above.

All the chemical reactions in living systems are catalyzed by enzymes. Understanding the fabulous catalytic activity of enzymes and their remarkably precise specificity constitutes one of the great scientific problems of all time. Since Sumner crystallized urease in 1926, considerable progress has been made in developing the pathways, and in some instances rather detailed mechanisms for the action of enzymes. Theory has led to quantitative estimates of the catalytic activation that can be anticipated by a number of means: by bringing substrates into intimate contact with one another, by orienting molecules properly for reaction, by desolvating reactants, by internal acid and base catalysis, by special complexing of metal ions, by introducing steric strains, and by other means. Whereas two decades ago, no one

0-8412-0514-0/80/33-191-017$05.00/0
© 1980 American Chemical Society

knew how any enzyme worked, the more optimistic among the biochemical community now look at the problems of enzyme mechanisms
as essentially solved.

Such, however, is regrettably not the case. Although we like to
believe that we "understand" some enzymes at a moderate level of
sophistication, we cannot really claim to have solved the problem until
we are able to design a structure, previously unknown, that will serve
as catalyst for a specific reaction, and that will be selective for a particular substrate; we must then synthesize that structure and test the
predictions. Further, we must be able to make at least an approximate
estimate—within two orders of magnitude—of the expected rate and
binding constant of the substrate for our "synthetic enzyme." If this
sounds like a tall order—it is. But that is the challenge, that is the
problem that chemists face in their attack on enzymology. Biomimetic
chemistry is concerned with building models that imitate enzyme activity and specificity. The final objective is to convert the problem to
one of molecular engineering, rather than one of answering fundamental questions. Even after someone has synthesized the first good
enzyme-like catalyst, much will remain to be done as we explore the
various ways in which the enormous catalytic activity of enzymes can
be achieved. The leading synthetic organic chemists can now make
any "small" molecule, but we are still interested in the ingenious
ways in which such syntheses are achieved; similarly, even after we
can synthesize appropriate catalysts, chemists will be interested in the
actual accomplishment in specific cases. Although such catalysts eventually may have important industrial and medical uses, we are still far
from the time when such catalysts can be designed and made.

Biomimetic chemistry has been dominated by Japanese and
American scientists; interestingly, a number of the joint publications
from U.S. laboratories are from teams with both Japanese and American collaborators. A joint Japanese–American symposium on
biomimetic chemistry is then especially appropriate. In this volume,
Tagaki's chapter concerns the effects of metal ions in promoting a
specific hydrolytic process. His work illustrates the effects of metal
ions in numerous enzymic systems. Kaiser's chapter is directed toward
an understanding of the role of protein structure in enzymology. Since
we cannot as yet predict how polypeptide chains will fold, he has
adapted to his purposes a natural protein that has folded to a stable
conformation, and is making an attempt to use its binding site to introduce a different catalytic activity than the one characteristic of the
native enzyme; specifically, he is trying to convert a hydrolytic enzyme to one for oxidation–reduction. Yoshikawa has examined a system that promotes a highly stereoselective process; such systems lead
us toward the objective of building catalysts with the complete

stereospecificity that is characteristic of enzymes. Finally, Breslow will describe chemical modifications of cyclodextrins, modifications designed to allow an understanding of the specificity of binding, and the connection between binding and catalytic activity.

These introductory remarks will have a twofold objective: first, to review the history of the field, in an attempt to put the following papers into prospective; second, to present some work from my own laboratory on a possible biomimetic system—the generation of monomeric metaphosphate, and its reactions that seem to parallel the action of pyruvate kinase and of amidotransferases.

Pyridoxal Phosphate

An accurate history of biomimetic reactions cannot easily be written. Emil Fischer's synthesis, in 1887, of fructose by the aldol condensation of a crude mixture of glyceraldehyde and dihydroxyacetone (1, 2) reasonably can be described as a biomimetic reaction; probably other, earlier syntheses could be cited with almost equal merit. The reactions mentioned here obviously have been selected somewhat arbitarily from the wealth of possibilities in the literature. Certainly one of the most elegant examples comes from the work of E. E. Snell and his collaborators. Although his work is modern by comparison with that of Fischer cited above, Snell reported in 1945 that he could transaminate glutamic acid and pyridoxal (3) by heating them together at about 120°C. Subsequently, in a series of papers with M. Ikawa, D. E. Metzler, and others, Snell reported (4–8) that the nonenzymic reactions of pyridoxal with amino acids in aqueous solutions, pH 4–10, are strongly promoted by polyvalent cations, such as Al^{+++}, Fe^{+++}, and Cu^{++}. Most amino acids tested under these conditions undergo reversible transamination; glycine is an exception. Here the equilibrium in transamination with pyridoxal lies far over the side of the amino acid; the reverse reaction, however, between pyridoxamine and glyoxylic acid, occurs readily in hot aqueous solutions in the presence of di- and trivalent cations.

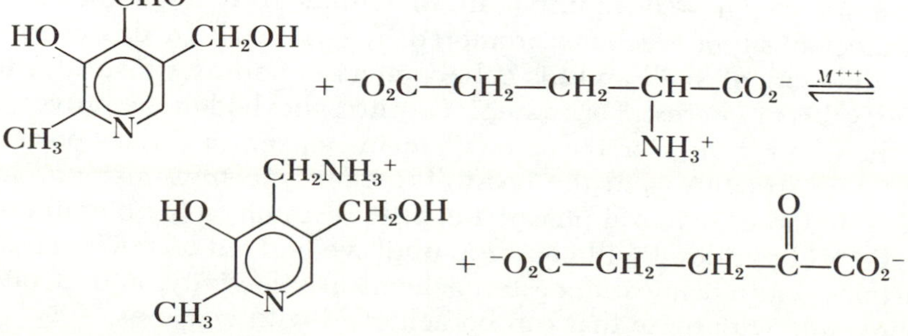

$$\underset{\substack{\text{HO}\\ \text{CH}_3}}{}\overset{\text{CH}_2\text{NH}_2}{\underset{\text{N}}{\bigcirc}}\text{CH}_2\text{OH} \quad + \; \text{H}\overset{\text{O}}{\underset{\|}{\text{C}}}\text{CO}_2^- \; \underset{}{\overset{M^{+++}}{\rightleftharpoons}}$$

(2)

$$\underset{\substack{\text{HO}\\ \text{CH}_3}}{}\overset{\text{CHO}}{\underset{\text{N}}{\bigcirc}}\text{CH}_2\text{OH} \quad + \; \overset{+}{\text{H}_3\text{N}}\text{—CH}_2\text{CO}_2^-$$

Other reactions that mimic the enzymic processes that require pyridoxal phosphate also have been realized. Werle and Koch reported the nonenzymic decarboxylation of histine (9). The racemization of alanine occurs in preference to its transamination when aqueous solutions with polyvalent cations are maintained at pH 9.5. Other amino acids are likewise racemized; the order of rates is Phe, Met > Ala > Val > Ileu. At lower pH, the dominant reaction is transamination, with pH maxima varying from 4.3–8 with the nature of the metal ion used as catalyst.

When serine-3-phosphate or cysteine or threonine is heated with pyridoxal and an appropriate metal ion, the compounds are degraded to the corresponding ketoacids. Similarly, cystathionine is decomposed to yield α-ketobutyric acid (4–8). Further, serine reacts with indole in hot aqueous solution at pH 5 in the presence of metal ions to yield tryptophane, in strict imitation of the corresponding enzymic synthesis. The biomimetic reactions promoted by pyridoxal are remarkable in that so many different reactions of pyridoxal-promoted enzymes can be imitated with a simple metal ion serving as catalyst. Of course, these do not duplicate the enzymic processes (10, 11). To begin with, the enzymes probably do not require metal ions. Furthermore, the metal-ion promoted reactions are not specific; many competing reactions of the same substrate occur concurrently, and of course, the nonenzymic processes are not stereospecific. Nevertheless, at a minimum, they have been important in helping to elucidate the chemical mechanism of reactions promoted by pyridoxal. In this connection, the work of Snell and his collaborators in finding substitutes for pyridoxal is of interest. For example, 4-nitrosalicylaldehyde proves an excellent substitute for the coenzyme in nonenzymic reactions; its electronic structure, and the possibilities for "electron-pushing" are similar to those pyridoxal phosphate. Still, research on the biomimetic reactions of pyridoxal will continue until we find out how to promote reactions with chemical and stereochemical selectivity, and at rates comparable with those that can be achieved with enzymes.

Thiamin

Another early success in biomimetic chemistry concerns reactions promoted by thiamin. In 1943, more than 35 years ago, Ukai, Tanaka, and Dokowa (*12*) reported that thiamin will catalyze a benzoin-type condensation of acetaldehyde to yield acetoin. This reaction parallels a similar enzymic reaction where pyruvate is decarboxylated to yield acetoin and acetolactic acid. Although the yields of the nonenzymic process are low, it is clearly a biomimetic process; further investigation by Breslow, stimulated by the early discovery of Ugai et al., led to an understanding of the mechanism of action of thiamin as a coenzyme.

In 1957, Breslow (*13*) showed that the hydrogen atom in the 2-position of the thiazolium-ion portion of thiamin is ionized readily; the electronic structure of the anion imitates that of cyanide ion. The chemistry of thiamin can then be explained; the decarboxylation of pyruvate and the acetoin condensation are processes that follow conventional mechanisms; in modern language, thiamin allows an acyl group to become an anion equivalent. Subsequent to Breslow's initial discovery, he and McNelis (*14*) synthesized 3,4-dimethyl-2-acetylthiazolium ion, and showed that in fact it is hydrolyzed rapidly.

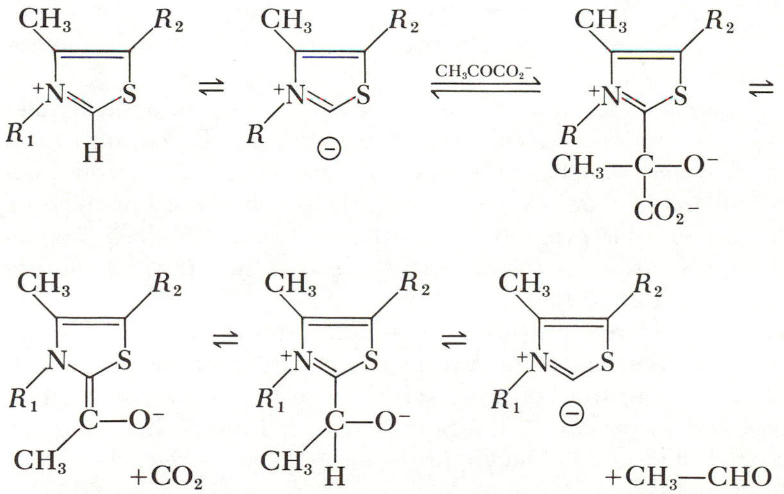

$$(3)$$

A number of investigators (*15, 16, 17*), working with pyruvate decarboxylase, actually isolated the pyruvate adduct to thiamin pyrophosphate, and demonstrated that the enzyme will then decarboxylate that adduct to yield CO_2 and acetaldehyde. In an elegant biomimetic study, Lienhard (*18, 19*) and his collaborators synthesized an analog of this adduct, and showed that it undergoes decarboxylation upon heating in water. Further, the rate is enhanced 10^5-fold by carrying out the reaction in ethanol at low ionic strength instead of in water at high ionic strength; a similar increase in rate resulted from the substitution of DMSO for water, although the actual rate constant wasn't measured. The authors estimated that the acceleration by the enzyme for the decarboxylation is about 10^6-fold, so that the model comes within a factor of ten of the enzyme. Of course, as Lienhard pointed out, the decarboxylation step may not be rate limiting, or the one with the greatest enzymic acceleration, but at least for this one step, the model, operating in an environment of low dielectric constant, simulates the enzyme fairly well. An environment of low dielectric constant similarly enhances the rates of other steps in the overall process.

Acetoacetate Decarboxylase

Another example where mechanism and model have been developed is that for the decarboxylation of acetoacetic acid; here no coenzyme is required, and the chemistry involves the enzyme itself. The mechanism for the enzymic decarboxylation with crystalline decarboxylase from *Clostridium acetobutylicum* has been worked out in some detail; it is presented below (20, 21). The initial work, carried out in the author's laboratory by G. Hamilton (22) and I. Fridovich (23, 24) proved that the essential intermediate is a ketimine; much of the subsequent development of the enzymic system resulted from the researches of W. Tagaki (25).

Investigations show that the active amino group of acetoacetate decarboxylase—the one that is concerned with the formation of the ketimine intermediate—has an especially low pK (26, 27). Model experiments revealed that amines of low pK are the best nonenzymic catalysts; in particular, cyanomethylamine led to a rate of decarboxylation that is only a few orders of magnitude less than that for the enzymic system (28, 29, 30).

Here, too, the process involves several steps, and the enzyme must accelerate them all. The rate of the decarboxylation step itself certainly is faster in the ketimine salt than in the ketoacid. Taguchi (31) succeeded in preparing the Schiff base, I. Prior to his work, no β-carboxy Schiff base was known, presumably because they decarboxylated too rapidly.

Compound I could be isolated because the decarboxylation of the ketimine must produce an intermediate with a badly twisted double bond. The twist is imposed by the bicyclic structure; the intermediate in the decarboxylation violates Bredt's rule. This slows the decarboxylation. Taguchi's Schiff base, nevertheless, undergoes decarboxylation about 10^6 times as fast as does the ketoacid from which it was made. The model accounts for the major part of the acceleration in the decarboxylation step for the overall process.

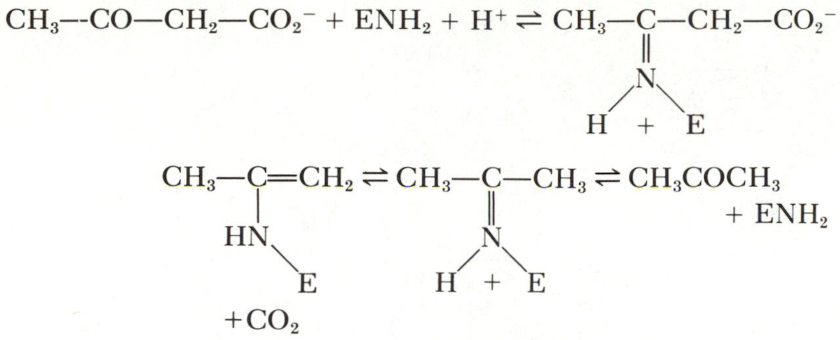

$$CH_3\text{--}CO\text{---}CH_2\text{---}CO_2^- + ENH_2 + H^+ \rightleftharpoons CH_3\text{---}\underset{\underset{\underset{H \quad + \quad E}{\diagdown N \diagup}}{\|}}{C}\text{---}CH_2\text{---}CO_2^- \rightleftharpoons$$

$$CH_3\text{---}\underset{\underset{\underset{+CO_2}{E}}{HN\diagdown}}{C}{=}CH_2 \rightleftharpoons CH_3\text{---}\underset{\underset{\underset{H \quad + \quad E}{\diagdown N \diagup}}{\|}}{C}\text{---}CH_3 \rightleftharpoons CH_3COCH_3$$

$$+ ENH_2$$

(4)

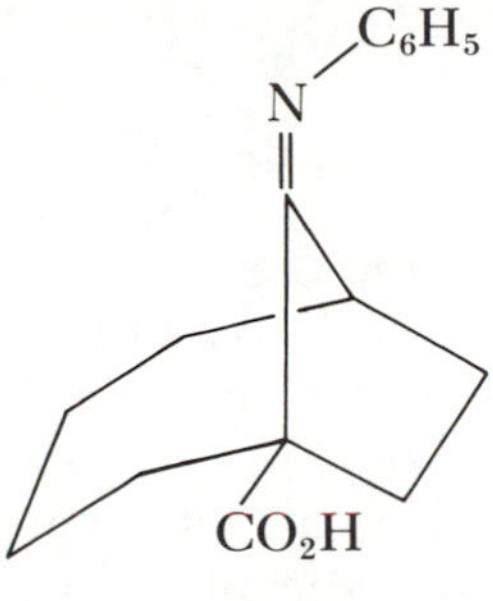

(I)

Dehydrogenases

Biomimetic chemistry has also been notably successful in imitating the oxidation–reduction processes mediated by the coenzyme system NAD–NADH. The many enzymes that utilize these coenzymes proceed with direct and stereospecific transfer of hydrogen between substrate and the 4-position of the coenzyme molecule (*32, 33*). A number of early attempts (*34, 35, 36*) to find model systems for the oxidation–reduction processes were only partially successful, but in 1971, D. J. Creighton and D. S. Sigman (*37*) found that they could mimic the action of alcohol dehydrogenase with a properly chelated metal ion. The dehydrogenases, or anyway most of them, are zinc enzymes (*38, 39, 40, 41*), and, on the basis of Sigman's model system, a reasonable mechanistic role can be assigned to the metal ion. Sigman treated the zinc complex of phenanthroline-2-carboxyaldehyde with N-propyldihydronicotinamide in acetonitrile as solvent; nonenzymic oxidation–reduction occurs with direct hydrogen transfer at a moderate rate at room temperature. The chelation of the zinc ion enormously increases the polarity of the carbonyl group, so as to promote the transfer of an incipient hydride ion from the dihydronicotinamide to the aldehydic carbon atom. The reaction fails in water, presumably because water hydrates and thus deactivates the zinc ion.

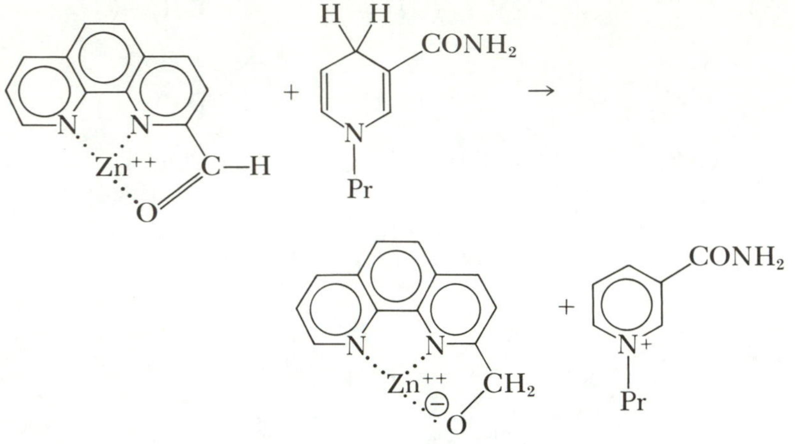

(5)

Several other models mimic the action of the oxidation–reduction enzymes that require NAD. Shinkai and Bruice (42) carried out a nonenzymic oxidation–reduction by using an aldehyde that combined a highly electronegative carbonyl group with hydrogen bonding; the latter served almost as well as a metal ion to further enhance the reactivity of the carbonyl. The effect of hydrogen bonding in accelerating such reactions had been noted previously (34).

More recently, Ohnishi, Kagami, and Ohno (43) have reduced ethyl benzoylformate with N-benzyldihydronicotinamide in acetonitrile as solvent, using magnesium perchlorate as catalyst; the reaction proceeded in 86% yield in 17 hr at room temperature. Although this rate falls far short of the corresponding enzymic ones, the example is especially noteworthy because the needed metal ion is not chelated to the substrate, as in the example provided by Creighton and Sigman, but was free in solution. Similar reductions with chiral dihydronicotinamides gave partially resolved product. A different approach has been taken by Van Bergen and Kellogg (44), who have chelated the metal ion to the reducing agent rather than to the substrate. In all of these examples, the rates are low and much remains to be discovered before we can duplicate the action of the enzymes.

Metal Ions

Other biomimetic reactions are based on the catalytic properties of metal ions. Many enzymes require metal ions that function, in one way or another, in oxidation–reduction processes. The wide range of such metal-ion reactions precludes mentioning more than a few; in addition to the iron-porphyrin class, and in addition to chlorophyll, a number of enzymes require cobalamin as cofactor; ferridoxin and high-potential iron proteins require iron–sulfur clusters, and nitrogenase requires an iron–sulfur–molybdenum cluster. Further, many hydrolytic enzymes require metal ions for activity, and the effects of these enzymes can in some instances be mimicked by polyvalent cations.

The mechanisms of action of enzymes that require a vitamin B_{12} coenzyme have been under intensive investigation; a start has been made toward understanding their mode of action (45). Recently, a number of compounds with covalent bonds between cobalt and organic substrates have been prepared that mimic reactions of B_{12} enzymes. In particular, Dowd and his collaborators (46, 47, 48) have reduced vitamin B_{12} to the highly nucleophilic B_{12s}, and allowed this to react with 2-methylene-3-carboalkoxy-4-bromobutanoic ester; the resulting organo-cobalt compound readily undergoes rearrangement and, on hydrolysis, gives mixtures containing rearranged as well as unrearranged acids. The system mimics the chemistry of the isomerase that equilibrates methyl itaconic acid with methylene glutaric acid.

The iron–sulfur cubes that have been detected by x-ray analysis (49) as constituents of the enzymes ferridoxin and high-potential iron protein have been extruded from these enzymes by replacing the sulfhydryl ligands of the enzymes with simple mercaptans, and these cubes identified with the corresponding synthetic compounds, **II** (50, 51, 52, 53). The latter have oxidation–reduction properties that closely mimic those of the enzymes. Similarly, an iron–sulfur–molybdenum double cube has been ejected from nitrogenase, and a similar double cube has been synthesized by Holm and his collaborators (54). It remains to be seen whether or not the iron–sulfur–molybdenum double cube can mimic the properties of nitrogenase (55).

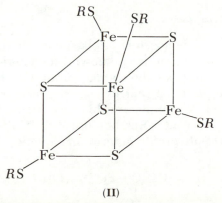

(**II**)

The effects of metal ions in promoting hydrolytic reactions have been noted by many investigators (56, 57). Probably the most spectacular of these biomimetic reactions come from the work of Sargeson, Buckingham, and their collaborators (58, 59). They have made a number of cobalt complexes of various esters, and found that the rates of hydrolysis were accelerated enormously. For example, the rate of hydrolysis of p-nitrophenyl phosphate was increased by a factor of 10^9 when it was coordinated with cobaltbistrimethylenediamine. Of course, the hydrolysis is not catalytic in this instance; the product remains coordinated to the cobaltic ion. But if the hydrolysis of the Complex **III** could be duplicated by that of a transitory complex with another metal (e.g., ferric, zinc, or cupric ion) where the rate of formation and cleavage of the bonds to metal are rapid, then the reaction would imitate those of metal-ion promoted hydrolytic enzymes (60, 61, 62, 63). Although the reaction is not truly a biomimetic one, it shows that enormous catalytic effects can be achieved with polyvalent cations.

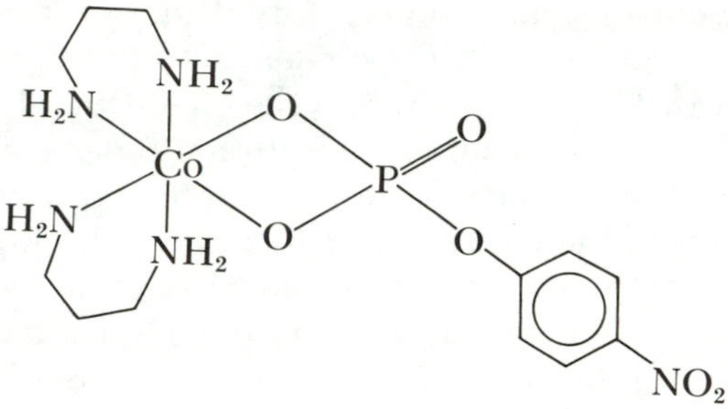

(**III**)

No list of biomimetic reactions could possibly be complete, and the list above obviously covers only a few of many examples that could be chosen for discussion. They do introduce, however, several important examples.

Monomeric Metaphosphates

The hydrolysis of monoesters of phosphoric acid presumably proceeds by way of the presently unknown monomeric metaphosphate ion, PO_3^-. These hydrolyses show a maximum rate at pH 4, where the monoester monoanion is present in largest concentration. More than two decades ago, C. A. Bunton and his co-workers (64) at University College, London, and Walter Butcher (65) at the University of Chicago interpreted the pH–rate profile in terms of Equation 6. The postulated

$$CH_3OPO_3H^- \rightleftharpoons CH_3\overset{+}{\underset{H}{O}}-PO_3^- \rightarrow CH_3OH + [PO_3^-] \qquad (6)$$

monomeric metaphosphate anion was assumed to react then with solvent. Whereas nitrate ion is stable, and orthonitric acid (H_3NO_4) is unknown, orthophosphate is stable, and metaphosphate is (essentially) unknown. The metaphosphate ion is isoelectronic with SO_3, and so would be expected to be a strong electrophile. If it is indeed a chemical intermediate, it might serve as a phosphorylating agent. Over the years, much indirect evidence has accumulated that such is indeed the fact; important contributions have arisen from the laboratories of W. P. Jencks (66), and others (67, 68, 69) and recently from those of J. Rebek and F. Gaviña (70) and ourselves.

A few years ago, Charles Clapp and collaborators (71, 72) pyrolyzed the vapor of methyl butenylphosphonate, and obtained a gas stream that contained a highly reactive phosphorylating agent, which presumably is monomeric methyl metaphosphate.

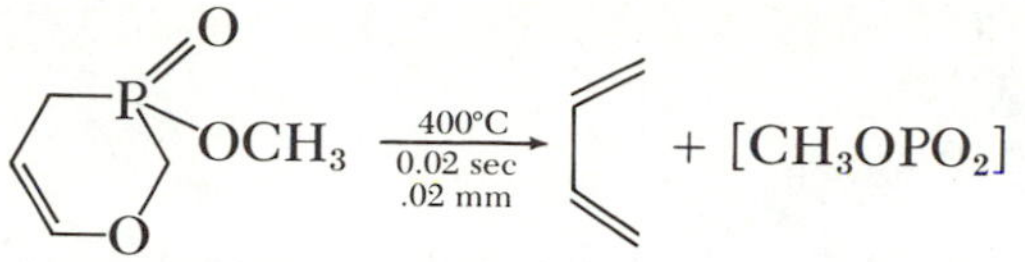

$$\text{(7)}$$

Arnold Satterthwait (73), continuing that work, also has prepared what appears to be monomeric methyl metaphosphate in solution. His preparative method utilizes the Conant–Swan fragmentation of β-halophosphonates (Equation 8). The material reacts with substi-

$$
\begin{array}{ccc}
C_6H_5-CBr-CHBr-CH_3 & & C_6H_5-CBr=CHCH_3 \\
\quad\quad | & \rightarrow & \\
CH_3O-PO_2^- & & + [CH_3OPO_2^-] + Br^-
\end{array}
\qquad (8)
$$

tuted anilines in part by electrophilic substitution to yield esters of aromatic phosphonic acids, as for example in the reaction depicted in Equation 9. The reaction in solution is carried out at 70°C in the presence of 2,2,6,6-tetramethylpiperidine, a sterically hindered base.

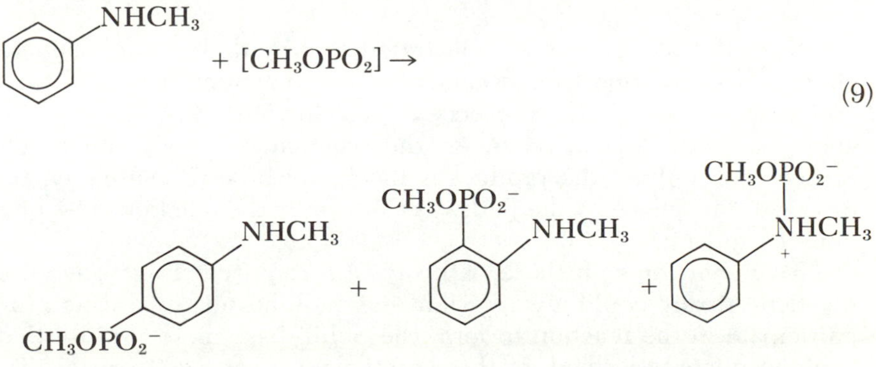

$$\text{(9)}$$

Satterthwait recently has found that the electrophilic substitution is sensitive to solvent effects. When the reaction with N-methylaniline is carried out in the absence of solvent, the product consists of about

40% of a mixture of methyl *o*- and *p*-methylaminophosphonic acid, and almost 60% of the phosphoramidate. When the reaction mixture is diluted 10 : 1 with chloroform, the fraction of aromatic substitution is reduced to about 30%, but when it is diluted 10 : 1 with either dioxane or acetonitrile, only a trace of aromatic substitution can be found (*see* Table I).

Table I.   Aromatic Substitution by "Monomeric Methyl Metaphosphate"

Experimental Conditions	*Yield of Aromatic Substitution*
No solvent	~40%
$CHCl_3$	~30%
CH_3CN	~ 2%
Dioxan	~ 2%

These data imply that dioxane and acetonitrile suppress the electrophilic character of the reagent. The simplest interpretation of these results is that methyl metaphosphate itself is the reagent for electrophilic substitution, but since it will react readily with any unshared pair of electrons, it is deactivated by dioxane or acetonitrile to produce Compounds **IV** and **V** shown below. Since SO_3 reacts with dioxane to form a crystalline adduct, **VI,** a parallel reaction of metaphosphate appears reasonable.

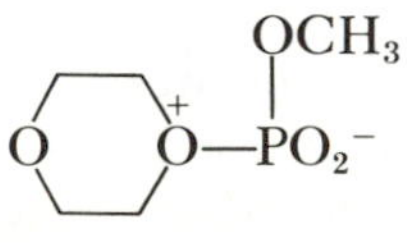

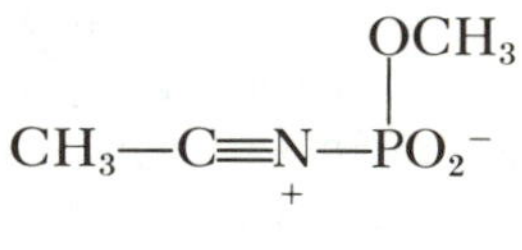

Satterthwait also has demonstrated that methyl metaphosphate, generated in solution by the Conant–Swan fragmentation, will react with acetophenone in the presence of base to yield the corresponding enol phosphate. When, however, the reaction is carried out in the presence of aniline, the product is the Schiff base. Presumably, the processes take place by the pathways shown by the reactions in Equations 10 and 11.

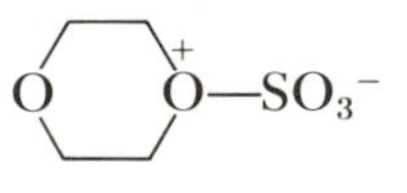

Two comments on these pathways are required. First, since the reaction mixture could always yield the enolphosphate, aniline must participate in the reaction to form the Schiff base more rapidly than enolphosphate is formed. If the equations are correct, then the first step of the reaction is the same in both cases, that is, methyl metaphosphate attacks the carbonyl group of acetophenone to produce an intermediate that is activated both for nucleophilic attack at the carbonyl group and for removal of a proton from the methyl group of

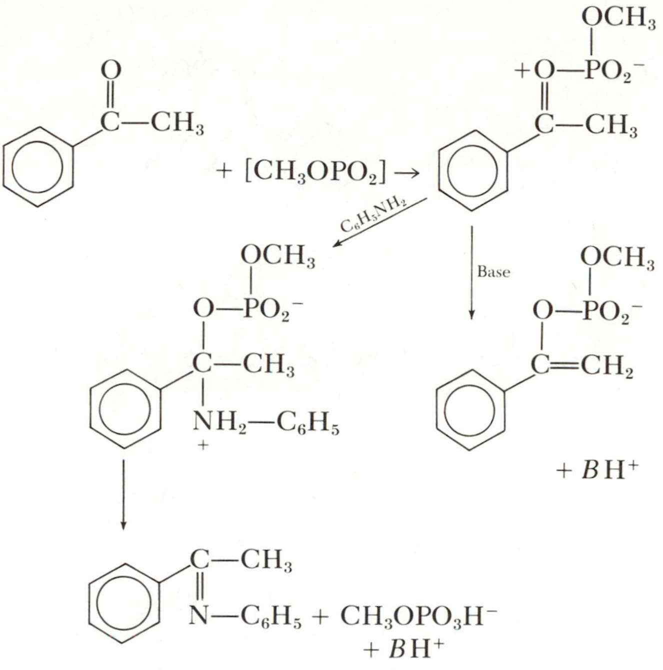

$$
\begin{array}{c}
\underset{\text{(acetophenone)}}{\overset{\displaystyle\overset{O}{\parallel}}{C6H5{-}C{-}CH_3}} + [CH_3OPO_2] \rightarrow \quad
\underset{}{\overset{\displaystyle\overset{OCH_3}{\mid}}{\overset{+O{-}PO_2^-}{\overset{\parallel}{C{-}CH_3}}}}
\end{array}
\tag{10}
$$

(equation 10: acetophenone + [CH$_3$OPO$_2$] →)

$$
\xleftarrow{\ C_6H_5NH_2\ }
\overset{OCH_3}{\underset{}{\mid}}\ O{-}PO_2^-\ \ \overset{\text{Base}}{\Big\downarrow}\ \ \overset{OCH_3}{\mid}\ O{-}PO_2^-
$$

Intermediate (left): ring–C(OCH$_3$)(O—PO$_2^-$)—CH$_3$ with $\overset{+}{N}H_2{-}C_6H_5$

Intermediate (right): ring–C(O—PO$_2^-$)=CH$_2$ $\;+ BH^+$

$$
\underset{}{\overset{C{-}CH_3}{\underset{N{-}C_6H_5}{\parallel}}} + CH_3OPO_3H^- \\ + BH^+
\tag{11}
$$

acetophenone. Apparently the attack of aniline at the carbonyl carbon atom is sufficiently rapid relative to the loss of a proton to the strong but sterically hindered base that is present, so that a Schiff base is formed preferentially to the enol phosphate. The second comment concerns control experiments. An alternative pathway for the formation of the Schiff base would involve the direct attack of aniline on acetophenone.

Measurements of the rate of formation of the Schiff base suggested that, indeed, two separate processes occur, with the metaphosphate-promoted process being much the faster. This conclusion was supported strongly by the observation that, with 2-trifluoromethylaniline, the difference in rates is so large that, for practical purposes, the only reaction that occurs is the one promoted by metaphosphate.

Finally, when methyl metaphosphate is generated in the presence of ethyl acetate and aniline, ethyl N-phenylbenzamidate is produced. Presumably, the reaction follows the course shown in Equation 12.

These reactions suggest the corresponding biochemical ones (*see* Equation 13). In the presence of pyruvate kinase, ATP and pyruvate are equilibrated with phosphoenolpyruvate and ADP (*74, 75, 76*). Furthermore, in the presence of formylglycinamide ribonucleotide amidotransferase, ATP reacts with 5-phosphoribosyl-N-formylglycinamide and glutamine to yield 5-phosphoribosyl-N-formylglycinamidine according to Equation 14 (*77*).

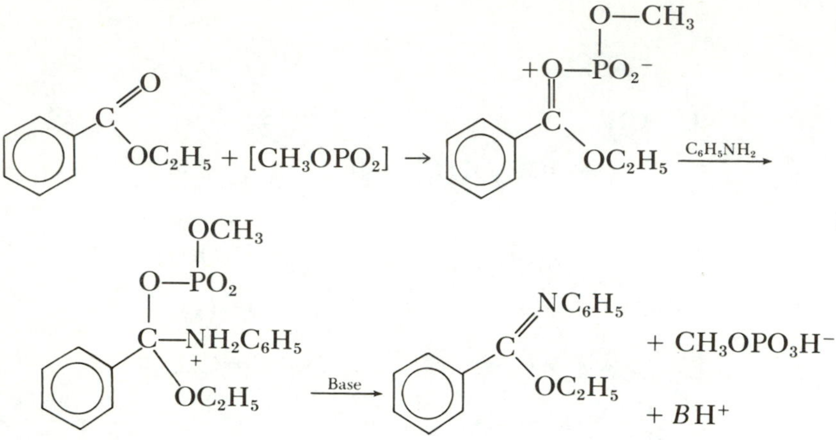

$$C_6H_5\text{-}C(\text{=}O)\text{-}OC_2H_5 + [CH_3OPO_2] \rightarrow C_6H_5\text{-}C(\overset{+O\text{-}PO_2^-}{\underset{O\text{-}CH_3}{})}\text{-}OC_2H_5 \xrightarrow{C_6H_5NH_2}$$

$$\text{(12)}$$

$$C_6H_5\text{-}C(\overset{O\text{-}PO_2}{\underset{OCH_3}{}})(\overset{+}{NH_2C_6H_5})\text{-}OC_2H_5 \xrightarrow{Base} C_6H_5\text{-}C(\text{=}NC_6H_5)\text{-}OC_2H_5 + CH_3OPO_3H^- + BH^+$$

$$CH_3COCO_2^- + ATP \underset{kinase}{\overset{Pyruvate}{\rightleftharpoons}} CH_2\text{=}C(\overset{OPO_3^=}{\underset{CO_2^-}{}}) + ADP \quad \text{(13)}$$

[Structure with $\overset{H}{N}$, CH_2, CHO, $O\text{=}C$, NH, Ribose phosphate] $+ ATP + glutamine \xrightarrow[Mg^{++}]{Amidotransferase}$

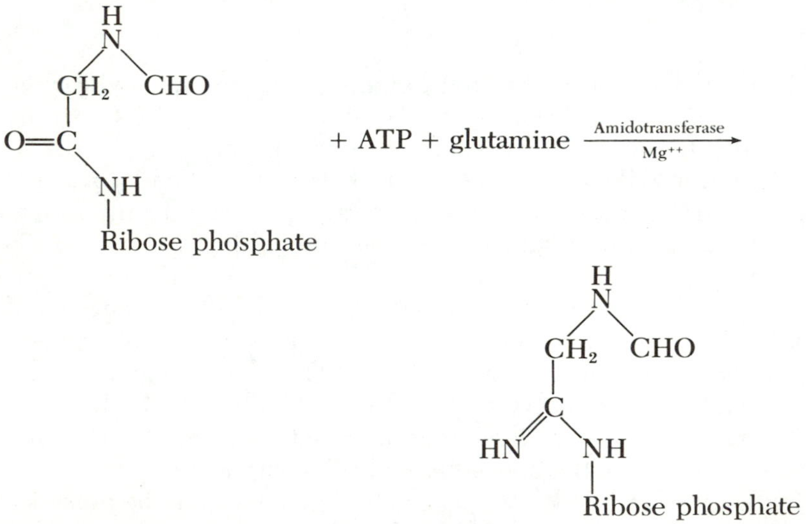

$$\text{(14)}$$

[Structure with $\overset{H}{N}$, CH_2, CHO, C, HN, NH, Ribose phosphate]

Of course, the mechanisms of these enzymic processes are at present unknown. G. Lowe and B. S. Sproat (78) have suggested that the action of pyruvate kinase involves monomeric metaphosphate as an intermediate. His evidence concerns the effect of the enzyme in scrambling the oxygen atoms of ATP in the absence of pyruvate, Equation 15. This scrambling process is enzyme catalyzed, and is interpreted most easily, but not exclusively, in terms of the reversible formation of monomeric metaphosphate.

$$\underset{\underset{\text{AMP}-\circledast}{|}}{\overset{\overset{\ominus}{\diagup}\overset{\circledast}{|}}{\circledast-\text{P}-\text{O}-\text{PO}_3^{=}}} \rightleftharpoons \underset{\underset{\text{AMP}-\circledast}{|}}{\overset{\overset{\circledast^{=}}{|}}{\circledast-\text{P}-\text{O}}} + [\text{PO}_3^-]$$

$$\Updownarrow \qquad\qquad (15)$$

$$\underset{\text{AMP}-\circledast}{\overset{\overset{\ominus}{\diagup}\overset{\circledast}{|}}{\text{O}-\text{P}-\circledast-\text{PO}_3^{=}}}$$

Although these data suggest a monomeric metaphosphate mechanism, they are far from conclusive. Furthermore, while the reactions shown for methyl metaphosphate apparently mimic the enzymic processes, the mechanisms are not necessarily parallel. In particular, Kuo and Rose (79) have shown that the action of pyruvate kinase probably involves the energetically unfavorable enolization of pyruvate, followed by phosphorylation of the enol. (More precisely, they studied the reaction of Equation 13 in the direction from right to left.) By contrast, the reaction of methyl metaphosphate on acetophenone almost certainly proceeds by initial attack of the monomeric metaphosphate on the carbonyl oxygen atom of the ketone. Nevertheless, an interesting speculation concerns the possibility that, in some instances at least, phosphorylation proceeds by pathways such as those shown in Equations 11 and 12; if this proves to be the case, then ATP is not only thermodynamically, but also kinetically important. F. Lipmann (80), in an early and classical advance in biochemistry, showed that thermodynamically unfavorable processes could be carried forward if they were coupled to the hydrolysis of ATP, or to that of another "high-energy phosphate". The net free energy for the overall process, for instance, of the amidation of 5-phosphoribosyl-N-formylglycinamide is negative, provided that in the process ATP is degraded to ADP and inorganic phosphate. Now it appears that ATP may serve a second important function. It could activate a carbonyl group by phosphorylating it, much in the fashion that a superacid might activate a carbonyl group by protonating it, converting the polar carbonyl compound to an even more polar intermediate with a positively charged oxygen atom. The molecule would then be primed for rapid reaction, even with a relatively mild nucleophile. The kinetic activation of carbonyl compounds by methyl metaphosphate may prove to be a biomimetic process, and to illuminate the kinetic importance of ATP.

Literature Cited

1. Fischer, E.; Tafel, J. *Ber. Dtsch. Chem. Ges.* **1887**, 20, 3384.
2. Fischer, E. *Ber. Dtsch. Chem. Ges.* **1890**, 23, 387.
3. Snell, E. E. *J. Am. Chem. Soc.* **1945**, 67, 194.

4. Metzler, D. E.; Snell, E. E. *J. Am. Chem. Soc.* **1952**, *74*, 979.
5. Olivard, J.; Metzler, D. E.; Snell, E. E. *J. Biol. Chem.* **1952**, *199*, 669.
6. Metzler, D. E.; Ikawa, M.; Snell, E. E. *J. Am. Chem. Soc.* **1954**, 76, 648.
7. Ikawa, M.; Snell, E. E. *J. Am. Chem. Soc.* **1954**, 76, 653.
8. Longnecker, J. B.; Snell, E. E. *J. Am. Chem. Soc.* **1957**, 79, 142.
9. Werle, E.; Koch, W. *Biochem. Z* **1949**, *319*, 305.
10. Braunstein, A. E. "The Enzymes"; Boyer, P. D., Lardy, H., Myrbäch, K., Eds.; Academic: New York, 1960; Vol. II, 113.
11. Snell, E. E. *Vitam. Horm. (N.Y.)* **1958**, *16*, 77.
12. Ukai, T.; Tanaka, R.; Dokowa, T. *J. Pharm. Soc. Jpn.* **1943**, *63*, 269.
13. Breslow, R. *J. Am. Chem. Soc.* **1957**, 79, 1762.
14. Breslow, R.; McNelis, E. *J. Am. Chem. Soc.* **1960**, *82*, 2394.
15. Carlson, G. L.; Brown, G. M. *J. Biol. Chem.* **1960**, 235, PC3.
16. Holzer, H.; Beaucamp, K. *Biochim. Biophys. Acta* **1961**, *46*, 225.
17. Krampitz, L. O.; Suzuki, I.; Greull, G. *Fed. Proc. Fed. Am. Soc. Exp. Biol.* **1961**, *20*, 971.
18. Lienhard, G. E. *J. Am. Chem. Soc.* **1966**, 88, 5642.
19. Crosby, J.; Lienhard, G. E. *J. Am. Chem. Soc.* **1970**, *92*, 5707.
20. Westheimer, F. H. *Proc. Chem. Soc., London* **1963**, 253.
21. Westheimer, F. H. *Robert A. Welch Foundation Conference on Chemical Research*, 15th, **1971**, 7.
22. Hamilton, G.; Westheimer, F. H. *J. Am. Chem. Soc.* **1959**, *81*, 6332.
23. Fridovich, I.; Westheimer, F. H. *J. Am. Chem. Soc.* **1962**, *84*, 3208.
24. Warren, S.; Zerner, B.; Westheimer, F. H. *Biochemistry* **1966**, 5, 817.
25. Tagaki, W.; Westheimer, F. H. *Biochemistry* **1968**, 7, 891, 895, 901.
26. Frey, P. A.; Kokesh, F. C.; Westheimer, F. H. *J. Am. Chem. Soc.* **1971**, *93*, 7266.
27. Kokesh, F. C.; Westheimer, F. H. *J. Am. Chem. Soc.* **1971**, *93*, 7270.
28. Guthrie, J. P.; Westheimer, F. H. *Fed. Proc. Fed. Am. Soc. Exp. Biol.* **1967**, *26*, 562.
29. Guthrie, J. P. *J. Am. Chem. Soc.* **1972**, *94*, 7024.
30. Guthrie, J. P.; Jordan, F. *J. Am. Chem. Soc.* **1972**, *94*, 9132, 9136.
31. Taguchi, K.; Westheimer, F. H. *J. Am. Chem. Soc.* **1973**, *95*, 7413.
32. Westheimer, F. H.; Fisher, H. F.; Conn, E. E.; Vennesland, B. *J. Am. Chem. Soc.* **1951**, 73, 2403.
33. Vennesland, B.; Westheimer, F. H. "The Mechanism of Enzymes"; McElroy, W. D., Glass, B., Eds.; The Johns Hopkins Press: Baltimore, 1954; 357.
34. Abeles, R. H.; Hutton, R. F.; Westheimer, F. H. *J. Am. Chem. Soc.* **1957**, 79, 712.
35. Norcross, B. E.; Kleindinst, P. E., Jr.; Westheimer, F. H. *J. Am. Chem. Soc.* **1962**, *84*, 797.
36. Dittmer, D. C.; Fouty, R. A. *J. Am. Chem. Soc.* **1964**, *86*, 91.
37. Creighton, D. J.; Sigman, D. S. *J. Am. Chem. Soc.* **1971**, *93*, 6314.
38. Brändén, C. I.; Jörnvall, H.; Eklund, H.; Furugren, B. *Enzymes* **1975**, 105.
39. Holbrook, J.; Liljas, A.; Steindel, S. J.; Rossmann, M. G. *Enzymes* **1975**, 191.
40. Smith, E. L.; Austen, B. M.; Blumenthal, K. M.; Nye, J. F. *Enzymes* **1975**, 294.
41. Banaszak, L. J.; Bradshaw, R. A. *Enzymes* **1975**, 369.
42. Shinkai, S.; Bruice, T. C. *Biochemistry* **1973**, *12*, 1750.
43. Ohnishi, Y.; Kagami, M.; Ohno, A. *J. Am. Chem. Soc.* **1975**, 97, 4766.
44. Van Bergen, T. J.; Kellogg, R. M. *J. Am. Chem. Soc.* **1977**, 99, 3882.
45. Abeles, R. H. "Biological Aspects of Inorganic Chemistry," Dolphin, D., Ed.; John Wiley: New York, 1978.
46. Dowd, P.; Shapiro, M.; Kang, K. *J. Am. Chem. Soc.* **1975**, 97, 4754.
47. Dowd, P.; Trivedi, B. K.; Shapiro, M.; Marwaha, L. K. *J. Am. Chem. Soc.* **1976**, *98*, 7875.
48. Dowd, P.; Shapiro, M. *J. Am. Chem. Soc.* **1976**, *98*, 3724.

49. Carter, C. W., Jr.; "Iron-Sulfur Proteins"; Lovenberg, W., Ed.; Academic: 1977; Vol. III, p. 157.
50. Wolff, T. E.; Berg, J. M.; Warrick, C.; Hodgson, K. O.; Holm, R. H.; Frankel, R. B. *J. Am. Chem. Soc.* **1978**, *100*, 4630.
51. Reynolds, J. G.; Laskowski, E. J.; Holm, R. H. *J. Am. Chem. Soc.* **1978**, *100*, 5315.
52. Laskowski, E. J.; Frankel, R. B.; Gillum, W. O.; Papaefthymiou, G. C.; Renaud, J.; Ibers, J. A.; Holm, R. H. *J. Am. Chem. Soc.* **1978**, *100*, 5322.
53. Johnson, R. W.; Holm, R. H. *J. Am. Chem. Soc.* **1978**, *100*, 5338.
54. Kurtz, D. M., Jr.; Wong, G. B.; Holm, R. H. *J. Am. Chem. Soc.* **1978**, *100*, 6777.
55. McKenna, C. E. "Molybdenum and Molybdenum-Containing Enzymes;" Coughlan, M. P., Ed.; Pergamon: Oxford, 1980; 14.
56. Lloyd, G. J.; Hsu, C. M.; Cooperman, B. S. *J. Am. Chem. Soc.* **1971**, *93*, 4889.
57. Hsu, C. M.; Cooperman, B. S. *J. Am. Chem. Soc.* **1976**, *98*, 5652, 5657.
58. Buckingham, D. A.; Marzilli, L. G.; Sargeson, A. M. *J. Am. Chem. Soc.* **1967**, *89*, 2772, 4539.
59. Anderson, B.; Milburn, R. M.; Harrowfield, J. MacB.; Robertson, G. B.; Sargeson, A. M. *J. Am. Chem. Soc.* **1977**, *99*, 2652.
60. Butler, L. G. *Enzymes* **1971**, 529.
61. Reid, T. W.; Wilson, I. B. *Enzymes* **1971**, 373.
62. Hartsuck, J. A.; Lipscomb, W. N. *Enzymes* **1971**, 1.
63. Delange, R. J.; Smith, E. L. *Enzymes* **1971**, 81.
64. Bernard, P. W. C.; Bunton, C. A.; Llewellyn, D. R.; Oldham, K. G.; Silver, B. L.; Vernon, C. A. *Chem. Ind. (London)* **1955**, 760.
65. Butcher, W. W.; Westheimer, F. H. *J. Am. Chem. Soc.* **1955**, 77, 2420.
66. DiSabato, G.; Jencks, W. P. *J. Am. Chem. Soc.* **1961**, 83, 4400.
67. Brown, D.; Hamer, N. *J. Chem. Soc. (London)* **1960**, 1155.
68. Samuel, D.; Silver, B. *J. Chem. Soc. (London)* **1961**, 4321.
69. Miller, D. L.; Ukena, T. *J. Am. Chem. Soc.* **1969**, *91*, 3050.
70. Rebek, J.; Gaviña, F. *J. Am. Chem. Soc.* **1975**, *97*, 1591, 3221.
71. Clapp, C. H.; Westheimer, F. H. *J. Am. Chem. Soc.* **1974**, *96*, 6710.
72. Clapp, C. H.; Satterthwait, A.; Westheimer, F. H. *J. Am. Chem. Soc.* **1975**, *97*, 6873.
73. Satterthwait, A. C.; Westheimer, F. H. *J. Am. Chem. Soc.* **1978**, *100*, 3197.
74. Rose, I. A. *J. Biol. Chem.* **1960**, *235*, 1170.
75. Robinson, J. L.; Rose, I. A. *J. Biol. Chem.* **1972**, *247*, 1096.
76. Levitzki, A.; Koshland, D. E., Jr. *Biochemistry* **1971**, *10*, 3365.
77. Buchanan, J. *Adv. in Enzymol.* **1973**, *39*, 91.
78. Lowe, G.; Sproat, B. S. *Chem. Commun.* **1978**, 783.
79. Kuo, D. J.; Rose, I. A. *J. Am. Chem. Soc.* **1978**, *100*, 6288.
80. Lipmann, F. *Adv. in Enzymol.* **1941**, *1*, 99.

RECEIVED May 21, 1979.

Studies on the Mechanism of Action and Stereochemical Behavior of Semisynthetic Model Enzymes

E. T. KAISER, HOWARD L. LEVINE, TETSUO OTUSKI, HERBERT E. FRIED, and ROSE-MARIE DUPEYRE

Searle Chemistry Laboratory, University of Chicago, 5735 South Ellis Avenue, Chicago, IL 60637

The objective of our research is the rational design of enzymatic catalysts. Toward this end, new oxidation–reduction enzymes have been prepared by the covalent modification of hydrolytic enzymes with coenzyme analogs. Among the semisynthetic enzymes that have been generated, the most effective one that we have studied to date is the covalent flavin–papain complex produced by the reaction of the sulfhydryl group of Cys-25 in the papain active site with 7 α-bromoacetyl-10-methylisoalloxazine. The kinetics of the oxidation of dihydronicotinamides observed using this flavopapain as the catalyst indicated that saturation occurs at low substrate concentrations and that the rate accelerations exceed an order of magnitude relative to models. Furthermore, examination of the stereochemistry of the oxidation of labelled NADH derivatives revealed that the flavopapain shows a marked preference for removing hydrogen from the 4A-position. The research we have performed has demonstrated the feasibility of tampering very significantly with an enzyme active site without destruction of the enzyme as a catalytic species.

The long range objective of our research is the rational design of enzymatic catalysts. In striving toward this goal, the two principal phenomena that must be considered are the acts of binding and of the subsequent intracomplex catalysis. Our research on the binding

0-8412-0514-0/80/33-191-035$05.00/0
© 1980 American Chemical Society

process encompasses work on peptides designed to simulate the major protein constituents of lipoproteins in their ability to bind lipids and to activate enzymes (1, 2, 3) and research on the preparation of polypeptide inhibitors of enzymes (4, 5). In this chapter, however, rather than discussing our research on modelling binding interactions, the focus is on the "chemical mutation" studies we have performed in which new and effective catalytic groups have been introduced into the active sites of existing enzymes. In particular, we have been examining the conversion of moderate-molecular-weight enzymes, which are relatively simple structurally (as enzymes go), into modified enzymes capable of catalyzing a variety of reactions important to both organic chemistry and biochemistry, including processes such as oxidation–reduction, transamination, and decarboxylation. Our experimental approach has involved the use of coenzyme analogs containing reactive functional groups—permitting them to be attached at (Equation 1) or near the periphery of the active sites of hydrolytic enzymes that are readily available, easily purified, stable, and can be immobilized on solid supports. As will be discussed, we have demonstrated that suitable coenzyme analogs can be attached covalently to such relatively simple enzymes in a manner that permits the binding sites of the enzyme to remain accessible to organic substrates. Our work has shown for the first time that it is possible to convert hydrolytic enzymes into modified semisynthetic enzymes capable of effective catalysis of oxidation–reduction reactions. We feel that this represents a significant scientific advance not only conceptually but also in terms of its potential implications for the use of enzymes in organic chemistry.

$$\text{Residue at active site} + \text{reactive coenzyme analog} \atop \downarrow$$

$$\text{covalently bound coenzyme analog} \atop \text{permitting substrate binding} \tag{1}$$

The Choice of the Enzyme and Residue to be Modified

In our initial research on semisynthetic enzymes, we examined briefly the modification of the serine proteinase α-chymotrypsin, perhaps the best understood of the proteolytic enzymes. A logical choice as a residue for alkylation in the active site of α-chymotrypsin is His-57. However, an examination of a three-dimensional model (Lab Quip) of chymotrypsin in which coenzyme analogs were covalently attached to His-57 suggested strongly that such modifications would block completely the enzyme's active site region and that the probability of new reactions being catalyzed by the modified enzyme would be low. Another possible site of modification of chymotrypsin that could be considered was Met-192. This residue, located on the periphery of the enzyme's active site, is alkylated readily by phenacyl bromides (6), yielding stabilized sulfonium ylids. Therefore, in our first attempt to prepare semisynthetic enzymes, we explored briefly the introduction of covalently bound coenzyme groups on the periphery of the active site of α-chymotrypsin by the reaction (under somewhat acidic conditions) of coenzyme analogs containing alkylating functional groups with the nonessential amino acid residue, Met-192 (7). Our early findings indicated that we could prepare semisynthetic enzymes in this manner, but we felt that reaction with an essential amino acid in the active site would make it more facile to follow the course of modification, aiding in the large-scale preparation of modified enzymes. Although in the case of α-chymotrypsin, alkylation of an essential amino acid residue would appear to preclude the possibility of preparing a catalytically viable semisynthetic enzyme, this is not the case for the cysteine proteinase papain.

Examination of the three-dimensional structure of papain (Lab Quip) indicated that because of the extended groove present in the vicinity of the active site residue Cys-25, it should be feasible to alkylate the sulfhydryl group with a coenzyme analog, still permitting the binding of potential substrates.

Development of methodology for the large-scale purification of papain for the preparation of coenzyme analog-modified enzymes has been important to our research. Blumberg et al. (8) reported that papain could be purified by taking advantage of the affinity of the enzyme for the immobilized peptide inhibitor, agarose–Gly–Gly–Tyr(Bz)–Arg. However, using pure tetrapeptide obtained commercially and standard coupling procedures, a significant purification of papain could not be achieved. M. O. Funk (9) found that both active and nonactivatable enzyme bound to a column prepared in this manner were eluted together by the use of deionized water. We showed that an affinity medium with properties similar to those reported by Blumberg et al. (8) was obtained by removal of the benzyl protecting group present on the Tyr residue prior to coupling with agarose. The deprotected tetrapeptide Gly–Gly–Tyr–Arg was also synthesized by an independent route, and this material also could be employed in the purification experiments (9).

The Choice of the Coenzyme Analog Modifying Agent

To give ourselves what we felt would be the best chance for success, most of our work to date has been concerned with the use of flavin derivatives as the enzyme modifying agents. There is already a great deal of information in the literature indicating that even model flavins can be quite effective catalysts (10). Since we could not be sure at the beginning of our research what would be the best enzyme template to which our coenzyme species might be attached, it seemed prudent to pick coenzyme systems like the flavin compounds, where the chances were reasonable that we could observe at least some catalysis with the enzyme-bound coenzyme. In our studies on the preparation of flavin-modified papain, we have concentrated on the species produced by the alkylation of the sulfhydryl group of Cys-25 in the active site by two flavin reagents, 8α-bromo-2′,3′,4′,5′-tetra-O-acetylriboflavin, I, and 7α-bromoacetyl-10-methylisoalloxazine, II. These flavins produce the semisynthetic oxidoreductases flavopapains III and IV, respectively. From the examination of a three-dimensional model of papain in which the flavin groups were attached covalently to the sulfhydryl of Cys-25, substantial differences in the activity of flavopapains III and IV were anticipated. In particular, the structures of three papain derivatives resulting from reaction of three chloromethyl ketone substrate analogs with the sulfhydryl of Cys-25 have been determined by x-ray diffraction (11). In each case, the carbonyl oxygen of what had originally been the chloromethyl ketone group was found to be near two potential hydrogen-bond donating groups, the backbone NH of Cys-25 and a sidechain NH_2 of Gln-19. In our model building, we oriented the flavin group in flavopapain IV so that the carbonyl group of what had been the halocarbon function in the modifying agent II was brought to a similar position. When this was done, it was seen easily that N-benzyl-1,4-dihydronicotinamide, a potential substrate, could be fitted snugly into the binding pocket of the enzyme in close proximity to the flavin group, suggesting that hydrogen transfer might be facilitated. On the other hand, in the case of flavopapain III, the carbonyl group attached to the flavin ring in IV is not present, and it seems probable that the flavin group would not be drawn into the active site region, as has been postulated for flavopapain IV. An additional point to consider is that the tetraacetyl ribose group at the 10-position of the flavin ring in flavopapain III is much larger than the methyl group at the 10-position in flavopapain IV; this steric difference also could contribute to the greater likelihood that the flavin group in IV would be drawn into the active site region than would be the case for III. The prediction based on model building that IV would be a more effective flavoenzyme than III has been borne out by our experimental work.

I   II

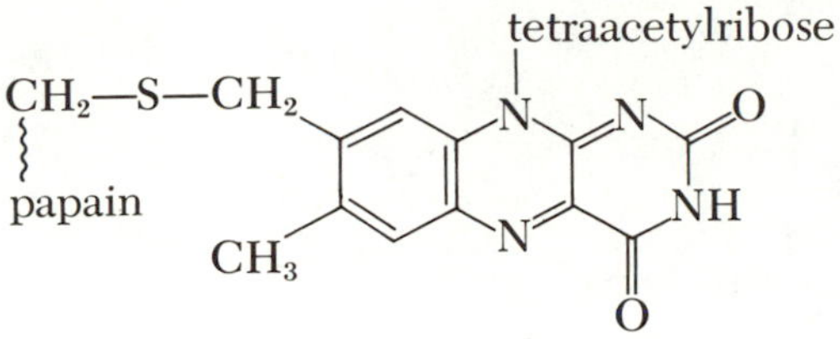

tetraacetylribose

CH_2—S—CH_2

papain

CH_3

III

CH_3

CH_2—S—CH_2—C

papain O

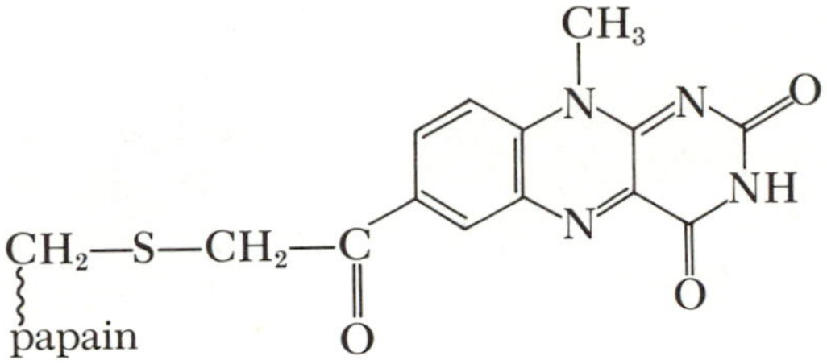

IV

Results Obtained with Flavopapain III and Related Derivatives

First, we will discuss the results obtained with flavopapain **III**. We have found that two discrete flavopapains differing in the oxidation state of the sulfur of the Cys-25 residue can be isolated on modification of papain using 8α-bromo-2′,3′,4′,5′-tetra-O-acetylriboflavin, **I**, under different conditions (*12*). In particular, when papain is treated with **I** at pH 7.5 and 25°C, modification produces **V**, the sulfone form of the modified enzyme, rather than **III**, the sulfide form. To obtain **III**, it was necessary to carry out the modification of papain under nitrogen, employing the reduced reagent dihydro-8α-bromo-2′,3′,4′,5′-tetra-O-acetylriboflavin, produced by in situ reduction. The assignment of the oxidation state of the sulfur in the flavopapains **III** and **V** was made by comparison of the spectral properties, oxidation behavior, and amino acid analyses determined for these modified enzymes to the properties determined for several naturally occurring enzymes and the corresponding glutathione derivatives. The mechanism by which flavopapain **V** is produced in the modification of papain by **I**, and why it is necessary to use the reduced (dihydro) form of **I** to give the modified enzyme III, remain to be clarified.

The specificity of flavopapains **III** and **V** has been examined by measuring the kinetics of oxidation of a series of dihydro-nicotinamides, including the N-substituted methyl-, ethyl-, n-butyl-, n-pentyl-, n-hexyl-, benzyl- and β-phenylethyl compounds. The variations made in the R substituent on the ring nitrogen of the dihydronicotinamides had less than a five-fold effect on k_{cat}/K_m. Also, the rate parameters measured indicated that flavopapains **III** and **V** were much less effective as catalysts for dihydronicotinamide oxidation than

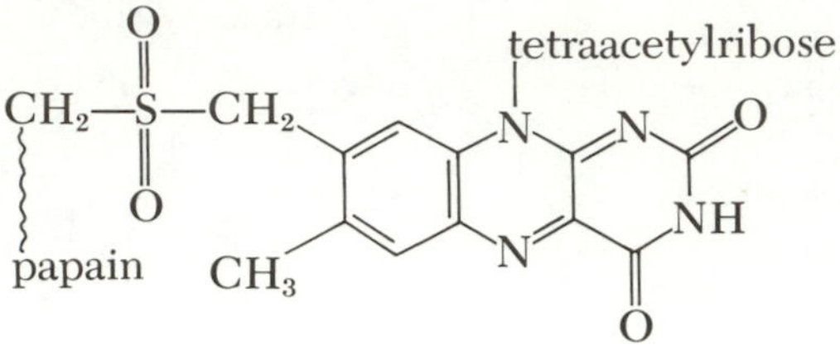

V

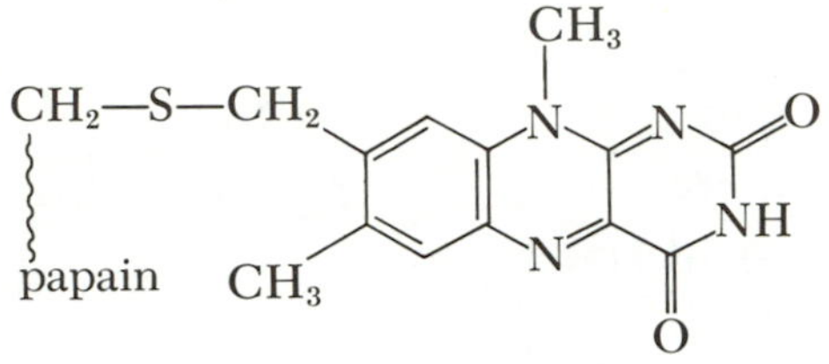

VI

was flavopapain **IV** (see below) and that the rate accelerations observed for the reactions carried out by **III** and **V**, as compared with the model system 2′,3′,4′,5′-tetra-*O*-acetylriboflavin, were modest, about a factor of two- or threefold. We have found that reduction in the size of the *N*-10 substituent, as in the case of the lumiflavin-modified papain species **VI**, has some effect in increasing the k_{cat}/K_m values for the flavopapain-catalyzed oxidation of dihydronicotinamides. However, the catalytic behavior of the modified enzyme **VI** is not particularly impressive. Indeed, our results to date with flavin derivatives bound to papain through a simple thioether linkage at the 8-benzylic position of the flavin system has been useful primarily in pointing out the importance of the carbonyl group in the acetyl sidechain of the flavin in the modified enzyme **IV**. In a sense, the results with the flavin-modified papains **III**, **V**, and **VI** calibrate our findings with flavopapain **IV**, indicating that the specific design of the flavin molecule attached to the enzyme is crucial for observing the production of an effective oxido-reductase.

Kinetic Results Obtained with Flavopapain IV

In contrast with our results with the flavopapains **III** and **V**, we have found that flavopapain **IV** can act as an effective catalyst for the oxidation of dihydronicotinamides (*13*, *14*). Using dihydronico-tinamide in excess, the oxidation reactions of *N*-benzyl-, *N*-ethyl-, *N*-propyl- and *N*-hexyl-1,4-dihydronicotinamide as well as that of NADH by flavopapain **IV** exhibit saturation kinetics at relatively low substrate concentrations. This behavior contrasts with the oxidation by model flavins or flavopapains **III** and **V**, where saturation kinetics are not observed at low dihydronicotinamide concentrations. Our kinetic measurements made under aerobic conditions for the oxidation of the dihydronicotinamides by **IV** have been explained by the scheme shown in Equation 2, where E_{ox} and EH_{2red} represent oxidized and reduced flavopapain, respectively, NRNH and NRN represent reduced and oxidized dihydronicotinamide, and ES represents a Michaelis complex formed between the substrate and the enzyme. If it is assumed that under the aerobic conditions used, the rate of oxidation of NRNH is independent of oxygen, that is, $k_0(O_2) >> k_{cat}$, then the scheme of Equation 2 leads to the expression of Equation 3 for the rate of oxidation of the dihydronicotinamide where $[E]_0$ is the total enzyme concentration. Computer fitting of the data for the oxidation of the dihydronicotinamides by flavopapain **IV** to Equation 3 gives the values of k_{cat} and K_m listed in Table I.

$$E_{ox} + NRNH \underset{K_m}{\rightleftharpoons} ES \xrightarrow{k_{cat}} EH_{2red} + NRN \qquad (2a)$$

$$EH_{2red} + O_2 \xrightarrow{k_0} E_{ox} + H_2O_2 \qquad (2b)$$

$$v = \frac{k_{cat}[E]_0[NRNH]}{K_m + [NRNH]} \qquad (3)$$

Table I. Rate Parameters[a] for the Oxidation of Dihydronicotinamides by Flavopapain IV[b] and the Model Flavin 7-Acetyl-10-Methylisoalloxazine

Dihydro-nicotin-amide	Enzymatic Reaction			Model Reaction Second-Order Rate Constant k $(M^{-1} sec^{-1})$
	$K_m \times 10^4$ (M)	k_{cat} (sec^{-1})	k_{cat}/K_m $(M^{-1} sec^{-1})$	
NEtNH[c]	1.3	0.72	5,540 (3,200)[d]	853
NPrNH[b]	1.0	0.81	8,100 (3,180)[d,e]	845
NHxNH[c]	0.42	0.44	10,480 (10,700)[d]	843
NBzNH	1.9	0.64	3,370 (3,430)[d,e]	185
NADH	2.2	0.015	68[d,e]	13

[a] We have discovered recently that k_{cat} and K_m increase dramatically upon addition of EDTA to the buffer solutions. (Identical results are observed when the buffer solutions are passed through a Chelex-100 column.) In contrast, little or no EDTA effect is seen in the model reactions involving 7-acetyl-10-methylisoalloxazine. Although the distilled water used was passed through a Continental Demineralizer, it is likely that trace metal-ion contaminants are responsible for the differences in kinetic parameters found for untreated flavopapain **IV** solutions compared with those measured for EDTA- (or Chelex-100-) treated enzyme solutions. Detailed studies are in progress to clarify the role of EDTA (or Chelex-100) in these systems.
[b] Measured at 25°C in 0.1M Tris–HCl containing 0.001M EDTA (or in buffer solutions passed through Chelex-100), pH 7.5, 0–5% ethanol (v/v).
[c] NEtNH, *N*-ethyl-1,4-dihydronicotinamide; NPrNH, *N*-propyl-1,4-dihydronicotinamide; and NHxNH, *N*-hexyl-1,4-dihydronicotinamide.
[d] Measured in buffer that was not passed through Chelex-100 and in the absence of EDTA.
[e] Data from Ref. *14*.

By means of stopped-flow spectrophotometry, the reaction of flavopapain IV with N-benzyl-1,4-dihydronicotinamide (NBzNH) has been studied anaerobically. Using conditions of excess substrate and following the decrease in absorbance of E_{ox} (at 427 nm) with time, we observed biphasic kinetic behavior. The experimental data could be fitted to a scheme using two consecutive first-order processes, and rate constants could be determined for the two phases. A reasonable interpretation of our results is that a labile intermediate is formed in the reaction, as illustrated in Equation 4, where ES' represents the intermediate. In this interpretation, the second, substrate-independent, slower phase of the reaction corresponds to the breakdown of the intermediate ES' (k_3 step in Equation 4). The initial, faster phase of the reaction corresponds to the formation of the intermediate from E_{ox} and N-benzyl-1,4-dihydronicotinamide. Using Equation 5, the calculated rate constants for this phase, k_f, can be related to K_s and k_2. Values of k_{cat} and K_m were calculated from the relationships of Equations 6 and 7, using the measured values of the kinetic parameters k_2, k_3, and K_s, and the numbers obtained were in reasonable agreement with the values obtained aerobically.

$$E_{ox} + NBzNH \underset{K_s}{\rightleftharpoons} ES \xrightarrow{k_2} ES' \xrightarrow{k_3} EH_{2red} + NBzN \qquad (4)$$

$$k_f = \frac{k_2[NBzNH]}{K_s + [NBzNH]} \qquad (5)$$

$$k_{cat} = \frac{k_2 k_3}{k_2 + k_3} \qquad (6)$$

$$K_m = K_s \left(\frac{k_3}{k_2 + k_3} \right) \qquad (7)$$

We have considered several explanations for the observation of biphasic kinetics in the anaerobic reduction of flavopapain IV by NBzNH. To test these explanations, kinetic measurements were performed under conditions of excess enzyme, following the disappearance of NBzNH. If the ϵ_{355} values for the free and enzyme-bound forms of the product N-benzylnicotinamide are different and if dissociation of this product from the enzyme was rate-determining, then the kinetics seen at 355 nm under conditions of excess flavopapain should be biphasic. The observed kinetics were pseudo first-order, however, providing no evidence for rate-limiting dissociation of N-benzylnicotinamide from the reduced form of flavopapain IV under these conditions. Another possibility we considered was that the species ES' may correspond to a chemically altered intermediate along the reaction pathway from oxidized-flavin–reduced-nicotinamide to reduced-flavin–oxidized-nicotinamide. However, whatever the nature of the intermediate, the species is unlikely to be one in which the double bond or the ring nitrogen of the dihydronicotinamide moiety is modified structurally, since changes in the absorbance caused by N-benzyl-1,4-dihydronicotinamide were not seen in stopped-flow spectral measurements on the formation of ES'. Thus, it is improbable that the intermediate is a type of flavin–nicotinamide radical pair.

The spectrum of flavopapain IV exhibits significant long-wavelength tailing extending beyond 600 nm. While the cause of this tailing is not certain, it may be the result of a charge-transfer interaction between a tyrosine or tryptophan residue in the active site with the flavin moiety. Similarly, the reduced flavopapain–N-benzylnicotinamide mixture has a spectrum that exhibits substantial wavelength tailing. In contrast, stopped-flow spectral measurements showed that the intermediate ES' observed in the anaerobic oxidation of N-benzyl-1,4-dihydronicotinamide by flavopapain IV exhibits far less long-wavelength tailing than either the starting flavopapain or the reduced-flavopapain–product mixture. A reasonable interpretation of this result is that if a charge-transfer complex exists between the flavin moiety and an aromatic amino acid in the enzyme's active site, then this complex is disrupted prior to the redox reaction with N-benzyl-1,4-dihydronicotinamide. The destruction of such a complex should result in a decrease in the long-wavelength absorption observed for the oxidized enzyme. It is important to emphasize that biphasic kinetics are observed anaerobically only when monitoring the reduction of E_{ox} at 427 nm. In aerobic solutions, only the decrease at 355 nm (NBzNH) can be monitored, since the absorbance at 427 nm remains unchanged as a result of the rapid oxidation of EH_{2red} by oxygen (i.e., $k_0(O_2) >> k_{cat}$). We hypothesize, therefore, that the transient intermediate ES', observed kinetically in the anaerobic reaction of flavopapain IV with N-benzyl-1,4-dihydronicotinamide, corresponds to a species in which the flavin moiety has moved to a distinctly different environment than it occupies in the Michaelis complex ES. In the step producing ES', the flavin remains in the oxidized state while the substrate, N-benzyl-1,4-dihydronicotinamide, remains reduced. After the realignment of the flavin moiety has taken place to give ES', the redox reaction proceeds via the normal mechanism.

Interpretation of the Kinetic Data for Flavopapain IV

As seen in Table I, the second-order rate constants for the reactions of the model flavin 7-acetyl-10-methylisoalloxazine (or, as described in Ref. *13*, the model-flavin–peptide derivative 7α-(S-

glutathionyl)-acetyl-10-methylisoalloxazine) with the dihydronicotin-amides examined in our studies compared with the values measured for k_{cat}/K_m for the corresponding flavopapain **IV**-catalyzed reactions show that the enzymatic reactions are more rapid. The rate accelerations measured in this way can exceed an order of magnitude, as illustrated in the cases of N-benzyl-, N-hexyl-, and N-propyl-1,4-dihy-dronicotinamides. In the enzymatic reactions, saturation kinetics were observed at low substrate concentrations, indicating that flavo-papain **IV** binds well to the dihydronicotinamide substrates. Since no binding of the dihydronicotinamides to the flavin–peptide derivative 7α-(S-glutathionyl)-acetyl-10-methylisoalloxazine (*14*) or to the flavoenzymes **III** and **V** was observed under comparable conditions, the binding of these substrates to flavopapain **IV** appears to be the result of specific enzyme–substrate interactions and not simply a phenomenon associated with having a peptide or even a polypeptide substituent attached to the flavin moiety. Furthermore, the variations in the k_{cat} and K_m values shown in Table I illustrate the important role of enzymatic specificity in the catalytic action of the semisynthetic enzyme **IV**.

In Table II, k_{cat}/K_m values for four naturally occurring flavoenzymes that utilize either NADH or NADPH as substrates are listed. Comparison of these data with those of Table I shows that the oxidation of N-hexyl-1,4-dihydronicotinamide by flavopapain **IV** proceeds faster than the reaction illustrated for one of the enzymes, somewhat slower than those illustrated for two of the enzymes, and much slower than that observed for bovine-heart NADH dehydrogenase. The k_{cat}/K_m values measured for the reaction of N-hexyl-1,4-dihydronicotinamide as well as those for the N-benzyl-, N-ethyl-, and N-propyl compounds

Table II.   Kinetic Parameters for Several Naturally Occurring Flavoenzymes

Enzyme	$k_{cat}(sec^{-1})$	$K_m(\mu M)$	$k_{cat}/K_m(M^{-1}\ sec^{-1})$
NADH-specific FMN oxido-reductase (B. Harveyi)[a]	15.5	47.5	3.26×10^5
NADPH-specific FMN oxido-reductase (B. Harveyi)[a]	34.0	40.0	8.50×10^5
Old yellow enzyme (Yeast)[b]	0.71	220	3.23×10^3
NADH dehydrogenase (bovine heart)[c]			$\sim 10^8$

[a] Data from Ref. *17*.
[b] Data from Ref. *18*.
[c] Data from Ref. *19*.

with flavopapain **IV** are comparable with those found for NADH or NADPH reactions with naturally occurring flavoenzymes. Our kinetic results, therefore, demonstrate that the semisynthetic enzyme flavopapain **IV** is capable of binding reduced nicotinamide substrates with considerable specificity as well as catalyzing their oxidation with modest, though appreciable, rate accelerations.

Stereochemical Studies

The next logical step in elucidating the enzymatic behavior of the flavopapain **IV** species was to investigate the stereochemistry of hydrogen transfer from the dihydronicotinamide to the flavoenzyme. The $C(4)$ position of the dihydronicotinamides is a prochiral center. Natural pyridine nucleotide-linked oxido-reductases are capable of distinguishing the diastereotopic methylene protons at this position of NAD(P)H, transferring stereospecifically a hydride ion to their substrate. Ideally, the stereochemistry of the hydride transfer reaction between flavopapain **IV** and the dihydronicotinamide substrates that we studied could be determined using dihydronicotinamides stereospecifically labelled with deuterium at the $C(4)$ position. However, when we started the work described in this chapter, no means was available to label stereospecifically model compounds like N-benzyl- and N-hexyl-1,4-dihydronicotinamide in a simple fashion. Consequently, we studied the reaction of flavopapain **IV** with (4A ^{2}H) NADH and (4B ^{2}H) NADH (15). Unfortunately, as can be seen from Table I, NADH was the poorest dihydronicotinamide substrate for which we had studied the kinetics in detail and is probably not an ideal choice for a stereochemical study of flavopapain action.

A complicating factor in studying the oxidation of deuterated NADH derivatives is the possibility that nonstereospecific exchange of $C(4)$ hydrogen between the product NAD$^+$ and the reactant NADH would take place. To minimize exchange under our reaction conditions, we used NADH concentrations less than 0.5mM. The results of our experiments are listed in Table III. It is quite clear from Table III that flavopapain **IV** showed considerable, but not complete, prefer-

Table III. Product Ratios and Rate Parameters for the Oxidation of NADH and Deuterated NADH Derivatives by Flavopapain IV

NADH Derivative	$(4^2H)NAD^+/NAD^{+a}$	$k_{cat}/K_m (M^{-1} sec^{-1})$
NADH	—	68.1
[4A ^{2}H]NADH	0.47	17.3
[4B ^{2}H]NADH	7.33	43.6
[4AB ^{2}H$_2$]NADH	—	3.2

^{a} Calculated from the percentage of NAD$^+$ observed in the NMR spectrum of the isolated product.

ence for the transfer of the $4A$ (pro-R) hydrogen of NADH to the covalently bound flavin. If we assume that the stereoselectivity exhibited by flavopapain **IV** results from different rates of hydride transfer from the A and B sides of the dihydronicotinamides, then the scheme of Equation 8 can be used to explain our results. In Equation 8, A represents the hydrogen (or ^2H) on the A side and B represents the group on the B side. The hydrogen (or ^2H) being transferred is denoted by the overbar on A or B, and the air oxidation of the reduced enzyme E′ back to the oxidized form E is postulated to occur rapidly relative to the hydride-transfer process under our experimental conditions.

By combining the results of rate measurements (with the various dihydronicotinamides—undeuterated, monodeuterated and dideuterated) and the results of product determinations (on the amount of hydrogen or deuterium transfer), the ratio for the rate of hydrogen transfer from the A side to that from the B side can be calculated, in principle (15). However, we could not obtain a meaningful solution in this way—either because of experimental problems or because of the possible presence of kinetically significant intermediates along the reaction pathway from ES to E′ + product (Equation 8). In view of this, we estimated the ratio of rate constants for hydrogen transfer from the A face vs. the B face from the relative values of $(k_{cat}/K_m)_{H\bar{D}}$ and $(k_{cat}/K_m)_{D\bar{H}}$, assuming that secondary isotope effects can be neglected. From this approach, we have found that there is an approximately seven-fold preference for transfer to flavopapain **IV** of hydrogen from the A face of NADH.

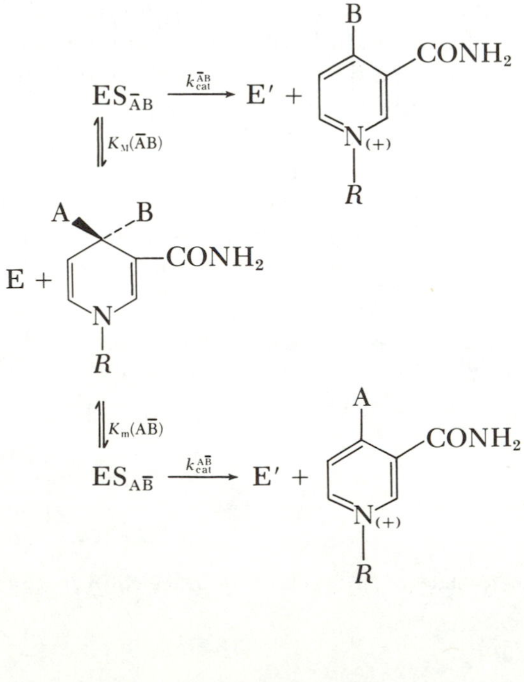

$$(8)$$

Examination of the three-dimensional model that we have constructed for flavopapain **IV** reveals two reasonable binding modes for the dihydronicotinamide moiety of NADH. In the first mode, the dihydronicotinamide ring lies near the α-carbon of Trp-26 with the carboxamide sidechain pointing out into the surrounding solvent. In the second mode, the plane of the dihydronicotinamide ring is rotated 180° so that the carboxamide sidechain is now lying in the interior of the protein. This binding mode could be stabilized by hydrogen bonds between the carboxamide sidechain and the carbonyl oxygen of either Cys-25 or Ala-16. In addition, a slight movement of the imidazole sidechain of His-159 would permit hydrogen bonding between this residue and the substrate. In light of the observation that the enzyme showed preference for removal of the 4A hydrogen rather than the 4B hydrogen from the dihydronicotinamide ring, this latter binding mode presumably is favored.

Since NADH is a relatively poor dihydronicotinamide substrate for flavopapain **IV**, we would have preferred to study the stereochemistry of the oxidation of a better dihydronicotinamide substrate (such as the N-ethyl-, N-propyl-, N-hexyl-, or N-benzyl-1,4-dihydronicotinamide). However, to the best of our knowledge, the preparation of these chirally labelled dihydronicotinamides has not been described, although we have been considering several possible routes. The only model dihydronicotinamide derivatives we have found in the literature that seem to be suitable for stereochemical studies are the compounds described by A. Ohno and co-workers (16). Ohno has supplied us with N-(S)-α-methylbenzyl-1-propyl-2-methyl-4-(S)-methyl-1,4-dihydronicotinamide as well as the corresponding RR- and SR-isomers. Our studies on these compounds are still in progress, but we have found already with the (S,S)-dihydronicotinamide derivative that flavopapain **IV** catalyzes the oxidation of this substrate with $k_{cat}/K_m = 4500M^{-1}sec^{-1}$, a value comparable with those for the **IV**-catalyzed oxidation of N-ethyl-, N-propyl-, and N-benzyl-1,4-dihydronicotinamides. Analysis of the reactions of flavopapain **IV** with the 4-(R)-methyl derivatives will be interesting and should provide more information about the possible binding modes for the dihydronicotinamides.

Finally, we are beginning to apply our flavopapain systems to the oxidation and reduction of a wider variety of substrates. Also, the approach we have employed to the development of semisynthetic enzymes is not confined to flavoenzyme systems, and research is currently in progress in our laboratory on the development of new semisynthetic enzymes, especially including NAD$^+$-analog systems. The research we have performed demonstrates the feasibility of tampering significantly with an enzyme's active site without destroying the enzyme as a catalytic species. We have shown that it is possible to design rationally, by a chemical method, a catalytic site on an enzyme that can carry out the desired type of reaction.

Acknowledgments

We wish to thank Y. Nakagawa for his help in the initial phase of the research described here and for many useful discussions. Our work was supported in part by National Science Foundation Grant APR 72-03577 and AER 77-14529 (E.T.K.) and by National Research Service Award HS-5-BM-07151 from the National Institutes of General Medical Sciences (H.L.L.).

Literature Cited

1. Kroon, D. J.; Kaiser, E. T. *J. Org. Chem.* **1978**, *43*, 2107.
2. Kroon, D. J.; Kupferberg, J. P.; Kaiser, E. T.; Kézdy, F. J. *J. Am. Chem. Soc.* **1978**, *100*, 5975.
3. Fukushima, D.; Kupferberg, J. P.; Yokoyama, S.; Kroon, D. J.; Kaiser, E. T.; Kézdy, F. J. *Abstracts, Medicinal Chemistry Division, American Chemical Society National Meeting, 1979.*
4. Tan, N. H.; Kaiser, E. T. *J. Org. Chem.* **1976**, *41*, 2787.
5. Tan, N. H.; Kaiser, E. T. *Biochemistry* **1977**, *16*, 1531.
6. Sigman, D. S.; Torchia, D. A.; Blout, E. R. *Biochemistry* **1969**, *8*, 4560.
7. Howe, N.; Ballesteros, A.; Delker, W.; Kaiser, E. T., unpublished data.
8. Blumberg, S.; Schechter, I.; Berger, A. *Eur. J. Biochem.* **1970**, *15*, 97.
9. Funk, M. O.; Nakagawa, Y.; Skochdopole, J.; Kaiser, E. T. *Int. J. Pept. Protein Res.* **1979**, *13*, 296.
10. Bruice, T. C. In "Progress in Bioorganic Chemistry"; Kaiser, E. T., Kézdy, F. J., Eds.; Wiley-Interscience: New York, 1976; Vol. 4, p. 1–87.
11. Drenth, J.; Kalk, K. H.; Swen, H. M. *Biochemistry* **1976**, *15*, 3731.
12. Otsuki, T.; Nakagawa, Y.; Kaiser, E. T. *J. Chem. Soc., Chem. Commun.* **1978**, 457.
13. Levine, H. L.; Nakagawa, Y.; Kaiser, E. T. *Biochem. Biophys. Res. Commun.* **1977**, *76*, 64.
14. Levine, H. L.; Kaiser, E. T. *J. Am. Chem. Soc.* **1978**, *100*, 7670.
15. Levine, H. L.; Kaiser, E. T. *J. Am. Chem. Soc.* **1980**, *102*, 342.
16. Ohno, A.; Ikeguchi, M.; Kimura, T.; Oka, S. *J. Chem. Soc., Chem. Commun.* **1978**, 328.
17. Jablonski, F.; De Luca, M. *Biochemistry,* **1977**, *16*, 2932.
18. Honma, T.; Ogura, Y. *Biochim. Biophys. Acta,* **1977**, *9*, 484.
19. Singer, T. P. In "Biological Oxidations"; Singer, T. P., Ed.; Interscience: New York, 1968; p. 339.

RECEIVED May 15, 1979.

Asymmetric Transformation of α-Substituted Carbonyl Compounds Via Enamine or Iminazoline Formation

S. YOSHIKAWA

Department of Synthetic Chemistry, University of Tokyo, Tokyo 113, Japan

H. MATSUSHITA and S. SHIBATA

Central Research Institute, The Japan Tobacco & Salt Public Corporation, 6-2-Umegaoka, Midori-ku, Yokohama, Kanagawa 227, Japan

Acid hydrolysis of iminium salts of enamines composed of optically active secondary amines and racemic α-substituted carbonyl compounds yielded corresponding optically active compounds. During these experiments, 1-(β-methylstyryl)-2-methylpiperidinium chloride was isolated and characterized. This enammonium salt was found to exchange deuterium and then change easily to the corresponding iminium salt. The relationships of chirality and deuteration are characterized. By analogy, the asymmetric transformation mechanisms of the above reaction was extended to iminazoline ring formation. (S)- and (R)-alanine were converted into optically labile iminazoline derivatives. When (S)-amino methylpyrrolidine was used as a diamine component, (S)-alanine was converted to (R)-alanine in an optical yield of 93.8% (e.e.).

Optical Activation via Enamine Hydrolysis Using Optically Active Acid

We observed optical rotation in the recovered carbonyl compound when an enamine, as the salt of an optically active acid, was hydrolyzed. An example of this is illustrated in Figure 1 (*1*), where the racemic carbonyl compound must be chiral.

D-10-camphorsulfonic acid is added to a benzene solution of the pyrrolidine enamine of α-phenylpropionaldehyde. Then water is

0-8412-0514-0/80/33-191-049$05.00/0
© 1980 American Chemical Society

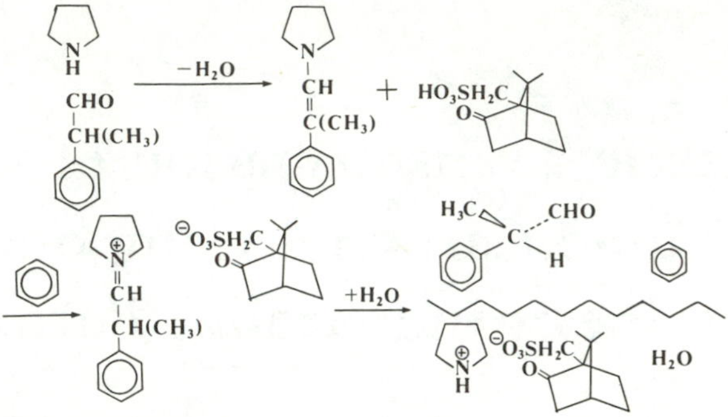

Figure 1.   Hydrolysis of the enamine salt of an optically active acid

added dropwise and hydrolysis is carried out by vigorously stirring the mixture. The original compound is found in the benzene layer, which exhibits optical rotation. Illustrated in Figure 2 are chiral α-carbonyl compounds formed by this method. These compounds usually produce optically active isomers, as illustrated in Table I. The optical yield depends upon the combination of enamine components.

The secondary amines of a five-membered ring and six-membered ring are used as the amine component. Since 2-methylpiperidine is chiral, this piperidine is believed to form diastereomeric iminium salts with chiral carbonyl compounds. (See section entitled "Asymmetric Transformation via Iminazoline Formation" for further discussion.)

No optical rotation is found in the recovered substances when achiral phenyl acetaldehyde or cyclohexanone is the carbonyl compound used.

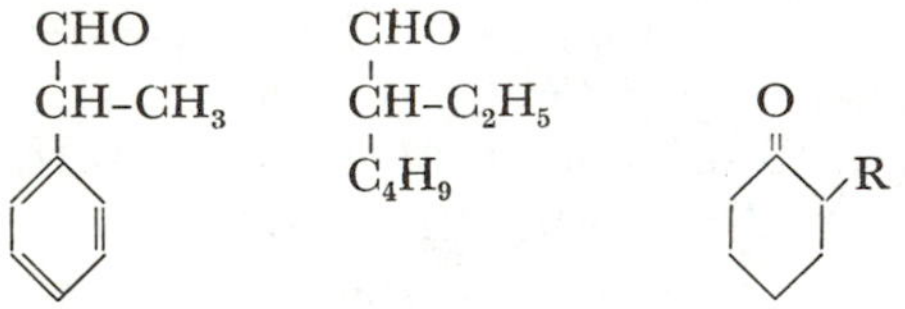

(1)   (2)   (3) R=CH$_3$
                                        (4) R=CH$_2$CH$_2$CN
                                        (5) R=CH$_2$CH$_2$COOCH$_3$

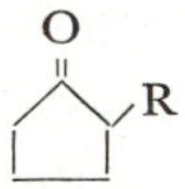

(6) R=CH$_2$CH$_2$CN
(7) R=CH$_2$CH$_2$COOCH$_3$

Bulletin of the Chemical Society of Japan

*Figure 2.   Chiral α-carbonyl compounds from the hydrolysis of
enamines (1)*

Table I.   Enamine Hydrolysis Products (*1*)

Carbonyl Compounds	Amines	$[\alpha]_D^{23}$	Recovery (%)
1	piperidine	+16.2 (c3.67)	89.6
	pyrrolidine	+13.8 (c7.10)	84.1
	2-methyl-piperidine	+15.9 (c1.15)	92.0
2	piperidine	+0.22 (c55.1)	82.5
	pyrrolidine	+0.18 (c20.5)	62.8
	2-methyl-piperidine	+0.11 (c28.8)	78.0
3	piperidine	+0.63 (c2.30)	73.1
	pyrrolidine	+0.39 (c10.8)	88.8
	2-methyl-piperidine	+0.89 (c20.1)	80.0
4	pyrrolidine	+0.058(neat)	51.5
5	pyrrolidine	+0.085(neat)	39.1
6	pyrrolidine	+0.018(neat)	42.2
7	pyrrolidine	+0.011(neat)	38.8
Phenyl-acetaldehyde	piperidine	+0.000(c11.5)	80.7
	pyrrolidine	+0.000(c22.4)	68.9
Cyclohexanone	piperidine	+0.000(c18.1)	62.3
	pyrrolidine	+0.000(c57.1)	71.2

Bulletin of the Chemical Society of Japan

Illustrated in Table II are the effects of different optically active acids used for producing the piperidine enamine salt of α-phenylpropionaldehyde. Based on the rate of hydrolysis and the asymmetric transformation of the recovered substances, a strong acid is most effective.

Illustrated in Table III are the solvent effects. The carbonyl compound used is α-phenylpropionaldehyde and the optically active acid is D-camphorsulfonic acid. The figure reveals that when hydrolysis is carried out, less miscible solvents are more effective suggesting that interfacial reactions are effective for stereoselectivity of asymmetric transformations.

Table II.   Results of Acid Hydrolysis to Form the Piperidine Enamine Salt of α-Phenylpropionaldehyde (*1*)

Acids	Time (min)	$[\alpha]_D^{23}$	Recovery (%)
L-Tartaric acid	80	−0.93(c7.3)	97.8
D-Camphoric acid	75	+2.61(c3.4)	53.3
D-Quinic acid	80	+0.56(c3.2)	29.8
D-10-Camphorsulfonic acid	35	+16.2(c3.67)	89.6

Bulletin of the Chemical Society of Japan

Table III. Solvent Effect on Optical Purity (*1*)

	Benzene	Dioxane	CH_3CN	EtOH	McOH	H_2O
Piperidine enamine						
$[\alpha]_D^{23}$	+16.2 (c3.67)	+10.0 (c1.81)	+2.81 (c2.10)	+0.091 (c23.3)	+0.100 (c31.9)	+0.000 (c50.8)
Recovery (%)	89.6	28.0	42.3	61.7	78.0	89.7
Pyrrolidine enamine						
$[\alpha]_D^{23}$	+18.3 (c6.77)	+11.8 (c3.88)	+1.08 (c11.3)	+0.026 (c35.5)	+0.082 (c41.3)	+0.000 (c67.1)
Recovery (%)	84.1	32.7	38.1	49.9	58.9	92.3

Bulletin of the Chemical Society of Japan

Optical Activation via Achiral Acid Hydrolysis of Enamines Containing Optically Active Secondary Amines

Next, illustrated in Figure 3, is the case when optically active (+)2-methylpiperidine is used and hydrolysis is carried out with an achiral hydrochloric acid solution (2).

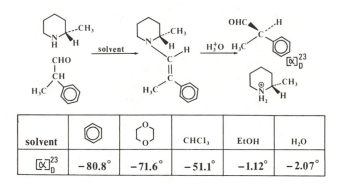

solvent			CHCl₃	EtOH	H₂O
$[\alpha]_D^{23}$	$-80.8°$	$-71.6°$	$-51.1°$	$-1.12°$	$-2.07°$

Figure 3. Acid hydrolysis of enamines derived from (+)2-methyl-piperidine

The solvent effects show almost the same tendencies as when hydrolysis is effected by optically active acids. The absolute value of the optical rotation of the recovered α-phenylpropionaldehyde is much larger in these examples. This indicates that the use of an optically active amine is more effective for asymmetric transformation than the use of an optically active acid. In addition, this method has been proved very effective for each type of carbonyl compound previously mentioned (3).

If a proton is added to enamine, diastereomers are possible products when an optically active secondary amine is used, assuming the iminium salt has been produced. To investigate this matter, the series of reactions shown in Figure 4 is performed. The salt is made from the acid using HCl, and D₂O hydrolysis of the resultant salt is carried out. Since deuterium is not incorporated into the recovered α-phenylpropionaldehyde, chirality is believed to be induced before the hydrolysis (4).

Illustrated in Figure 5 is the ¹H NMR spectra of two types of iminium salt together with that of the enamine. The resonance signal shown in Figure 5(b) denotes two types of chemical shift, which reveals the existence of diastereomers whose concentration ratio is different than one. The resonance signal in Figure 5(c) is the iminium salt that is made with DCl, then hydrolyzed with H₂O.

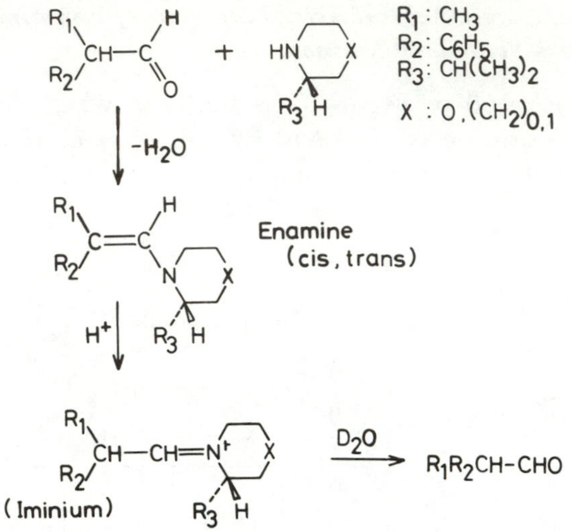

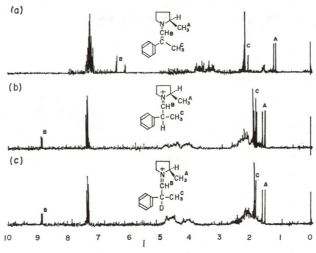

Figure 4. Formation and hydrolysis of an enamine in the presence of D_2O

Bulletin of the Chemical Society of Japan

Figure 5. ^1H NMR spectra of (a, b) two types of iminium salt and (c) the iminium salt prepared as described in Figure 7 (4)

Mechanism of the Optical Activation

While an experiment with the proton addition to the enamine was being carried out, a very significant observation was made (5) when an enammonium salt was isolated. A white precipitate was produced when the enamine, synthesized from α-phenylpropionaldehyde and optically active 2-methylpiperidine, was dissolved in anhydrous benzene and dry HCl gas was added to it (cooling with an ice–salt bath). This compound is isomeric with the iminium salt produced earlier; however, it exhibits a different NMR spectrum. Moreover, when the NMR spectrum is recorded after the salt is warmed to 60°C, the compound exhibits nearly the same spectral characteristics as the iminium salt. Shown in Figure 6 is the NMR spectrum taken of salt produced by DCl. Even in this case, the pattern is different from the one shown before (Figure 5) (6).

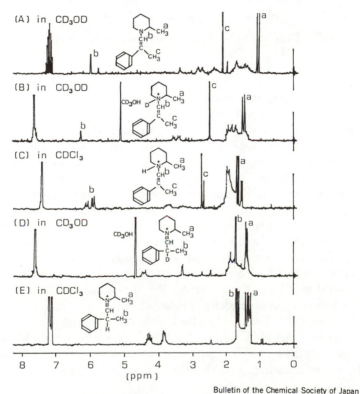

Bulletin of the Chemical Society of Japan

Figure 6.   100MHz ¹H NMR spectra, in CD₃OD and CDCl₃, of enamine and enammonium salts derived from α-phenylpropionaldehyde and 2-methylpiperidine (6)

When acidic hydrolysis is carried out with the material supposed to be enammonium salt, (+)α-phenylpropionaldehyde is produced in excess. When the same substance is heated at 60°C for several hours, a levorotatory material is produced in excess.

Moreover, when hydrolysis is carried out in DCl, a deuterated aldehyde is recovered (*see* Figure 7).

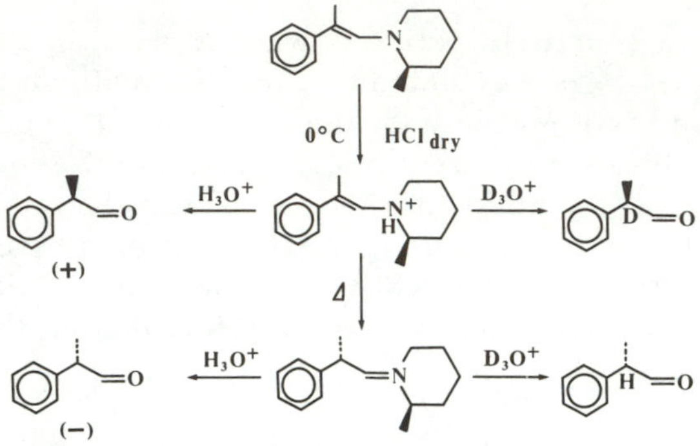

Figure 7. Reaction scheme of the hydrolysis of an enammonium salt
with DCl

To corroborate this phenomenon, enammonium salt is made using
DCl and is decomposed in light water. Hydrogen replaces deuterium
in the recovered aldehyde. When the enammonium salt is heated before hydrolysis, the resultant aldehyde is deuterated (*see* Figure 8).

From these data, we have proposed the following scheme.

The chirality is decided when a porton is added to the enamine
nitrogen on the amine group. Furthermore, when rearrangement to the
iminium salt occurs, the chirality of the iminium proton is decided by
this first proton. Upon hydrolysis of the enammonium salt, water is
added from the opposite side of the enammonium proton and asym-
metry of the recovered carbonyl compounds results. When an iminium
salt with established chirality is hydrolyzed, the resultant carbonyl
compound will have the reversed chirality because the hydrolysis occurs at the carbon–nitrogen double bond.

This result is believed to be brought about entirely by chance by
the fact that the HCl salt of 2-methyl-1-(α-methylstyryl) piperidine
was separated as a solid outside the benzene solution system. Several
trials were conducted with every type of enamine composed of

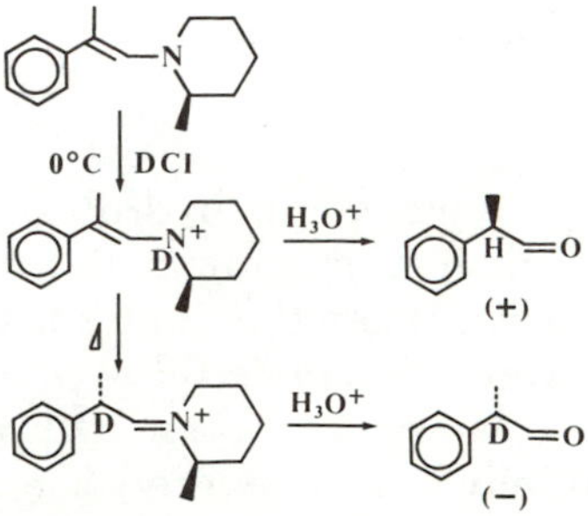

Figure 8. Reaction scheme for the
hydrolysis of an enammonium salt
after heating

numerous carbonyl compounds and secondary amines (7). However, no clear result has been obtained as yet.

Illustrated in Table IV is the asymmetric transformation of predominant chirality in the pyrrolidine that has the same chirality (8). With regard to I, hydrolysis is carried out at 10°C; II indicates hydrolysis at 50–60°C. Different chiralities are produced at the two temperatures.

These results indicate that reaction conditions can be adjusted to yield preferentially a product with the desired optical activity.

Table IV. Asymmetric Transformation in Recovered α-Phenylpropionaldehyde[a]

Amine component		Chirality of Recovered
	A	α-Phenylpropionaldehyde
	CH$_3$	S
	COOC$_2$H$_5$	S
	CONH$_2$ { I	R
	II	S
	CON⟨⟩  I	R
	II	S
	CH$_2$N⟨⟩	R

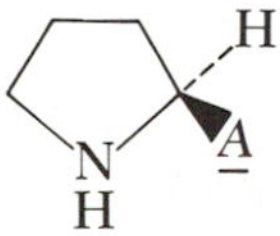

[a] Hydrolysis temperature: I, 10°C; II, 50–60°C.

Asymmetric Transformation via Iminazoline Formation

Reference was made earlier to a study to expand the idea of the asymmetric transformation. Figure 9 is a summary of previous work carried out on the determination of the activity of Bacitracin A (9). The activity of Bacitracin A is decreased at pH 4.5–6.5, of which is believed to be based upon epimerization of carbon. The activity is recovered in 3% acetic acid.

A comparison of enamine and iminazoline structures is shown in Figure 10. There are two types of structural isomerism in the proton-added type of enamine. Differences between the enamine and iminazoline arise because the former is stabilized as the enamine type, whereas the latter is stabilized as the imine type.

Cyclization of S-aminopyrrolidine induced from S(−)-proline and the iminoether of carboxylic acid were used to produce an iminazole ring. Alanine with protected groups is used as the carboxylic acid component. The yield of iminazoline, based on iminoether, was 86% (Figure 11). The absolute configuration and optical purity of an alanine residue contained in a iminazoline derivative were determined by comparison with a sample of alanine after hydrolysis was carried out (6N HCl, 110°C, 20 hr).

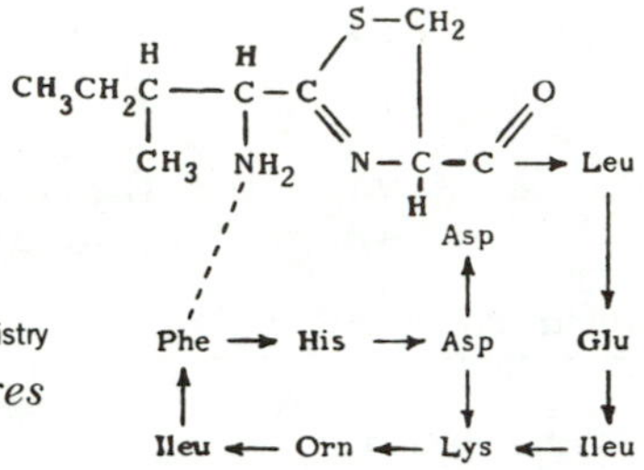

Journal of Organic Chemistry

Figure 9.   Main structural features
                of Bacitracin A (9)

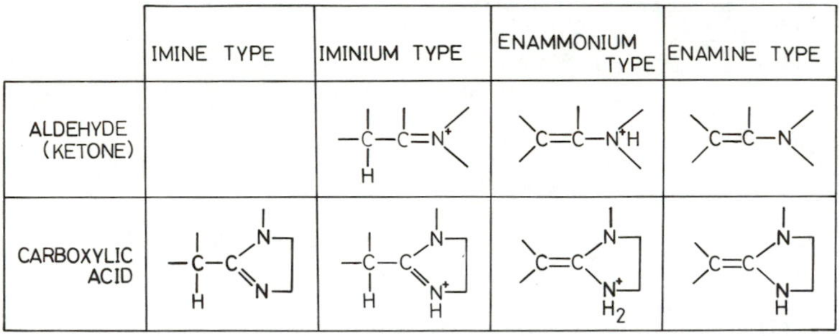

Figure 10.   Enamine and iminazole structures

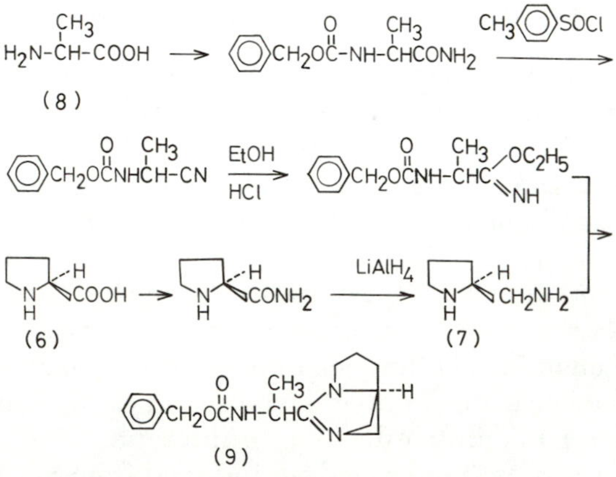

Figure 11.   Cyclization of S-aminopyrrolidine to form an iminazole
ring

In Table V, the yields of hydrolysis and optical purity of amino acids that are recovered from iminazoline are shown. These iminazolines are isolated from the reaction mixture of the scheme shown in Figure 11.

Differences were observed with the types of diamine used. In the case of aminomethylpyrrolidine, iminazoline yield is 86%, from which approximately 90% alanine is recovered; its optical purity is approximately 93% R-enantiomeric excess regardless of whether S-alanine or R-alanine is the starting material. Thus, we are confident that the alanine has been transformed asymmetrically (*10*).

On the other hand, in the case of isopropylpropylene diamine, the configuration of the recovered alanine is retained. Also, the hydrolysis is difficult and the iminazoline yield is very low.

To ascertain the presence of asymmetric transformation, mutarotation in solution was determined. Figure 12 shows the mutarotation of the above products of alanyl iminazolines in methanol solution.

Table V. Asymmetric Transformation via Iminazoline

Diamine Component	Chirality of starting Alanine	Recovered Alanine Chirality	e.e. %	yield
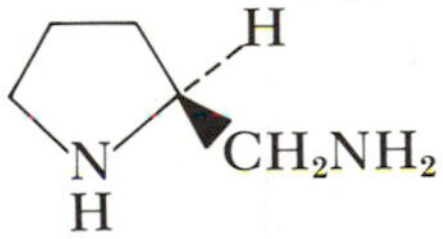 (pyrrolidine, H, CH₂NH₂ structure)	S	R	93.8	89.2
	R	R	93.2	90.1
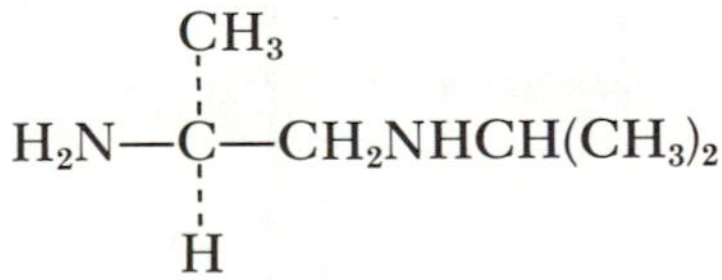 $H_2N-\overset{CH_3}{\underset{H}{C}}-CH_2NHCH(CH_3)_2$	S	S	91.9	5.6
	R	R	91.8	4.3

In the case of S-aminopyrrolidine, equilibrium is established when the R content reaches 35.5 (e.e.)%, regardless of whether S-alanine or R-alanine is the starting material. The S-propylene diamine derivative of S-alanine is almost a racemate (*see* Figure 12).

It follows from this that although epimerization occurs in the methanol solution, the activation energy to change the chirality of the alanine methine carbon is very different.

The mutarotation when an amino-acid component is changed from alanine to proline is illustrated in Figure 13. In the case of amino pyrrolidine, mutarotation occurs, whereas with N-isopropylpropylene diamine, no mutarotation is observed. The deuterium-incorporation velocity of the methine proton in heavy methanol (CH_3OD) parallels the mutarotation velocity.

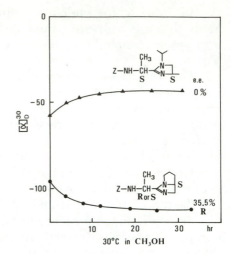

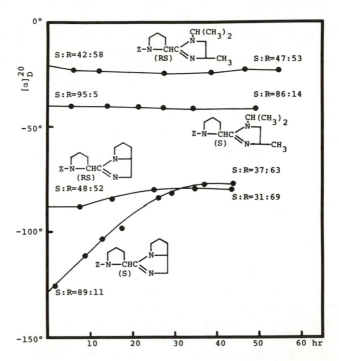

Figure 12. Mutarotation of iminazolines

30°C in CH₃OH

Figure 13.   Mutarotation of imisazoline derivatives of proline; imid-
azoline derivatives were refluxed in isopropanol.

For these reactions to occur, we believe that an iminazoline ring
and an α-branched carbon must be coplanar. A reaction sequence is
depicted in Figure 14. An environment conductive to asymmetric
transformation is required.

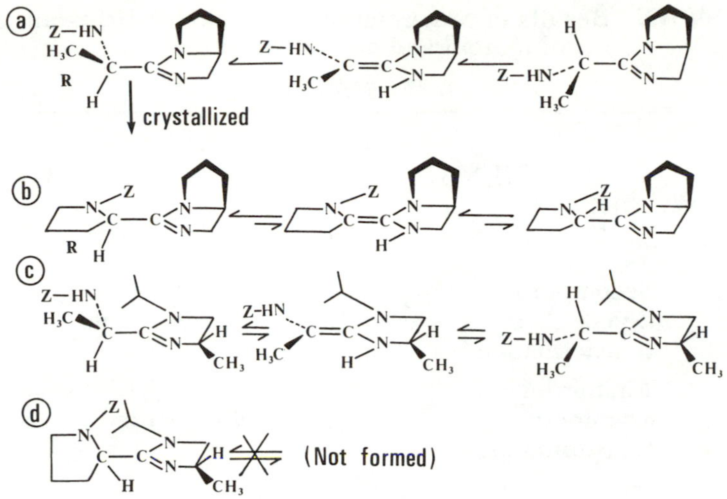

Figure 14.   Mutarotation of an iminazoline: (a) removal from the system by crystallization; (b) asymmetric transformation, but no preferential crystallization; (c) racemization; and (d) no epimerization

The optical purity of the iminazoline consisting of alanine and 2-aminopiperidine can be increased to 93.8 (e.e.)% by crystallizing it. In Figure 14(a), once a crystal is removed to the outside of the system, it forms the solution in which asymmetric transformation is carried out; the second-order asymmetric transformation where further transformation develops is obtained. In the case of crystallization of a diastereomer, 100 (e.e.)% might be possible. In this instance, the alanine appears to be racemized when the iminazoline ring is decomposed with strong acid.

In Figure 14(b), the combination of proline and 2-aminomethyl pyrrolidine, the asymmetric transformation was observed but no preferential crystallization occurred. In Figure 14(c) only racemization was observed. Moreover, in Figure 14(d), non-coplanarity of the iminazoline and α-branched carbon prevented the double-bond shift and epimerization did not occur.

Table VI is a summarization of our results; amino acid and amine components are arranged into four representative cases.

In the mutarotation experiment of iminazoline from alanine and isopropylpropylenediamine, the recovered alanine residue racemized entirely in methanol solution. Why, then, can the iminazoline containing the optically retained amino acid residue be synthesized?

To answer this question, the iminazole mutarotation was measured in various achiral solvents. Figure 15 and 16 show progressive mutarotation in solvents from which iminazole derivatives are isolated as solids.

**Table VI. Results of Mutarotation Experiments Using an
Amino Acid and an Iminazoline**

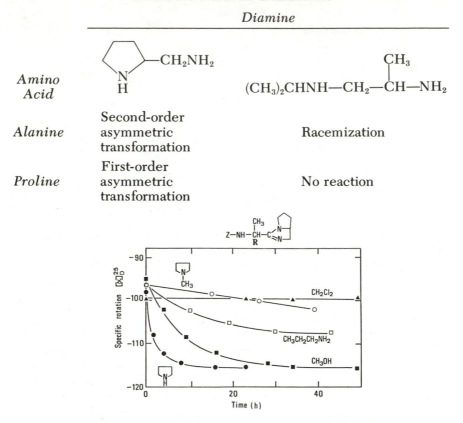

Figure 15.   Progressive mutarotation of an iminazole

In methylene dichloride solution, no mutarotation was observed. Methanol has moderate ability in regard to mutarotation. Amine solutions used in this experiment are all one molar (in methylene dichloride).

In both instances, the mutarotation velocity with pyrrolidine was faster than with the other amines. When iminazoline synthesis was conducted between alanyliminoether and 2-aminomethylpyrrolidine, excess pyrrolidine in the reaction media probably accelerated the epimerization reaction. Racemization is faster than crystallization of this diastereomer, which is less soluble than the other diastereomers. In the case of isopropylpropylenediamine, the excess amine racemizes poorly, so the crystallization was faster than the racemization.

Figure 17 shows the results of this experiment with α-phenylbutyric acid replacing alanine. In this reaction, the asymmetric transformation is first order.

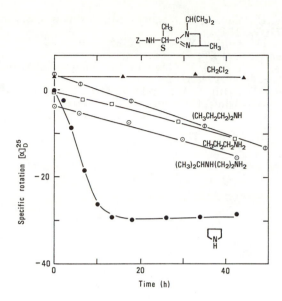

Figure 16. *Progressive mutarotation of an iminazole*

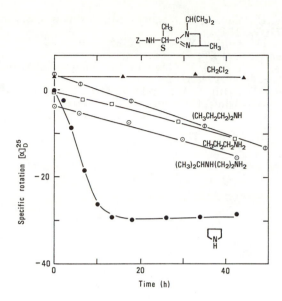

Figure 17. *Asymmetric transformation using α-phenylbutyric acid*

Acknowledgments

This work was partially supported by a Grant-in-Aid for Scientific Research from the Ministry of Education, Japan (No. 335211).

Literature Cited

1. Matsushita, H.; Noguchi, M.; Saburi, M.; Yoshikawa S. *Bull. Chem. Soc. Jpn.* **1975**, 48, 3715.
2. Matsushita, H.; Noguchi, M.; Yoshikawa, S. *Chem. Lett.* **1975**, 1313.
3. Matsushita, H.; Tsujino, Y.; Noguchi, M.; Yoshikawa, S. *Bull. Chem. Soc. Jpn.* **1976**, 49, 3629.
4. Matsushita, H.; Noguchi, M.; Yoshikawa, S. *Bull. Chem. Soc. Jpn.* **1976**, 49, 1928.
5. Matsushita, H.; Tsujino, Y.; Noguchi, M.; Yoshikawa, S. *Chem. Lett.* **1976**, 1087.

6. Matsushita, H.; Tsujino, Y.; Noguchi, M.; Yoshikawa, S. *Bull. Chem. Soc. Jpn.* **1977,** *50,* 1513.
7. Matsushita, H.; Tsujino, Y.; Noguchi, M.; Saburi, M.; Yoshikawa, S. *Bull. Chem. Soc. Jpn.* **1978,** *51,* 862.
8. Matsushita, H.; Tsujino, Y.; Noguchi, M.; Saburi, M.; Yoshikawa, S. *Bull. Chem. Soc. Jpn.* **1978,** *51,* 201.
9. Koningsberg, W.; Hill, R. H.; Craig, L. C. *J. Org. Chem.* **1961,** *26,* 3867.
10. Shibata, S.; Matsushita, H.; Noguchi, M.; Saburi, M.; Yoshikawa, S. *Chem. Lett.* **1978,** 1305.

RECEIVED May 21, 1979.

Model Studies and the Biochemical Function of Coenzymes

DAVID DOLPHIN

Department of Chemistry University of British Columbia Vancouver, B.C., Canada V6T 1W5

Our understanding of the roles of coenzymes in the catalytic function of enzymic systems stems, in great part, from knowledge that has been gained from in vitro studies with coenzymes and model systems that mimic them. Moreover, as our knowledge increases it becomes apparent that the chemistry of the coenzymes, in the absence of proteins, often parallels that of the enzymic system, albeit usually at a slower rate and less stereospecifically. Two coenzyme systems that have been recalcitrant in allowing their mechanisms of action to be elucidated are the flavins and the vitamin B_{12} coenzyme. It is our purpose here to review the progress that has been made, principally from model studies, in elucidating their modes of action and to set the stage for the more detailed studies reported on these systems in this volume (1, 2, 3).

The Vitamin B_{12} Coenzyme

In 1926 Minot and Murphy (4) announced that whole liver was effective in the treatment of pernicious anemia. The initial assay methods, which were clinical (5), coupled with what we now know are the exceptionally small amounts of B_{12} (even in a relatively rich source such as liver) required that two more decades pass before Folkers (6) and Smith (7) in 1948 simultaneously isolated crystalline vitamin B_{12} (1, R = CN). A further decade passed before it was realized that the so-called vitamin (cyanocobalamin) was an artifact of the isolation procedure and that the enzymatically active species is the vitamin B_{12} coenzyme (5′-deoxyadenosylcobalamin, 1, R = 5′-deoxyadenosyl). This initial observation arose during Barker's study on the conversion

0-8412-0514-0/80/33-191-065$05.75/0
© 1980 American Chemical Society

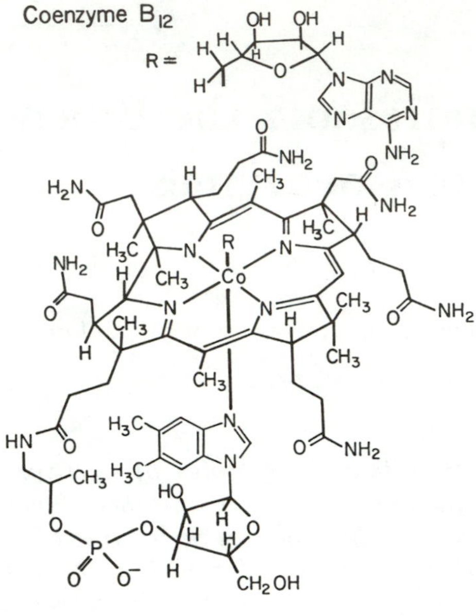

1

of glutamate to β-methyl aspartate (8) and since then the enzymatic rearrangements listed in Figure 1 have all been shown to require the vitamin B_{12} coenzyme as a cofactor.

These rearrangements can, with the exception of ribonucleotide reductase, be generalized as the migration of a hydrogen from a carbon atom to an adjacent one, coupled with the migration of a group X (an alkyl, acyl, or electronegative group) from the adjacent carbon atom to the one to which the hydrogen was bound. In a series of elegant experiments Abeles (9) showed that the migration of hydrogen could be, but was not exclusively, intramolecular. It was found that the conversion of ethylene glycol to acetaldehyde by the enzyme dioldehydrase in the presence of propylene glycol, which was converted to propionaldehyde, allowed tritium to be transferred from the C_2 substrate to the C_3 product. These observations were followed by the even more significant observation that tritium from the substrate could accumulate in the $C(5)'$ methylene group of the coenzyme and that either of the diasteriotopically labelled positions could transfer tritium to the product (9) (Scheme 1). This apparent lack of stereoselectivity in what was an otherwise typically stereochemically well behaved enzyme has since been explained. Homolytic cleavage of the Co–C bond is well established as an initial event in some B_{12}-catalyzed rearrangements (10, 11, 12, 13) and this area recently has been reviewed (14). Homolytic cleavage generates B_{12r} (Co(II)) and a 5'-deoxyadenosyl radical.

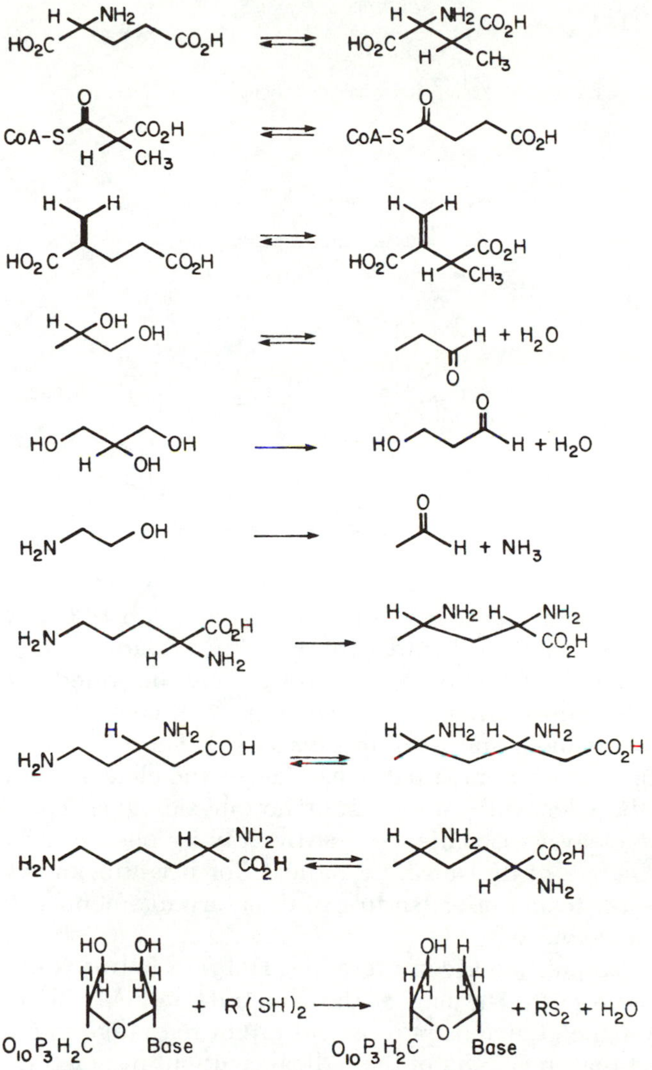

Figure 1. Enzymatic rearrangements controlled by coenzyme B_{12}

Scheme 1.

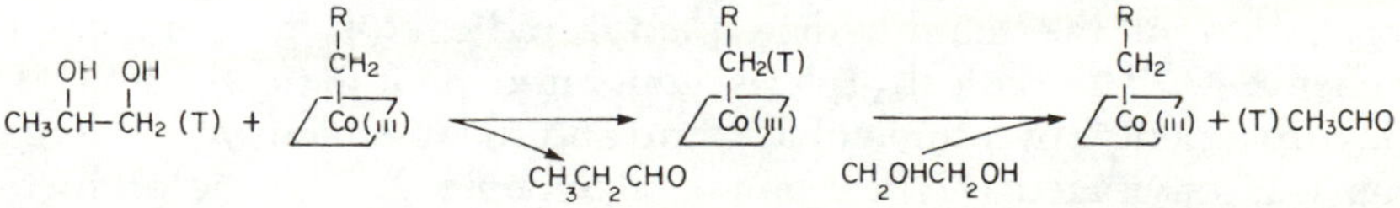

Scheme 2.

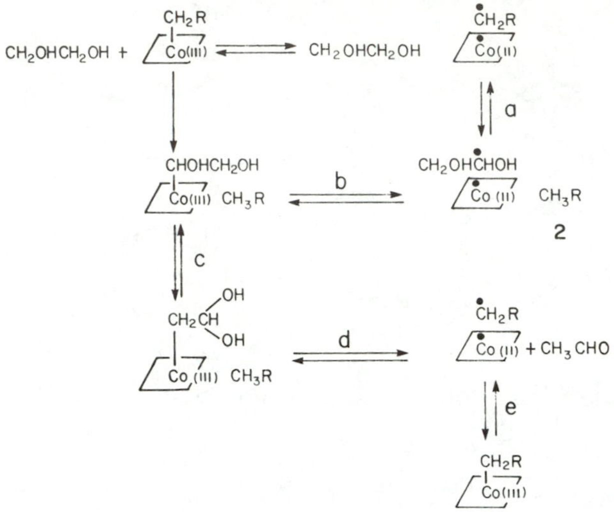

Abstraction of a hydrogen atom from the substrate by the 5'-deoxyadenosyl radical would generate 5'-deoxyadenosine (**2**) and a substrate radical (Reaction a, Scheme 2). The intermediacy of **2** has been established during the conversion of ethanolamine to acetaldehyde by ethanolamine ammonia lyase (*15*). These observations raise two important questions. First, what causes the cleavage of the Co–C bond in the coenzyme, since under nonphysiological conditions the homolytic cleavage is achieved readily only by photolysis? Secondly, what is the fate of the substrate radical; for it is this knowledge that holds the key to an understanding of the rearrangements catalyzed by the B_{12} coenzyme.

The first question is considered by Halpern in this volume (*1*) and I shall move on to the fate of the substrate radical. There are four routes by which the substrate radical might rearrange to the product. The direct rearrangement of the radical (without interacting with B_{12r}, Route b Scheme 2) is possible, and was demonstrated for the conversion of ethylene glycol to acetaldehyde. Thus, the glycol radical (**3**), generated using Fenton's reagent, rearranges to acetaldehyde; the mechanism in Scheme 3 has been proposed for this reaction (*16*). Another example closer to that of the B_{12}-catalyzed rearrangements is the conversion of the 4,5-dihydroxypentyl radical (**4**) to pentanal (**5**) (Scheme 4). The radical (**4**) was generated photolytically from 4,5-dihydroxypentyl(pyridine)cobaloxime and the mechanism of the reaction was envisaged (*17*) as shown in Scheme 4. The yields in this reaction were low ($\leq 0.5\%$) and this is consistent with the energetically

Scheme 3.

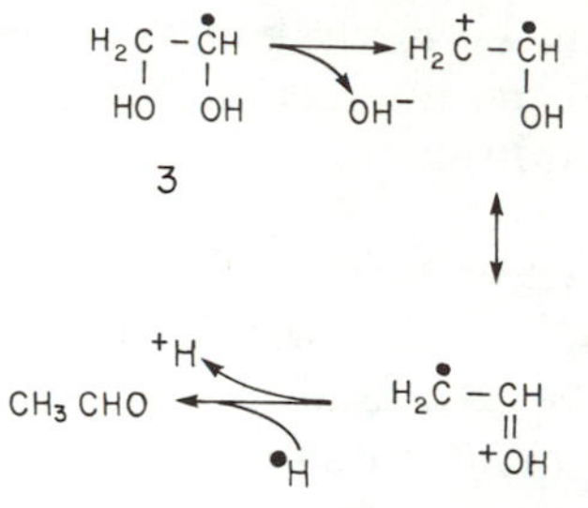

Scheme 4.

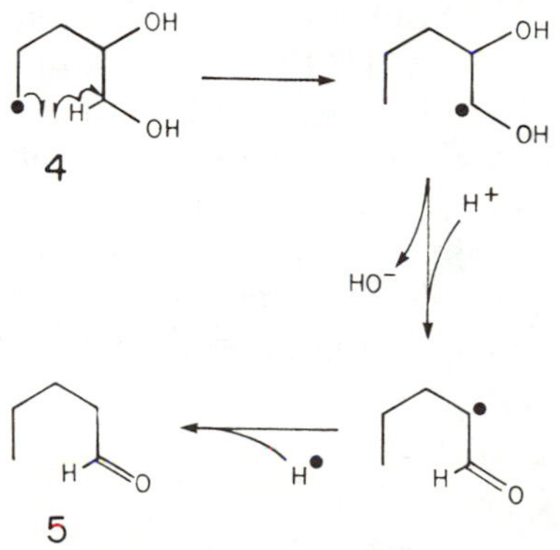

unfavorable 1,2-rearrangement of radicals. While the direct intramolecular 1,2-rearrangements of carbonium ions can be facile, the corresponding radical rearrangement is seen rarely in chemistry. The transition state for the carbonium-ion rearrangement places all electrons in
bonding orbitals, while the corresponding radical rearrangement requires that a single electron be placed in an antibonding orbital. It is
only when a low-energy antibonding orbital is available (as in the case
of π^* orbitals for an aryl migrating group or an empty d orbital of a
halogen atom) that radicals rearrange. A well known example of this is
conversion of **6** to **7**, which was discovered by Urry and Kharasch (*18*).
It seems unlikely then that the unassisted rearrangement of substrate
radical will account for product formation.

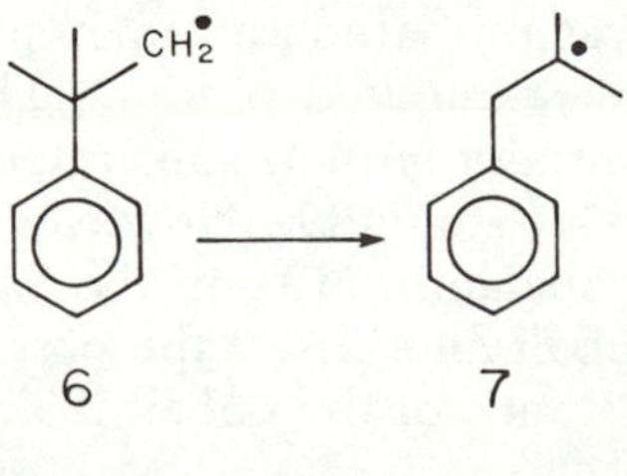

Golding and Radom (19), using ab initio calculations, have suggested that protonation of the transition state may facilitate the radical
rearrangement. This can be envisaged as lowering the energy of the
transition state through the interaction of the unpaired electron with
the new σ* orbital resulting from the protonation of the migrating
group.

While the rearrangement of the substrate radical either protonated
or otherwise "modified" by the protein may be found eventually to
accurately describe the B_{12}-catalyzed rearrangements, other rearrangements under the influence of the cobalt can be envisaged. In a
series of closely related experiments, Bidlingmaier et al. (20) showed
that the cobaloxime (8) upon photolysis gave rearranged product (9)
while Dowd and Shapiro (21) observed the analogous cobalamin (10),
upon standing in the dark, underwent Co–C bond cleavage and rearrangement to give succinic acid (11) upon work up. Again the yields of
rearranged product were very low and Flohr et al. (22) suggested that,
"soon after the homolytic cleavage of the cobalt–carbon bond the
malonic ester "substrate" [from (8)] loses contact with the central
cobalt atom and, thus deprived of that atom's catalytic effect, is no
longer able to rearrange." This suggestion could account for both the
low and variable yields of rearranged product observed in these model
systems.

To test the possible role of the metal, the cobaloxime (12) with an
anchored "substrate" was prepared and upon photolysis and work up
gave, in high yield, the rearranged product. This suggested that the
rearrangement of the initially formed substrate radical could, indeed,
be promoted by the cobalt atom (22). However, rearrangement of the
thiol-ester-containing cobalamin (13), in the dark, to (14) once again
obscured the role of cobalt and raised the possibility that the radical
rearranges without assistance of the cobalt (23).

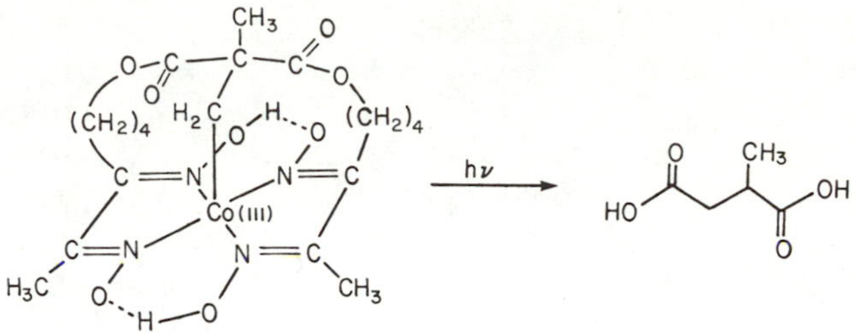

12

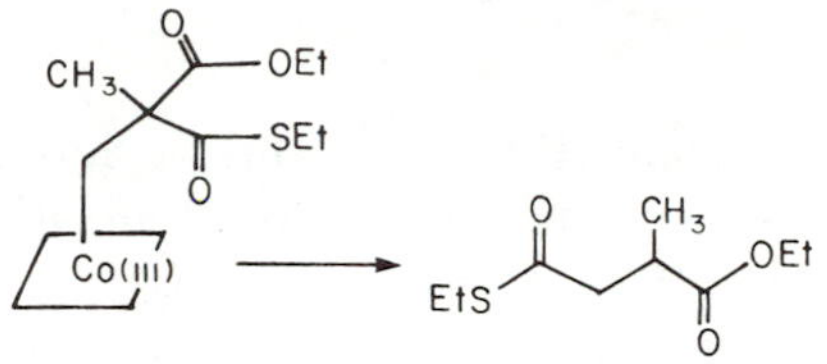

13   14

By way of mimicking the α-methylene glutarate rearrangement, which is B₁₂-coenzyme dependent, Dowd et al. (*24*) found that the cobalamin (**15**) rearranges in the dark to (**16**). The intermediacy of an allylcarbinyl group in this rearrangement was suggested, and such intermediates were further implicated, when it was observed that the two cobaloximes (**17**) and (**18**) are in equilibrium (*25*). On the basis of this facile equilibrium, the mechanism outlined in Scheme 5 has been suggested for the B₁₂-catalyzed rearrangements (*25*).

Even though an initial homolytic cleavage, as outlined in Scheme 1, is well established during some of the B₁₂-coenzyme catalyzed rearrangements, the ease with which electron transfers may occur always leaves open the possibility that subsequent electron transfer between the substrate radical and B₁₂ could generate either a carbonium ion or

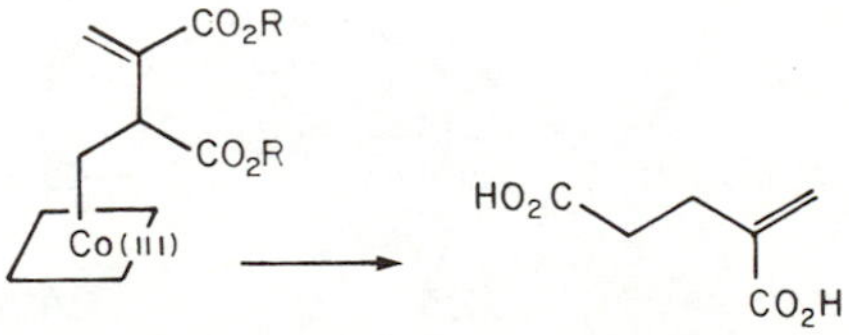

15   16

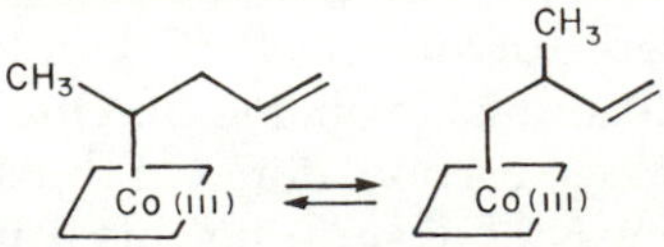

17   18

Scheme 5.

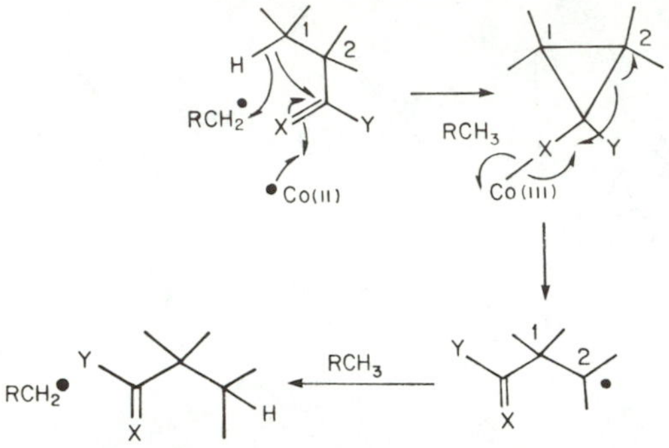

carbanion of the substrate (Routes b and c, Scheme 2). One of the earliest proposed mechanisms for the B_{12}-controlled rearrangements suggested the intermediacy of carbanions (26), and carbonium ions can, of course, undergo 1,2-migration (27). As yet, though, no model systems have been forthcoming for either of these ionic routes.

An area that we have explored involves the rearrangement of the substrate σ-bonded to cobalt. After the initial homolysis of the Co–C bond and hydrogen-atom abstraction from substrate, the substrate radical and B_{12r} could reform a Co–C bond (Reaction b, Scheme 2). This is merely the reversal of the initial homolysis, and the reaction of carbon radicals with Co(II) to form Co–C bonds is well documented (1). The net result would be a transalkylation of 5′-deoxyadenosyl for substrate. Our studies in this area were prompted by Golding's observation (28) that 2-acetoxylethylcobaloxime (19) reacted with ethanol under pseudo first-order kinetics to give the corresponding 2-ethoxyethylcobaloxime. If one assumes that the reaction proceeds via an initial loss of acetate, then the resultant carbonium ion can be represented in the extreme electronic configurations (22, 23, 24). Of

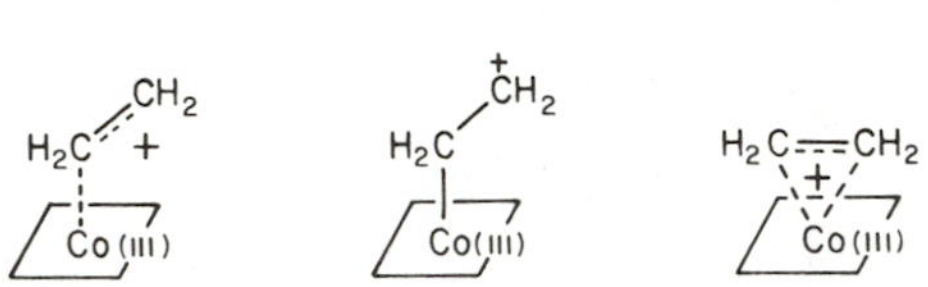

| 22 | 23 | 24 |

these three forms, only the olefin π-complex (24) is symmetric and reaction with alcohol would require that the carbon atoms in the product be scrambled; indeed, solvolysis of ^{13}C-labelled 19 in methanol gave equal amounts of 20 and 21 (Scheme 6). One must conclude that during the solvolysis the two carbon atoms of the alkyl ligand became equivalent; the simplest way of describing this is via the intermediacy

Scheme 6.

$$Co\,(III) \; —^{13}CH_2 — CH_2 — OAc \xrightarrow{CH_3OH} Co\,(III) — CH_2 \; —^{13}CH_2 — OCH_3$$

$$+ \quad \textbf{20}$$

19

$$Co\,(III) \; —^{13}CH_2 — CH_2 — OCH_3$$

21

of the cobalt(III)–olefin π-complex (29). Since our initial observations, additional evidence supporting the existence of such π-complexes have been reported (30, 31).

Transition-metal–olefin π-complexes exhibit synergic bonding, which involves interaction of the filled olefin π-orbitals with empty metal d orbitals, and backbonding from the filled metal d orbitals to the empty π^* olefin orbitals. The net result is that stable metal–olefin π-complexes are formed when the metal is electron-rich (low oxidation state) and the olefin is electron deficient (32). The opposite situation would occur in π-complex **24** and the bonding mode may be different since the metal is in a high oxidation state. This suggested to us that olefin π-complexes of trivalent cobalt might be stablized by an electron-rich olefin.

Reaction of electron-rich olefins such as vinyl ethers have not enabled us to observe spectroscopically a complex with cobalt(III) cobaloximes. Nevertheless, the reaction between a cobal(III)oxime, ethyl-vinyl ether, and ethanol gives complete alkylation of the cobalt (33). When the ethanol is not vigorously anhydrous, some formyl-methylcobaloxime (**25**) (Scheme 7) is formed also and these observations are consistent (34) with the reactions outlined in Scheme 7. When the water concentration was increased, the amount of formylmethyl-cobaloxime also increased at the expense of the acetal. Thus, we anticipated that the analogous reaction with vitamin B_{12}, which is impossible to obtain in an anhydrous condition and is best manipulated in

Scheme 7.

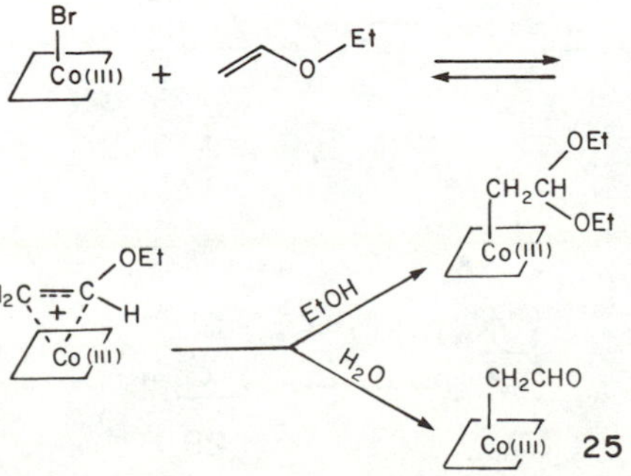

water, would give principally the formylmethylcobalamin that we anticipated (and later showed) would be unstable. To prevent the interception of the cobalamin π-complex by water, we used hydroxyethylvinyl ether (26) in the hope that intramolecular collapse of the olefin π-complex would occur. Indeed, when an aqueous solution of hydroxocobalamin (B_{12b}) was treated with a large excess of 26 in the presence of triethylamine, the cyclic acetal (27) was formed quantitatively as outlined in Scheme 8 (33, 34).

Scheme 8.

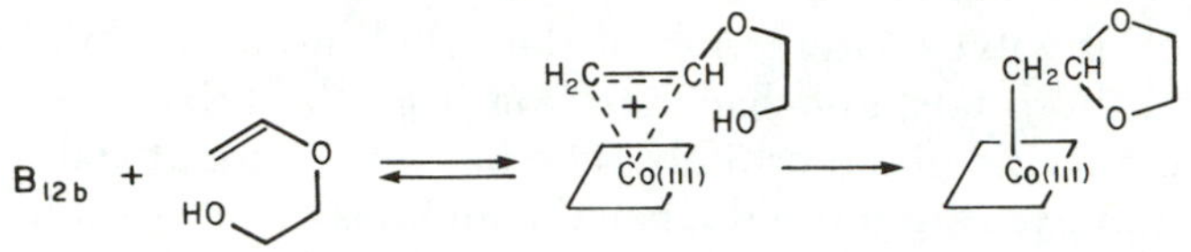

26   27

All of the above chemistry suggests the intermediacy of cobalt(III)–olefin π-complexes and one can then ask what role might such π-complexes play in B_{12}-coenzyme catalyzed rearrangements? To be more specific, can such chemistry account for the conversion of substrate σ-bound to cobalt to product σ-bound to cobalt (Reaction c, Scheme 2)? In Scheme 9, we suggest how ethylene glycol as substrate σ-bound to cobalt (28), via the transalkylations described above, can be converted by the loss of the β-hydroxy group to the π-complex (29). Readdition of the migrating water to 29 would, by analogy to the chemistry described above, give product σ-bound to cobalt (30). The

Scheme 9.

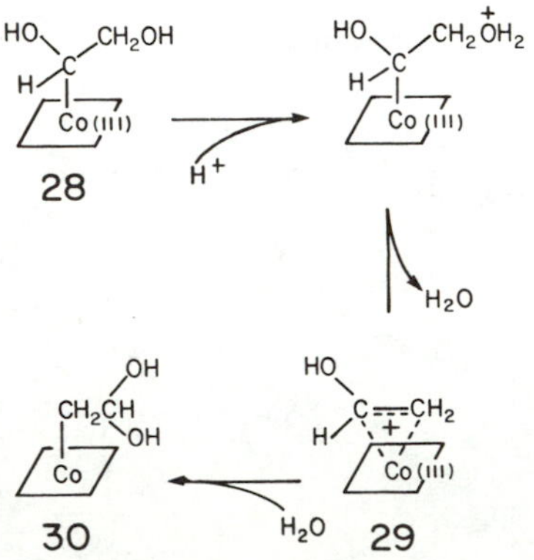

28   30

final step in forming acetaldehyde and generating a new catalytic site is shown in Reactions d and e, Scheme 2. To mimic more closely the reaction proposed in Scheme 9, we have prepared the cobaloxime **31** where the OH leaving group of **28** has been replaced by acetoxy and the hydroxy group at C(1) of **28** is methoxy. When **31** was passed through a silica gel column, it rearranged to formylmethylcobalamin (**25**); the mechanism for this reaction (outlined in Scheme 10) is consistent with the chemistry described above and once again supports the intermediacy of a cobalt(III)–olefin π-complex (**35**).

Scheme 10.

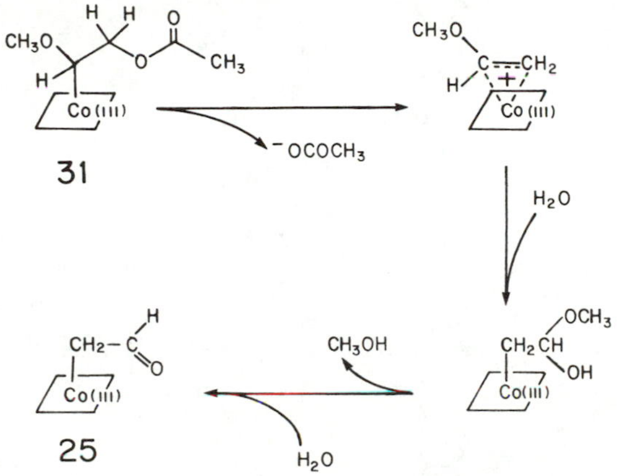

Whether or not the model systems described here bear any relevance to the B_{12}-coenzyme catalyzed rearrangements must await the test of time. Nevertheless, it is intriguing to speculate that the $\sigma \rightleftarrows \pi \rightleftarrows \sigma$ rearrangement described above, which is but one of many such examples that dominate the catalytic activity of transition-metals complexes understood by chemists for the past three decades, may have been appreciated and used by nature in the B_{12}-coenzyme catalyzed rearrangements for the past three billion years!

Flavoproteins

Flavoproteins play varied roles as coenzymes in biochemical processes. They serve as key intermediates in electron transport between those systems, such as the cytochromes, that carry one electron at a time and the two-electron transfers involved in the redox chemistry of organic-substrate metabolism. Flavins such as FMN or FAD (Figure 2) are designed structurally and electronically to stabilize both one- and two-electron redox intermediates. Thus, the one-electron reduction of

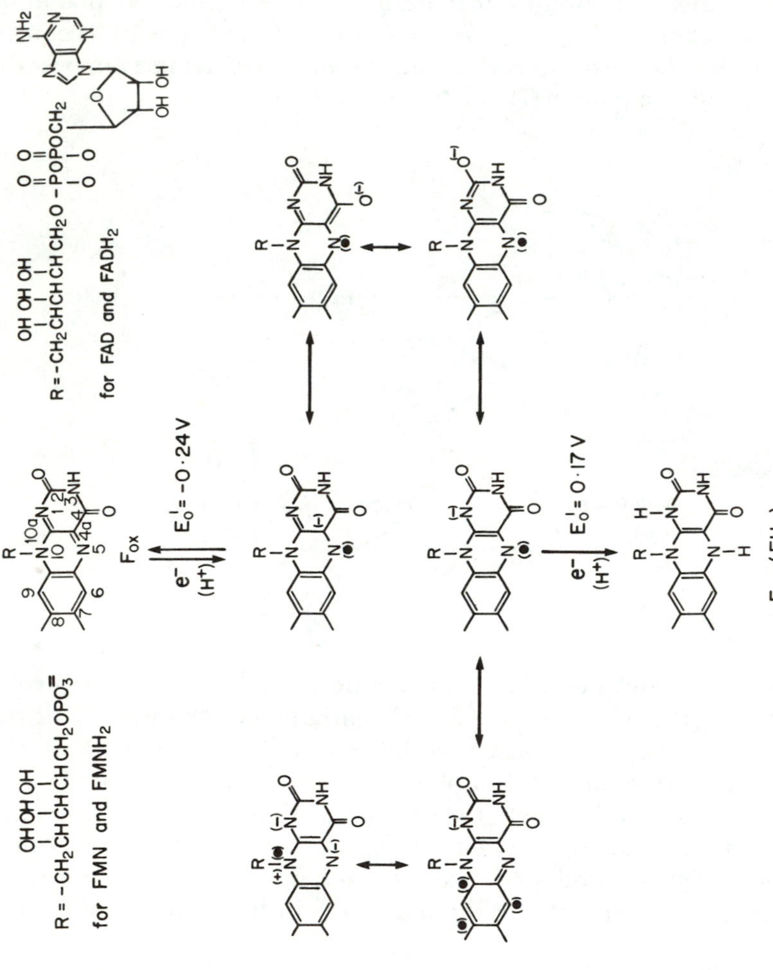

Figure 2. Flavins and their redox chemistry

the oxidized coenzyme (F_{ox}) gives the semiquinone anion radical where the unpaired electron of the free radical and the nonbonding (before protonation) lone pair can be delocalized over the whole chromophore. From some of the resonance contributions shown in Figure 2, the extensive delocalization of the electrons explains how the isoalloxazine ring stabilizes both one- and two-electron intermediates. In addition, and often as part of these electron transfers, the flavin coenzymes function as dehydrogenases and catalyze the dehydrogenations (and hydrogenations) of a variety of substrates as outlined in Figure 3.

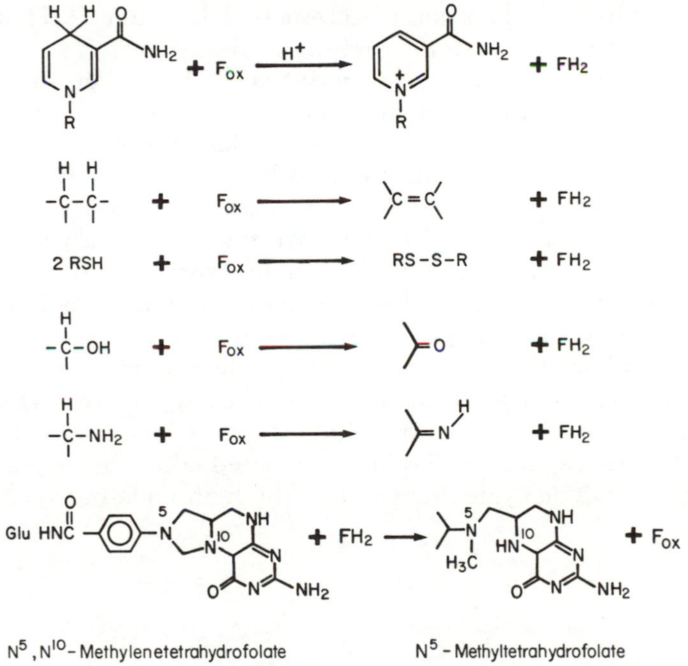

$$N^5, N^{10}\text{-Methylenetetrahydrofolate} \qquad N^5\text{-Methyltetrahydrofolate}$$

Figure 3. Hydrogenations and dehydrogenations controlled by flavoproteins

While the isoalloxazine ring is a characteristic feature of the flavins, several naturally occurring modifications, especially in the benzene ring, are known and these modifications (Figure 4) often are employed in the covalent binding of the flavin to its apoprotein. Recently, several synthetic analogs of flavins have been prepared (36). These have been principally deaza analogs where the nitrogen atom at the 1, 3, and 5 positions have been replaced by CH (Figure 5). As can be seen from the chapters by Bruice (2) and Walsh (3) (this volume),

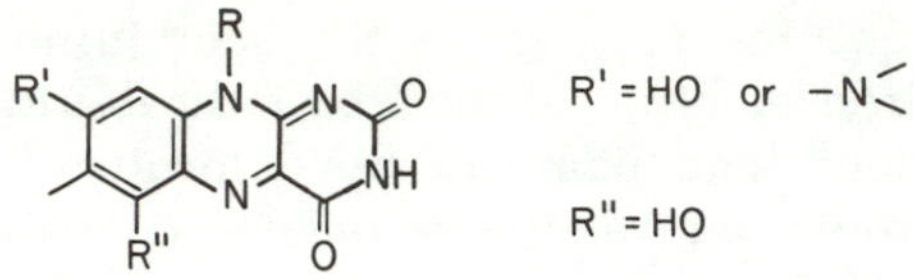

R' = HO or −N⟨

R" = HO

Figure 4.   Naturally occurring modifications of the flavins

these analogs are active as coenzymes in many flavin-requiring systems and their study already has allowed for numerous speculations of the modes of action of flavoproteins to be abandoned or modified. In addition, the characterization of a naturally occurring 8-hydroxy-5-deazaflavin, from methanogenic bacteria by Eirich et al. (37), makes a study of the synthetic systems even more significant.

Our principal interest in the flavins arises from their interactions with dioxygen, and the function of flavoproteins as monooxygenases. Examples of oxidase reactions mediated by flavoproteins are shown in Figure 6 and include the oxidative decarboxylation of α-hydroxy acids with internal mixed-function monooxygenase, the bacterial luciferases that chemiluminesce when aldehydes are oxidized to carboxylic acids, and the external monooxygenases, which hydroxylate activated aromatic rings (38). It is this latter class of reactions that attracted our attention, for in these systems ground-state triplet oxygen is activated for oxidation without the help of a metal. At first sight, these aromatic hydroxylations seem reminiscent of those catalyzed by cytochrome P-450. There are, however, some important differences when the active oxygenating species of P-450 is associated with a heme porphyrin; during the catalytic cycle (Scheme 11) the hydroxylating species (**32**)

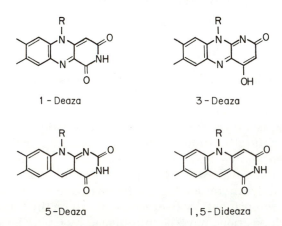

1 - Deaza

3 - Deaza

5 - Deaza

1,5 - Dideaza

Figure 5.   Synthetic analogs of the flavins

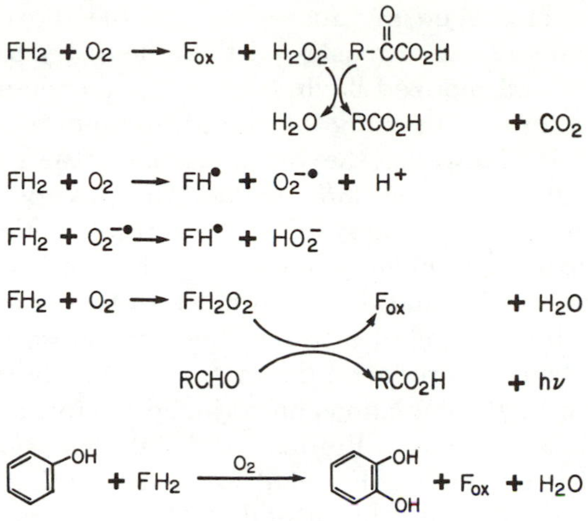

$$FH_2 + O_2 \longrightarrow F_{ox} + H_2O_2$$

$$R-\overset{O}{\overset{\|}{C}}CO_2H$$

$$H_2O \quad RCO_2H \quad + CO_2$$

$$FH_2 + O_2 \longrightarrow FH^\bullet + O_2^{-\bullet} + H^+$$

$$FH_2 + O_2^{-\bullet} \longrightarrow FH^\bullet + HO_2^-$$

$$FH_2 + O_2 \longrightarrow FH_2O_2$$

$$F_{ox} \quad + H_2O$$

$$RCHO \quad RCO_2H \quad + h\nu$$

(phenol + FH_2) $\xrightarrow{O_2}$ (catechol) $+ F_{ox} + H_2O$

Figure 6.   *Reactions of reduced flavins with dioxygen and oxidations controlled by flavoproteins*

undoubtedly is associated with the iron atom (*see* the chapter by J. Groves, this volume (39) for detailed discussion). In the case of flavoproteins, however, no assistance can be provided by a transition metal, and one must ask how does the flavin, by itself, activate molecular oxygen?

The first question that arises is how does triplet oxygen interact with reduced flavin in what is a spin-forbidden reaction? We originally noted the similarity, both electronic and structural, between reduced flavin and tetrakis(dimethylamino)ethylene (33). In the presence of triplet oxygen, 33 reacts rapidly to give the corresponding urea. This

Scheme 11.

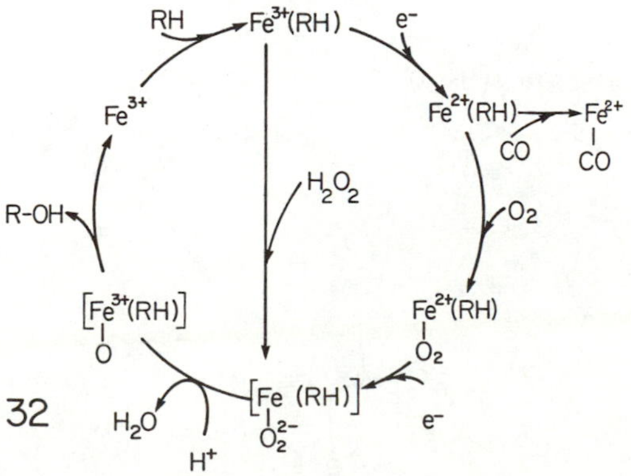

reaction is best envisaged as proceeding via the dioxetane (34), as shown in Scheme 12. An extension of this chemistry to the reaction between oxygen and reduced flavin to give the dioxetane (35) would allow for an alternate opening of the dioxetane to give the 4a-hydroperoxide (36). However, the unimportance of the 10a-position for the reaction with dioxygen (38) makes the involvement of the dioxetane less likely in natural systems. Nevertheless, the existence of a 4a-hydroperoxide has been provided by the work of Kemal and Bruice (40) (Scheme 13) and the importance of the 4a-hydroperoxide in the enzymic functioning of external flavin monooxygenases seems secured. How then does reduced flavin react with triplet oxygen?

Molecular orbital calculations on reduced flavins, such as the extended Hückel calculations of Pullman and Pullman (41) and the self-consistent field calculations of Fox et al. (42), suggest that the energy of the highest occupied molecular orbital (HOMO) varies only slightly with conformational changes about the $N(5)$–$N(10)$ axis, and all of these calculations show the energy of the HOMO to be nonbonding to

Scheme 12.

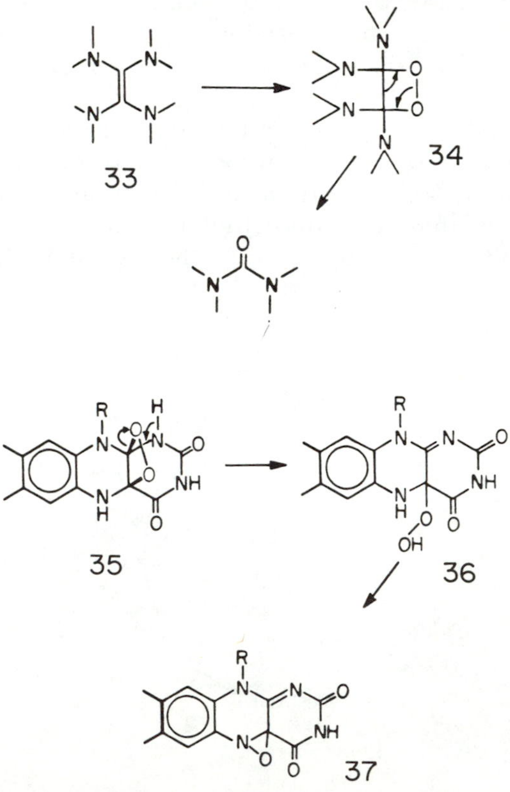

Scheme 13.

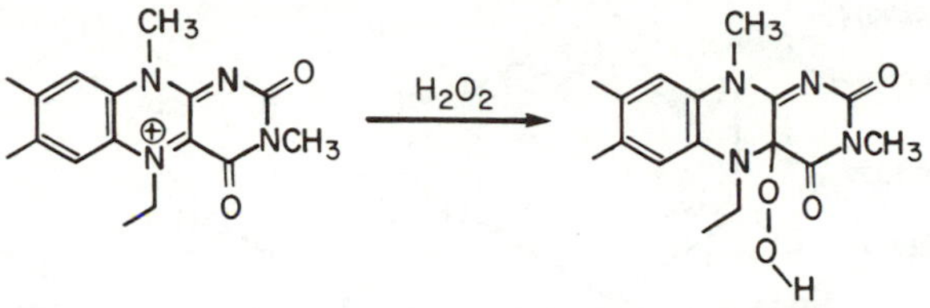

slightly antibonding. In addition, our extended Hückel calculations (*43*) (Figure 7) show that the HOMO electron density distribution, the π-charge distribution and the total-charge distribution favor attack at 4a. Indeed, a redox reaction with the loss of an electron from the antibonding HOMO of the electron rich, and easily oxidizable, reduced flavin to triplet oxygen to generate a flavin radical and superoxide ion has been suggested by Bruice (*38*). This electron transfer overcomes the spin forbiddenness of the reaction and collapse of the resultant ion pair would give the 4a-hydroperoxide.

How can such a peroxide cause hydroxylation of a phenolic substrate when hydrogen peroxide and simple alkyl hydrogen peroxides do not attack phenols unless catalyzed by metal ions via radical pathways (*44*)?

Entsch et al. (*45*) have suggested the mechanism outlined in Scheme 14 for hydroxylation via the 4a-hydroperoxide, and they suggest that by ring opening the isoalloxazine ring the peroxide might be a better electrophile. This mechanism suffers from the complexity of the ring opening, but it must be noted that Kemal et al. (*46*) have shown, at least for oxygen-atom transfer to sulfur, the 4a-hydroperoxide to be around 10^5 times more efficient than simple alkyl hydrogen peroxides.

An early suggestion of Hamilton's (*44*), once again involving the intermediacy of the 4a-hydroperoxide and a ring opening, is outlined in Scheme 15. It was postulated, on the basis of the contributing resonance forms of **38**, that the terminal oxygen of the carbonyl oxide should be electrophilic. However, Hückel calculations that we have carried out on this system indicate (*43*) that there is a high π-electron density at the terminal oxygen. In addition, the O–O bond order is similar to that of a carbonyl group, suggesting that the terminal oxygen will not be electrophilic. It must be pointed out, however, that these are arguments based only on ground-state considerations. Nevertheless, they do not encourage us in accepting the carbonyl-oxide model, which also suffers from the fact that generation of the highly oxidizable *o*-phenylenediamine moiety could lead to irreversible destruction of the flavin chromophore.

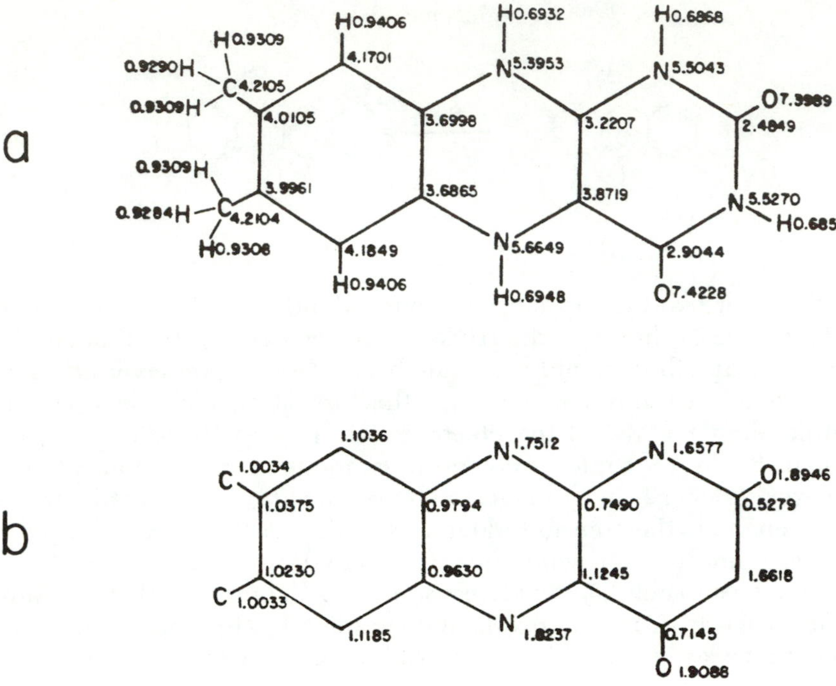

Figure 7. Extended Hückel calculations: a) total electronic charge distribution; b) total π-electronic charge distribution

We have suggested (*43*) that collapse of the 4a-hydroperoxide via intramolecular nucleophilic attack of $N(5)$ on the peroxide could generate the three-membered oxaziridine ring (*37*) (Scheme 12). Should such an oxaziridine prove to be electrophilic, the oxidation of the phenolic substrate would be envisaged proceeding as in Scheme 16.

Scheme 14.

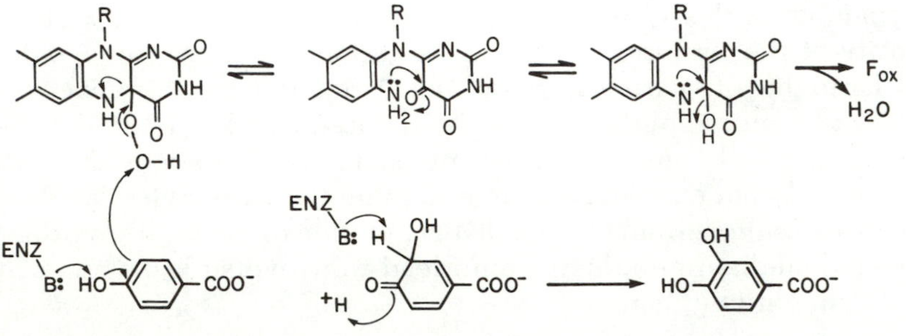

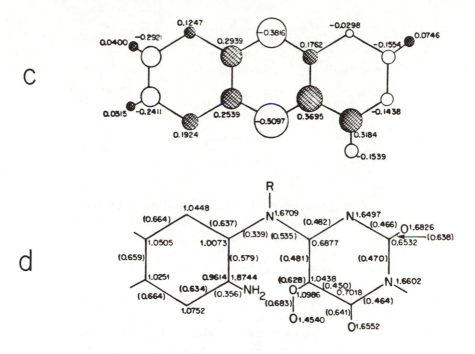

Figure 7.   Extended Hückel calculations: c) HOMO electron density distribution; d) π-electron densities (π-bond orders) for carbonyl oxide (38, Scheme 15)

We based the possible oxidizing power of the oxaziridine on the observation that photolysis of pyridine *N*-oxides apparently generates oxaziridines, and that the photolysis of pyridine *N*-oxide itself gener-

Scheme 15.

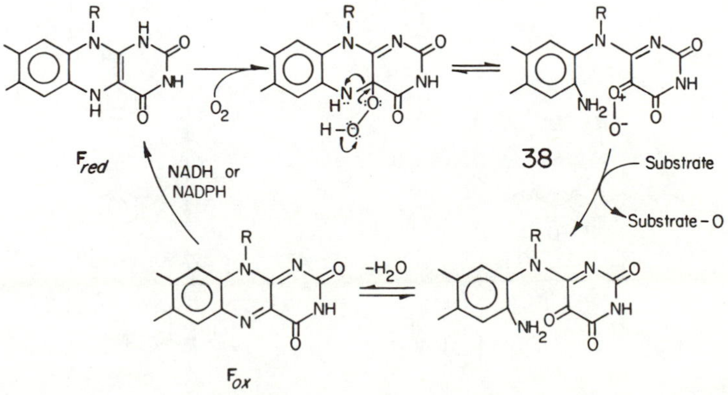

Scheme 16.

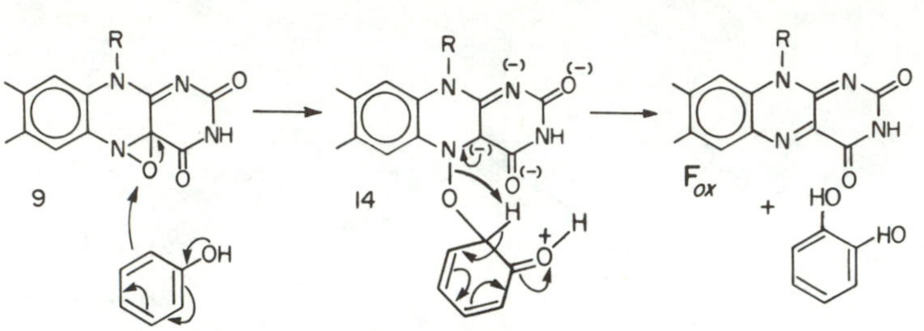

ates a species capable of oxidizing ethanol to acetaldehyde and naphthalene to 1-naphthol (47). This oxaziridine mechanism for flavin-mediated oxidations has been criticized since nonphotochemical oxygen insertions with oxaziridines is unknown. An even more serious criticism raised by Rastetter et al. (48) is that the photochemically generated oxaziridines that we suggested as models for oxidation are not oxaziridines (49). Nevertheless, the postulation of oxaziridines as intermediates in the enzymic oxidations by flavins is still viable as shown by Rastetter's group, who have shown that upon photolysis the flavin $N(5)$-oxide (39) oxidizes phenols to hydroquinones by the mechanisms suggested in Scheme 17. Clearly, the enzymic mechanism does not involve photochemistry of the flavin N-oxide. However, Rastetter et al. (48) point out that homolysis of the oxaziridine C–O bond

Scheme 17.

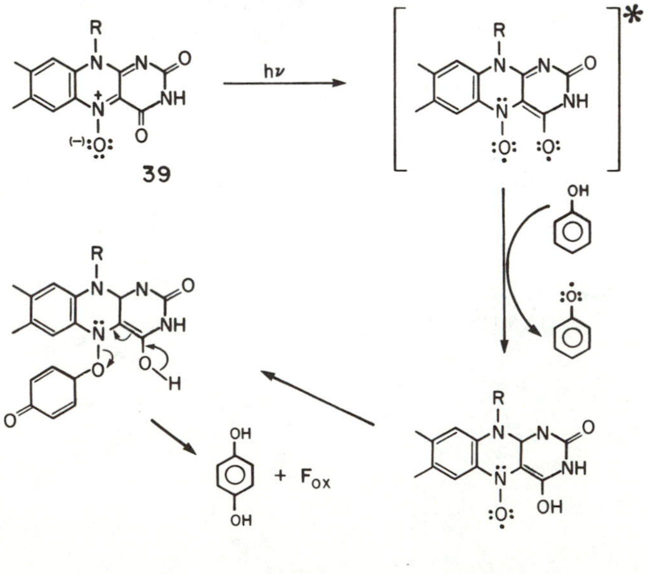

with the concerted hydrogen-atom abstraction from the phenol (Scheme 18) would generate the phenoxy radical and nitroxyl radical (**40**), which could react as outlined in Scheme 17 to give F_{ox} and hydroxylated substrate.

Scheme 18.

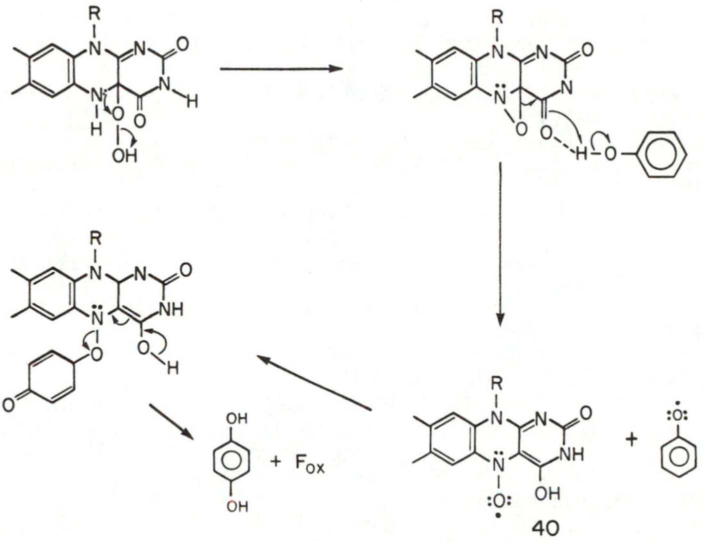

Conclusions

Clearly, the chemistry carried out on model systems for both B_{12}-coenzyme dependent and flavin-dependent enzymes have suggested a number of mechanisms by which the natural systems might function. As with all biomimetic chemistry, the information given by the in vitro studies must be compared with and extended to the in vivo system to assess whether or not the model chemistry bears any relationships to the natural systems.

Acknowledgment

Acknowledgment is made to the donors of the Petroleum Research Fund, administered by the American Chemical Society, for partial support of this research. This work was a contribution from the Bioinorganic Chemistry Group and is supported partially by the United States National Institutes of Health (AM-17989).

Literature Cited

1. Halpern, J. "Some Mechanistic Aspects of Vitamin B_{12} Coenzyme Dependent Rearrangements. Homolytic Cleavage of Metal-Carbon Bonds"; Chapter 9 in this book, pp. 165–177.
2. Bruice, T. "Carbon Acid Oxidations and Oxygen Activation by Flavins"; Chapter 6 in this book, pp. 89–118.
3. Walsh, C.; Jacobson, F.; Ryerson, C. C. "Flavin Coenzyme Analogs As Probes of Fluorenzyme Reaction Mechanisms"; Chapter 7 in this book, pp. 119–138.
4. Minot, G. R.; Murphy, W. P. *J. Am. Med. Assoc.* **1926**, *87*, 470.
5. Smith, E. L. "Vitamin B_{12}"; Methuen and Co. Ltd.: London; John Wiley & Sons Inc.: New York, 1960.
6. Rickes, E. L.; Brink, N. G.; Konioszy, F. R.; Wood, T. R.; Folkens, K. *Science* **1948**, *107*, 396.
7. Smith, E. L.; Paker, L. F. J. *Biochem. J.* **1948**, *43*, viii.
8. Barker, H. A.; Weissbach, H.; Smyth, R. D. *Proc. Natl. Acad. Sci. USA.* **1958**, *44*, 1093.
9. Abeles, R. H. In "Bioinorganic Chemistry," *Adv. Chem. Ser.* **1971**, *100*, p. 346.
10. Babior, B. M.; Moss, T. H.; Orme–Johnson, W. H.; Beinert, H. *J. Biol. Chem.* **1974**, *249*, 4537.
11. Finlay, T. H.; Valinsky, J.; Mildvan, A. S.; Abeles, R. H. *J. Biol. Chem.* **1973**, *248*, 1285.
12. Cockle, S. A.; Hill, H. A. O.; Williams, R. J. P.; Davies, S. P.; Foster, M. A. *J. Am. Chem. Soc.* **1972**, *94*, 275.
13. Valinsky, J. G.; Abeles, R. H.; Fee, J. A. *J. Am. Chem. Soc.* **1974**, *96*, 4709.
14. Pilbrow, J. R. In "Proc. Eur. Symp. Vitamin B_{12} Intrinsic Factor, 3rd," Zagalak, B., Ed. Walter de Gruyter: Berlin and New York, 1979, p. 505.
15. Babior, B. M.; Carty, T. J.; Abeles, R. H. *J. Biol. Chem.* **1974**, *249*, 1689.
16. Walling, C.; Johnson, R. A. *J. Am. Chem. Soc.* **1975**, *97*, 2405.
17. Golding, B. T.; Sell, C. S.; Sellars, P. J. *Chem. Commun.* **1976**, 773.
18. Urry, W. H.; Kharasch, M. S. *J. Am. Chem. Soc.* **1944**, *66*, 1438.
19. Golding, B. T.; Radom, L. *J. Am. Chem. Soc.* **1976**, *98*, 6331.
20. Bidlingmaier, G.; Flohr, H.; Kempe, U. M.; Traute, K.; Rétey, J. *Angew Chem. Int. Ed. Engl.* **1975**, *14*, 822.
21. Dowd, P.; Shapiro, M. *J. Am. Chem. Soc.* **1976**, *98*, 3724.
22. Flohr, H.; Pannhorst, W.; Rétey, J. *Angew. Chem. Int. Ed. Engl.* **1976**, *15*, 561.
23. Scott, A. I.; Kang, K. *J. Am. Chem. Soc.* **1977**, *99*, 1997.
24. Dowd, P.; Trivedi, B. K.; Shapiro, M.; Marwaka, L. K. *J. Am. Chem. Soc.* **1976**, *98*, 7875.
25. Bury, A.; Ashcroft, M. R.; Johnson, M. D. *J. Am. Chem. Soc.* **1978**, *100*, 3214.
26. Ingraham, L. L. *Ann. N. Y. Acad. Sci.* **1964**, *112*, 713.
27. Rétey, J.; Umani–Ronchi, A.; Seible, J.; Arigoni, D. *Experientia.* **1966**, *22*, 502.
28. Golding, B. T.; Holland, H. L.; Horn, U.; Sakrikav, S. *Angew. Chem. Int. Ed. Engl.* **1970**, *9*, 959.
29. Silverman, R. B.; Dolphin, D.; Babior, B. *J. Am. Chem. Soc.* **1972**, *94*, 4028.
30. Golding, B. T.; Sakrikav, S. *Chem. Commun.* **1972**, 1183.
31. Brown, K. L.; Ingraham, L. L. *J. Am. Chem. Soc.* **1974**, *96*, 7681.
32. Herberhold, M. "Metal π-Complexes"; Elsevier: New York, 1972; Vol. 2, Part 1.
33. Silverman, R. B.; Dolphin, D. *J. Am. Chem. Soc.* **1976**, *98*, 4626.
34. Silverman, R. B.; Dolphin, D. *J. Am. Chem. Soc.* **1976**, *98*, 4633.

35. Dolphin, D.; Banks, A. R.; Cullen, W. R.; Cutler, A. R.; Silverman, R. B. In "Proceedings Eur. Symp. Vitamin B_{12} Intrinsic Factor, 3rd," Zagalak, B., Friedrich, W., Eds.; Walter de Gruyter: Berlin and New York, 1979, p. 575.
36. Walsh, C.; Fisher, J.; Spencer, R.; Graham, D. W.; Ashton, W. T.; Brown, J. E.; Brown, R. D.; Rogers, E. F. *Biochemistry* **1978**, *17*, 1942.
37. Eirich, D.; Balch, W.; Wolfe, R. *Biochemistry* **1972**, *17*, 4583.
38. Bruice, T. C. In "Progress in Bioorganic Chemistry," Kaiser, E. T., Kezdy, F. J., Eds.; Wiley Interscience: New York, 1976; Vol. 4, p. 1.
39. Groves, J. T. Chapter 15 in this book.
40. Kemal, C.; Bruice, T. C. *Proc. Natl. Acad. Sci., USA.* **1976**, *73*, 995.
41. Pullman, A.; Pullman, B. "Quantum Biochemistry"; Interscience: 1963.
42. Fox, J. L.; Laberge, S. P.; Nishimoto, K.; Forster, L. S. *Biochim. Biophys. Acta* **1967**, *136*, 544.
43. Orf, H. W.; Dolphin, D. *Proc. Natl. Acad. Sci. USA.* **1974**, *71*, 2646.
44. Hamilton, G. A. In "Progress in Bioorganic Chemistry"; Kaiser, E. T., Kedzy, T. J., Eds.; Wiley Interscience: New York, 1971; Vol. 1, p. 83.
45. Entsch, B.; Ballou, D. P.; Massey, V. *J. Biol. Chem.* **1976**, *251*, 2550.
46. Kemal, C.; Chan, T. W.; Bruice, T. C. *Proc. Natl. Acad. Sci., USA.* **1977**, *74*, 405.
47. Alkaitis, A.; Calvin, M. *Chem. Commun.* **1968**, 292.
48. Rastetter, W. H.; Gadek, T. R.; Tane, J. P.; Frost, J. W. *J. Am. Chem. Soc.* **1979**, *101*, 2228.
49. Kaneko, L.; Yamada, S.; Yokoe, I.; Ishikawa, M. *Tetrahedron Lett.* **1967**, 1873.

RECEIVED May 23, 1979.

Carbon Acid Oxidations and Oxygen Activation by Flavins

THOMAS C. BRUICE

Department of Chemistry, University of California, Santa Barbara, CA 93106

Evidence is presented in support of free-radical mechanisms for the oxidation of ionizable carbon acids by oxidized flavin. Activation of molecular oxygen by reduced flavin is shown to occur through the formation of a 4a-hydroperoxyflavin that may, dependent upon conditions and substrate, transfer one or two oxygen atoms. Examples of all reactions are provided in the text. The present state of knowledge concerning the chemiluminescent oxidations of aldehydes by 4a-hydroperoxyflavin is considered.

The isolloxazine nucleus of the flavins [3-(R or H)-7,8-dimethyl-10-R'-isoalloxazines] may exist in the fully reduced (1,5-dihydro-), the radical (semiquinone), and the fully oxidized (quinone) states. Because of acid–base equilibria, each of these oxidation states

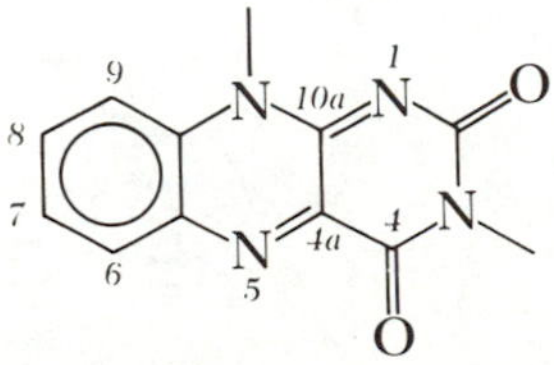

Isoalloxazine

may, dependent upon pH, exist as one or another of three species (Scheme 1) (*1*). Reduced and oxidized flavins disproportionate to provide the flavin radical, which is present at readily determinable concentrations at high and low pH.

0-8412-0514-0/80/33-191-089$07.50/0
© 1980 American Chemical Society

Scheme 1

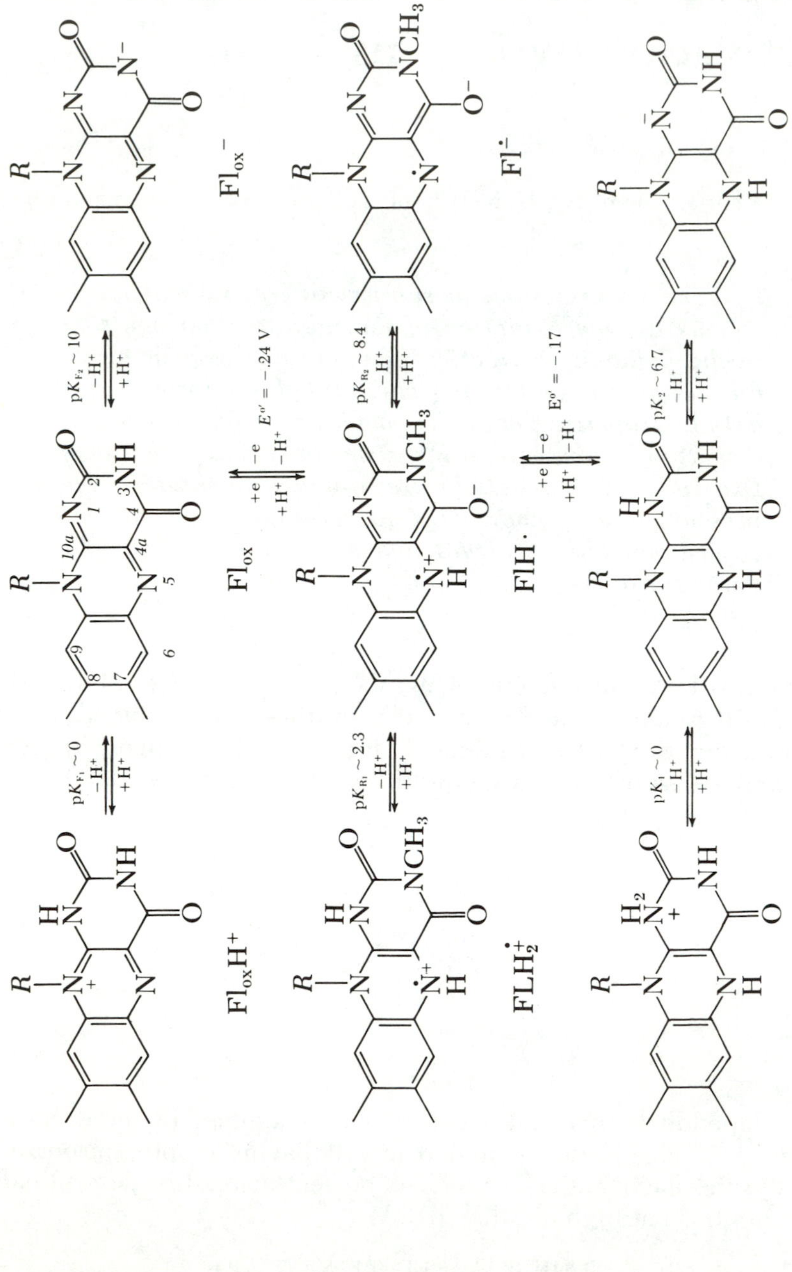

$$Fl_{ox} + FlH_2 \rightleftarrows 2\ FlH \cdot \tag{1}$$

There are over one hundred recognizable flavoproteins. It is not a great oversimplification to state, however, that the biological roles played by most flavoproteins are as: (i) electron carriers (flavodoxins) that undergo reversible one-electron oxidation and reduction between oxidized and radical states; (ii) dehydrogenating agents through the transfer of the elements of H_2 from substrate to oxidized flavoenzyme, yielding reduced flavoenzyme; and (iii) activators of molecular oxygen by reaction of reduced flavoenzyme with molecular oxygen to provide a species capable of inserting or adding O or O_2 to substrate. Of these three activities, the dehydrogenases and oxygenases provide mechanistic challenges to the organic chemist. The flavoenzymes that carry out dehydrogenation and monooxygenation reactions are not true catalysts since they do not catalyze their respective reactions in both directions. They are, in essence, regeneratable reagents.

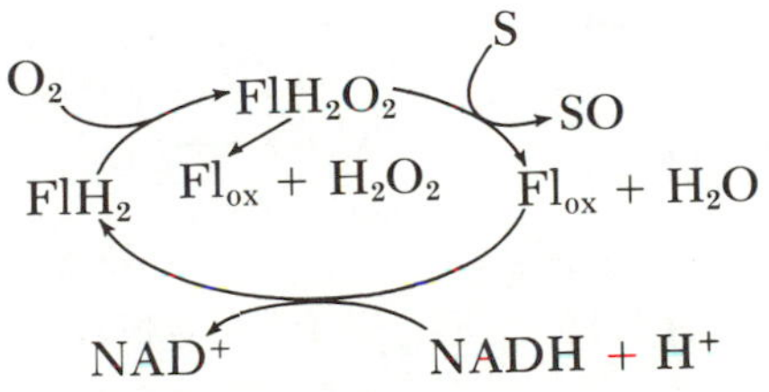

$$
\begin{array}{c}
O_2 \quad \longrightarrow FlH_2O_2 \overset{S}{\underset{}{\diagdown}} SO \\
FlH_2 \quad Fl_{ox} + H_2O_2 \quad Fl_{ox} + H_2O \\
NAD^+ \qquad\qquad NADH + H^+
\end{array}
\tag{2}
$$

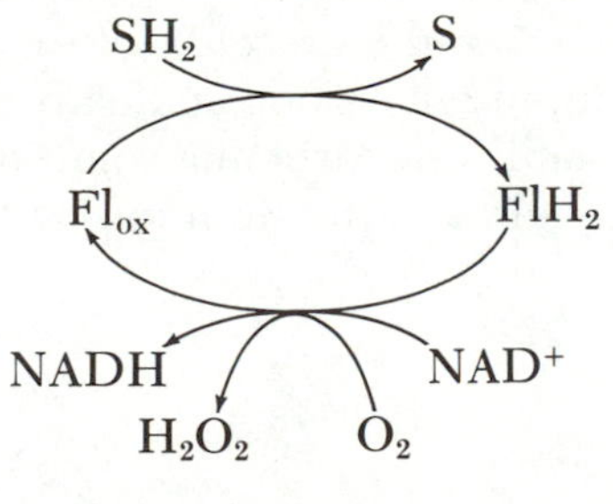

$$
\begin{array}{c}
SH_2 \qquad\qquad S \\
Fl_{ox} \qquad\qquad FlH_2 \\
NADH \qquad NAD^+ \\
H_2O_2 \qquad O_2
\end{array}
\tag{3}
$$

Reduced and oxidized isoalloxazines, sans protein, are also capable of serving as reagents for dehydrogenation and oxygen activation reactions. What follows is a concise overview of our mechanistic studies of the dehydrogenation of carbon acids by oxidized flavin and activation of molecular oxygen by reduced flavin. Our investigations of the chemistry of flavins began in 1971 (2) and our progress was summarized within a general review of the mechanisms of flavin reactions published in 1976 (3).

Carbon Acid Dehydrogenation

The carbon acids that serve as the normal substrates for flavoenzymes are divisible into two groups, dependent upon the acidity of the C—H bond that ultimately gives up its bonding electrons to the flavin cofactor. In the first group, the pK_a of the C—H function of the substrate is so great that oxidation of a derived carbanion cannot be involved. The substrates for flavoenzymes NADH dehydrogenase, glucose oxidase, and monoamine oxidase fall into this category.

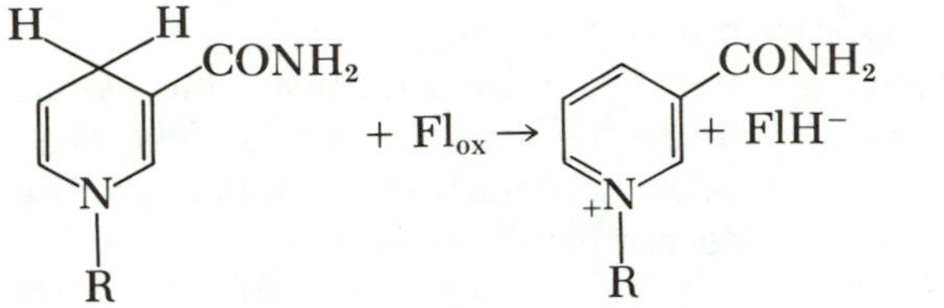

$$\text{(4)}$$

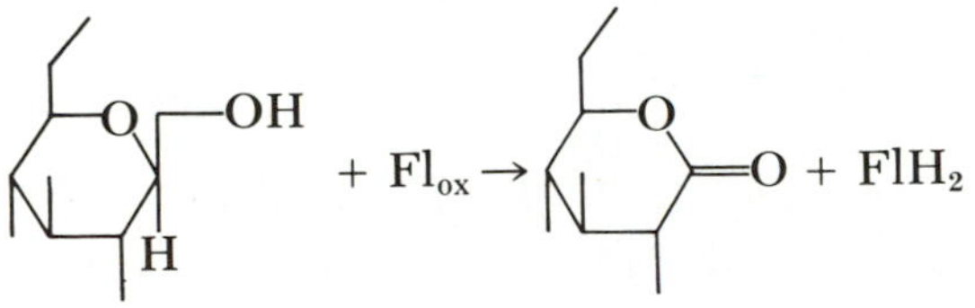

$$\text{(5)}$$

$$RCH_2NH_2 + Fl_{ox} \rightarrow RCH{=}NH + FlH_2 \qquad \text{(6)}$$

Though the reduction of Fl_{ox} by NADH is a facile process in solution, the reactions of Equations 5 and 6 are characterized by large kinetic barriers. The second grouping of carbon acid substrates are those that dissociate to resonance-stabilized carbanions and that, due to internal electron release, do not yield a carbonium ion on two-electron oxidations, for example:

Amino Acid Oxidases

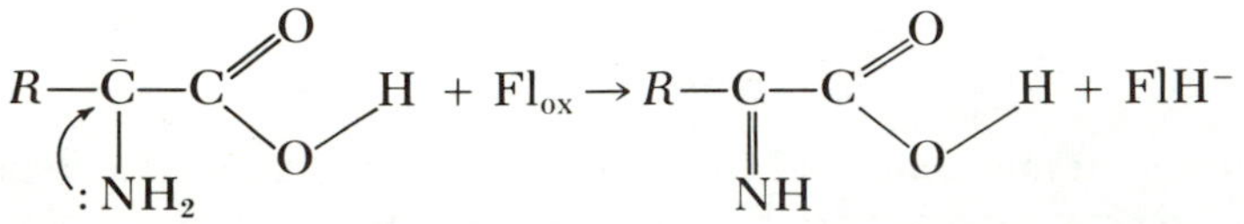

$$\text{(7)}$$

Lactic Acid Oxidase

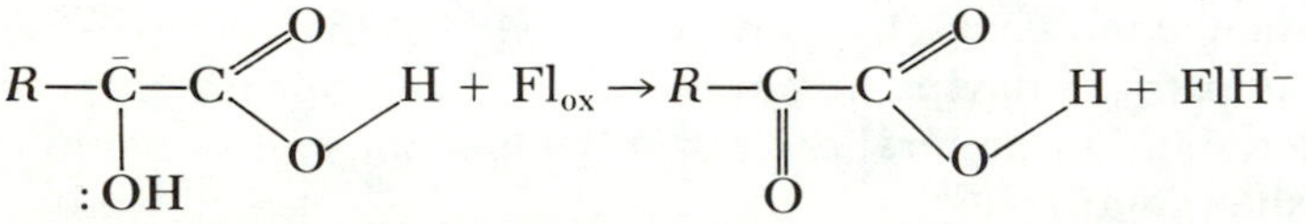

$$\text{(8)}$$

Succinic Acid Oxidase

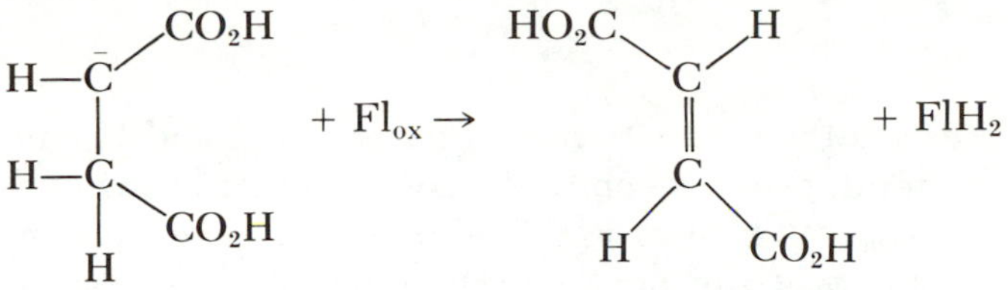

$$\underset{\underset{H}{\overset{|}{H-\underset{CO_2H}{\overset{CO_2H}{\overset{|}{C}}}}}{\overset{\overset{CO_2H}{/}}{H-\overset{-}{C}}} + Fl_{ox} \rightarrow \underset{H \quad CO_2H}{\overset{HO_2C \qquad H}{C}} + FlH_2 \qquad (9)$$

The third category of carbon acid substrates is nitroalkanes that readily dissociate to resonance-stabilized carbanions. The carbanions cannot be oxidized by loss of two electrons because of the instability of the resultant carbonium ion. Oxidation of nitroalkanes occurs through loss of nitrite ion.

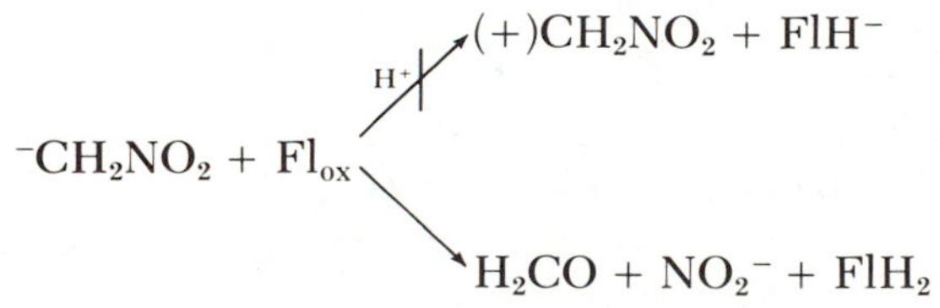

$$^-CH_2NO_2 + Fl_{ox} \underset{\underset{H_2CO + NO_2^- + FlH_2}{\searrow}}{\overset{\overset{(+)CH_2NO_2 + FlH^-}{\nearrow}}{\overset{H^+}{\cancel{}}}} \qquad (10)$$

The implication of carbanion intermediates in the mechanisms of flavoenzyme oxidations of substrates of the second class has been dependent upon the determination of the competition between β-halide release and oxidation (e.g., Equation 11) (*4, 5, 6, 7*). In model studies,

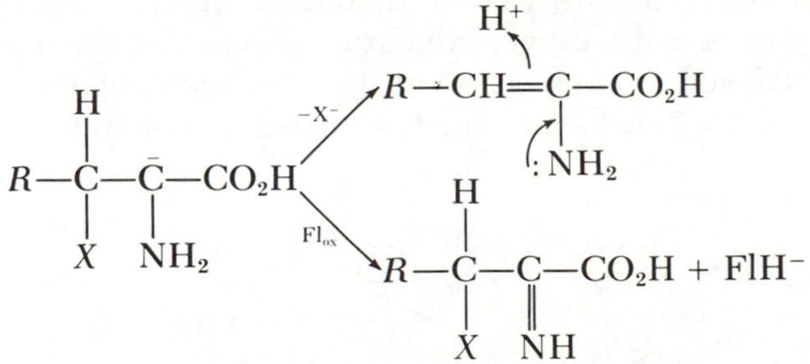

$$R-\underset{\underset{X}{\overset{|}{C}}}{\overset{\overset{H}{\overset{|}{C}}}{C}}-\underset{\underset{NH_2}{\overset{|}{C}}}{\overset{|}{C}}-CO_2H \underset{\underset{Fl_{ox}}{\searrow}}{\overset{\overset{-X^-}{\nearrow}}{}} \begin{matrix} \overset{H^+}{\underset{R-CH=C-CO_2H}{\searrow}} \\ \underset{:NH_2}{\overset{H}{}} \\ R-\underset{\underset{X}{\overset{|}{C}}}{\overset{H}{\overset{|}{C}}}-\underset{\underset{NH}{\overset{||}{C}}}{\overset{|}{C}}-CO_2H + FlH^- \end{matrix} \qquad (11)$$

we have shown that the flavin oxidation of carbon acids of this class occurs through the formation of carbanions. In these instances, carbanion formation is rate limiting at low Fl_{ox} concentrations, but Fl_{ox} dependent at even lower concentrations. The rates of disappearance of Fl_{ox} from solution on oxidation of a number of substrates follow accurately the kinetic expression of Equation 12 (*8–12*). In Equation 12,

$$\frac{d[Fl_{ox}]}{dt} = \frac{k_1' k_2 [CH][Fl_{ox}]}{k_{-1}' + k_2[Fl_{ox}]} \tag{12}$$

[CH] represents the concentration of carbon acid and k_1' and $k_{-1}'{}^1$ are pH- and buffer-dependent constants. By computer fitting of the absorbance (Fl_{ox} $\lambda_{max} \cong 442$ nm) vs. time plots we have determined the values of k_1' and k_2/k_{-1}'. From the dependence of these constants upon $[HO^-]$ and [buffer], it has been possible to show that $k_1' = k_{HO}[HO^-] + k_{gb}[B]$ and $k_2/k_{-1}' = k_2/(k_{H2O}[H_2O] + k_{ga}[BH])$. These findings are most simply assigned to the sequential reactions of Equation 13 where B

$$C\!-\!H \underset{\substack{k_{H_2O}[H_2O] \\ k_{ga}[BH]}}{\overset{\substack{k_{HO}[HO^-] \\ k_{gb}[B]}}{\rightleftarrows}} C^{(-)} \xrightarrow{k_2[Fl_{ox}]} C_{ox} + FlH^- \tag{13}$$

and BH are general base and general acid respectively. A steady state assumption in carbanion concentration (i.e., $[C^-]$) leads to Equation 14.

$$\frac{k[Fl_{ox}]}{dt} = \frac{(k_{HO}[HO^-] + k_{gb}[B])k_2[CH][Fl_{ox}]}{k_{H2O} + k_{ga}[BH] + k_2[Fl_{ox}]} \tag{14}$$

That the expression $(k_{HO}[HO^-] + k_{gb}[B])[CH]$ pertains to the rate of carbanion formation is established under conditions of high $[Fl_{ox}]$ when carbanion formation is rate determining. Under this condition, the rate constant for the flavin oxidation of dimethyl *trans*-1,2-dihydrophthalate to dimethyl phthalate is identical to the rate constant for base-catalyzed isomerization of the substrate to its 1,4-isomer (Equation 15) (8). This flavin oxidation serves as a biomimetic reaction

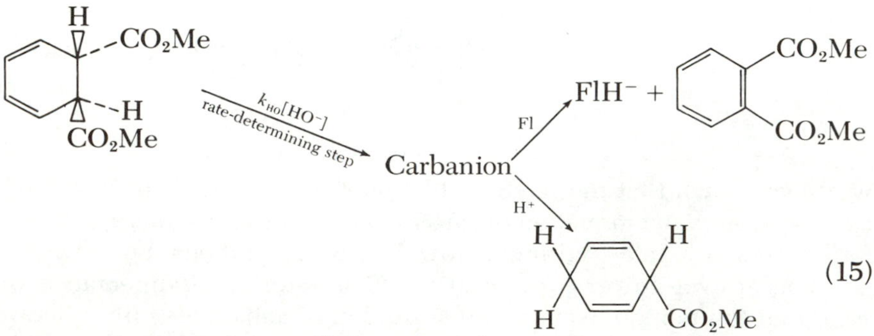

$$\tag{15}$$

for the enzyme succinic acid oxidase. In like manner, the general-base-catalyzed flavin oxidation and the general-base-catalyzed racemization of (+)-benzoin possess the same rate constant (*10*). The oxidation of α-ketols as benzoin and furoin serve as models for lactic acid oxidase.

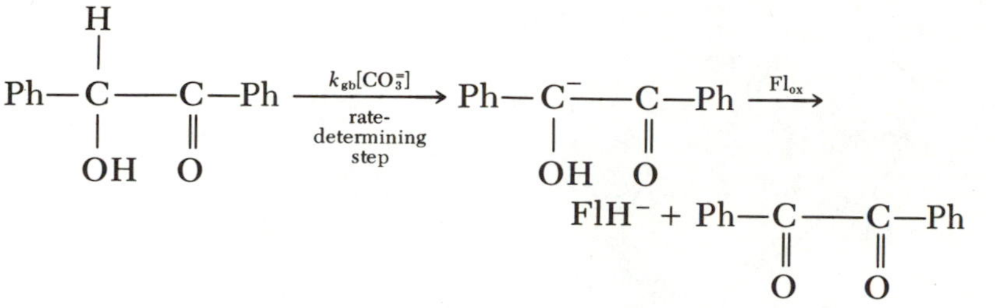

$$\underset{\substack{| \\ OH}}{Ph-\overset{\overset{\displaystyle H}{|}}{C}}-\underset{\substack{\| \\ O}}{C}-Ph \xrightarrow[\substack{rate-\\determining\\step}]{k_{gb}[CO_3^-]} Ph-\underset{\substack{| \\ OH}}{\overset{-}{C}}-\underset{\substack{\| \\ O}}{C}-Ph \xrightarrow{Fl_{ox}}$$

$$FlH^- + Ph-\underset{\substack{\| \\ O}}{C}-\underset{\substack{\| \\ O}}{C}-Ph \quad (16)$$

Having shown that general-base-catalyzed carbanion formation precedes the oxidation step in flavin oxidation of the second class of carbon acids, the question arises as to how the electron pair moves from the carbanion to flavin. Covalent addition of carbanion to Fl_{ox} followed by a base-catalyzed elimination reaction is one possibility (*13*). Addition to the 4a- (Equation 17) and 5-position (Equation 18) would appear to be feasible and, a priori, it would seem reasonable to expect that these adducts could undergo an elimination to yield oxidized substrate and reduced flavin. Nucleophilic addition of $SO_3^=$ to

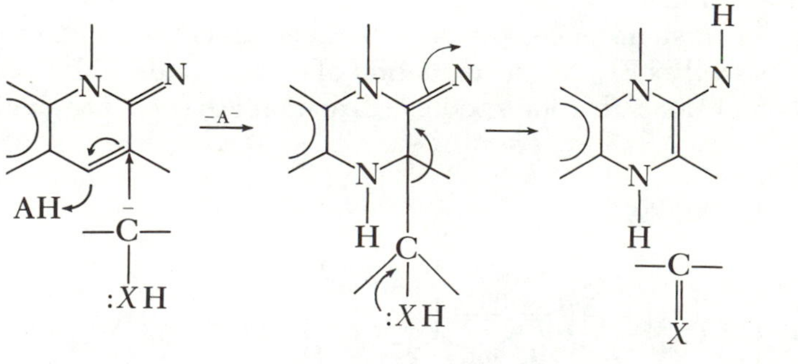

$$(17)$$

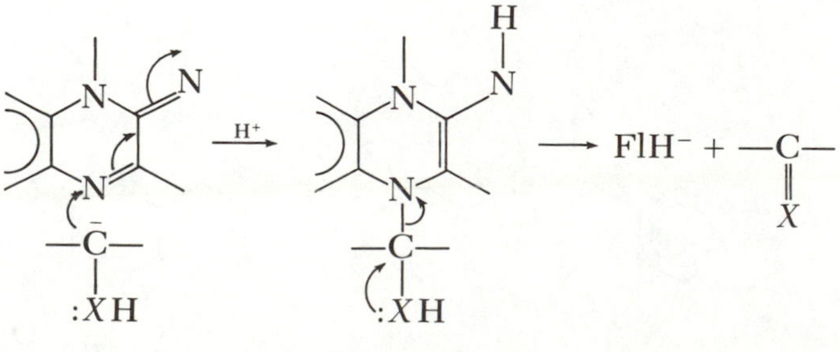

$$FlH^- + -\underset{\substack{\| \\ X}}{C}- \quad (18)$$

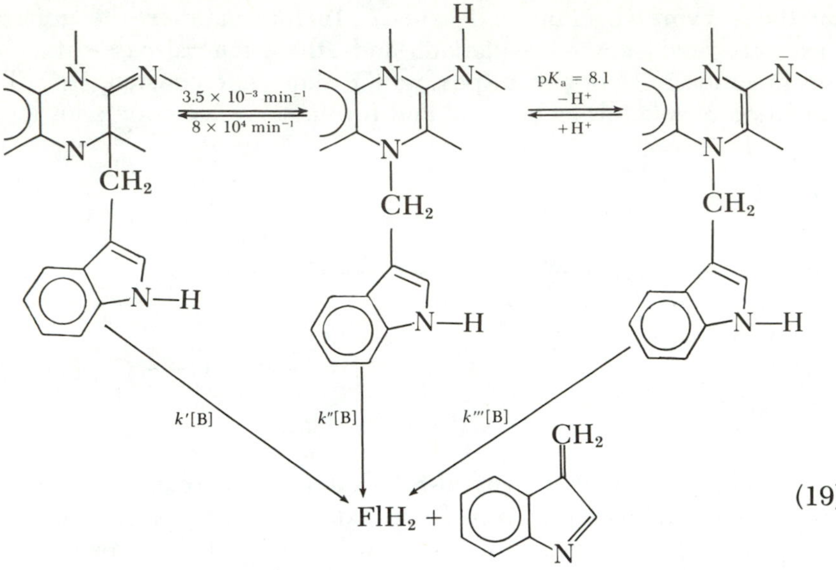

$$(19)$$

both the 4a- and 5-position (*14, 15, 16*) of the isoalloxazine ring is known. We have synthesized photochemically appropriate 4a- and 5-adducts and studied the kinetics of such elimination reactions (Equation 19) (*17*). However, several arguments may be advanced that disfavor 4a- and 5-adducts as general intermediates in the Fl_{ox} oxidation of carbanions.

In the first instance, the $N(5)$-blocked flavinium cation (Fl_{ox}^+Et) serves as well as Fl_{ox} for the oxidation of carbon acids and proceeds by oxidation of the carbanion species (Equation 20) (*11*). The availability

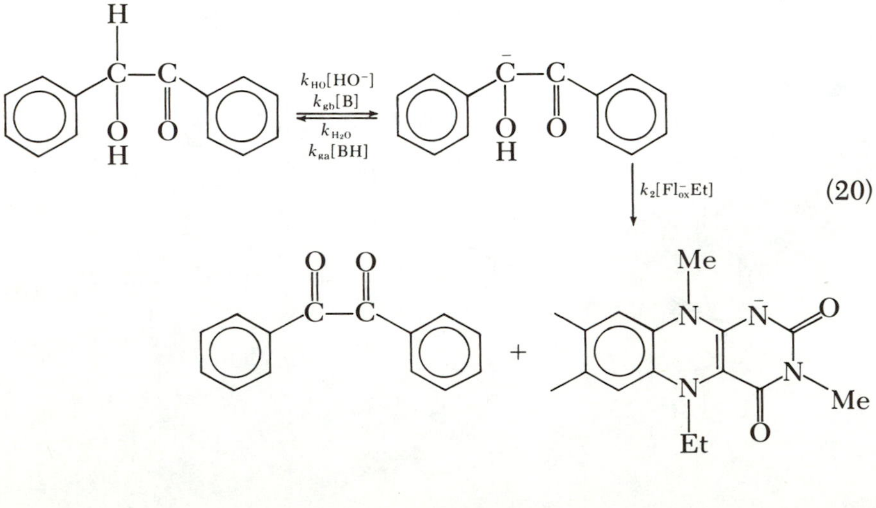

$$(20)$$

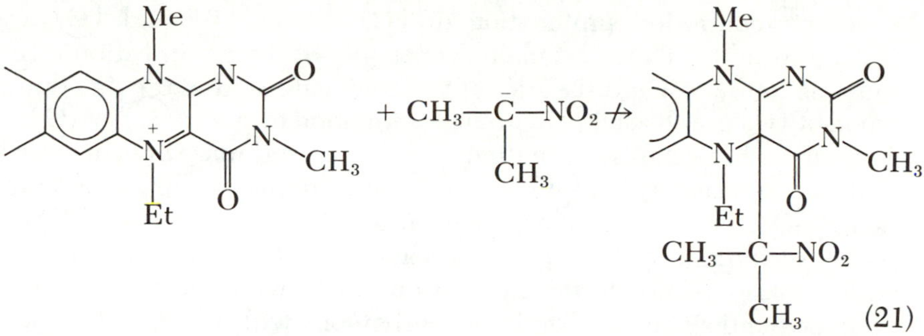

$$+ \; CH_3 - \overset{\text{\textemdash}}{\underset{CH_3}{C}} - NO_2 \nrightarrow \qquad (21)$$

of the 5-position is therefore not required in the Fl_{ox} oxidation of carbanion. Addition to the 4a-position also appears highly unlikely. For example, addition to the 4a-position of Fl_{ox}^+Et by a carbanion with the steric requirements of benzoin endiolate should be highly unfavorable. In aqueous solution, nitromethane anion forms a stable 4a-adduct with Fl_{ox}^+Et while the 4a-adduct of nitromethane is much less stable and 2-nitropropane anion does not add to Fl_{ox}^+Et (Equation 21). An additional argument in disfavor of a 4a-adduct mechanism may be offered.

The possible stepwise mechanisms for 4a-addition of carbanion to Fl_{ox} (A) to yield a 4a-adduct (D) are shown in the MAR diagram of Scheme 2. The route (A) → (B) → (D) represents specific acid catalysis and the sequence (A) → (C) → (D) would represent an

Scheme 2.

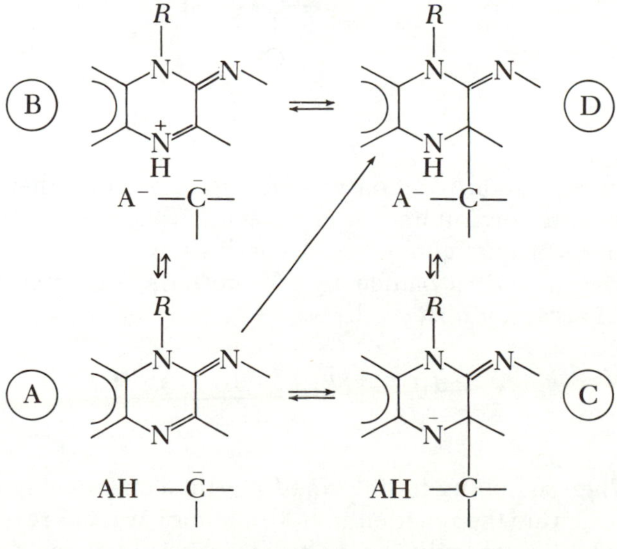

encounter-controlled protonation of Ⓒ. States Ⓑ and Ⓒ are highly unstable. Thus, the microscopic pK_a of the flavin cation of Ⓑ must be below -3 and the pK_a of the conjugate acid of the 4a-adduct anion of Ⓒ is at least 20. To obviate formation of states Ⓑ or Ⓒ, a general-acid mechanism is anticipated (concerted nucleophilic attack and proton transfer, i.e. Ⓐ → Ⓓ). An example of this is the addition of $SO_3^=$. 4a-Addition of $SO_3^=$ is catalyzed by the general acid HSO_3^- while N^5-addition of $SO_3^=$ is not so catalyzed (14). We conclude that addition of carbanion to the 4a-position of Fl_{ox} would be general-acid catalyzed and since the reaction of carbanions with Fl_{ox} is not subject to catalysis (see Equation 13), 4a-addition is not expected. The dehydrogenation of dimethyl trans-dihydrophthalate (Equation 15) follows the kinetic equations established for the oxidation of other ionizable carbon acids (Equation 14), which supports the general sequence of Equation 13. Examination of Equation 13 reveals that reaction of the carbanion species with Fl_{ox} is neither acid nor base catalyzed. However, the 4a-addition should be general-acid catalyzed and the ensuing elimination reaction from any 4a- or 5-adduct to provide a carbon–carbon double bond should be general-base catalyzed (Equation 22).

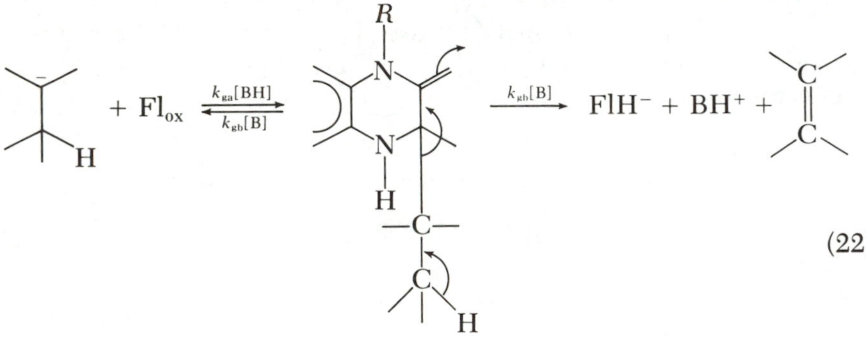

$$
\begin{array}{c}
\text{(structure)} + Fl_{ox} \xrightleftharpoons[k_{gb}[B]]{k_{ga}[BH]} \quad \text{(structure)} \xrightarrow{k_{gb}[B]} FlH^- + BH^+ + \quad \text{(structure)}
\end{array}
\tag{22}
$$

If the elimination reaction was rate determining, then conversion of carbanion to unsaturated product would follow the rate expression of Equation 23 and general-acid catalysis would be required in the oxidation of the carbanion by Fl_{ox} to form a carbon–carbon double bond. This is not observed.

$$
\frac{d[C^-]}{dt} = \frac{k_{ga}k'_{gb}[BH][Fl_{ox}][CH]}{(k_{gb} + k'_{gb})}
\tag{23}
$$

When searching for intermediates in a chemical reaction it is useful to examine the reaction in both the forward and reverse directions. The reduction of pyruvic acid by dihydroflavin is the retrograde of the

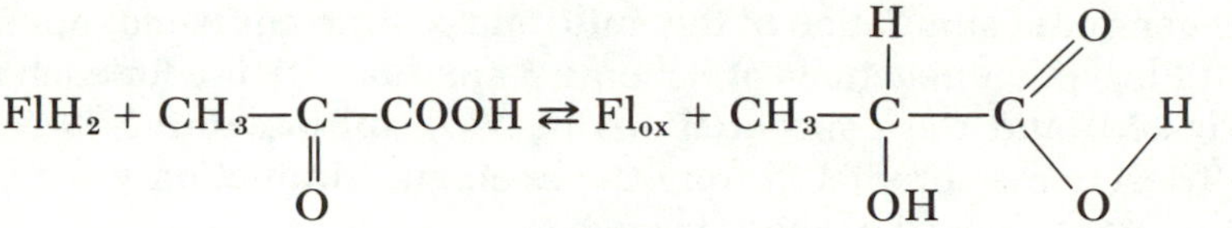

$$FlH_2 + CH_3-\underset{\underset{O}{\|}}{C}-COOH \rightleftarrows Fl_{ox} + CH_3-\underset{\underset{OH}{|}}{\overset{\overset{H}{|}}{C}}-C\overset{\nearrow O}{\underset{\searrow O}{}}H \quad (24)$$

oxidation of lactic acid by oxidized flavin (Equation 24). The reactions that occur upon mixing aqueous solutions of $CH_3COCOOH$, CH_3COCOO^-, $CH_3COCOOC_2H_5$, $CH_3COCONH_2$ (*18*), or CH_2O (*19*) with dihydroflavin under anaerobic conditions are provided in Scheme 3. The validity of Scheme 3 rests upon the findings that:

Scheme 3.

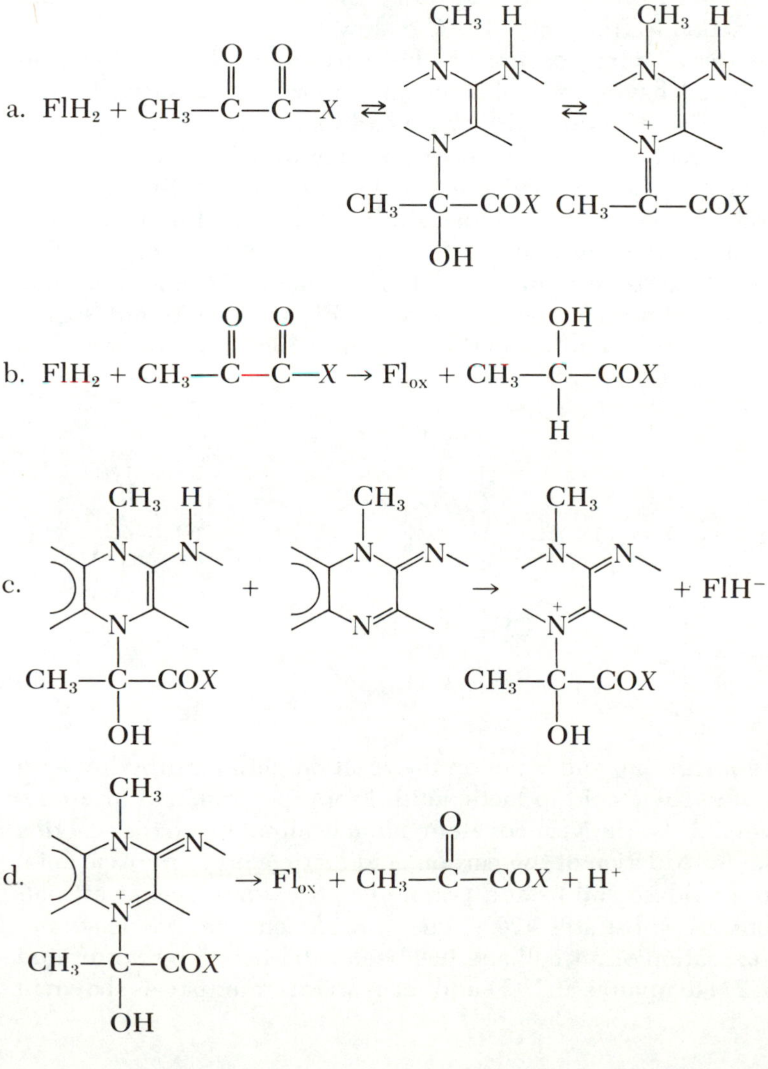

a. $FlH_2 + CH_3-\overset{\overset{O}{\|}}{C}-\overset{\overset{O}{\|}}{C}-X \rightleftarrows$

b. $FlH_2 + CH_3-\overset{\overset{O}{\|}}{C}-\overset{\overset{O}{\|}}{C}-X \rightarrow Fl_{ox} + CH_3-\underset{\underset{H}{|}}{\overset{\overset{OH}{|}}{C}}-COX$

c.

d. $\rightarrow Fl_{ox} + CH_3-\overset{\overset{O}{\|}}{C}-COX + H^+$

analog-computer simulation of the multiphasic time course for appearance of Fl_{ox} via the reactions of Scheme 3 are quantitative for each pH investigated; and the concentrations of $N(5)$-carbinolamine + $N(5)$-imine (Reaction a) predicted from the analogue simulation were verified through trapping of these species.

Reactions were studied under the pseudo first-order condition of [substrate] much greater than [initial dihydroflavin]. Under these conditions, the reactions are characterized by a "burst" in the production of Fl_{ox} followed by a much slower rate of Fl_{ox} formation until completion of reaction. The initial burst is provided by the competition between parallel pseudo first-order Reactions a and b of Scheme 3. These convert dihydroflavin and carbonyl compound to an equilibrium mixture of carbinolamine and imine (Reaction a), and to Fl_{ox} and alcohol (Reaction b), respectively. The slower production of Fl_{ox}, following the initial burst, occurs by the conversion of carbinolamine back to reduced flavin and substrate and, more importantly, by the disproportionation of product Fl_{ox} with carbinolamine (Reaction c followed by d). Reactions c and d constitute an autocatalysis by oxidized flavin of the conversion of carbinolamine back to starting dihydroflavin and substrate. In the course of these studies, the contribution of acid–base catalysis to the reactions of Scheme 3 were determined. The significant feature to be pointed out here is that carbinolamine does not undergo an elimination reaction to yield Fl_{ox} and lactic acid (Equation 25). The carbinolamine ($N(5)$-covalent adduct) is formed in a

$$FlH_2 + CH_3{-}CO{-}COX \rightleftharpoons \quad \not\rightarrow Fl_{ox} + CH_3{-}\overset{\overset{\displaystyle H}{|}}{\underset{\underset{\displaystyle OH}{|}}{C}}{-}COX \tag{25}$$

dead-end equilibrium and is not on the reaction path for dihydroflavin reduction of pyruvic acid to lactic acid. From the principal of microscopic reversibility, the $N(5)$-covalent adduct cannot be on the reaction path for flavin oxidation of the carbon acid lactic acid to pyruvic acid.

Glucose oxidase and D- and L-amino acid oxidase accept nitroalkane anions as substrates (20). The mechanism for flavoenzyme-catalyzed oxidation of nitroalkane has been established by Porter and Bright (20, 21) to involve an $N(5)$-adduct as an intermediate as shown in

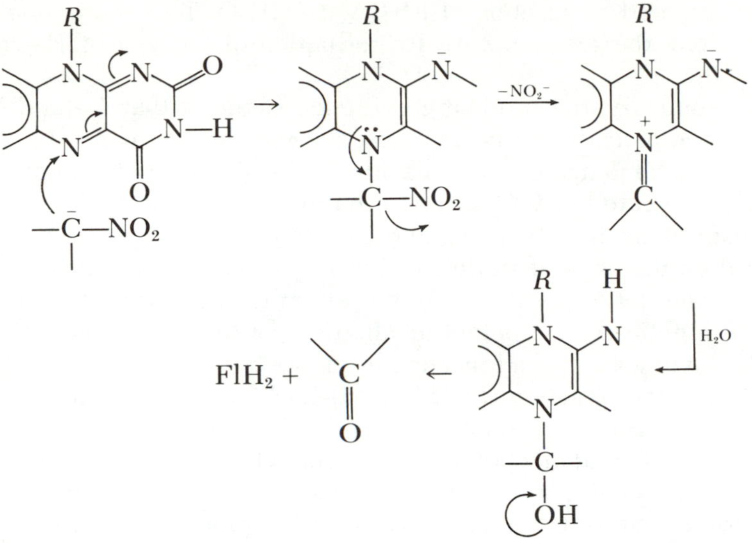

(26)

Equation 26. Electron-deficient flavins will also oxidize nitroalkane anions in model reactions (*12*). The observation (*11*) that nitromethane anion and Fl_{ox}^+Et yield a stable 4a-adduct is evidence that 4a-adducts are not on the reaction path for nitroalkane oxidation. That the blocking of the $N(5)$-position of flavin (i.e., Fl_{ox}^+Et) prevents oxidation of nitromethane would, however, be in accord with the requirement for an $N(5)$-adduct (*11*). The nitroalkane reaction with flavoenzyme has been used to implicate $N(5)$-adducts as intermediates in the oxidation mechanism of amino acid oxidases. However, it must be understood that nitroalkane anions differ significantly from the carbanions generated from a normal substrate. The nitroalkane anion on loss of its pair of electrons would provide an impossibly unstable carbonium ion, whereas in the case of the amino acid anion an internal electron release obviates carbonium ion formation.

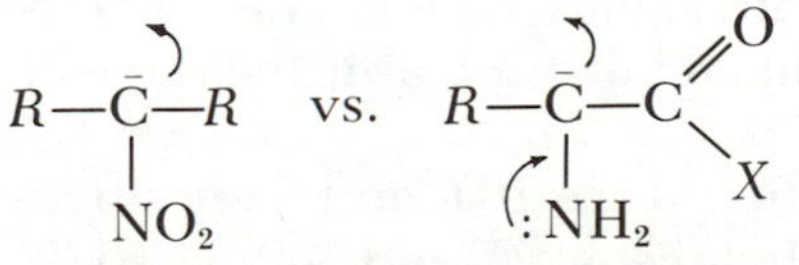

The nitroalkane anion cannot undergo a direct two-electron oxidation. However, once the nitroalkane $N(5)$-adduct is formed the internal displacement of the stable NO_2^- species (Equation 26) should be favorable energetically. The imine species formed on loss of NO_2^- from the condensation product of nitromethane and Fl_{ox} is the same imine obtained on reaction of FlH_2 with CH_2O. This imine does not occur along the reaction path for reduction of CH_2O to CH_3OH by FlH_2 (19).

In summary, the following evidence suggests that 4a- and $N(5)$-covalent-adducts do not occur along the reaction paths for the dehydrogenation of ionizable carbon acids: $N(5)$-substituted flavin oxidizes benzoin anion to benzyl though formation of an $N(5)$-covalent adduct of substrate, flavin is blocked, and 4a-addition would be strongly disfavored on the basis of steric considerations. General-acid catalysis, a requirement for 4a-addition, is not seen on reaction of carbanion with Fl_{ox}. (A kinetically competent mechanism for thiol anion oxidation by Fl_{ox} involves a 4a-thiol adduct and evidence exists for the general-acid catalysis of its formation (12, 22). $N(5)$-Carbinolamine adducts, when formed from carbonyl compounds and dihydroflavin, do not go onto oxidized flavin and alcohol, though carbonyl compounds are reduced readily to alcohols by dihydroflavin. Certain synthetic $N(5)$-covalent adducts (Equation 19) have provided dihydroflavins on elimination. This is to be expected. Inspection of the reactions of Equation 27

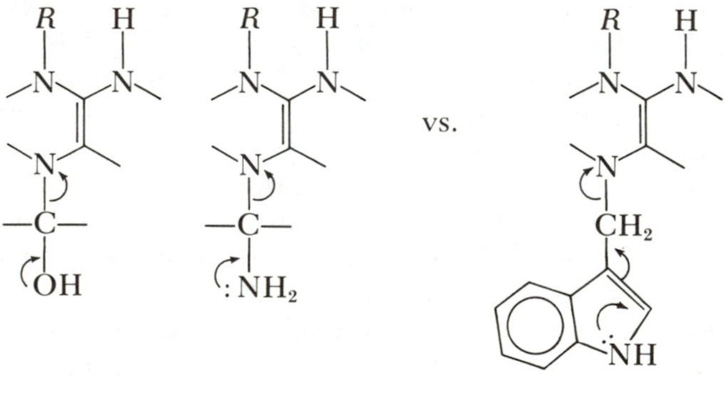

$$\text{(27)}$$

establishes their vinylogous relationship. The $N(5)(3$-methylene indole) adduct should and does exhibit the same chemistry as the $N(5)$-carbinolamine.

Covalent addition at the 10a- and 9a-positions are not likely since the 10-(2′,6′-dimethylphenyl) (I) and 5-ethyl-10-(2′,6′-dimethylphenyl) (II) isoalloxazines behave as ordinary flavins in all the reactions described in this chapter. With these analogues, the 2′- and 6′-methyl groups shield the 9a- and 10a-positions from nucleophilic attack. For instance, the 10a- and 4-positions are opened on alkaline hydrolysis of $N(3)$-substituted isoalloxazines (23). In the case of I, hydrolysis only occurs at the 4-position because the 10a-position is shielded.

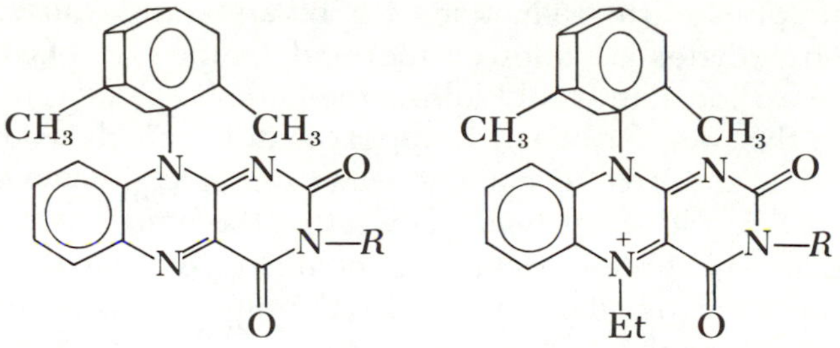

I   II

Pathways for oxidation reactions not involving covalent adducts or metal ions are radical in nature or involve hydride transfer. We have proposed the radical mechanisms of Scheme 4 (*18, 19*). Electrochemical calculations establish that the standard free energy of formation of the radicals FlH· and ·CH$_2$OH from FlH⁻ and CH$_2$O does not exceed the determined $\Delta G\ddagger$ for reduction of CH$_2$O by FlH⁻. The same con-

Scheme 4.

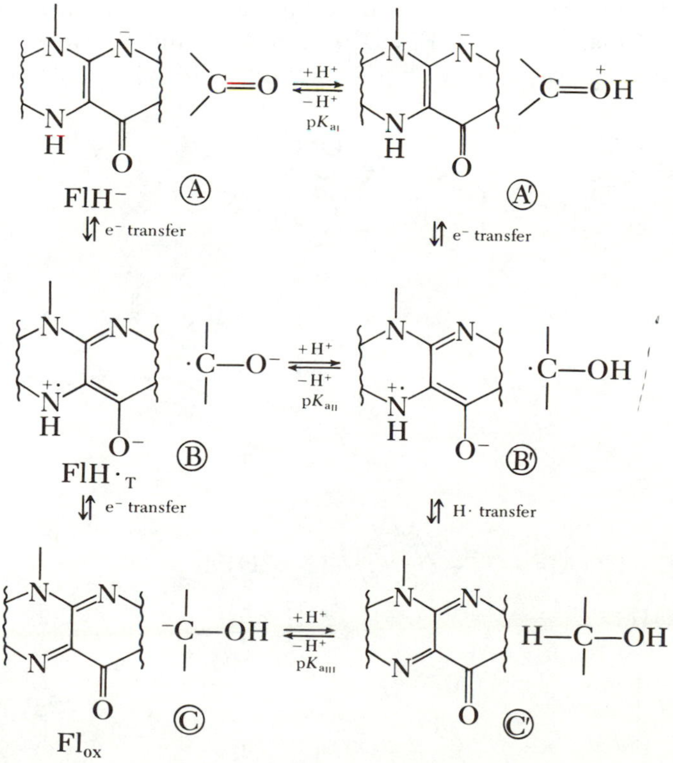

sideration applies to the reduction of pyruvate and derivative. Thus, radical intermediates are allowed thermodynamically. The oxidation of ionizable carbon acid would follow the path $\text{(C')} \rightarrow \text{(C)} \rightarrow \text{(B)} \rightarrow \text{(A)}$. Thus, the pK_{a_I} for benzil is approximately -7, the reduction of benzil by FlH_2 and FlH^- is not acid catalyzed; $pK_{a_{III}}$ has been determined to be 5.5, and it has been shown that the benzoin carbanion is the immediate substrate in the reduction of Fl_{ox} by benzoin. The oxidation of carbon acids whose derived carbanions are very unstable in water would follow the path $\text{(C')} \rightarrow \text{(B')} \rightarrow \text{(A')} \rightarrow \text{(A)}$. The oxidation of methanol is an example of this. The large free-energy content of $\cdot CH_2O^-$ and $(-)CH_2OH$ (i.e., $\Delta G°$ of formation of these species $> \Delta G\ddagger$) precludes the existence of states (B) and (C) . Specific-acid catalysis has been found for the reduction of CH_2O by FlH_2 as anticipated in steps $\text{(A)} \rightarrow \text{(A')}$ of Scheme 4.

Radical mechanisms account for the stoichiometry for reduction of triketohydrindane by $N(5)$-ethyldihydroflavin and reduction of triphenylmethyl carbonium ion species by dihydroflavin, (24). One-electron reduction of quinone by $N(5)$-ethyldihydroflavin also has been shown. These results are not surprising since the substrates and flavin support reasonably stable radical states. Radical species also can be established as intermediates in the oxidation of 9-hydroxyfluorene and methyl mandelate by Fl_{ox} (Equations 28 and 29, respectively). The reactions of Equations 28 and 29 are facile when carried out in

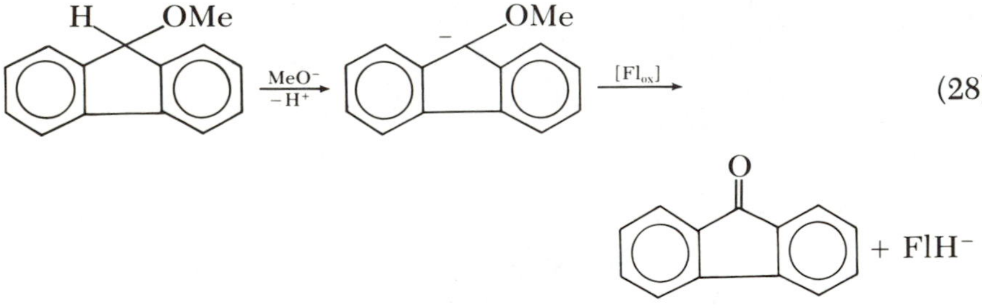

$$\text{(28)}$$

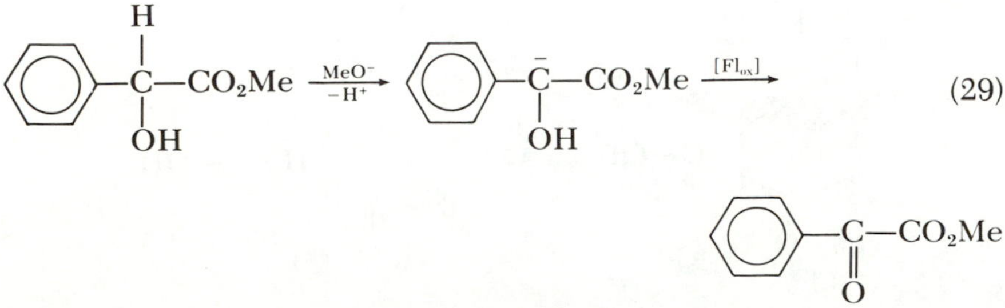

$$\text{(29)}$$

absolute methanol. One-electron transfer to Fl_{ox} from the resonance-stabilized carbanions formed from 9-hydroxyfluorene and methyl mandelate generated the radical species **III** and **IV**.

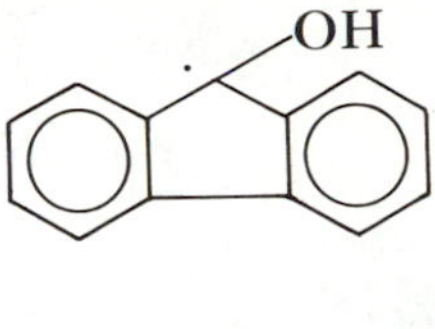

III 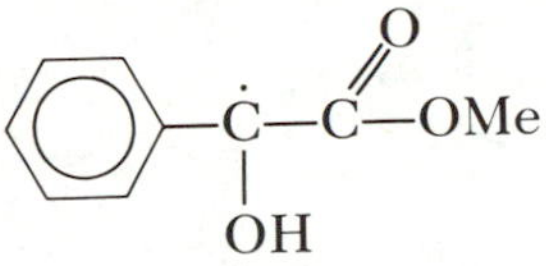   IV

Guthrie (25) has shown that radical oxidation of 9-methoxyfluorene anion by nitrobenzene under anaerobic conditions provides the dimeric coupling product **VI** of Equation 30, while under

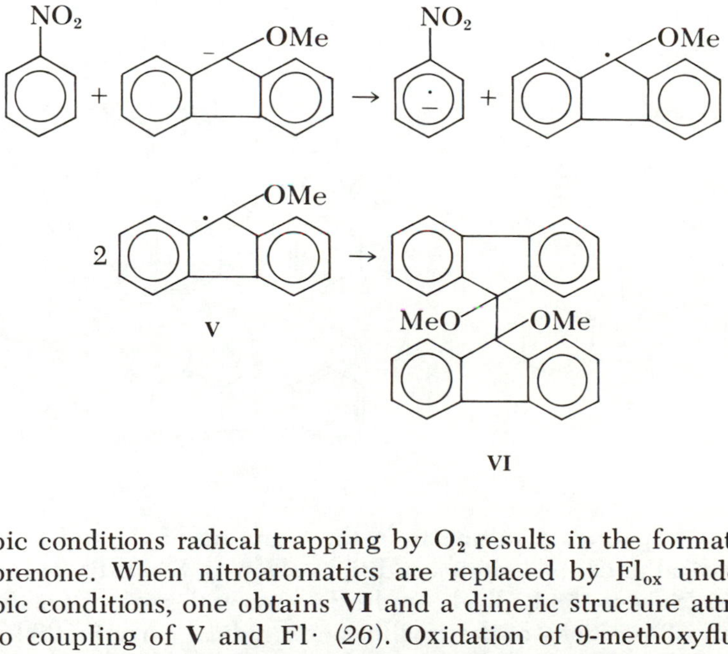

(30)

VI

aerobic conditions radical trapping by O_2 results in the formation of 9-fluorenone. When nitroaromatics are replaced by Fl_{ox} under anaerobic conditions, one obtains **VI** and a dimeric structure attributable to coupling of **V** and Fl· (26). Oxidation of 9-methoxyfluorene anion by Fl_{ox} in the presence of either O_2 or a nitroxide results in the formation of 9-fluorenone (Scheme 5). Studies of the oxidation of the carbanion of methyl α-methoxyphenyl acetate by Fl_{ox} under anaerobic conditions and in the presence of O_2 and/or a nitroxide provide results completely analogous to those obtained with 9-methoxyfluorene (27). These carbanion oxidations by Fl_{ox} are unquestionably free radical in nature.

Scheme 5.

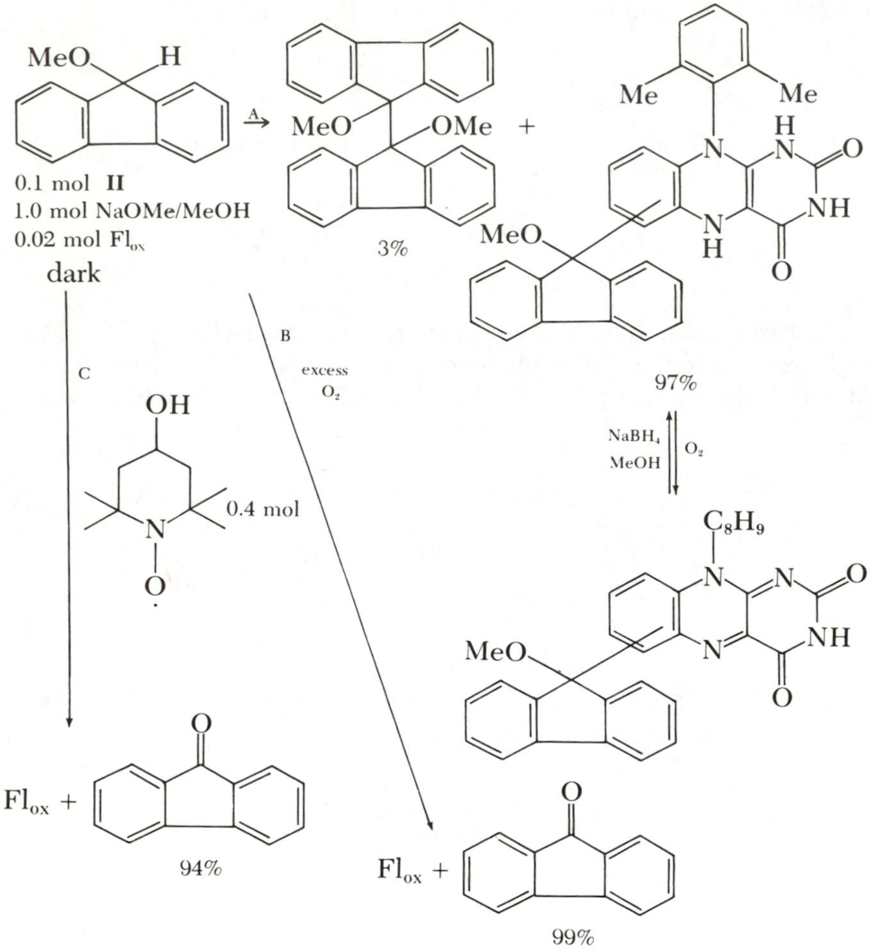

The compounds resulting from the coupling of **III** and **IV** with flavin radical under anaerobic conditions are 6- or 8-substituted isoalloxazines (Scheme 5). Radical coupling occurs at the 4a-position when the 7- and 8-positions of the isoalloxazine radical carry methyl substituents (flavins) and when the reaction involves species of general structure $R—CH_2\cdot$. Hemmerich (28) has employed the photosynthetic procedure of Equation 31 to provide 4a-substituted flavins. These reactions have been investigated in our laboratory using EPR spectroscopy and spin traps and have been shown to involve the coupling of flavin radical with $X—CH_2\cdot$ (27). Homolysis of the 4a-adducts (obtained as in Equation 31) does not occur.

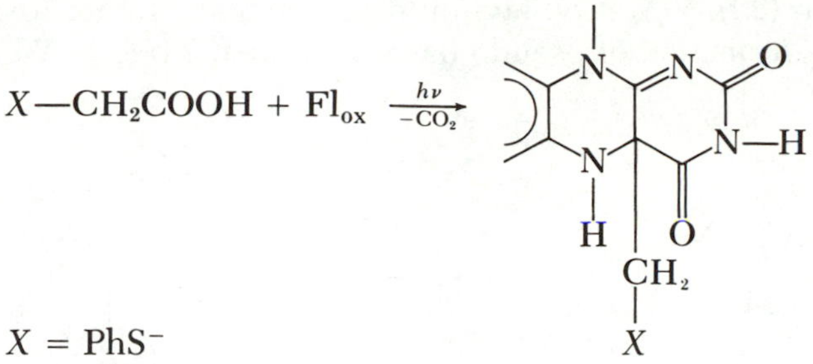

$$X—CH_2COOH + Fl_{ox} \xrightarrow[-CO_2]{h\nu}$$

(31)

X = PhS⁻
PhO⁻
Ph⁻

Mono- and Dioxygenation Reactions

Of the various proposals for the mechanism of activation of molecular oxygen by flavoprotein oxygenases, we have chosen to believe that oxygen reacts with dihydroflavin to yield a 4a-hydroperoxyflavin (Equation 32). The inability to detect 4a-FlHOOH in the reaction of

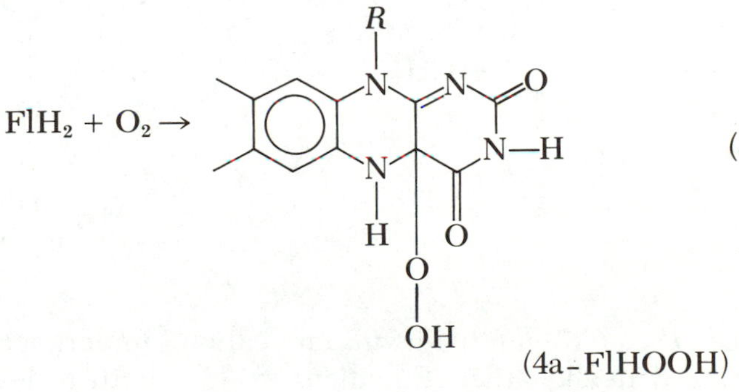

$$FlH_2 + O_2 \rightarrow$$

(32)

(4a–FlHOOH)

dihydroflavin with oxygen in solution would then be attributable to the expected facile elimination of hydrogen peroxide from 4a-FlHOOH with formation of Fl_{ox} (Equation 33). It was reasoned that N-alkyl substitution would greatly inhibit this hydrogen peroxide

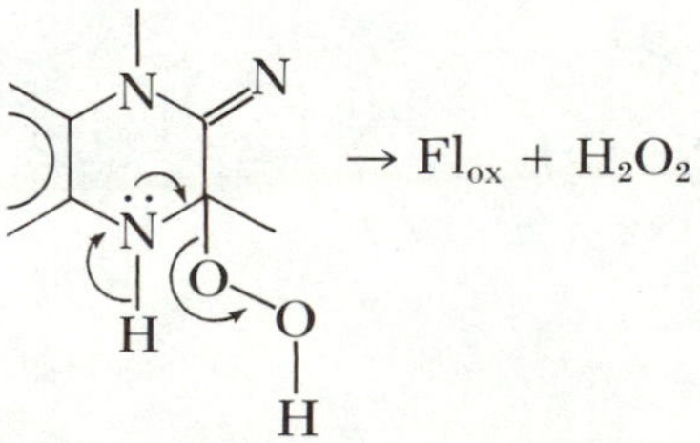

$$\rightarrow Fl_{ox} + H_2O_2$$

(33)

elimination (29). $N(5)$-Akylflavinium cations have rather low pK_a values for the formation of pseudo base (Equation 34) (30). We expected

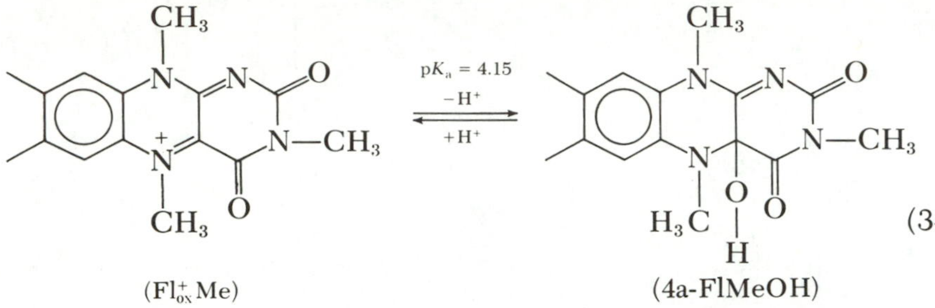

$$(34)$$

(Fl$_{ox}^+$Me)   (4a-FlMeOH)

that the 4a-hydroperoxy-$N(5)$-alkylflavin could be prepared by the addition of hydrogen peroxide to the flavinium cation (Equation 35).

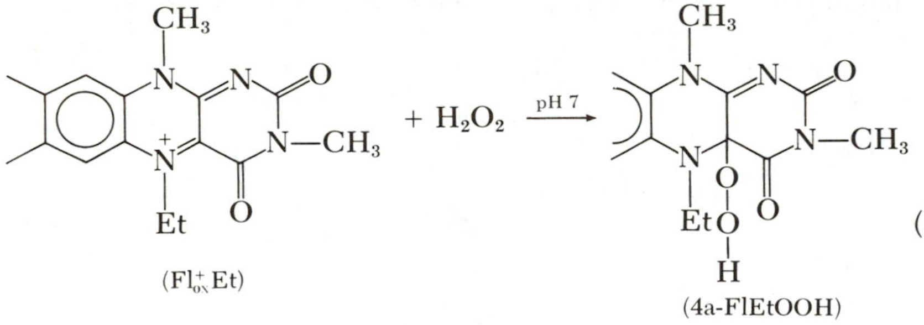

$$(35)$$

(Fl$_{ox}^+$Et)   (4a-FlEtOOH)

Since the $N(5)$-methylflavinium cation undergoes general-base-catalyzed dealkylation (Equation 36) (31) with a deuterium kinetic isotope effect ($N(5)$—CH$_3$/$N(5)$—CD$_3$) of greater than 10, we chose —CD$_3$ and more simply —CH$_2$CH$_3$ as the $N(5)$-blocking groups. The compound 4a-FlEtOOH is prepared routinely in our laboratory in

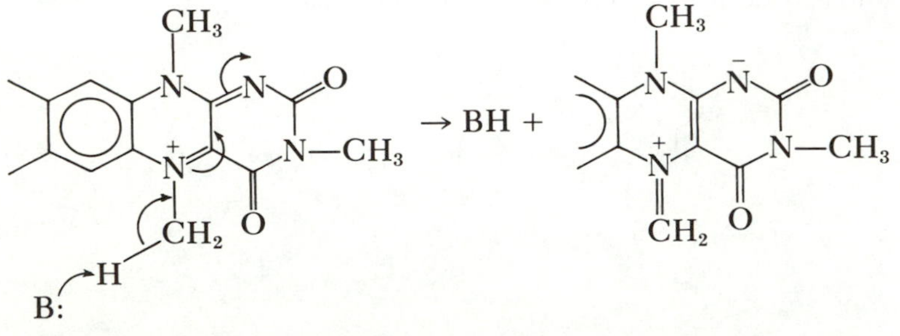

$$(36)$$

85%–96% purity. The identity of the UV–visible spectras of 4a-FlEtOOH and the intermediate formed from a reduced flavomono-oxygenase and oxygen provides us with the knowledge (29) that the enzyme-bound dihydroflavin oxygen adduct is indeed a 4a-hydro-peroxyflavin. With knowledge of the spectral characteristics of 4a-FlEtOOH, it has been possible to show that molecular oxygen reacts directly with N(5)-ethyl-1,5-dihydrolumiflavin to yield 4a-FlEtOOH (Equation 37). The formation of 4a-FlEtOOH on reaction of

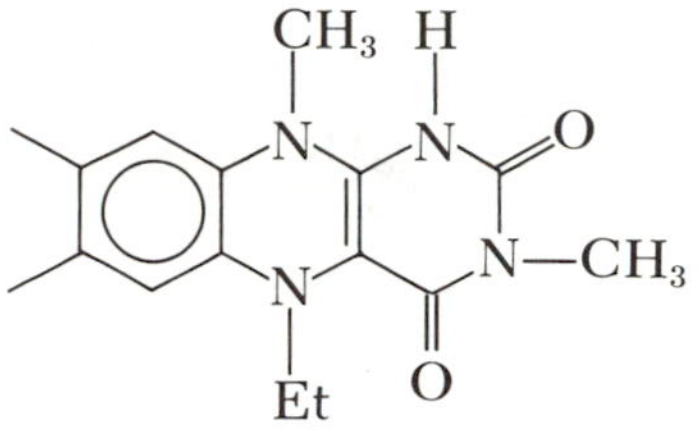

$$+\ ^3O_2 \rightarrow \text{4a-FlEtOOH} \qquad (37)$$

oxygen with dilute aqueous solutions of dihydroflavin is not observed because of the rapid exchange of HO^- and HOO^- to provide the 4a-pseudo base (Equation 38). The formation of 4a-FlEtOOH can be

$$\text{4a-FlEt—OOH} + H_2O \rightarrow \text{4a-FlEt—OH} + H_2O_2 \qquad (38)$$

studied in absolute methanol where exchange of the —OOH moiety for —OMe to provide 4a-FlEtOMe is slow. It can be studied better in absolute t-butyl alcohol where there is no perceptible exchange of HO_2^- for t-BuO$^-$.

The kinetics for the reaction of FlEtH and FlEt$^-$ with oxygen in water are quite similar to the kinetics for reaction of FlH$_2$ and FlH$^-$ with oxygen. The major differences are attributable to the greater stability of FlEt· compared with FlH· and that the radical anion obtained from flavin (Fl·$^-$) has no N(5)-alkyl counterpart (32). With neither FlEtH + FlEt$^-$ nor FlH$_2$ + FlH$^-$ can an oxygen adduct be detected in water. The major reactions occurring on reaction of N(5)-ethyl 3-methyl-1,5-dihydrolumiflavin with oxygen in methanol are provided in Scheme 6. The rate constants for the various reactions were determined either from the reaction of FlEtH with oxygen or from the rates of reactions of 4a-FlEtOOH (32). In Scheme 6, Reactions 1 and 2 are of most importance. At high ratios of [O$_2$] to [FlEtH] the yield of 4a-FlEtOOH is maximized and only a few percent of FlEt· is formed. When the [O$_2$] is decreased or [FlEtH] is increased, the yield of 4a-FlEtOOH is decreased while the yield of FlEt· is increased. These findings have been quantified and are in accord with competition of O$_2$ and 4a-FlEtOOH for FlEtH.

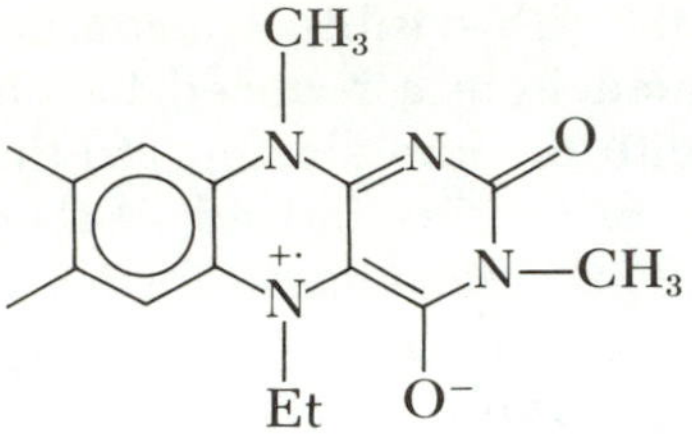

(FlEt·)   vs.   (FlH·)

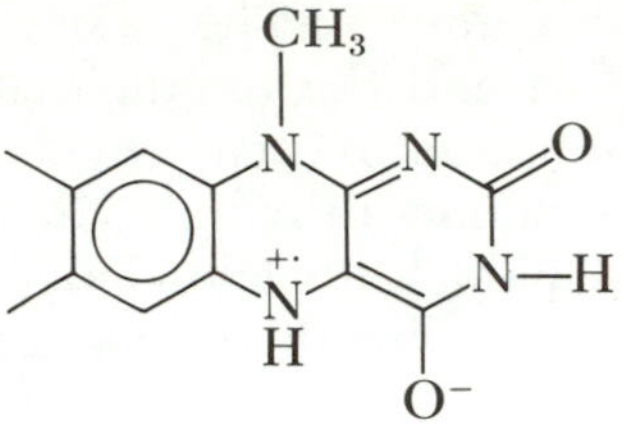

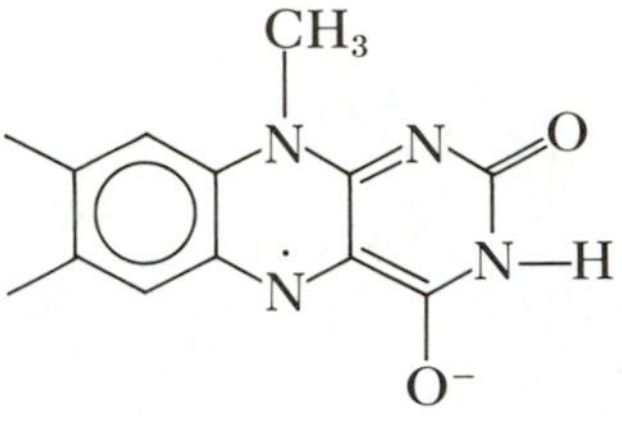

(FlEt⁻)

Scheme 6.

$$\text{FlHCH}_3 + {}^3\text{O}_2 \xrightarrow{k_1 = 10\text{–}30 \text{ mol}^{-1} \text{ sec}^{-1}} 4a\text{-FlCH}_3\text{—OOH} \qquad [1]$$

$$\text{FlHCH}_3 + 4a\text{-FlCH}_3\text{—OOH} \xrightarrow{k_2 = 500 \text{ mol}^{-1} \text{ sec}^{-1}}$$
$$\text{FlCH}_3\cdot + 4a\text{-FlCH}_3\text{—OH} + \text{HO}\cdot \qquad [2]$$

$$\text{FlCH}_3\cdot + 4a\text{-FlCH}_3\text{—OOH} \xrightarrow{k_3 \cong 20 \text{ mol}^{-1} \text{ sec}^{-1}}$$
$$\text{Fl}_{\text{ox}}{}^+\text{CH}_3 + 4a\text{-FlCH}_3\text{—OH} + \text{HO}\cdot \qquad [3]$$

$$4a\text{-FlCH}_3\text{—OOH} \xrightarrow{k_4 = 3.2 \times 10^{-4} \text{ sec}^{-1}} \text{Fl}_{\text{ox}}{}^+\text{CH}_3 + \text{HO}_2{}^- \qquad [4]$$

$$4a\text{-FlCH}_3\text{—OH} \xrightarrow{k_5 = 1.6 \times 10^{-4} \text{ sec}^{-1}} \text{Fl}_{\text{ox}}{}^+\text{CH}_3 + \text{HO}^- \qquad [5]$$

$$\text{FlCH}_3\cdot + {}^3\text{O}_2 \xrightarrow{k_6 \cong 0.4 \text{ mol}^{-1} \text{ sec}^{-1}} \text{Fl}_{\text{ox}}{}^+\text{CH}_3 + \text{O}_2{}^- \qquad [6]$$

$$\text{FlHCH}_3 + \text{Fl}_{\text{ox}}{}^+\text{CH}_3 \xrightarrow{k_7 > 10^7 \text{ mol}^{-1} \text{ sec}^{-1}} 2 \text{ FlCH}_3\cdot \qquad [7]$$

Investigations of the chemistry of 4a-FlR—OOH compounds is a current project in our laboratory. We view the availability of the flavin hydroperoxides as an opportunity to investigate what is in essence either the activated oxygen species or the precursor to the activated oxygen species of flavoenzyme monooxygenases and the one known flavin dioxygenase (33). The microsomal monooxygenases consist of the inducible and numerous cytochrome P-450-type enzymes as well as flavomonooxygenases. The hepatic microsomal monooxygenase apparatus is responsible for the conversion (or initiation of conversion) of xenobiotic materials into water-soluble excretable products. The N-oxidation of tertiary amines is a role played by microsomal flavoen-

zymes (34). In absolute *t*-butyl alcohol and under anaerobic conditions, 4a-FlEtOOH reacts with N,N-dimethylaniline to yield N-oxide
and flavin pseudo base. The bimolecular reaction of Equation 39 is

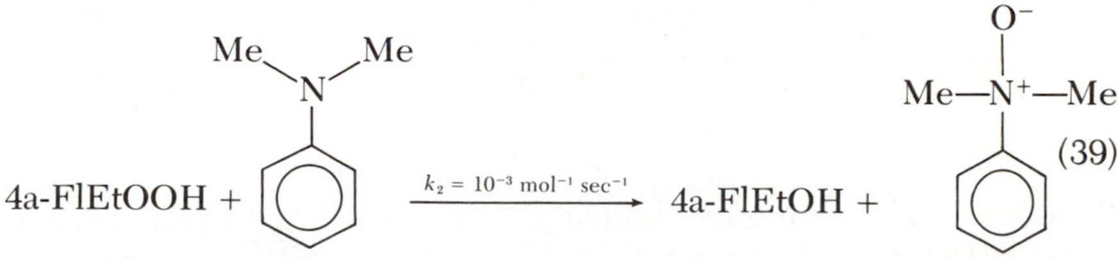

$$
\text{4a-FlEtOOH} + \underset{\text{[dimethylaniline]}}{\boxed{}} \xrightarrow{k_2 = 10^{-3}\ \text{mol}^{-1}\ \text{sec}^{-1}} \text{4a-FlEtOH} + \underset{\text{[N-oxide]}}{\boxed{}} \tag{39}
$$

quantitative (35). Under the conditions where 4a-FlEtOOH readily
monooxygenates dimethylaniline there is no detectable formation of
N-oxide with either H_2O_2 or *t*-BuOOH. When the concentration of
H_2O_2 was monitored it was found that the addition of dimethylaniline
did not increase its rate of disappearance from solution. Since the
second-order rate constants for reaction of dimethylaniline with H_2O_2
and *t*-BuOOH could not be obtained, it was not possible to determine
just how much better 4a-FlEtOOH is than these reagents as a
monooxygenation reagent. An answer to this question was available,
however, from a study of the related monooxygenation of thioxane
(Equation 40) (36). The reaction of Equation 40 is quantitative in absolute methanol and the relative rate constants in Table I could be determined.

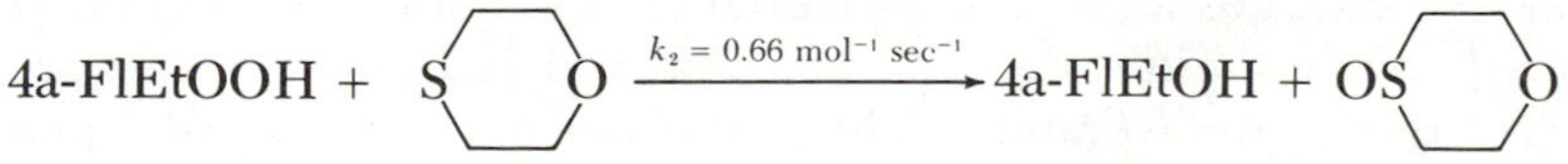

$$
\text{4a-FlEtOOH} + \underset{\text{[thioxane]}}{\boxed{}} \xrightarrow{k_2 = 0.66\ \text{mol}^{-1}\ \text{sec}^{-1}} \text{4a-FlEtOH} + \text{O}\underset{\text{[S-oxide]}}{\boxed{}} \tag{40}
$$

Table I.   Comparison of the Relative Rate Constants for S-Oxidation of Thioxane by Hydroperoxides

Hydroperoxide	k_{rel}
$(CH_3)_3COOH$	1
H_2O_2	2.4×10^1
4a-FlEtOOH	1.8×10^5

Dankleft et al. (37) have proposed that the S-oxidation of sulfide
by hydroperoxides requires the presence of a proton source. Their
conclusions were based on the finding that in the aprotic solvent dioxane
the S-oxidation of thioxane is second order in alkyl peroxides. The
mechanism suggested by them is shown in Equation 41. The

$$2\text{ROOH} + \text{S} \underset{\smile}{\bigcirc} \text{O} \longrightarrow \left| \begin{array}{c} \text{O-}R \\ | \\ \text{O} \\ \text{H} \overset{\frown}{\underset{\displaystyle \text{O-O}}{}} \text{H} \\ \overset{..}{\text{S}} \quad R \\ / \backslash \end{array} \right| \rightarrow \text{ROOH} + \text{ROH} + \text{OS} \underset{\smile}{\bigcirc} \text{O} \tag{41}$$

S-oxidation of thioxane by 4a-FlEtOOH in anhydrous dioxane is first order in the flavin hydroperoxide (35). Apparently the "oxygen atom transfer" potential of 4a-FlEtOOH is such that general-acid-catalyzed assistance of oxygen transfer is not required.

A number of proposals have been offered to explain the oxygen-transfer potentials of the flavin monooxygenases. These have been designed with the objective of creating a positively charged oxygen (an oxene) intermediate. The various proposals are summarized in Scheme 7. There is no experimental evidence to support the existence of these oxene structures. The machinations of Scheme 7 are not required to explain the relative rate constants of Table I. 4a-FlEtOOH may contain a highly polarizable oxygen–oxygen bond. This feature may have been deduced from the very low pK_a for pseudo base formation with the flavinium cation (Equation 34). In considering the proposals of Müller et al. (1976) (39), Hemmerich (1976) (40), and Dimitrienko (1977) (41) (Scheme 7), it is worth noting that the N(5)-ethylisoalloxazine II forms a 4a-hydroperoxy adduct that is comparable with 4a-FlEtOOH as an S- and N-oxidizing agent (43). As mentioned previously, nucleophilic addition to the 9a- and 10a-positions of I are sterically quite unfavorable. The mechanisms of the N- and S-oxidation of 4a-FlEtOOH are best ascribed to nucleophilic displacements (perhaps with a radical character, Equation 42).

$$\text{4a-FlEt-O} \underset{\text{O}}{\diagdown} \text{H} + \text{S} \overset{\diagup}{\diagdown} \longrightarrow \left| \begin{array}{c} \text{4a-FlEt-O}^- \overset{\text{H}}{\diagup} \\ \diagdown_{\text{O}} \\ \diagdown_{\text{S}^+} \end{array} \right| \rightarrow \text{4a-FlEtOH} + \text{OS} \overset{\diagup}{\diagdown} \tag{42}$$

The bacterial luciferases obtained from *Beneckae harveyi* and *Photobacterium fisheri* are a very intriguing class of flavoenzyme monooxygenases. The study of these enzymes is associated intimately with J. W. Hastings of Harvard University, who has recently reviewed

Scheme 7.

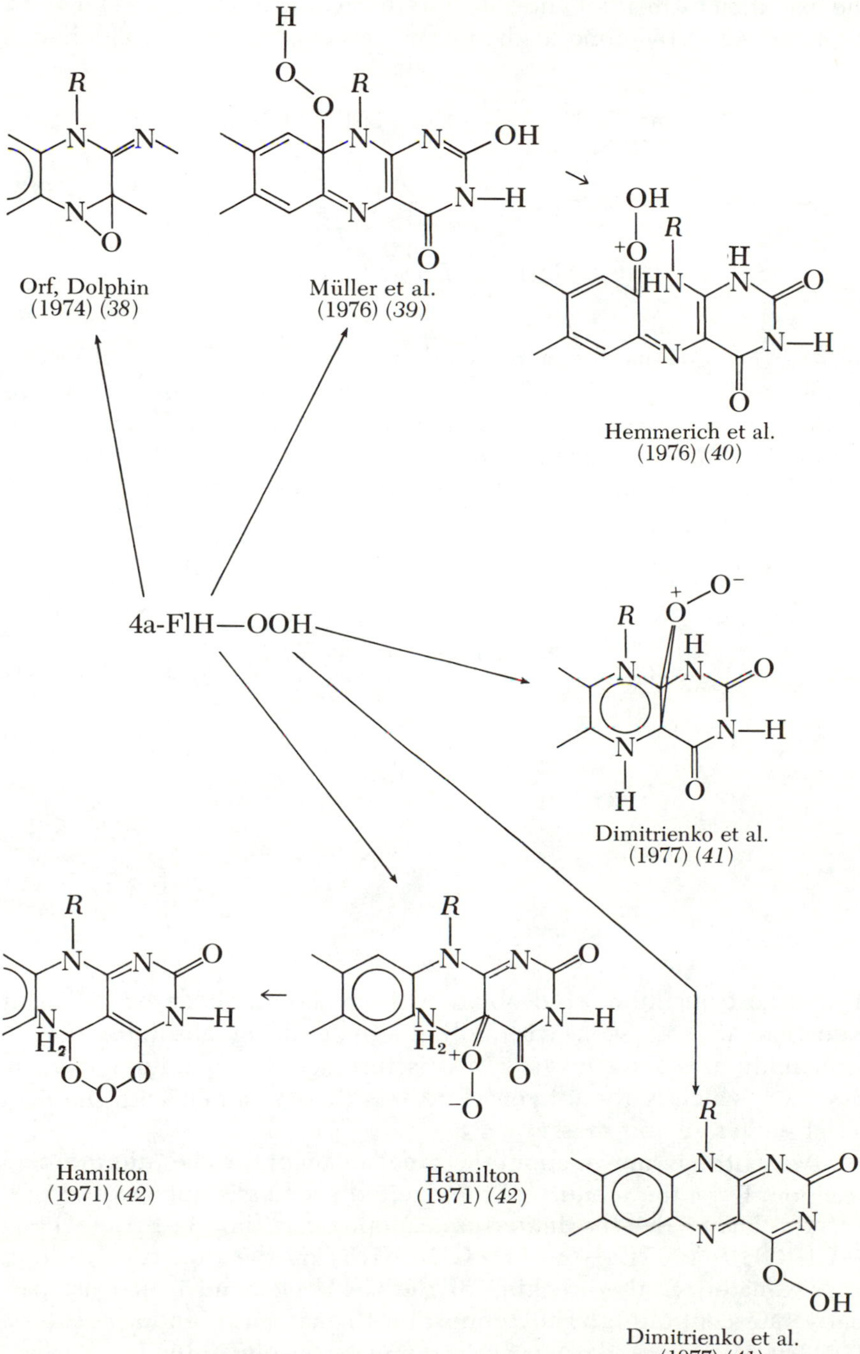

Orf, Dolphin
(1974) (*38*)

Müller et al.
(1976) (*39*)

Hemmerich et al.
(1976) (*40*)

4a-FlH—OOH

Dimitrienko et al.
(1977) (*41*)

Hamilton
(1971) (*42*)

Hamilton
(1971) (*42*)

Dimitrienko et al.
(1977) (*41*)

UNIVERSITY LIBRARIES
CARNEGIE-MELLON UNIVERSITY
PITTSBURGH, PENNSYLVANIA 15213

their chemistry (*44*). The light-producing reaction is associated with the oxidation of an aldehyde substrate to a carboxylic acid as shown in Equation 43. A biomimetic chemiluminescent oxidation of aldehydes

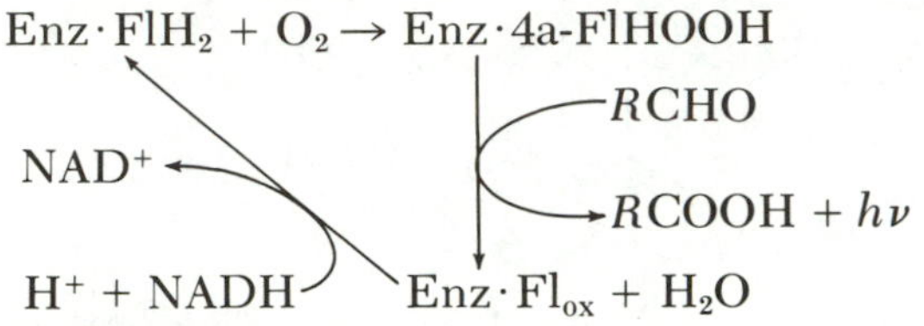

Enz·FlH₂ + O₂ → Enz·4a-FlHOOH
RCHO
NAD⁺
RCOOH + hν
H⁺ + NADH
Enz·Fl_ox + H₂O

(43)

by 4a-FlEtOOH has been realized in this laboratory (*29, 36*). The quantum efficiency of the chemiluminescent aldehyde oxidations are greatest when hydroperoxy hemiacetal is added to $Fl_{ox}^{+}Et$ in an aprotic solvent. The mixed peroxide intermediate (4a-FlEt—O—O—CH(OH)*R*) is formed in the stopped-flow time range (Equation 44).

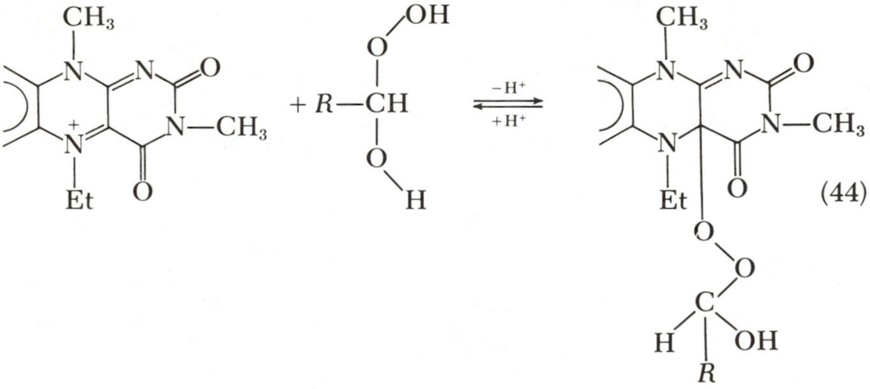

CH₃
N
N
O
N—CH₃
N
Et
O
OH
O
+ R—CH
O
H
−H⁺
+H⁺
CH₃
N
N
O
N—CH₃
N
Et
O
O
O
C
H
OH
R

(44)

The largest quantum yield obtained from the aldehyde oxidation of Equation 45 was seen with tolualdehyde using absolute diethyl formamide as a solvent (*45*). In this instance, the quantum yield is 6×10^{-3}, which is about twentyfold less than obtained with the bacterial luciferase enzymes.

Very little is known about the mechanism of the chemliminescent reaction. Even the identity of the excited species is unknown at preent. It is known that the deuterium isotope effect (i.e., 4a-FlEt—O—O—CH(OH)*R*/4a-FlEt—O—O—CD(OH)*R*) on the quantum yield is approximately 2; the breaking of the C—H(D) bond is at least partially rate controlling. The proposal (*50*) that the pseudo base (i.e., 4a-FlEtOH) is formed as an excited species seems unlikely.

At least five detailed mechanisms have been proposed for bacterial luciferase within the last six years (*28, 47, 48, 49, 50*). With the assumption that our biomimetic model relates to the enzyme-catalyzed reaction, we conclude that these five proposals of mechanisms are incorrect. Four (*28, 47, 48*) are inconsistent with the intermediacy of the mixed peroxide (4a-FlEt—O—O—CH(OH)R') and cannot be applied to $N(5)$-alkylflavins. In addition, two (*28, 47*) of these mechanisms as well as another (*49*), require the hydroxyl group of the aldehyde-peroxide adduct. No mechanism can be taken seriously without the identification of the excited species. We do know that a hydrogen substitutent is required on the carbon that is converted to a carbonyl group (*47*).

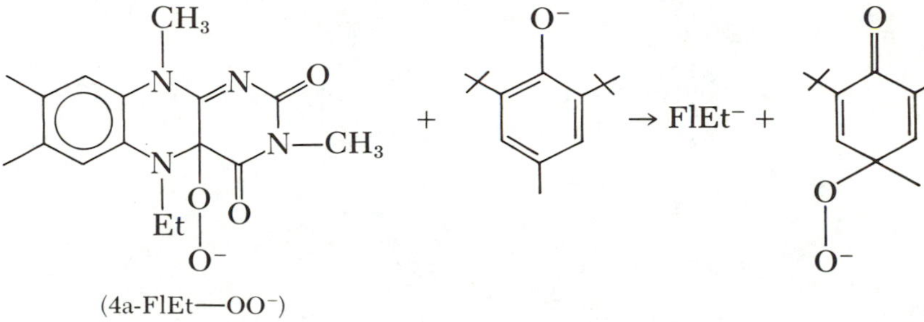

(4a-FlEt—OO⁻)   (45)

The peroxy-anion transfer reaction of Equation 45 has been established (*51*). This finding was intriguing because we had shown previously that FlEt⁻ reacts with molecular oxygen to provide 4a-FlEtOO⁻ (Equation 37). Combination of Equations 37 and 45 yields Equation 46, establishing that the dihydroflavin anion is a catalyst for the reaction of oxygen with the phenolate ion.

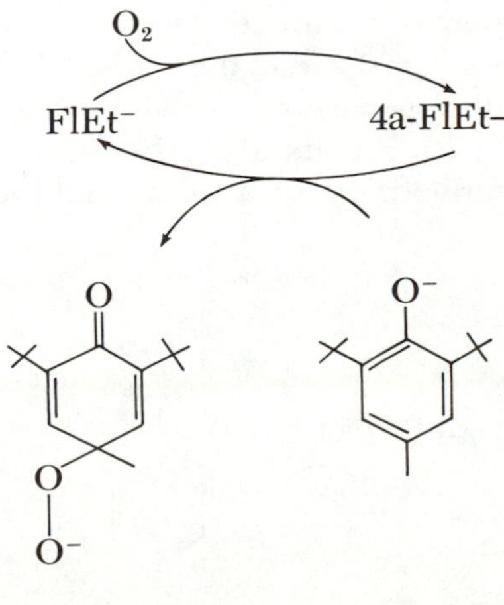

(46)

The mechanism for the peroxide transfer reaction of Equation 45 is not established. We know from kinetic studies that 4a-FlEtOO$^-$ undergoes conversion to a peroxidizing species before reaction with the phenolate ion. The species enumerated in Scheme 7 would have no propensity to deliver both oxygen atoms of the 4a-FlEtOO$^-$ to a substrate. Dissociation of 4a-FlEtOO$^-$ to FlEt$^-$ and O$_2$ and reaction of the phenolate ion with O$_2$ can be ruled out because the bimolecular rate constant for reaction of O$_2$ with phenolate is one hundredfold too small to account for the reaction of Equation 45. The mechanism of Equation 47 has been proposed (*51*). In favor of this mechanism is

a. 4a-FlEtOO$^-$ $\rightleftarrows$ FlEt$\cdot$ + O$_2^{\cdot-}$

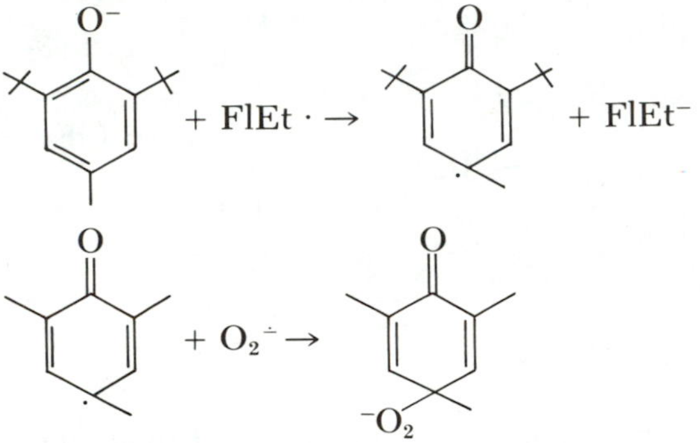

(47)

the finding that Reaction b is associated with a minimum rate constant of 10^7 mol^{-1}sec^{-1}. Disfavoring the mechanism is the report that O$_2^-$ does not couple with the radical species as shown in Equation 47, but reduces the radical to phenolate ion with the production of oxygen (*52*).

Regardless of the peroxide transfer reaction mechanism of Equation 45, its occurrence suggests that the 4a-FlEtOO$^-$ species might serve as a biomimetic dioxygenase reagent. This has been found to be the case (Equation 48) (*53*). Presumably, the reaction of Equation 48 occurs through the intermediacy of a peroxy adduct (Equation 49).

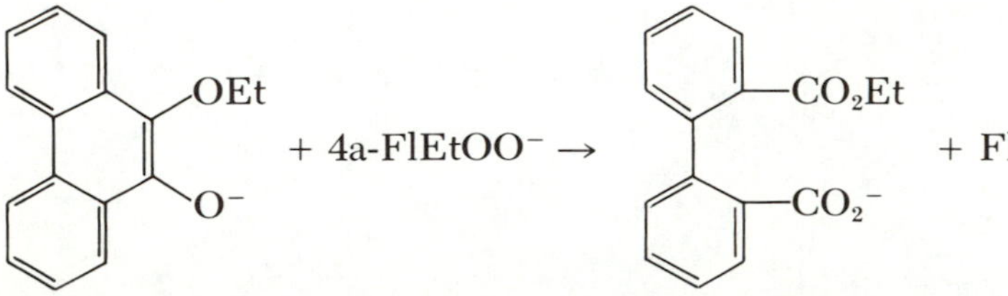

$$(49)$$

Acknowledgments

Studies leading to this review were supported by grants from the National Science Foundation and the National Institutes of Health. The identity of the graduate students and postdoctoral fellows who carried out the investigations leading to the review are provided in the cited references.

Literature Cited

1. Bruice, T. C. "Progress in Bioorganic Chemistry"; Kaiser, E. T., Kezdy, F. J., Eds.; Wiley Interscience: New York, 1976; Vol. 4, p. 1.
2. Bruice, T. C.; Main, L.; Smith, S.; Bruice, P. Y. *J. Am. Chem. Soc.* **1971**, 93, 7327.
3. Bruice, T. C. "Progress in Bioorganic Chemistry"; Kaiser, E. T., Kezdy, F. J., Eds.; Wiley Interscience: New York, 1976; Vol. 4, p. 1.
4. Tober, C. L.; Nicholls, P.; Brodie, J. D. *Arch. Biochem. Biophys.* **1970**, 138, 506.
5. Chening, Y. F.; Walsh, C. *Biochemistry* **1976**, 15, 2432.
6. Walsh, C. T.; Lockridge, O.; Massey, V.; Abeles, R. H. *J. Biol. Chem.* **1973**, 348, 7049.
7. Ghisla, S.; Massey, V. *J. Biol. Chem.* **1977**, 245, 6729.
8. Main, L.; Kasperek, G. J.; Bruice, T. C. *Biochemistry* **1972**, 11, 3991.
9. Shinkai, I.; Kunitake, T.; Bruice, T. C. *J. Am. Chem. Soc.* **1974**, 96, 7140.
10. Bruice, T. C.; Taulane, J. P. *J. Am. Chem. Soc.* **1976**, 98, 7769.
11. Chan, T. W.; Bruice, T. C. *Biochemistry* **1978**, 17, 4784.
12. Yokoe, I.; Bruice, T. C. *J. Am. Chem. Soc.* **1975**, 97, 450.
13. Hamilton, G. A. "Progress in Bioorganic Chemistry"; Kaiser, E. T., Kezky, F. J., Eds.; Wiley Interscience: New York, 1971; 1, p. 83.
14. Bruice, T. C.; Hevesi, L.; Shinkai, S. *Biochemistry* **1973**, 12, 2083.
15. Hevesi, L.; Bruice, T. C. *Biochemistry* **1973**, 12, 290.
16. Muller, F.; Massey, V. *J. Biol. Chem.* **1970**, 54, 1295.
17. Clerin, D.; Bruice, T. C. *J. Am. Chem. Soc.* **1974**, 96, 5571.
18. Williams, R. F.; Bruice, T. C. *J. Am. Chem. Soc.* **1976**, 98, 7752.
19. Williams, R. F.; Shinkai, S.; Bruice, T. C. *J. Am. Chem. Soc.* **1977**, 99, 921.
20. Porter, D. J. T.; Bright, H. J. *J. Biol. Chem.* **1977**, 252, 4361.
21. Porter, D. J. T.; Voet, J. G.; Bright, H. J. *J. Biol. Chem.* **1973**, 248, 4400.
22. Loechler, E. L.; Hollacher, T. C. *J. Am. Chem. Soc.* **1975**, 97, 3235.
23. Smith, S. B.; Bruice, T. C. *J. Am. Chem. Soc.* **1975**, 97, 2875.
24. Bruice, T. C.; Yano, Y. *J. Am. Chem. Soc.* **1975**, 97, 5263.
25. Guthrie, R. D.; Wesley, D. P.; Pendygraft, G. W.; Young, A. T. *J. Am. Chem. Soc.* **1976**, 98, 5870.
26. Novak, M.; Bruice, T. C. *J. Am. Chem. Soc.* **1977**, 99, 8079.
27. Novak, M.; Bruice, T. C. *J. Chem. Soc. (Chem. Comm.)*, in press.
28. Hemmerich, P. *Progr. Chem. Org. Nat. Prod.* **1976**, 33, 451.

29. Kemal, C.; Bruice, T. C. *Proc. Natl. Acad. Sci. USA* **1976,** 73, 995.
30. Ghisla, S.; Hartman, U.; Hemmerich, P.; Muller, F. *Justus Liebigs Ann. Chem.* **1973,** 1388.
31. Kemal, C.; Bruice, T. C. *J. Am. Chem. Soc.* **1976,** 98, 3955.
32. Kemal, C.; Chan, T. W.; Bruice, T. C. *J. Am. Chem. Soc.* **1977,** 99, 7272.
33. Sparrow, L. G.; Ho, P. P. K.; Sunderam, T. K.; Zach, D.; Nyns, E. J.; Snell, E. E. *J. Biol. Chem.* **1969,** 244, 2590.
34. Gorrod, J. W. "Biological Oxidation of Nitrogen"; Elsevier: North Holland, 1978.
35. Ball, S.; Bruice, T. C. *J. Am. Chem. Soc.* **1979,** 101, 4017.
36. Kemal, C.; Chan, T. W.; Bruice, T. C. *Proc. Natl. Acad. Sci. USA* **1977,** 74, 405.
37. Dankleft, M. A. P.; Curci, P.; Edwards, J. O.; Pyun, H. Y. *J. Am. Chem. Soc.* **1968,** 90, 3209.
38. Orf, W. H.; Dolphin, D. *Proc. Natl. Acad. Sci. USA* **1974,** 71, 2646.
39. Müller, F.; Grande, H. J.; Jarbandhau, T. In "Flavins and Flavoproteins"; Singer, J. P., Ed.; Elsevier: New York, 1976; p. 38.
40. Hemmerich, P.; Wessiak, A. In "Flavins and Flavoproteins"; Singer, J. P., Ed.; Elsevier: New York, 1976; pg. 9.
41. Dimitrienko, I.; Snieckus, S.; Viswanatha, T. *Bioorg. Chem.* **1977,** 6, 421.
42. Hamilton, G. A. "Progress in Bioorganic Chemistry"; Kaiser, E. T., Kezky, F. J., Eds.; Wiley Interscience: New York, 1971.
43. Monahan-Miller, A.; Bruice, T. C. *J. Chem. Soc. (Chem. Comm.)* **1979,** 896.
44. Hastings, J. W.; Nealson, K. H. *Annu. Rev. Microbiol.* **1977,** 31, 549.
45. Shepherd, P.; Bruice, T. C. unpublished data.
46. Kemal, C.; Bruice, T. C. *J. Am. Chem. Soc.* **1977,** 99, 7064.
47. McCapra, F.; Hysert, D. W. *Biochem. Biophys. Res. Commun.* **1973,** 52, 298.
48. Lowe, J. N.; Ingraham, L. L.; Alspach, J.; Rasmussen, R. *Biochem. Biophys. Res. Commun.* **1976,** 73, 465.
49. Keay, R. E.; Hamilton, G. A. *J. Am. Chem. Soc.* **1975,** 97, 6876.
50. Eberhard, E.; Hastings, J. W. *Biochem. Biophys. Res. Commun.* **1972,** 47, 348.
51. Kemal, C.; Bruice, T. C. *J. Am. Chem. Soc.* **1979,** 101, 1635.
52. Nishinaga, A.; Itahara, T.; Shimizu, T.; Tomita, H.; Nishizawa, K.; Matsuura, T. *Photochem. Photobiol.* **1978,** 28, 687.
53. Muto, S.; Bruice, T. C. unpublished data.

RECEIVED May 14, 1979.

Flavin Coenzyme Analogs as Probes of Flavoenzyme Reaction Mechanisms

C. WALSH, F. JACOBSON, and C. C. RYERSON

Departments of Chemistry and Biology, Massachusetts Institute of Technology, Cambridge, MA 02139

The use of riboflavin coenzyme analogs as biomimetic probes of flavoenzyme reaction mechanisms has yielded new information about these biological redox processes. Substitution of carbon at the 1 or 5 positions for the nitrogens normally present in the flavin isoalloxazine ring system leads to dramatic alterations in redox properties but does not affect strongly recognition and binding by a variety of apoflavoproteins. 5-Carba-5-deazaflavins are catalytically competent coenzymes only with flavoenzymes that catalyze redox conversions involving two-electron transfer steps. Thus some dehydrogenases but no flavoprotein oxidases or monooxygenases are functional for turnover. In contrast, 1-carba-1-deazaflavin analogs, with accessible one-electron chemistry, function catalytically in all classes of flavoenzymes examined. The 1-deaza system is more difficult to reduce, so lower V_{max}'s with some enzymes may reveal that the redox steps are rate determining. When monooxygenases are reconstituted with 1-deazaFAD, some still show NADH oxidation coupled to specific substrate hydroxylation. Studies with synthetic 8-hydroxy-7-demethyl-5-deazariboflavin are presented to confirm the identity of this 5-deazaflavin chromophore in the novel redox coenzyme, Factor 420, from methanogenic bacteria.

The central topic of this volume is biomimetic chemistry. In this chapter we describe our recent studies using synthetic analogs of riboflavin-based coenzymes with specific atom substitutions in the

0-8412-0514-0/80/33-191-119$05.00/0
© 1980 American Chemical Society

tricyclic isoalloxazine ring system to dissect and analyze the chemical features that permit the utilization of flavin coenzymes in a wide range of biological oxidation–reduction processes.

An initial summary of the nature and scope of redox processes involving flavin coenzymes will be presented to set the framework for evaluation of the utility of the specific coenzyme analogs. Analogs altered at $N(5)$, at $N(1)$, or at $N(1)$ and $N(5)$ by synthetic substitution of carbon will be discussed before analysis of 8-demethyl-8-hydroxy analogs either with nitrogen or carbon at the 5 position.

Roles of Flavin Cofactors in Redox Enzymology

The functional end of the flavin coenzymes FMN and FAD is the tricyclic isoalloxazine system, with the numbering system shown in structure **I**, the air-stable, yellow, oxidized form. The other two functionally important redox states are the one-electron-reduced semiquinone, **II** ($pK_a = 8.4$ for dissociation at $N(5)$), and the two-electron-reduced, colorless dihydroflavin, **III**. In the dihydro form $N(5)$, $C(4a)$, $C(1a)$, and $N(1)$ form a diaminoethylene system and it was anticipated that nitrogen at the 5 and 1 positions would be key to coenzymatic function.

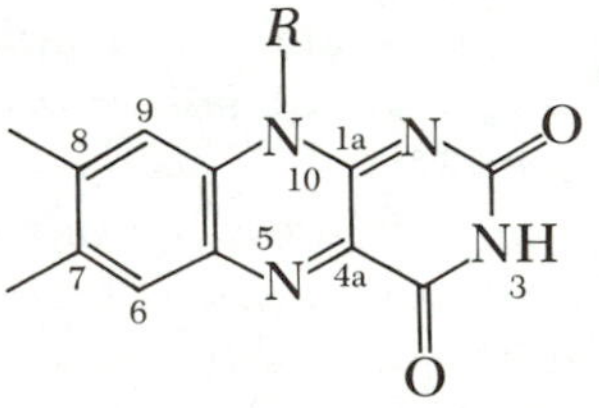

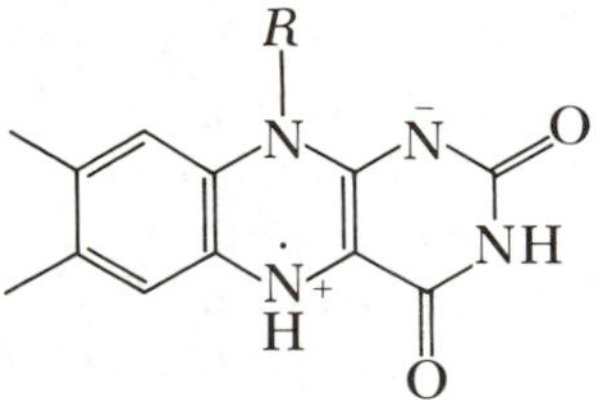

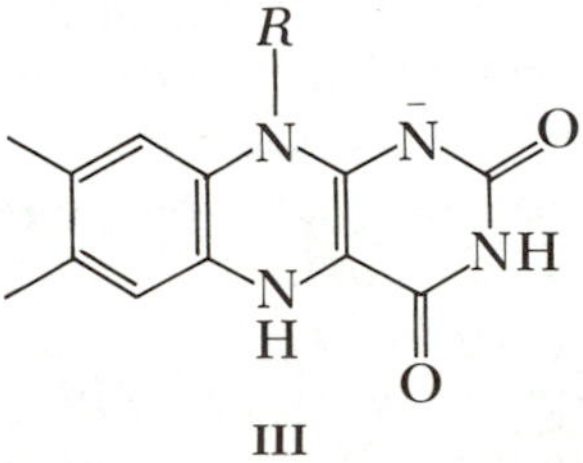

The ability to form a stable one-electron-reduced radical (semi-quinone) allows flavin cofactors to sit at the crossroads of two-electron and one-electron transfer chains. That is, they can be reduced by organic substrates two electrons at a time and be reoxidized by either obligate one-electron acceptors such as cytochromes (e.g., yeast cytochrome b_2 or cytochrome b_5 reductase/cytochrome b_5) and iron–sulfur cluster proteins (adrenodoxin reductase/adrenodoxin) or by facultative one-electron acceptors such as benzoquinones (coenzyme

Q) and naphthaquinones (vitamin K) (*1, 2*). These enzymes, in keeping with the membraneous nature of the reoxidants (small molecules or proteins), are often membrane-associated and are classified as dehydrogenases (e.g., succinate dehydrogenase). Other flavoenzymes are soluble and can be reoxidized by apparent two-electron acceptors such as NAD (e.g., transhydrogenase).

A second major category of flavoenzymes (soluble) are those where the enzyme-bound dihydroflavin is reoxidized by molecular oxygen (*1, 2, 3*). Net two-electron transfer yields **I** and hydrogen peroxide, H_2O_2. These are the flavoprotein oxidases (e.g., D- and L-amino-acid oxidases, amine oxidase, glucose oxidase). A variant of the flavoenzymes reacting with O_2 are certain bacterial enzymes that carry out monooxygenation, transfer of one oxygen atom to product. These flavoenzyme monooxygenases thus split O_2 and show activity towards two kinds of substrates: activated aromatic rings (phenol, *p*-hydroxybenzoate, salicylate) and ketones (cyclohexanone). The sole mammalian flavoprotein monooxygenase reported is a microsomal *N*-oxidase, converting, for example, tertiary amines to amine-*N*-oxides (*4*). Typical oxidase and monooxygenase reactions are shown in Table I.

The mechanisms for passage of electrons out of the cosubstrate undergoing oxidation and into the flavin undergoing reduction have been scrutinized carefully in the past decade, as have those for reoxidation of flavin and passage of electrons to the cosubstrate that experiences net reduction. Depending on substrate structure a variety of covalent intermediates have been postulated or demonstrated directly: thus O_2 reacts with dihydroflavins to form 4a flavin hydroperoxides **IV**

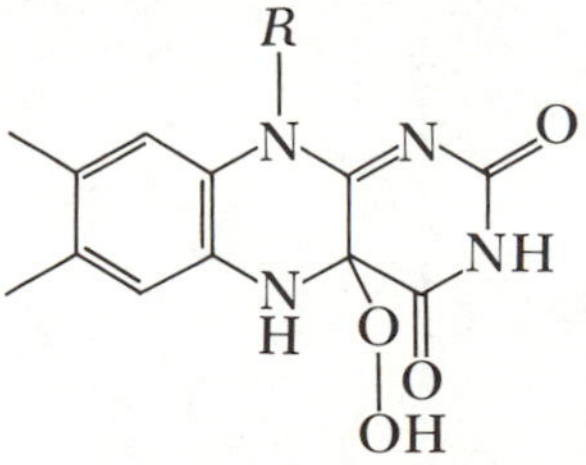

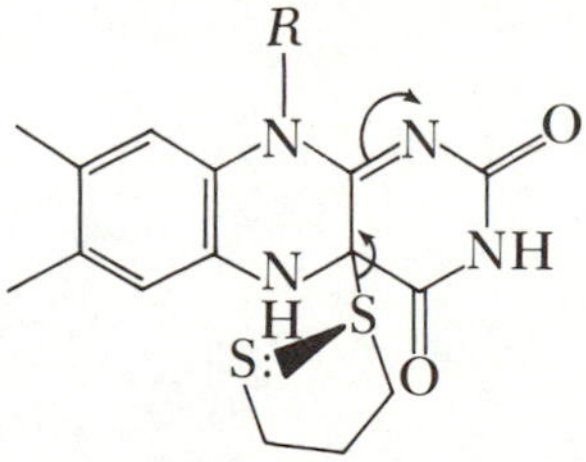

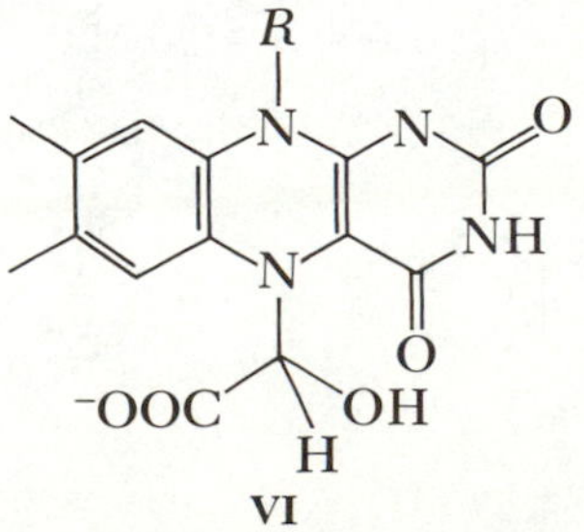

Table I. Types of Flavoenzymes that Reduce O_2 Catalytically

Oxidases

Substrates	Products	Enzyme
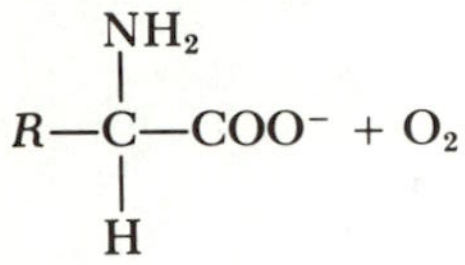 $\underset{H}{\overset{NH_2}{R-C-COO^-}} + O_2$	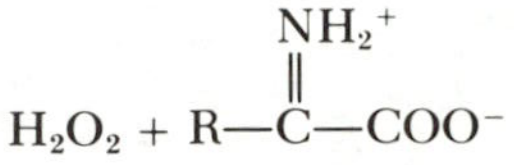 $H_2O_2 + R-\overset{NH_2^+}{\underset{\parallel}{C}}-COO^-$	D-amino-acid oxidase
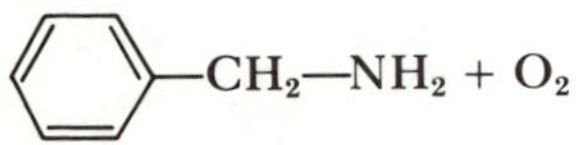 (C₆H₅)CH₂—NH₂ + O₂	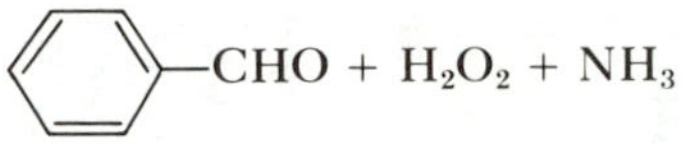 (C₆H₅)CHO + H₂O₂ + NH₃	amine oxidase

Oxygenases

Substrates	Products	Enzyme
 HO—(C₆H₄)—COO⁻ + O₂ + NADPH	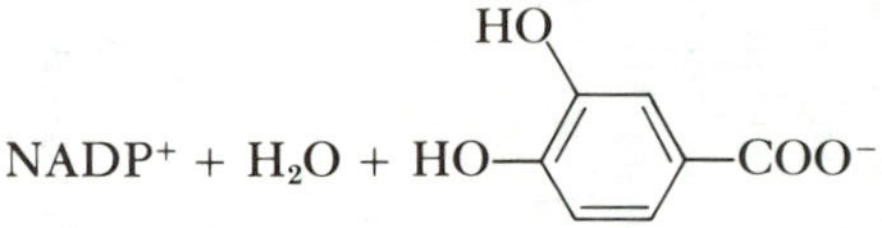 NADP⁺ + H₂O + HO—(C₆H₃)(OH)—COO⁻	p-hydroxybenzoate hydroxylase
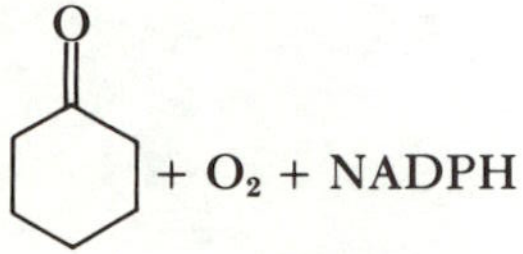 cyclohexanone + O₂ + NADPH	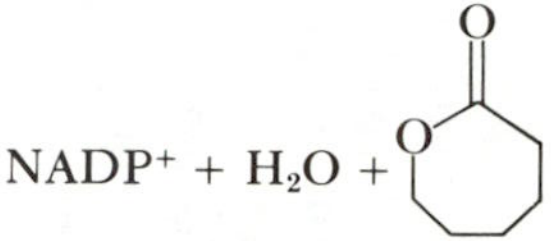 NADP⁺ + H₂O + (ε-caprolactone)	cyclohexanone monooxygenase
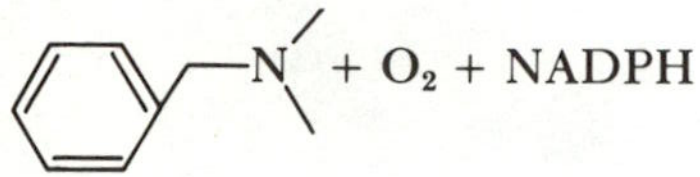 (C₆H₅CH₂)N(CH₃)₂ + O₂ + NADPH	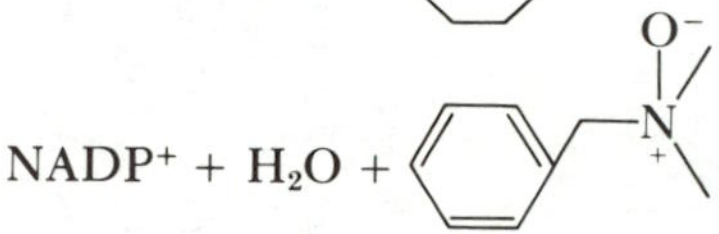 NADP⁺ + H₂O + (C₆H₅CH₂)N⁺(CH₃)₂O⁻	microsomal-amine N-oxidase

(5, 6) and dithiols undergo oxidation to disulfides by initial additions at the 4a position also as indicated in V, from model and enzymic data (7, 8). In oxidation of α-amino acids and α-hydroxy acids, α-carbanionic intermediates have been postulated to add to $N(5)$, as in VI (from glycolate) (9). In reactions with NADH (or NADPH), direct hydride transfers from $C(4)$ of dihydronicotinamide to $N(5)$ of flavin seem likely.

Flavin Coenzyme Analogs

We began studies with flavin coenzyme analogs in 1972 to probe what structural features in the flavin ring system were requisite for specific aspects of the enzymic catalyses noted above. In particular, evaluations of the 5-carba-5-deazaflavin and the 1-carba-1-deazaflavin system were selected, given the pivotal role of these nitrogens in the redox transformations.

5-Deazaflavins

5-Deazariboflavin, VII, was synthesized by Cheng and colleagues a decade ago (10), initially evaluated at the FMN level by Edmondston, Barman, and Tollin in 1972 for binding to and reactivity with A. vinlandii flavodoxin (11), and then extensively analyzed for coenzymatic function in our laboratories (12, 13, 14, 15) and those of Hersh (15–19). The chemistry of this system also has been analyzed by Massey and Hemmerich (20). The patterns of chemical and enzymatic reactivity have been summarized elsewhere (15, 20, 21) and only relevant points are raised here. As noted in Figure 1 and Table II, carbon replacement for nitrogen leads to retention of the two absorption maxima characteristic of flavins but a 50-nm hypsochromic shift occurs with no change in extinction coefficient. The 5-carba substitution lowers the reduction potential, $E^{\circ\prime}$, by about 100 mV, making the system much less easy to reduce (K_{eq} favors oxidized over reduced by $\approx 10^3$ compared with riboflavin values): it has an $E^{\circ\prime}$ now approximating the NADH/NAD couple (-320 mV). Fourier-transform ^{1}H NMR (14) of BD_4^- and NADD-reduced 5-deazaflavin confirmed deuteride transfer to $C(5)$, now stable to protonic exchange processes. In seven different 5-deazaflavin-reconstituted enzymes, direct hydrogen transfer occurs from oxidizing substrate to this $C(5)$ locus, proving it is the site of electron entry (15).

A consequence of carbon substitution at the 5 position is conversion of the central ring from a pyrazine to a pyrimidine (Figure 2). This change dramatically alters the kinetics and thermodynamics of redox processes. The dihydro form, VIII, is now a dihydropyridine not a dihydropyrazine and bears clear analogy to dihydronicotinamides

Table II. Comparative Properties of Flavins, 5-Deazaflavins, and 1-Deazaflavins

Property	Flavins	5-deazaFlavins	1-deazaFlavins
Accessible semiquinone	Yes	No	Yes
Redox reactivities	1 and 2 e$^-$	2 e$^-$ only	1 and 2 e$^-$
Rate of O$_2$ oxidation of dihydro form	1.9 sec^{-1}	<2 × 10^{-6} sec^{-1}	3.9 sec^{-1}
$E^{\circ\prime}$ (pH 7)	−208 mV	−280 mV	−311 mV
Stable incorporation of ^{3}H from substrate or solvent	No	Yes (at C(5))	No (None at C(1))
λ maximum (nm)	445,375	400,340	535,365

Table III. Functions of 5-Deazaflavin Reconstituted Flavoenzymes

Enzyme	Function
Dehydrogenases (SH$_2$ → S + 2e$^-$ + 2H$^+$) NADH/flavin oxidoreductase N-methylglutamate synthetase	Catalytic turnover V_{max} = 0.76% that of riboflavin V_{max} = 10% that of FMN
Oxidases (SH$_2$ + O$_2$ → S + H$_2$O$_2$) D-amino-acid oxidase Glucose oxidase	Coenzyme reduction only—(No reoxidation by O$_2$) Reduction rate 10^{-6}–10^{-7} that of FAD Reduction rate 10^{-3} that of FAD
Monooxygenases (SH + 2e$^-$ + O$_2$ → S—OH + H$_2$O) Orcinol hydroxylase p-Hydroxybenzoate hydroxylase	Coenzyme reduction only—(No O$_2$ transfer) Reduction is very slow Reduction is very slow

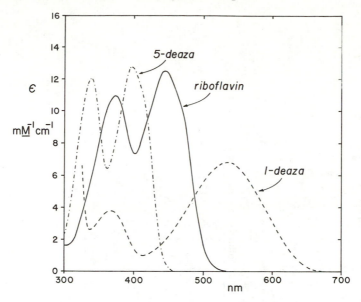

Figure 1.   Electronic spectra of riboflavin and 5-deaza and 1-deaza analogs

(NADH,NADPH). Like them, **VIII** is not reoxidized rapidly by O_2 nor is the one-electron-reduced (or one-electron-oxidized, coming from **VIII**) semiquinone obtainable in enzymatic systems. 5-Deazaflavins are restricted to the two-electron transfer cycles and so have only half the versatility of flavins (*15, 20, 21*). 5-Deazaflavin coenzymes in general are reduced only sluggishly when reconstituted with apoflavoenzymes (Table III) and function rapidly and catalytically only when the specific reoxidant is a two-electron acceptor (*15*). The ability to detect direct hydrogen transfer to $C(5)$ in these reconstituted enzymes is mechanistically informative, and could allow determination of 5*R* vs. 5*S* chirality in dihydro 5-deazaflavin (**IX**) of these reconstituted enzymes (in analogy to 4*R* and 4*S* forms of NAD²H (e.g., **IX**)), if the rapid disproportionation problem could be solved (*13*).

1-Deazaflavins

Given the disappearance of the one-electron-transfer capabilities of flavins caused by carbon substitution at the 5 position, it was of interest to determine the properties of carbon substitution at the 1 position. This is the other end of the key diaminoethylene moiety in the flavin middle. Compound **X**, 1-deazariboflavin, was synthesized by Ashton et al. at Merck (*22*), and the functional consequences of the switch from

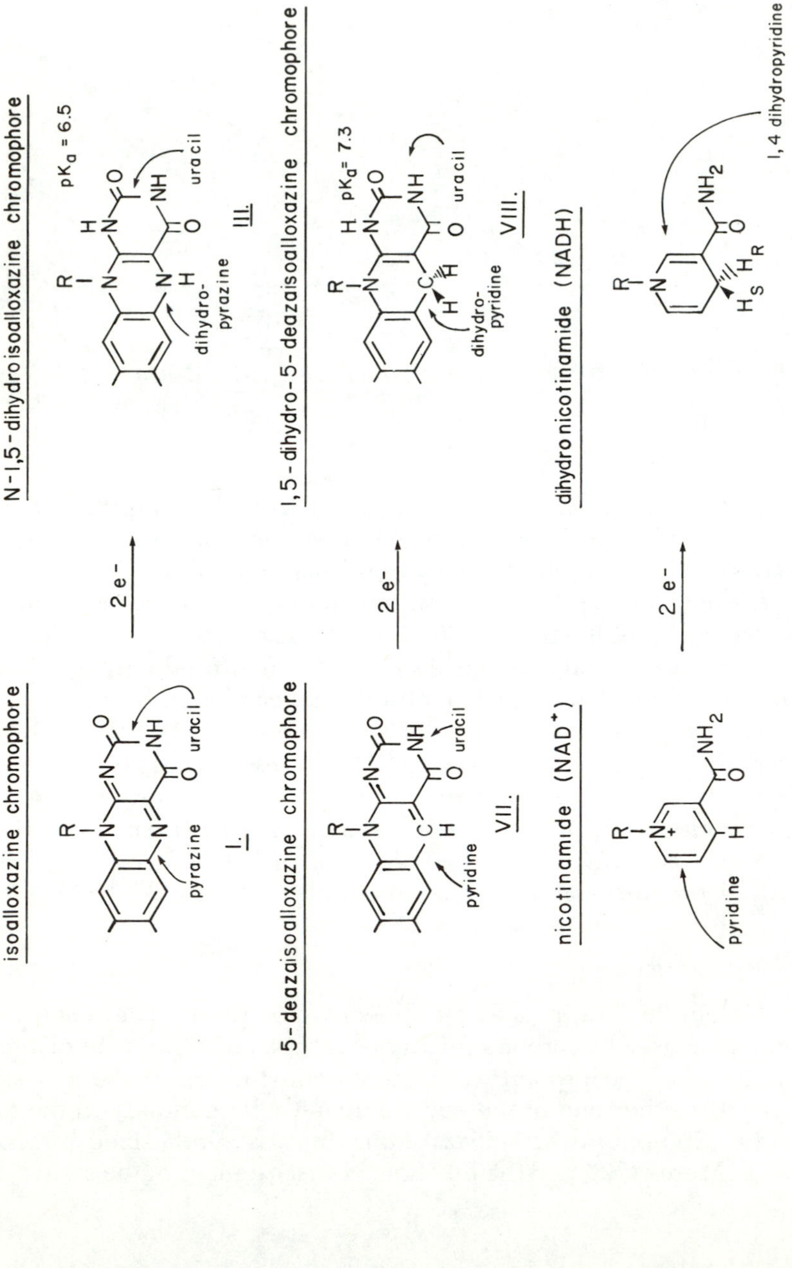

isoalloxazine chromophore

I.

pyrazine

uracil

N-1,5-dihydroisoalloxazine chromophore

III.

$pK_a = 6.5$

uracil

dihydro-pyrazine

5-deazaisoalloxazine chromophore

VII.

pyridine

uracil

1,5-dihydro-5-deazaisoalloxazine chromophore

VIII.

$pK_a = 7.3$

uracil

dihydro-pyridine

nicotinamide (NAD$^+$)

pyridine

dihydronicotinamide (NADH)

IX.

1,4 dihydropyridine

H_S H_R

2 e$^-$

2 e$^-$

2 e$^-$

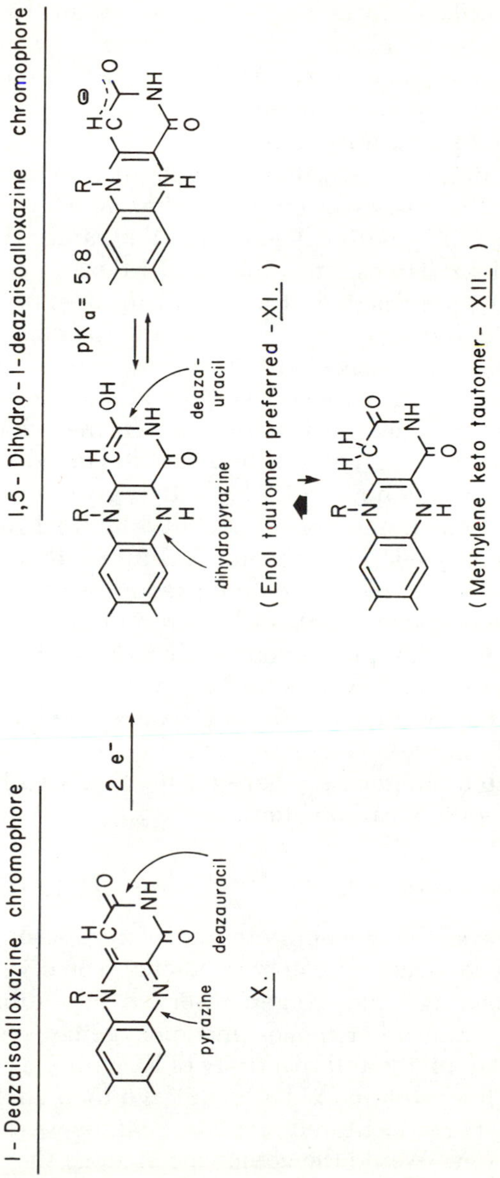

Figure 2.   *Analysis of the effect of carbon substitution at the 1 and 5 positions of the flavin nucleus*

uracil to deazauracil group (Figure 2) in the isoalloxazine system on chemical and coenzymatic properties was evaluated in our laboratories (23, 24). The 1-deazaflavins are purple (Figure 1, Table II) and again more difficult to reduce, by 80 mV, than the parent flavin system. This 80-mV change is exceedingly useful since in 1-deazaflavin-reconstituted enzymes, V_{max} comparisons with native FAD enzyme thus may give immediate indication whether or not the redox steps are rate determining in catalysis (24).

The central ring of 1-deazaflavins remains a pyrazine in **X**, a dihydropyrazine in the two-electron-reduced form, **XI**, and continues to dominate the chemistry with oxygen. Like the parent riboflavins, and unlike the 5-deazaflavins, the dihydro-1-deaza system, **XI**, is reoxidized by O_2 in a fraction of a second in air-saturated solutions (Table II); the semiquinone is accessible and 1-deazaFAD enzymes show full catalytic competence with flavoprotein dehydrogenases and oxidases (24). Turnover numbers vary from about 1% to 100% that of cognate FAD-enzymes but this variation reflects the -280 mV vs. -200 mV $E^{o\prime}$ values, respectively, for 1-deazariboflavin vs. riboflavin. The redox steps may or may not limit V_{max} with a given enzyme (15, 24).

In principle, one might have expected dihydro 1-deazaflavin, **XI**, to exist as the methylene keto species **XII** rather than the enol indicated, but this is not the case. The enol predominates as it does in the simple deazauracil systems (25), and the rapidly established equilibrium concomitant with rapid solvent exchange vitiates easy determination of whether or not substrate hydrogen is ever transferred to $C(1)$ as 1-deazaflavins are reduced. This would have been most interesting with the olefin-forming enzymes such as succinate dehydrogenase or dihydroorotate dehydrogenase where each of the two hydrogens lost are initially bound to a carbon atom.

1-Deazaflavins with Monooxygenases: Aromatic Ring Hydroxylases

In many ways, the flavoprotein monooxygenases comprise the most interesting flavoenzyme category mechanistically since the mode of O–O bond cleavage and oxygen insertion into substrate are still unclear for these and other mono- and dioxygenases (i.e., Fe or Cu dependent) (1). Given the full reactivity of **XI** with O_2, chemically and when bound to flavoprotein oxidases (24), we then analyzed whether or not 1-deazaFAD-reconstituted monooxygenases were competent for oxygen transfer (28). Would the coenzyme analogs show properties of uncoupling NADPH oxidation from substrate hydroxylation? This was not clear since substrate analogs may act as effectors, which uncouple NADPH oxidation from hydroxylation and result in O_2 going to H_2O_2 (26) (Figure 3).

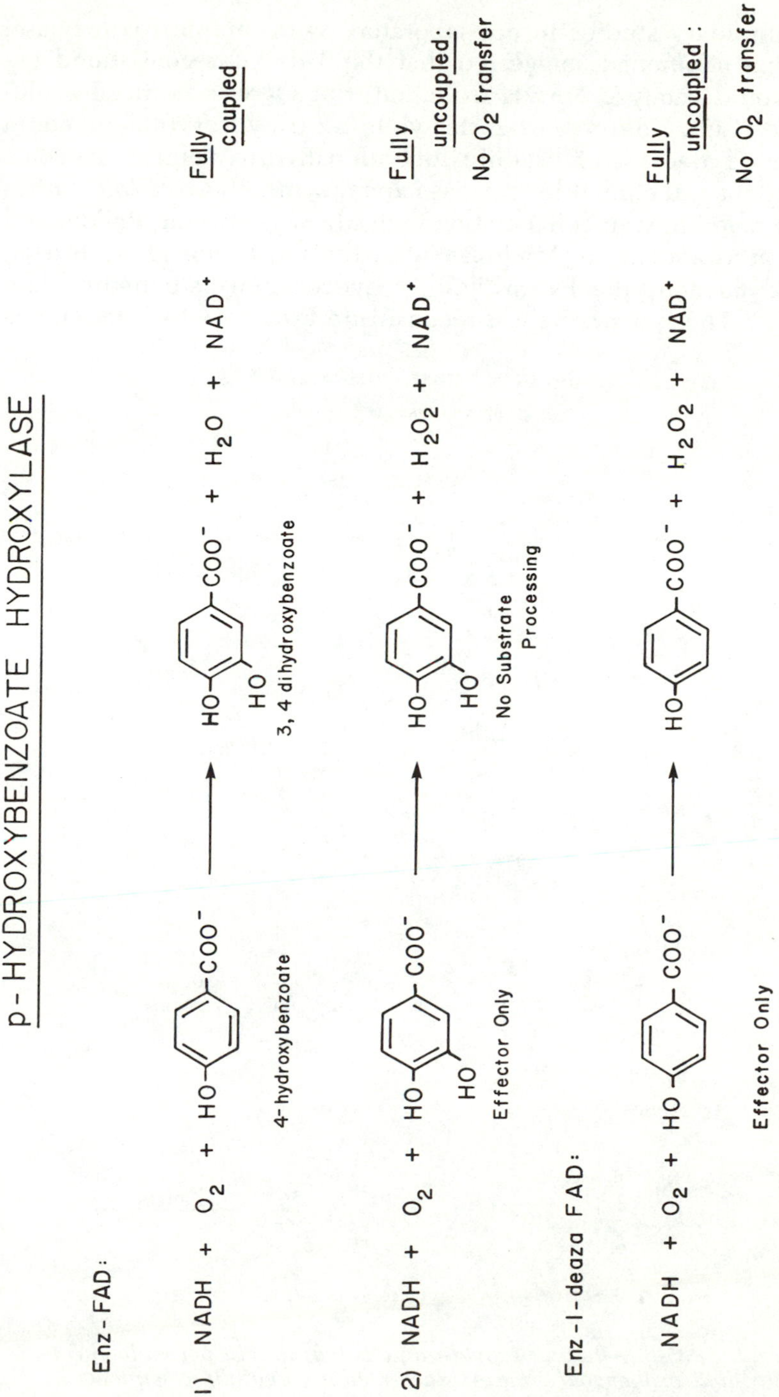

Figure 3. *Reaction stoichiometries for p-hydroxybenzoate hydroxylase with substrate or coenzyme analogs*

Preliminary studies in our laboratory with orcinol hydroxylase, from a pseudomonad, suggested that the 1-dFAD-reconstituted enzyme would catalyze NADH oxidation, but O_2 was reduced exclusively to H_2O_2. No oxygen transfer yielding trihydroxytoluene could be detected. Because of instability of both trihydroxytoluene oxygenation product and orcinol hydroxylase apoenzyme, the 1-dFAD studies were pursued next in collaboration with Massey, Husain, Ballou, and Entsch at University of Michigan with the most well characterized bacterial flavoenzyme hydroxylase, p-hydroxybenzoate hydroxylase (2, 26, 27). The apoenzyme was reconstituted stably with 1-deazaFAD

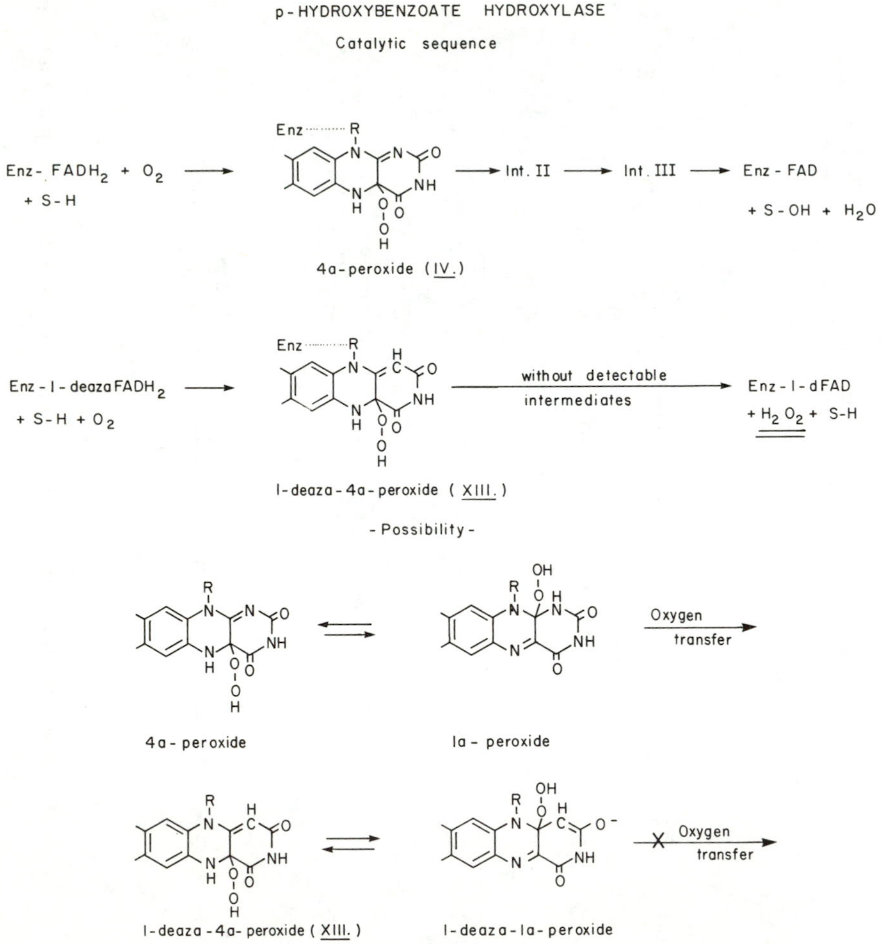

Figure 4.   Possible flavin peroxide and 1-deazaflavin peroxide species in p-hydroxybenzoate-hydroxylase-mediated catalytic sequence

and NADH oxidation is 25% that of FAD-holoenzyme (t.o. ≈ 667 sec^{-1}) (28). Rigorous analysis for 3,4-dihydroxybenzoate by HPLC failed to demonstrate O_2 transfer (Figure 4). The basis for this uncoupling is as yet unclear. Fast-reaction kinetic studies with normal FAD-enzyme reveal up to three spectroscopically detectable E-FAD derivatives (27), the first of these the 4a-hydroperoxy-FAD, **IV**, the latter two less well assigned. Similar fast-reaction kinetic analyses with 1-dFAD-p-hydroxybenzoate hydroxylase reveal an intermediate essentially identical to that of the FAD-4a-OOH, presumably 1-dFAD-4a-OOH, **XIII**, but no subsequent species are detectable, consistent with direct breakdown of **XIII** to 1-dFAD$_{ox}$ + H_2O_2 (Figure 4). The chemical nature of the subsequent intermediates (Int. II and Int. III of Figure 4) in normal holoenzyme catalysis are unclear. The electronic spectral similarities of FAD-4a-OOH and 1-dFAD-4a-OOH, each has λ_{max} at 370 nm (28), suggest similar reactivities and one might argue that the 4a-OOH species are not the proximal oxygen-transfer agents. These could be in equilibrium with 1a-OOH derivatives (via cyclic dioxetanes), which could show differential reactivity between FAD and 1-deazaFAD (Figure 4).

However, recent x-ray studies on p-hydroxybenzoate–p-hydroxybenzoate hydroxylase binary complex crystals clearly show the aromatic substrate is bound at the flavin 4a-5 edge and orthogonal to the isoalloxazine plane (29). Unless this binary complex structure is highly misinformative, it can be inferred that in the O_2, p-hydroxybenzoate, enzyme ternary active complex, oxygen transfer is in the 4a,5 region, not the 1a, 1 region of the bound FAD, which rules out 1a-OOH derivatives as important oxygenating intermediates for this enzyme.

1-Deazaflavins with Monooxygenases: Ketone Monooxygenases

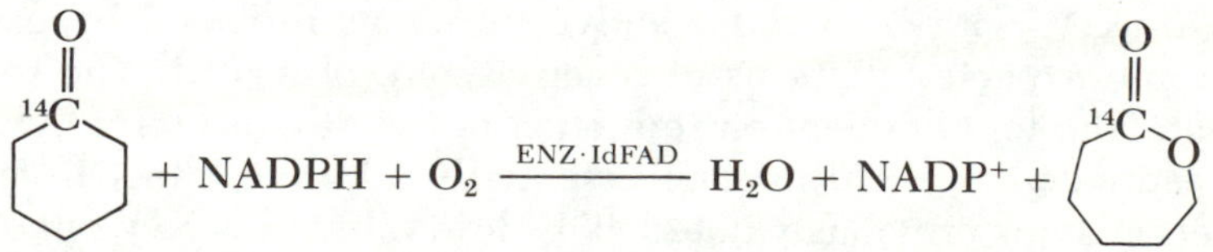

A second category of bacterial flavin monooxygenases can be compiled. This is the set of ketone monooxygenases, exemplified by pseudomonad cyclohexanone monooxygenase (30), which carries out an oxygen-insertion, ring-expansion sequence characteristic of a Baeyer–Villiger transformation. This type of monooxygen transfer differs importantly from the phenol hydroxylations in that it requires transfer of a nucleophilic oxygen equivalent to an electrophilic carbonyl carbon (2); the aromatic ring hydroxylations generally are envisioned as transfer of electrophilic oxygen to a carbanionic equivalent. This apparent reversal of polarity has prompted us to examine the ketone-monooxygenase transfer capacity after purification, apoenzyme formation, and reconstitution with 1-deazaFAD. The V_{max} for NADPH oxidation by 1-dFAD-cyclohexanone monooxygenase is 5% that of FAD-enzyme, possibly reflecting the more negative $E°'$ of 1-dFAD (Table II), but this rate is tenfold over the apoenzyme background rate (0.5%) and so readily detectable as a real 1-deazaFAD-dependent process. Oxygen transfer was analyzed by use of ^{14}C-cyclohexanone as substrate and isolation (by HPLC) of product ^{14}C-ε-caprolactone. Results from incubations with FAD-enzyme, apoenzyme, and 1-deazaFAD-enzyme respectively, show that both the FAD-enzyme and the 1-deazaFAD enzyme are 100% coupled: every NADPH oxidation is accompanied by oxygen transfer to substrate and yields a molecule of product ε-caprolactone. In this type of oxygen transfer, possibly requiring the distal 4a-OOH oxygen to be nucleophilic rather than electrophilic (2), this enzyme 1-deazaFAD species is a fully competent oxygen-transfer catalyst.

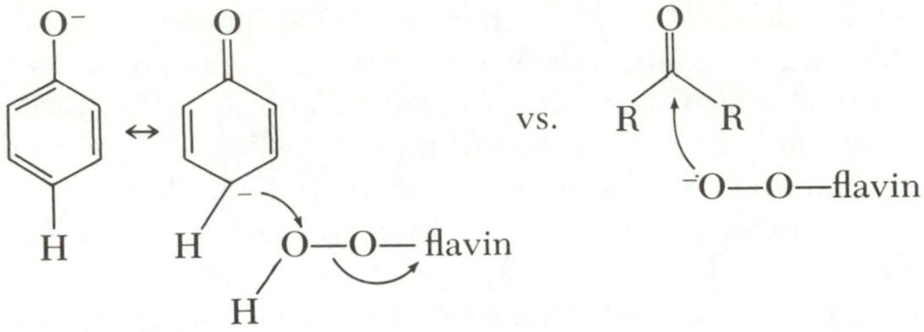

Model studies to generate $N(5)$-blocked-4a-OOH derivatives of 1-deazaflavins, and to test their capacity to deliver electrophilic and nucleophilic oxygen will be pursued.

1,5-Dideazaflavins

Subsequent to the synthesis of 5-carba-5-deaza- and 1-carba-1-deazariboflavin, the Merck group synthesized 1,5-dideazariboflavin, **XIV**, with both redox-active nitrogens replaced by carbon (22). Some expected patterns of reactivity are observed. The reduction potential for two-electron reduction to the dihydro form, **XV**, has been estimated at −370 mV, an essentially additive effect from the two deazaflavins previously noted. This low value puts **XIV** out of the

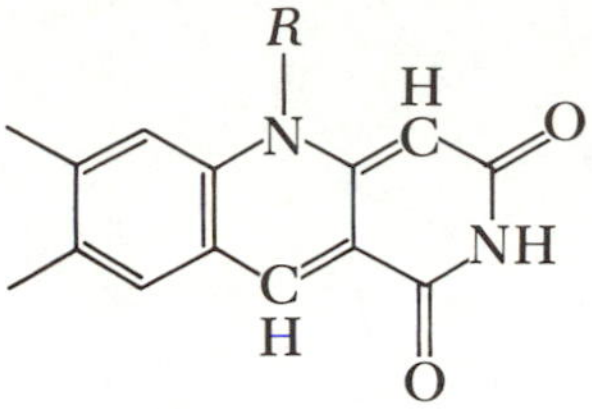

XIV 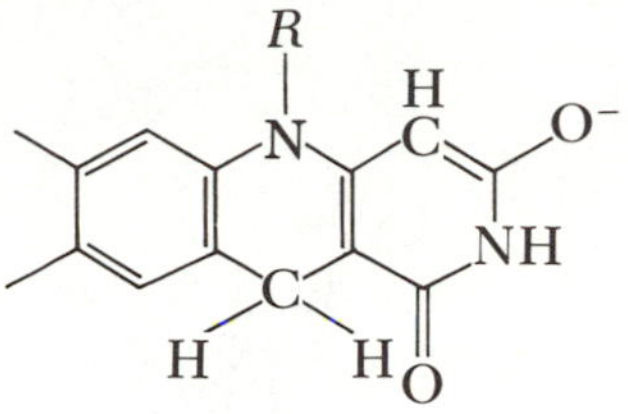   XV

realm of most biological reductants, and we have shown that it is not a substrate for the hydrogenase from methanogenic bacteria. This is in contrast to the equal catalytic competence of both 1- and 5-deazariboflavin in this system. The central dihydropyridine of 1,5-dideazariboflavin dominates the redox chemistry and only two-electron processes occur readily, as with 5-deazariboflavin itself. The two features of low redox potential and restriction to two-electron chemistry limit the utility of 1,5-dideazariboflavin as a functional flavin coenzyme analog.

8-Hydroxyflavins

In studies with an FAD-dependent D-lactate dehydrogenase from *Megaspheria elsdenii*, Ghisla and Mayhew (*31*) noted that some of the enzyme molecules purified as an inactive, orange form. The color was imparted by an altered FAD derivative, an 8-hydroxyFAD, **XVI**,

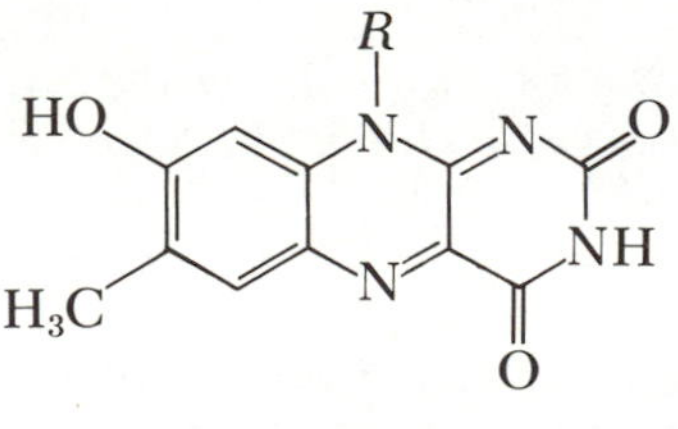

XVI $pK_a = 4.8$

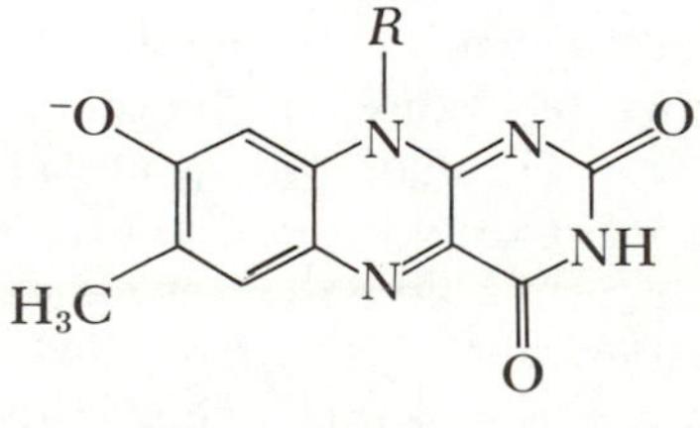

XVII 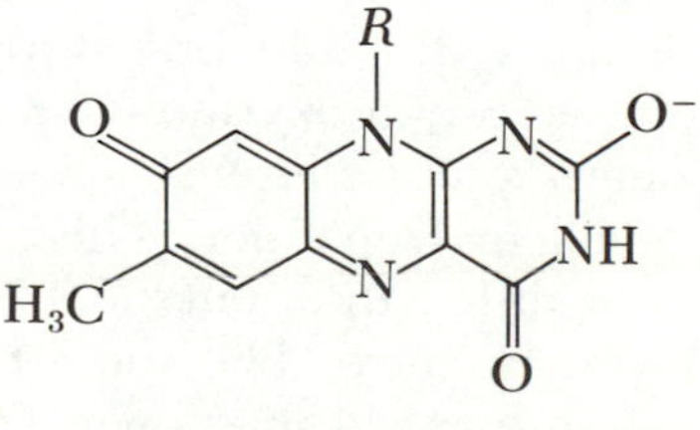   *p*-quinoid resonance
                             contributor

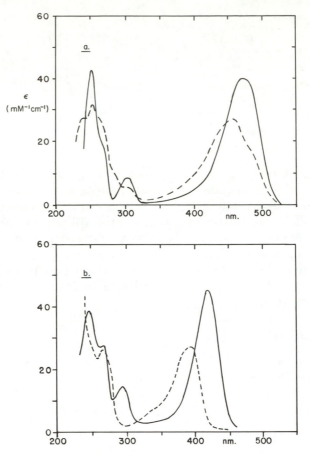

*Figure 5.   Spectra of 8-hydroxy-riboflavin and 8-hydroxy-5-deazaribo-
flavin: (a) 8-hydroxyriboflavin, (———) pH 8.3, (---) pH 4.5; (b) 8-hy-
droxy-5-deazariboflavin, (———) pH 8.3, (---) pH 4.5*

(mechanism of formation as yet unclear) which has an acidic pK_a of 4.8
for dissociation of the 8-hydroxyl proton to yield an anion, **XVII**. This
anion is stabilized as a para-quinoidal resonance contributor that is
essentially the sole form at pH 7. The spectrum of the acidic and basic
forms of 8-hydroxyriboflavin are shown in Figure 5a; the orange color
comes from the intense absorbance ($\epsilon = 41,000 M^{-1}cm^{-1}$) at 471 nm of
the delocalized anion. This anion is not readily reducible (*31, 32*) and
it has quite a different electronic distribution than the normal oxidized
flavin structure. The route of formation of the 8-hydroxyFAD molecules are as yet unknown. For the purposes of this article, this flavin
derivative serves as a standard of comparison for the 8-hydroxy-
5-deazaflavins to be discussed below.

8-Hydroxy-5-deazaflavins

In an incisive set of experiments (*33, 34, 35*), R. Wolfe and his colleagues recently have demonstrated for the first time the biological occurrence of a 5-deazaflavin, an 8-hydroxy one, as a natural product in anaerobic methanogenic bacteria. Following their discovery of a yellow, highly blue-fluorescent small molecule with absorbance λ_{max} at 420 nm, Factor 420 (F_{420}), as a characteristic marker molecule for bacteria-catalyzing methane formation, Wolfe's group demonstrated that F_{420} was an obligate redox coenzyme in the complex, eight-electron reduction of CO_2 by H_2, functioning as an initial reoxidant of hydrogenase, and, in turn, reducing cellular NADP to act as electron

$$4H_2 + CO_2 \xrightarrow[\text{transfer}]{8e^-} CH_4 + H_2O$$

donor in the reductive CO_2 conversion steps. They used the spectroscopic information accumulated for both 8-hydroxyflavin and 5-deazaflavin compounds in assignment (*35*) of the structure shown for F_{420}, **XVIII**, an 8-hydroxy, 7-demethyl-5-deazaFMN in phosphodiester linkage to a lactyl diglutamate sidechain. Mass spectra, ^{1}H NMR, and

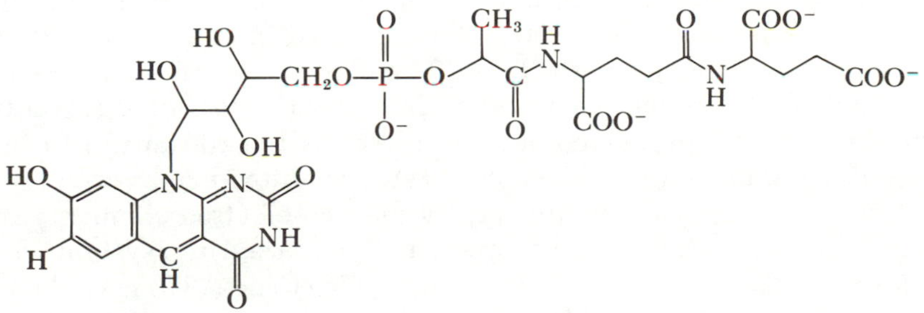

FACTOR 420, XVIII

NMR were consistent with the proposed structure (*35*). The UV–visible spectrum of the anionic and neutral forms of F_{420} are shown in Figure 5b and look very similar to the 8-hydroxyriboflavin spectra of Figure 5a except for the approximately 50-nm blue shift. This is the same shift seen on substitution of carbon for nitrogen at the 5 position in the 5-aza → 5-deazariboflavin series (Figure 1). Thus, the spectrum is consistent with what one might anticipate by additive effects of carbon substitution at the 5 position and replacement of 8-methyl by 8-hydroxyl.

The authentic 7-methyl-8-hydroxy-5-deazariboflavin, **XIX,** has been synthesized by Ashton and Brown at Merck and compared by us with the riboflavin level (Fo) of F_{420} obtained by controlled acid hydrolysis (36). This synthetic compound, **XIX**, has λ_{max} at 426 nm, 6 nm off the 420-nm absorbance of both Fo and F_{420}. The 7-demethyl-8-hydroxy-5-deazariboflavin nucleus, **XX**, also has been synthesized (at Merck) and examined in our laboratory and it is indeed

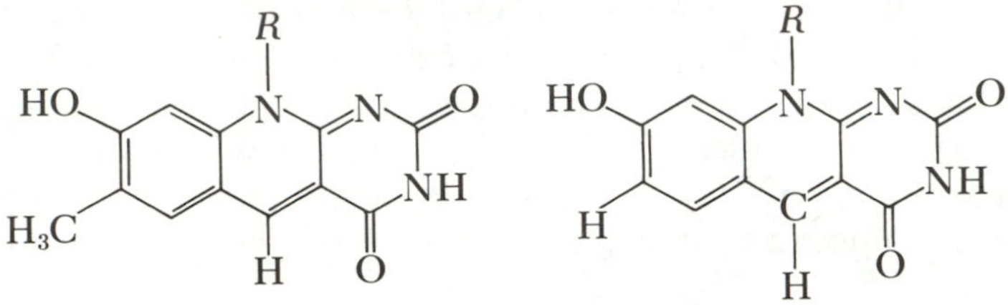

XIX   XX

identical to Fo and readily distinguishable from the 7-methyl compound, **XIX** (36). This synthesis unambiguously confirms Wolfe's proposed structure for F_{420}, and investigations are continuing into its chemistry and interaction with methanogen enzymes.

Wolfe and colleagues (35) demonstrated that methanogen-hydrogenase-mediated reduction of F_{420} or Fo by H_2 gas yields the 1,5-dihydro species, determined by NMR and UV comparison with synthetic 1,5-dihydro-5-deazariboflavin, **VIII**, noted above. When we reduce synthetic **XX** with D_2 and Pt and then allow reoxidation, the Ft-NMR of the product shows complete (>95%) loss of signal attributed to the $C(5)$ proton, confirming direct hydrogen transfer to $C(5)$ (36). The reduced methanogen coenzyme is autoxidized very slowly and in general appears dominated by the two-electron chemistry imposed by the pyridine/dihydropyridine 5-deaza ring system. The redox potential of F_{420} has been estimated (35) to be -370 mV, placing it neatly between the $H_2/(2H^+ +$ two electron) couple (-420 mV) and the NADPH/NADP couple (-320 mV). Thus the structural alterations from the riboflavin skeleton (-200 mV) alter the reduction potential to enable F_{420} to function thermodynamically as a redox shuttle between H_2 and NADP.

At present we are conducting experiments with **XIX, XX,** and other flavins with crude hydrogenase (quite air labile) kindly provided by Wolfe. Initial results demonstrate ready reduction of biological and synthetic Fo (**XX**) at approximately 200 nmol/min/mg while the 7-methyl compound, **XIX**, is reduced tenfold more slowly (at equivalent $20\mu M$ concentration). Studies are underway to investigate rate-determining steps, flavin specificity, stereochemistry, and mechanism of redox transfer between H_2 and F_{420} as well as its reoxidation by NADP.

Conclusion

The recent strategy of using synthetic mimics of flavin coenzymes by replacement of $N(5)$, $N(1)$, or $N(5)$ and $N(1)$ by carbon has begun to unravel the contributions of these atoms and loci to the diverse chemistry and the biological functions of these redox coenzymes. The three questions of sites of substrate-derived hydrogen transfer, control of two-electron vs. one-electron transfers, and modes of O_2 activation for two-electron reduction or reductive fragmentation and oxygen transfer already have been resolved partially by such studies. An unexpected dividend to flavin mimics is the finding, aided by the synthetic analog studies, that a natural 8-hydroxy-5-deazaflavin is a key redox coenzyme in biological production of natural gas. The availability of authentic synthetic samples in quantity coupled with continued analysis of the chemistry open to this heterocyclic system (and its analogs) will dictate what coenzymatic mechanisms we can anticipate in our studies with the specific methanogen enzymes.

Literature Cited

1. Walsh, C. *Annu. Rev. Biochem.* **1978**, *45*, 881–931.
2. Walsh, C. "Enzymatic Reaction Mechanisms"; W. H. Freeman: San Francisco, 1979; Chapters 11–12.
3. Massey, V.; Hemmerich, P. "The Enzymes," 3rd ed.; Boyer, P., Ed.; Academic: New York, 1976; Vol. 12, p. 191.
4. Prough, R.; Ziegler, D. *Arch. Biochem. Biophys.* **1977**, *180*, 363–373.
5. Kemal, C.; Chan, T.; Bruice, T. C. *Proc. Natl. Acad. Sci. USA* **1977**, *74*, 405–409.
6. Ghisla, S.; Entsch, B.; Massey, V.; Husain, M. *Eur. J. Biochem.* **1977**, *76*, 149–156.
7. Loechler, E.; Hollocher, T. *J. Am. Chem. Soc.* **1975**, *97*, 3235–3236.
8. Thorpe, C.; Williams, C. H. *J. Biol. Chem.* **1976**, *251*, 7726–7728.
9. Massey, V.; Ghisla, S. *Proc. 10th Fed. Eur. Biochem. Soc. Mtg.* **1975**, p. 145.
10. O'Brien, P.; Weinstock, L.; Cheng, C. *J. Heterocycl. Chem.* **1970**, *7*, 99–107.
11. Edmondson, D.; Barman, B.; Tollin, H. *Biochemistry* **1972**, *11*, 1133–1138.
12. Fisher, J.; Walsh, C. *J. Am. Chem. Soc.* **1974**, *96*, 4345–4346.
13. Spencer, R.; Fisher, J.; Walsh, C. *Biochemistry* **1976**, *15*, 1043–1053.
14. Fisher, J.; Spencer, R.; Walsh, C. *Biochemistry* **1976**, *15*, 1054–1064.
15. Walsh, C.; Hersh, L. "Methods in Enzymology," in press.
16. Jorns, M.; Hersh, L. *J. Am. Chem. Soc.* **1974**, *96*, 4012–4014.
17. Jorns, M.; Hersh, L. *J. Biol. Chem.* **1975**, *250*, 3620–3628.
18. Jorns, M.; Hersh, L. *J. Biol. Chem.* **1976**, *251*, 4872–4881.
19. Hersch, L.; Jorns, M.; Peterson, J.; Curie, M. *J. Am. Chem. Soc.* **1976**, *98*, 865–867.
20. Massey, V.; Hemmerich, P. *Biochemistry* **1978**, *17*, 1–15.
21. Walsh, C.; Jacobson, F.; Fisher, J.; Spencer, R. "Flavins and Flavoproteins," Yagi, K., Yamano, T., Eds., in press.
22. Ashton, W. T.; Graham, D. W.; Brown, R. D.; Rogers, E. F. *Tetrahedron Lett.* **1977**, 2551–2554.
23. Spencer, R.; Fisher, J.; Walsh, C. *Biochemistry* **1977**, *16*, 3586–3594.

24. Spencer, R.; Fisher, J.; Walsh, C. *Biochemistry* **1977**, *16*, 3594–3602.
25. Spinner, E.; White, J. *J. Chem. Soc.* **1966**, *B*, 991.
26. Flashner, M.; Massey, V. "Molecular Mechanisms of Oxygen Activation";
    Hayaishi, O., Ed.; Academic: New York, p. 245.
27. Entsch, B.; Ballou, D.; Massey, V. *J. Biol. Chem.* **1976**, *251*, 2550–2563.
28. Entsch, B.; Husain, M.; Ballou, D.; Massey, V.; Walsh, C. *J. Biol. Chem.*
    1980, 1420–1429.
29. Wieringer, W., Ph.D. Dissertation, University of Delft, Netherlands, 1978.
30. Griffin, M.; Trudgill, P. *Biochem. J.* **1972**, *129*, 595.
31. Ghisla, S.; Mayhew, S. *Eur. J. Biochem.* **1976**, *63*, 373–390.
32. Walsh, C.; Fisher, J.; Spencer, R.; Graham, D.; Ashton, W.; Brown, J.;
    Brown, R.; Rogers, E. *Biochemistry* **1978**, *17*, 1942–1951.
33. Tzeng, S.; Wolfe, R.; Bryant, M. *J. Bacteriol.* **1975**, *121*, 184.
34. Gunsalus, R.; Eirich, D.; Romessar, J.; Balch, W.; Shapiro, S.; Wolfe, R. S.
    "Microbial Production and Utilization of Gases"; Goltze: Gottingen,
    1976.
35. Eirich, D.; Vogels, G.; Wolfe, R. *Biochemistry* **1978**, *17*, 4583–4592.
36. Ashton, W.; Brown, R.; Jacobson, F.; Walsh, C. *J. Am. Chem. Soc.* **1979**,
    101, 4419–4420.

RECEIVED May 15, 1979.

Structure–Function Relationship of Vitamin B_{12} Coenzyme (Adenosylcobalamin) in the Diol-Dehydrase System

TETSUO TORAYA and SABURO FUKUI

Laboratory of Industrial Biochemistry, Department of Industrial Chemistry, Faculty of Engineering, Kyoto University, Sakyo-Ku, Kyoto 606, Japan.

The structure–function relationship of vitamin B_{12} coenzyme (adenosylcobalamin) was investigated with diol dehydrase through the use of a number of analogs of the coenzyme to gain information concerning the mechanism of cleavage of the C–Co bond of the coenzyme by the apoprotein. Modifications were introduced to the corrin ring or its amide sidechains, the Coα (lower) nucleotide ligand, or the Coβ (upper) nucleoside ligand. It was suggested strongly that interactions between enzyme and coenzyme at the adenosyl portion as well as at the peripheral amide sidechains of the corrin ring play essential roles in activation of the C–Co bond of the coenzyme. Some aspects of enzyme–coenzyme interactions also were studied by the affinity chromatography technique using several corrinoid derivatives immobilized on agarose gel through various positions of the molecule.

Adenosylcobalamin (vitamin B_{12} coenzyme) is a corrinoid compound that is distributed widely in living organisms. This coenzyme, first discovered by Barker and co-workers in 1958 (*1, 2*), is not only the most complicated nonpolymer organic compound of known structure (Figure 1), but also the first known naturally occurring organometallic compound containing a stable C–Co sigma bond. The coenzyme is known to participate as coenzyme in the ten enzymatic rearrange-

0-8412-0514-0/80/33-191-139$05.00/0
© 1980 American Chemical Society

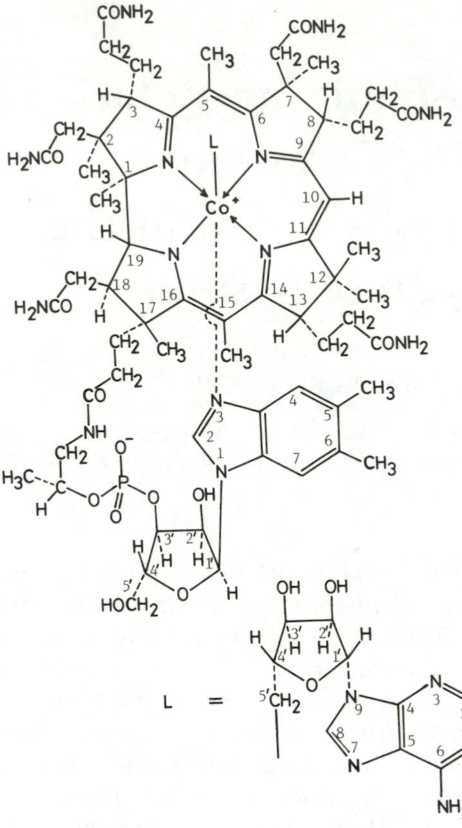

Figure 1. Structure of vitamin B_{12} coenzyme (adenosylcobalamin):
L = adenosyl group

ment and reduction reactions (Table I). The unifying feature of the
seemingly quite different chemical reactions catalyzed by adenosylcobalamin-requiring enzymes is that the coenzyme serves as
an intermediate carrier for a hydrogen transfer. It is accepted that an
early event in all of these reactions is the cleavage of the C–Co bond of
the coenzyme, which leads to generation of the catalytic center. How
does the enzyme catalyze this cleavage? This is one of the most important and interesting questions for biochemists and chemists in this
field. To answer this question, the structure–function relationship of
adenosylcobalamin was studied in detail with diol dehydrase (DL-
1,2-propanediol hydro-lyase, EC 4.2.1.28), one of the representative
B_{12} enzymes, through the use of a number of analogs of the coenzyme.
Some aspects of the enzyme–coenzyme interaction were investigated

Table I. Enzymatic Reactions in Which Adenosylcorrinoids Are Involved as Coenzyme

Reaction	Enzyme
Carbon–Carbon Bond Cleavage	

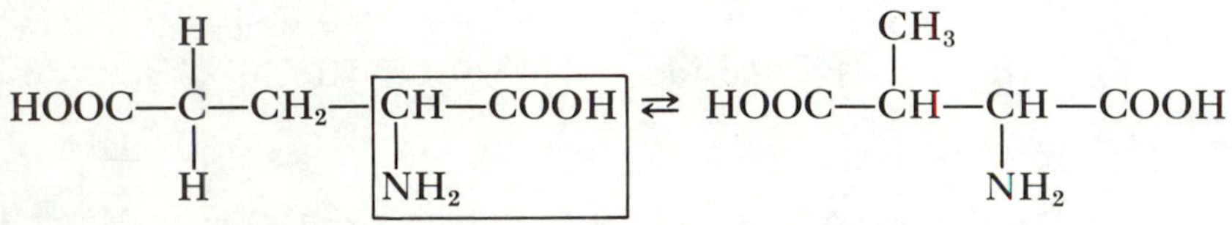

Glutamate mutase:

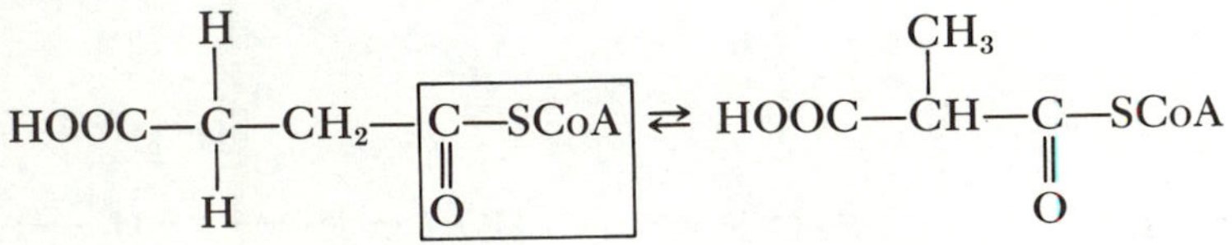

$$\text{HOOC–C(H)(H)–}\boxed{\text{CH(NH}_2\text{)–COOH}} \rightleftarrows \text{HOOC–CH}_2\text{–CH(CH}_3\text{)–CH(NH}_2\text{)–COOH}$$

Methylmalonyl-CoA mutase:

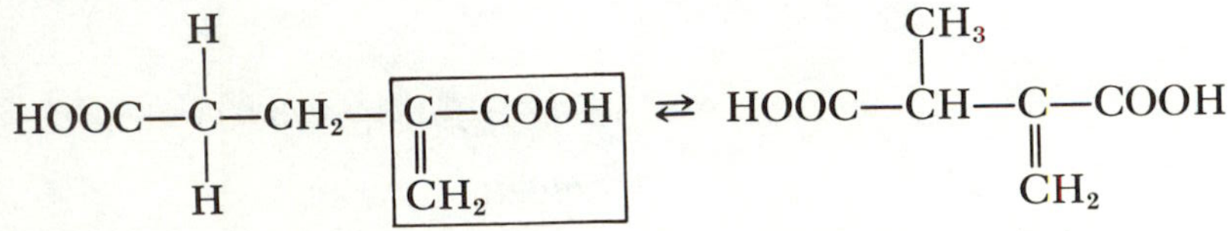

$$\text{HOOC–C(H)(H)–}\boxed{\text{C(=O)–SCoA}} \rightleftarrows \text{HOOC–CH}_2\text{–CH(CH}_3\text{)–C(=O)–SCoA}$$

α-Methyleneglutarate mutase:

$$\text{HOOC–C(H)(H)–}\boxed{\text{C(=CH}_2\text{)–COOH}} \rightleftarrows \text{HOOC–CH}_2\text{–CH(CH}_3\text{)–C(=CH}_2\text{)–COOH}$$

Table I. Continued

Reaction	Enzyme
Carbon–Oxygen Bond Cleavage	
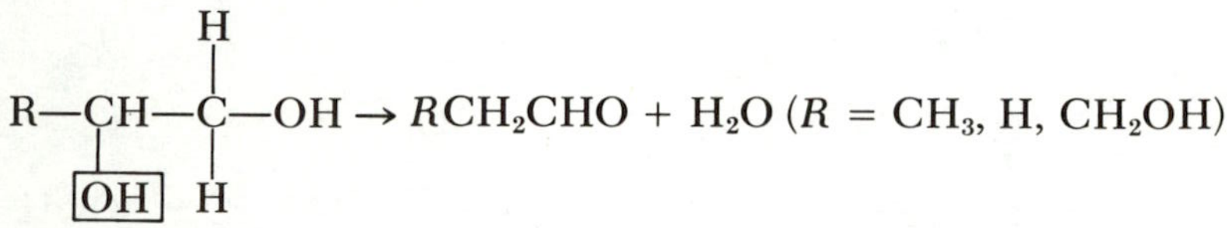 $R-CH-C-OH \rightarrow RCH_2CHO + H_2O \ (R = CH_3,\ H,\ CH_2OH)$	Diol dehydrase
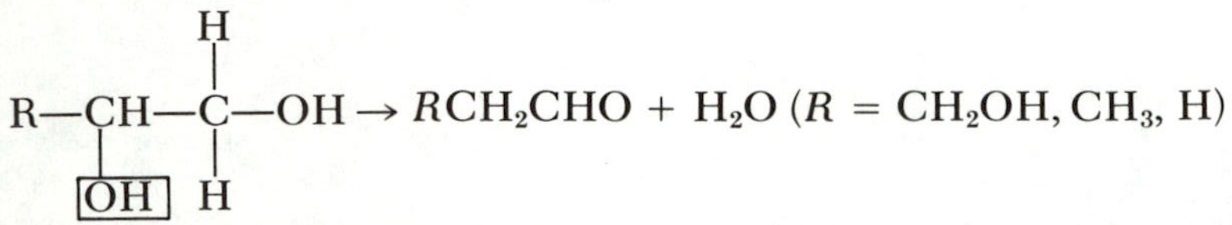 $R-CH-C-OH \rightarrow RCH_2CHO + H_2O \ (R = CH_2OH,\ CH_3,\ H)$	Glycerol dehydrase
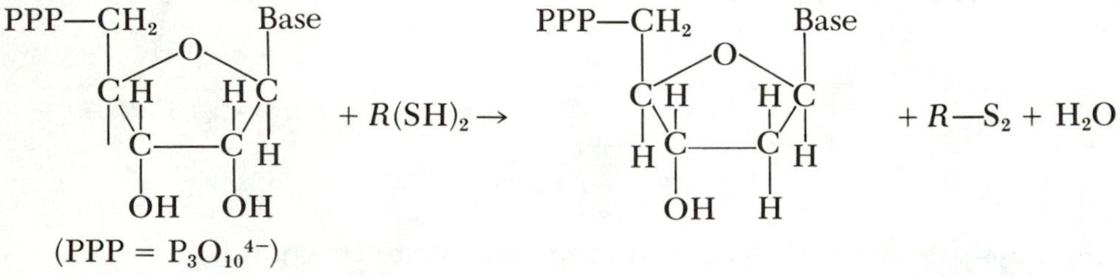 $+ R(SH)_2 \rightarrow$ $+ R{-}S_2 + H_2O$ $(PPP = P_3O_{10}{}^{4-})$	Ribonucleotide reductase

Enzyme	Reaction
	Carbon–Nitrogen Bond Cleavage
Ethanolamine ammonia-lyase	$CH_2(\boxed{NH_2})-\overset{H}{\underset{H}{C}}-OH \rightarrow CH_3CHO + NH_3$
L-β-Lysine mutase	$CH_2(\boxed{NH_2})-\overset{H}{\underset{H}{C}}-CH_2-CH_2-CH(NH_2)-COOH \rightleftarrows CH_3-CH(NH_2)-CH_2-CH(NH_2)-CH_2-COOH$
D-α-Lysine mutase	$CH_2(\boxed{NH_2})-\overset{H}{\underset{H}{C}}-CH_2-CH_2-CH(NH_2)-COOH \rightleftarrows CH_3-CH(NH_2)-CH_2-CH_2-CH(NH_2)-COOH$
Ornithine mutase	$CH_2(\boxed{NH_2})-\overset{H}{\underset{H}{C}}-CH_2-CH(NH_2)-COOH \rightleftarrows CH_3-CH(NH_2)-CH_2-CH(NH_2)-COOH$
Leucine 2,3-aminomutase	$CH_3-\overset{H}{\underset{CH_3}{C}}-CH_2-CH(\boxed{NH_2})-COOH \rightleftarrows CH_3-CH(CH_3)-CH(NH_2)-CH_2-COOH$

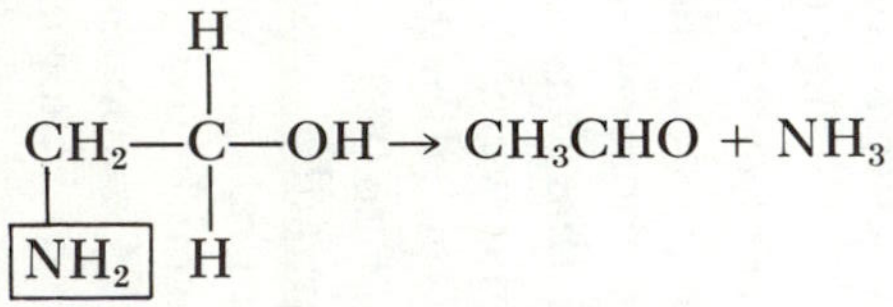

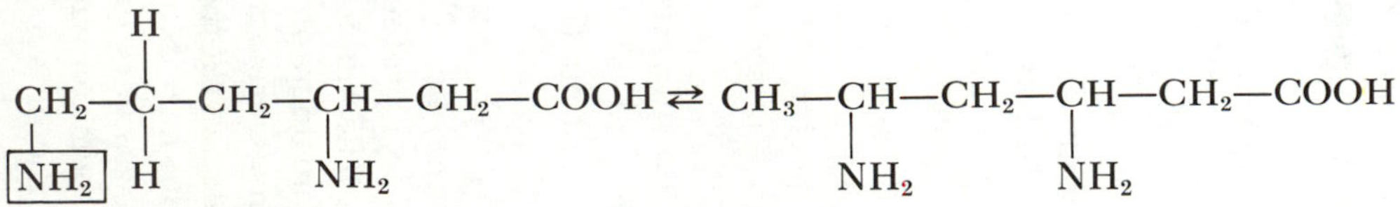

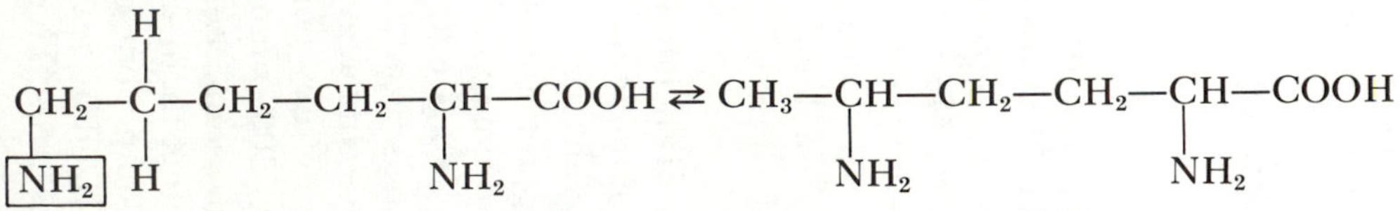

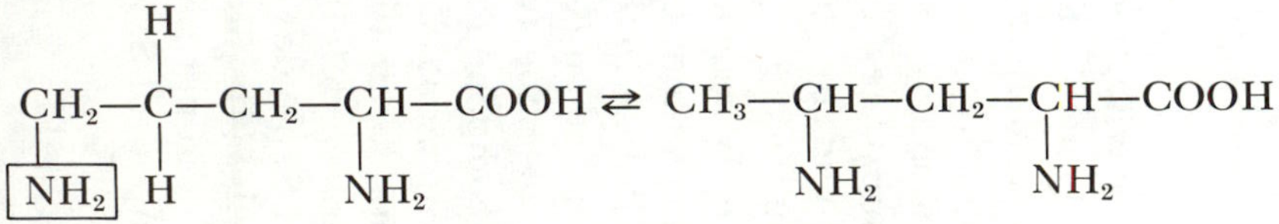

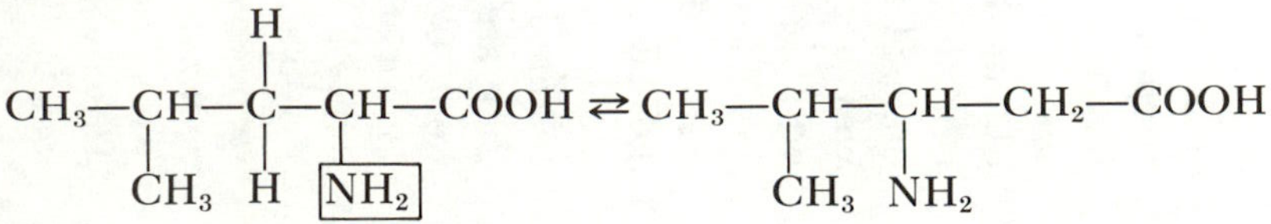

by affinity chromatography using several different types of agarose-gel-bound corrinoid derivatives as affinity adsorbents. Some enzymological properties also are described here.

Enzymological Properties and Mechanism of Action of Diol Dehydrase

Protein-chemical Properties of Apoenzyme. Adenosylcobalamin-dependent diol dehydrase was discovered and isolated first by Abeles and co-workers (3, 4) in the cells of *Klebsiella pneumoniae* (formerly known as *Aerobacter aerogenes*) ATCC 8724 grown without aeration in a glycerol or glycerol-1,2-propanediol medium. This enzyme catalyzes the conversion of 1,2-propanediol, 1,2-ethanediol, and glycerol to propionaldehyde, acetaldehyde, and β-hydroxypropionaldehyde, respectively (4, 5). Adenosylcobalamin and K^+ or other monovalent cations of a similar size are required for catalysis. Recently, the au-

$$R—CHCH_2OH \rightarrow R—CH_2CHO + H_2O \qquad (R = CH_3, H, \text{ or } HOCH_2)$$
$$\underset{\displaystyle OH}{\displaystyle |}$$

thors have developed a new procedure for purifying this enzyme, in cooperation with Soda, Tanizawa (Institute for Chemical Research, Kyoto University, Kyoto, Japan) and Poznanskaja (All-Union Research Vitamin Institute, Moscow, USSR) (6). The purified enzyme obtained by this procedure was homogeneous by the criteria of ultracentrifugation ($s_{20,w} = 8.9$ S) and disc gel electrophoresis in the presence of substrate. The molecular weight of approximately 230,000 was obtained by gel filtration and ultracentrifugal sedimentation equilibrium.

This enzyme was dissociated into two dissimilar protein components (designated Components F and S) by chromatography on DEAE-cellulose in the absence of substrate (7, 8). Neither component alone possessed any appreciable activity, and the enzyme activity was restored when the two components were combined, indicating that they are components of a single enzyme. The more acidic component, Component S, was a sulfhydryl protein that was sensitive to iodoacetamide, an alkylating agent. The molecular weights of Components F and S, determined by chromatography on calibrated Bio-Gel P-100 and Sephadex G-200 columns, were about 26,000 and 200,000, respectively (6). Both components were thermally unstable. The coenzyme did not protect them from heat denaturation. As compared with Component S, Component F was very unstable even at 0°C (8). However, it was stabilized by reassociation with Component S, which was facilitated by substrates. That the presence of 1,2-propanediol in the eluting buffer retarded the dissociation of the enzyme on DEAE-cellulose chromatography also suggests that the substrate strongly promotes association of the components.

Although a purified enzyme migrated as a single protein band upon disc gel electrophoresis in the presence of substrate, several protein bands appeared in the absence of substrate (6). This result is consistent with the above-mentioned observation that the substrate protected the enzyme from dissociation into the components. Upon electrophoresis in the absence of substrate, the mixture of Components F and S also gave several bands, as did the native enzyme. In the presence of substrate, they reassociated to give a single protein band, which corresponded to a purified enzyme. Subunit structures of Components F and S were further investigated by SDS-disc gel electrophoresis. Component F treated with SDS migrated as a single protein band, whereas Component S dissociated into four bands. This suggests that Component S is composed of at least four different polypeptide chains. The apparent molecular weights of the subunits of Component S were estimated from the calibration curve to be 60,000, 23,000, 15,500, and 14,000. The subunit heterogeneity of Component S was examined also by terminal amino-acid analyses. Methionine, leucine, cysteine, and lysine were identified as NH$_2$-terminal residues, and aspartate, glutamate, leucine, and histidine as COOH-terminals of these four subunits. These subunits seem to be bound together by electrostatic interactions, because Component S was separated into subunits by disc gel electrophoresis but not by Sephadex G-200 gel filtration. The possibility that Component S was nicked by proteolytic enzymes during purification could be ruled out by immunoelectrophoresis of freshly prepared, cell-free extracts (6).

Interaction with Adenosylcobalamin. It has been considered generally that adenosylcobalamin or its analogs binds to the apoprotein of diol dehydrase or other adenosylcobalamin-dependent enzymes almost irreversibly (4). However, we found that the holoenzyme of diol dehydrase was resolved completely into intact apoenzyme and adenosylcobalamin when subjected to gel filtration on a Sephadex G-25 column in the absence of K$^+$ (9, 10). Among the inactive complexes of diol dehydrase with irreversible cobalamin inhibitors, those with cyanocobalamin and methylcobalamin also were resolved upon gel filtration on Sephadex G-25 in the absence of both K$^+$ and substrate, yielding the apoenzyme, which was reconstitutable into the active holoenzyme (11). The enzyme–hydroxocobalamin complex, however, was not resolvable under the same conditions. The enzyme–cobalamin complexes were not resolved at all by gel filtration in the presence of both K$^+$ and substrate. When gel filtration of the holoenzyme was carried out in the presence of K$^+$ only, the holoenzyme was dissociated according to the first-order kinetics ($k = 1.4 \times 10^{-2}$ min^{-1} at 0°C). These data indicate that K$^+$ plays an essential role in the binding of cobalamins to diol dehydrase. Only hydroxocobalamin was able to bind to the enzyme even in the absence of both K$^+$ and substrate.

Another important factor affecting the cobalamin binding is the SH group(s) of enzyme. Cobalamins were not bound to the apoenzyme whose SH group(s) were modified by organic mercurials (11). On the other hand, once the enzyme bound cobalamins, it became insusceptible to SH-modifying agents (4, 12).

Both Components F and S were required for irreversible cleavage of the C–Co bond of adenosylcobalamin by oxygen upon aerobic incubation with the coenzyme in the absence of substrate. This suggests that activation of the C–Co bond of the coenzyme is dependent on both components. Sephadex G-25 filtration experiments showed that neither adenosylcobalamin nor cyanocobalamin was bound by the individual components, F or S. Both of them were necessary for the cobalamin binding (8).

Mechanism of Action. Diol dehydrase is one of the enzymes whose mechanism of action has been studied extensively (13, 14). It has been established that in general the rearrangement reactions catalyzed by adenosylcobalamin-requiring enzymes involve the migration of a hydrogen from one carbon atom of substrate to an adjacent carbon atom in exchange for a group X that moves in the opposite direction. In the case of the diol-dehydrase reaction, X represents a

$$\underset{\overset{|}{\underset{}{}}}{\overset{H}{\underset{|}{C_1}}} \underset{\overset{|}{\underset{}{}}}{\overset{X}{\underset{|}{C_2}}} \rightarrow \underset{\overset{|}{\underset{}{}}}{\overset{X}{\underset{|}{C_1}}} \underset{\overset{|}{\underset{}{}}}{\overset{H}{\underset{|}{C_2}}}$$

hydroxyl group, and the product (aldehyde) is formed by elimination of H$_2$O from the gem-diol. Scheme 1 shows a mechanism proposed by Abeles and co-workers (15) for diol dehydrase that is accepted by many, although not all, investigators (16, 17). Salient features of the mechanism are as follows. The interaction between enzyme and coenzyme leads to the activation and cleavage of the C–Co bond of the coenzyme to form cob(II)alamin and an adenosyl radical. In the absence of substrate, only a small fraction of the bound coenzyme is dissociated. Addition of substrate to the complex shifts the equilibrium so that a major fraction of the coenzyme is now present in the dissociated form. The adenosyl radical that results from the dissociation of the C–Co bond then abstracts a hydrogen atom from the substrate, producing a substrate-derived radical and 5'-deoxyadenosine. The substrate-derived radical rearranges to the product radical. The mechanism of

Scheme 1. *Mechanism of action of diol dehydrase: RCH₂— = adenosyl; [Co] = cobalamin; SH = substrate; PH = product*

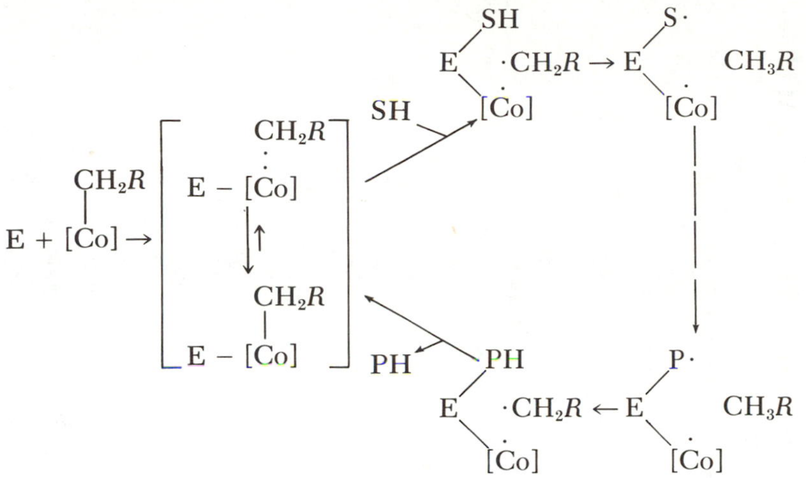

this rearrangement remains unclear. The product radical then abstracts a hydrogen atom from 5'-deoxyadenosine. This leads to the formation of the final product and regeneration of the coenzyme.

Studies of the Structure–Function Relationship of Adenosylcobalamin through the Use of Coenzyme Analogs

Corrin Nucleus and Its Amide Sidechains. As shown in Table II, 10-chloro and 10-bromo analogs of the coenzyme were active partially as coenzymes in the diol-dehydrase system (*18*). Since the affinity for the enzyme was not lowered by this substitution, it is likely that their lower coenzyme activity is attributable to the effect of halogen substitution on the electronic structure of the corrin nucleus, which possibly affects the electronic structure (reactivity) of the C–Co bond via Co–N$_{equatorial}$ bonds. This seems to be important evidence that modification of the corrin ring causes a change in reactivity of axial ligands (*cis*-effect). Adenosyl-13-epicobalamin is an analog in which the adenosyl moiety in the Coβ position is positioned differently above the corrin ring, because the inverted *e*-propionamide sidechain blocks its normal location above $C(13)$ (*19*). Furthermore, the lower pK_a of the "base on" ⇄ "base off" conversion of the 13-epimer coenzyme (2.8) compared with that of the regular coenzyme (3.5) suggests that the inversion at $C(13)$ also affects the electronic character of the cobalt atom. This epimerization at $C(13)$ causes a change in the conformation of ring C (*20*). These facts suggest that both electronic and steric effects are responsible for lower activity and lower affinity of this

Table II. Coenzyme Activity of Analogs in Which the Corrin Nucleus or Its Side Chains Are Modified

Part(s) modified	Analog[a]	Apparent K_m (μM)	Relative activity (%)	Reference
	AdoCbl	0.3–0.9	(100)	(18, 21, 23)
Corrin nucleus	AdoCbl(10-Cl)	0.6	40	(18)
	AdoCbl(10-Br)	0.4	20	(18)
	AdoCbl(13-epi)	13	14	(21)
	AdoCbl(8-NH-c-lactam)	—	Inactive	(22)
	AdoCbl(8-O-c-lactone)	—	Inactive	(22)
Sidechains	AdoCbl(b-OH)	2.3	86	(23)
	AdoCbl(e-OH)	4.5	11	(23)
	AdoCbl(d-OH)	8.0	66	(23)
	AdoCbl(b-OCH₃)	3.1	14	(23)
	AdoCbl(e-OCH₃)	1.9	7	(23)
	AdoCbl(d-OCH₃)	—	14	(23)
	AdoCbl(b-NHCH₃)	2.1	16	(23)
	AdoCbl(e-NHCH₃)	0.94	41	(23)
	AdoCbl(d-NHCH₃)	1.5	43	(23)
	AdoCbl(b-NHC₂H₅)	—	Inactive	(25)
	AdoCbl(e-NHC₂H₅)	—	19	(25)
	AdoCbl(OH)₂	—	Inactive	(24)
	AdoCbl(OH)₃	—	Inactive	(24)

[a] Abbreviations: AdoCbl, adenosylcobalamin or Coα-[α-(5,6-dimethylbenzimidazolyl)]-Coβ-adenosylcobamide; AdoCbl(b-, d-, or e-OH), adenosylcobalamin b-, d-, or e-carboxylic acid; AdoCbl(b-, d-, or e-OCH₃), adenosylcobalamin b-, d-, or e-methyl ester; AdoCbl(b-, d-, or e-NHCH₃), adenosylcobalamin b-, d-, or e-methylamide; AdoCbl(OH)₂ or AdoCbl(OH)₃, adenosylcobalamin di- or tricarboxylic acid, respectively.

analog in the diol-dehydrase system (*21*). *c*-Lactam and *c*-lactone analogs of the coenzyme have been reported by Rapp to be inactive in this system (*22*). Detailed discussion on these analogs is impossible, since no kinetic parameters were reported.

Coenzyme analogs with modified sidechains also were prepared. Detailed studies on the function of peripheral amide sidechains of the coenzyme were carried out by Toraya et al. (*23*). Isomers of adenosyl-cobalamin in which one of the three amide groups of the pro-pionamide sidechain of the corrin ring was converted to —COOH, —COOCH$_3$ or —CONHCH$_3$ were tested for coenzyme activity with diol dehydrase. The coenzyme activity of these nine isomers ranged from 86%–7% that of the normal coenzyme (Table II). With the *b*-site modifications of the coenzyme, the relative activity decreases in the order —COO$^-$≫—CONHCH$_3$, —COOCH$_3$, suggesting inhibition by steric effects, while at the diagonal *e* site the relative activity decreases in the order —CONHCH$_3$≫—COO$^-$, —COOCH$_3$, suggesting the im-portance of hydrogen-bond donation from coenzyme to enzyme. The binding of the analogs to enzyme was reversible, with apparent K_m values much higher than that of the unmodified coenzyme. In the presence of 1,2-propanediol, the *e*-carboxylic acid, *e*-methyl ester, *b*- and *e*-methylamide coenzymes brought about irreversible inactivation of the enzyme. All of the four binary complexes of the inactivating analogs with the apoenzyme were relatively stable in the absence of substrate. In contrast, the complexes with the normal coenzyme and with the analogs that do not inactivate are unstable in the absence of substrate. This stability of the inactivating complexes suggests insuffi-cient activation of the coenzyme. From the data obtained, Toraya et al. concluded that the interactions of the coenzyme propionamide side-chains with the apoprotein facilitate the homolytic cleavage of the C–Co bond and contribute to the stabilization of the radical intermedi-ates (*23*).

Di- and tricarboxylic-acid analogs did not show coenzyme activity (*24*). Rapp and Hildebrand have reported that the *e*-ethylamide coen-zyme is 19% as active as the normal coenzyme, while the *b*-ethylamide analog is inactive (*25*). This is consistent with the data of Toraya et al. on methylamide analogs.

Coα (Lower) Ligand. Cobinamide coenzyme, which lacks the nucleotide loop, was found to be partially active as a coenzyme (*26*). This indicates that the nucleotide portion in the Coα axial position does not participate directly in the catalysis. The relatively high K_m value (6μM) for this analog suggests the important contribution of interactions at the nucleotide portion to the tight binding of the normal coenzyme.

2'-*O*-Succinyl and 5'-*O*-succinyl derivatives of cyanocobalamin were not inhibitory for diol dehydrase even when enzyme was preincubated with them prior to the addition of adenosylcobalamin (27). This result suggests that the affinity of the analog for the enzyme is lowered by modification of 2'-OH or 5'-OH of the ribose in the nucleotide loop, presumably due to the steric and/or electrostatic effects.

Several analogs in which 5,6-dimethylbenzimidazole is replaced by different heterocyclic bases have been tested for coenzyme activity with diol dehydrase by Abeles and Lee (3) and Kamikubo and co-workers (28, 29, 30). It is noteworthy that relative activity of these analogs do not differ so much as do their apparent K_m values (Table III). This suggests that the base moiety is important for the recognition of the coenzyme by the enzyme molecule, but not directly for the catalytic action.

Friedrich et al. have reported that the analog with 1,1-dimethyl-1-amino-2-ethanol in place of 1-amino-2-propanol is inactive as a coenzyme in the diol-dehydrase system, although it is bound to the enzyme as tightly as is cyanocobalamin (31).

Coβ (Upper) Ligand.   The adenosyl group in the Coβ axial position is the most essential part of the coenzyme for the catalytic function. Before the present study, Hogenkamp and Oikawa have reported that 2'-deoxyadenosylcobalamin is a partially active coenzyme for diol dehydrase, while thymidylcobalamin is quite inactive (32). Abeles and co-workers have shown that a carbocyclic analog of the coenzyme serves as coenzyme, which eliminated the possibility that the ribosyl oxygen of the coenzyme plays an essential role in the catalytic process (33). Recently, a number of new coenzyme analogs with modified nucleoside groups were synthesized and tested for coenzyme activity with diol dehydrase (34).

As shown in Table IV, coenzyme analogs whose adenine part is modified slightly were still active coenzymes, although their relative activity decreased very much (3-isoadenosylcobalamin and nebularylcobalamin). Neither inosylcobalamin nor cobalamins with pyrimidine or benzimidazole bases served as coenzymes. This suggests that for the coenzyme activity a purine ring is required absolutely, and that 6-NH$_2$ is very important. It is of much interest that 3-isoadenosylcobalamin acted as an active coenzyme for this enzyme as well as for glycerol dehydrase (35) and ribonucleotide reductase (36). This may not be surprising, if the adenine ring of the coenzyme interacts with the apoprotein through hydrogen bonding at $N(1)$ and 6-NH$_2$ positions upon binding. Similar hydrogen bonding would be possible at $N(7)$ and 6-NH$_2$ of the 3-isoadenosine.

The coenzyme activity of analogs with a modified ribose are listed in Table V. Again, only small modifications were permitted without complete loss of activity (2'-deoxyadenosylcobalamin, ara-adenosylcobalamin and aristeromycylcobalamin). However, the

Table III. Coenzyme Activity of Analogs in Which the Coα (lower) Ligand Is Modified

Part(s) modified	Analog[a]	Apparent K$_m$ (μM)	Relative Activity (%)	Reference
	AdoCbl = (Me$_2$Bza)AdoCba	0.2–0.4	(100)	(3, 26, 28, 29, 30)
Base, ribose, phosphate	(aq)AdoCbi	6	Partially active	(26)
Base	(Bza)AdoCba	0.2	100	(3)
	(Ade)AdoCba	0.2	100	(3)
	(2-SHAde)AdoCba	3.5	63	(28)
	(2-ClAde)AdoCba	0.1		(29)
	(6-MeSPur)AdoCba	8.9	117	(30)
1-Amino-2-propanol	AdoCbl[—NHC(CH$_3$)$_2$CH$_2$O—]	(I$_{50}$ = 1.8)[b]	Inactive	(31)

[a] Abbreviations: Cbl, cobalamin; Cba, cobamide; Cbi, cobinamide; Ado, adenosyl; Bza, benzimidazolyl; Ade, aden-7-yl; Pur, purinyl; Me, methyl.
[b] I$_{50}$ represents a 50%-inhibition index.

Table IV. Coenzyme Activity of Analogs in Which

Analog[a]		Apparent K_m (μM)
Adenosyl-Cbl	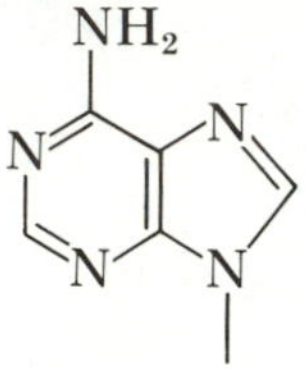	0.8
3-Isoadenosyl-Cbl	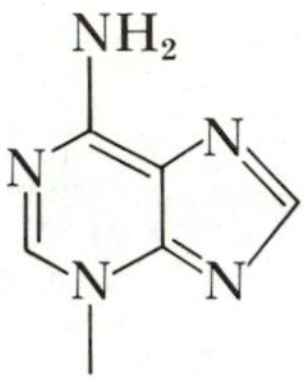	1.4
Nebulary-Cbl	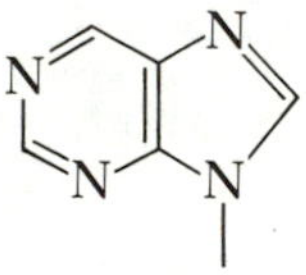	1.7
Inosyl-Cbl	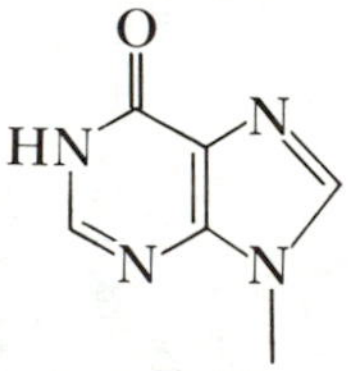	
Cytidyl-Cbl	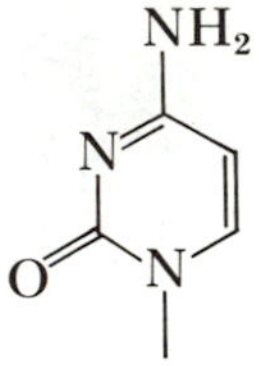	
Thymidyl-Cbl	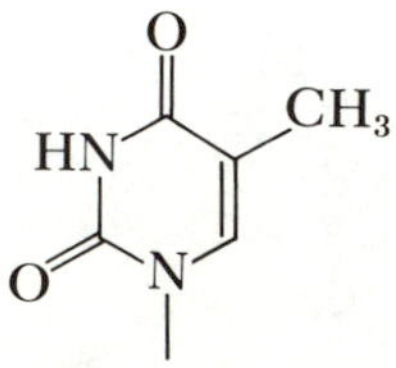	
Benzimidazoleribosyl-Cbl	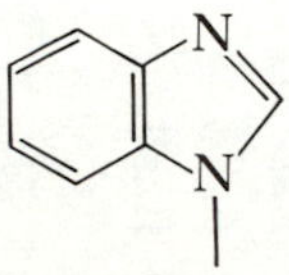	

[a] Cbl = cobalamin.

the Adenine Moiety of the Coβ (upper) Ligand is Modified.

Relative Activity (%)	Apparent K$_i$ (μM)	Inhibition[b] (%)	Reference
(100)	—	(0)	(34)
36	—	—	(34)
9	—	—	(34)
Inactive	22	72	(34)
Inactive	6.1	95	(34)
Inactive	—	—	(32)
Inactive	3.3	38	(34)

[b] After incubation of the apoenzyme with each analog (25μM) at 37°C for 10 min, the activity was assayed by 10-min reaction with 5μM adenosylcobalamin added.

Table V.   Coenzyme Activity of Analogs in Which the

Analog[a]

Adenosyl-Cbl

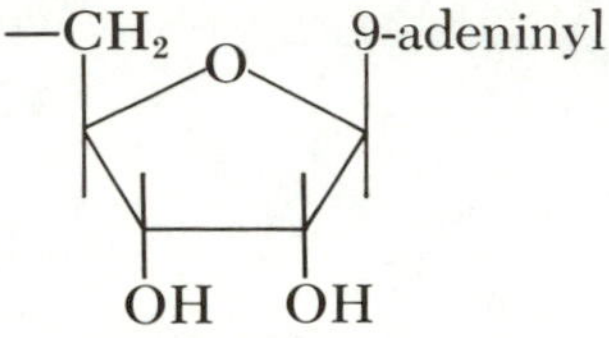

$-CH_2$ O   9-adeninyl

OH   OH

2'-Deoxyadenosyl-C

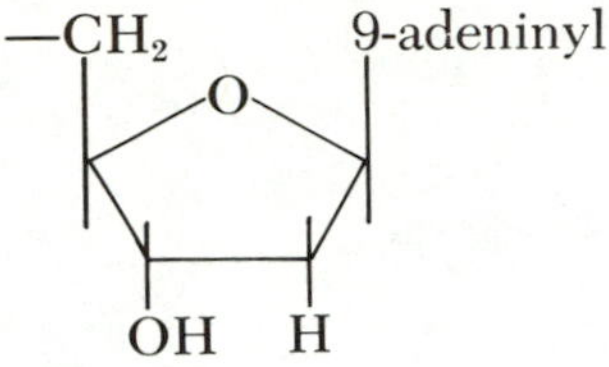

$-CH_2$ 9-adeninyl

O

OH   H

Ara-adenosyl-Cbl

$-CH_2$ 9-adeninyl

O

HO

OH

Aristeromycyl-Cbl

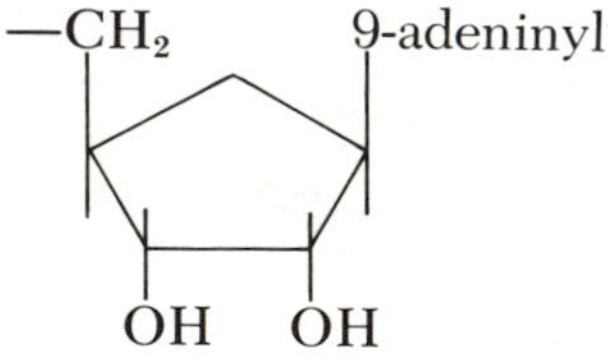

$-CH_2$ 9-adeninyl

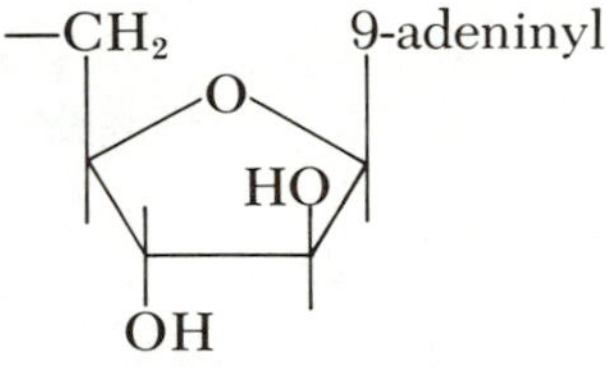

OH   OH

L-Adenosyl-Cbl

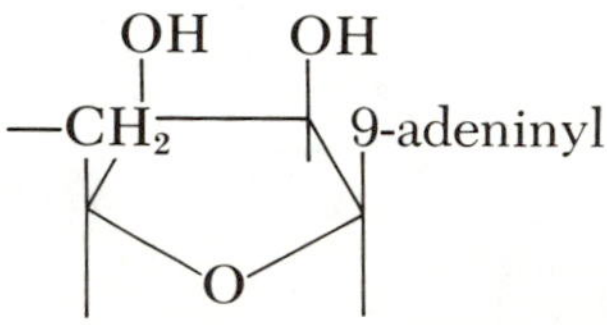

OH   OH

$-CH_2$ 9-adeninyl

O

Adenosylethyl-Cbl $-CH_2CH_2CH_2$ 9-adeninyl

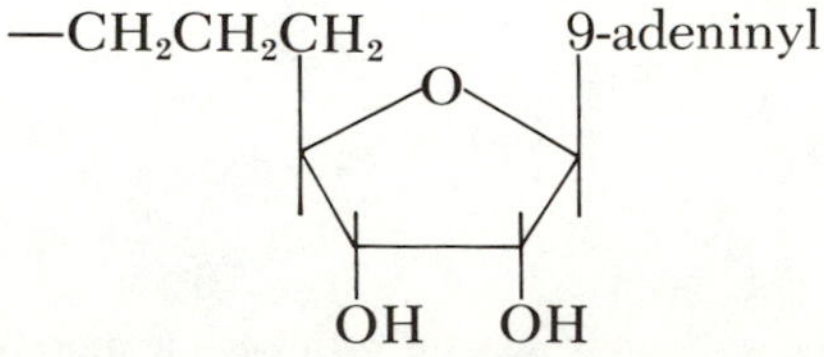

O

OH   OH

Ribose Moiety of the Coβ (upper) Ligand is Modified

Apparent K$_m$ (μM)	*Relative* Activity (%)	*Apparent* K$_i$ (μM)	*Inhibition*[b] (%)	*Reference*
0.1–0.8	(100)	—	(0)	(32, 34)
0.3	50	—	—	(32)
1.4	61[c]	—	—	(34)
1.1	42	—	—	(34)
	Inactive	1.2	100	(34)
	Inactive	0.6	100	(34)

Table V.

Analog[a]	*Apparent* $K_m(\mu M)$
Adeninylhexyl-Cbl	—$(CH_2)_6$-9-adeninyl
Adeninylpentyl-Cbl	—$(CH_2)_5$-9-adeninyl
Adeninylbutyl-Cbl	—$(CH_2)_4$-9-adeninyl
Adeninylpropyl-Cbl	—$(CH_2)_3$-9-adeninyl
Adeninylethyl-Cbl	—$(CH_2)_2$-9-adeninyl
Methyl-Cbl	—CH_3
Cyano-Cbl	—CN
Hydroxo-Cbl	—OH

[a] Cbl = cobalamin.
[b] After incubation of the apoenzyme with each analog (25μM) at 37°C for 10 min, the activity was assayed by 10-min reaction with 5μM adenosylcobalamin added.

analogs in which the D-ribose moiety is replaced by L-ribose or by alkyl chain of two to six carbons were quite inactive as coenzymes, but acted as competitive inhibitors with extremely high affinity for the enzyme.

It is not surprising that the analogs in which both adenine and ribose moieties are modified were inactive (methylcobalamin, cyanocobalamin, and hydroxocobalamin).

Spectroscopic properties of the complexes of diol dehydrase with various coenzyme analogs were compared (*34*). The spectrum of the enzyme–adenosylcobalamin complex in the absence of substrate was very similar to that of the free coenzyme. The addition of 1,2-propanediol to the complex caused a diminution of the absorption peak at 525 nm and the appearance of a new peak at 478 nm. This spectral change suggests that the C–Co bond of the coenzyme is cleaved homolytically to yield cob(II)alamin and an adenosyl radical. Like the native holoenzyme (enzyme–adenosylcobalamin complex), complexes with *ara*-adenosylcobalamin, 3-isoadenosylcobalamin, aristeromycyl-cobalamin, and nebularylcobalamin showed a cob(II)alamin-like absorption peak or shoulder at 478 nm in the presence of substrate ("reacting complex"). The intensity of the shoulder, which represents a steady state concentration of cob(II)alamin in the reaction, may correspond to the coenzyme activity of these analogs (except aristeromycylcobalamin). Spectra of the inactive analogs did not change appreciably upon binding to the enzyme even in the presence of substrate (1,2-propanediol). Upon anaerobic photolysis, the spectra of complexes with inosylcobalamin, cytidylcobalamin, and benzimidazoleribosylcobalamin changed to that of cob(II)alamin. Subsequent aeration did not cause a spectral change indicating that the enzyme-bound cob(II)alamin is unusually resistant to oxidation. On

Continued

Relative Activity (%)	*Apparent* $K_i(\mu M)$	*Inhibition*[b] (%)	*Reference*
Inactive		100	(34)
Inactive	0.3	100	(34)
Inactive	0.5	100	(34)
Inactive		100	(34)
Inactive		100	(34)
Inactive	0.7	100	(34)
Inactive	1.9	100	(34)
Inactive	0.6	100	

[c] Obtained on the assumption that the reaction rate with this analog was linear for at least 5 min.

the other hand, complexes with adenosylethylcobalamin, adeninyl-propylcobalamin, and methylcobalamin were converted to the hydroxocobalamin–enzyme complex by photolysis and subsequent aeration. Complexes with adeninylpentylcobalamin and L-adenosyl-cobalamin were stable to light under the same conditions.

Relationship between the Structure of the Coβ (Upper) Ligand and Activation of the C–Co Bond and the Binding to the Enzyme. The above-mentioned data demonstrate that both the adenine and the sugar portions of the nucleoside ligand are essential for coenzyme activity. The analogs can be grouped in four categories. Type 1 cobalamins have coenzyme activity and show high affinity for the enzyme ($K_m \sim 1\mu M$). These analogs contain an adenine or a purine and an intact or slightly modified D-ribose moiety. Type 2 cobalamins are the analogs that contain bases other than adenine. They do not serve as coenzymes and are weak competitive inhibitors with respect to adenosylcobalamin. The bound analogs of this type can be at least partially displaced by the normal coenzyme, suggesting reversibility of the binding of these analogs to the enzyme. Type 3 cobalamins are the analogs whose D-ribose is replaced by L-ribose or other groups. They are inactive as coenzymes but function as very strong competitive inhibitors ($K_i < 1\mu M$) with respect to adenosylcobalamin. Type 4 cobalamins are the analogs that are modified on both adenine and ribose moieties. They show fairly high affinity for the enzyme ($K_i \sim 1\mu M$). Both Type 3 and Type 4 cobalamins form an irreversible complex with enzyme. From the relationship with structural features of the analogs, a possible mechanism was postulated for interactions of the coenzyme and its analogs with enzyme and for activation of the C–Co bond in the resulting complex (Figure 2). This mechanism is based on the assumption that apoprotein has the sites at which it inter-

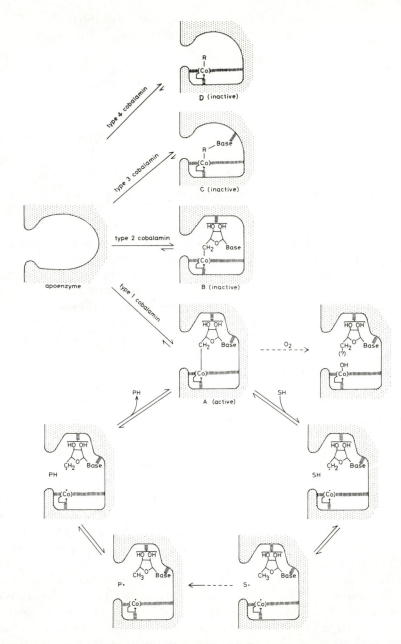

Figure 2. Proposed mechanism of the binding of adenosylcobalamin and its analogs to diol dehydrase and of activation of the C–Co bond in the resulting complex: SH = substrate; PH = product

acts with the cobalamin moiety (corrin ring, its sidechains, and nucleotide loop) and the adenosyl moiety (adenine and ribose) of the coenzyme. Type 1 cobalamins bind tightly to the enzyme by interaction at both the D-ribose and base moieties in addition to the cobalamin moiety. These interactions lead to activation and cleavage of the C–Co bond. In the absence of substrate, the activated C–Co bond is cleaved by reaction with oxygen resulting in irreversible inactivation of the holoenzyme. With Type 2 analogs, the affinity for the enzyme is relatively weak owing to inadequate interaction at the base moiety. As a result, the C–Co bond is not sufficiently activated. Type 3 analogs are able to interact with the enzyme only via the adenine and cobalamin moieties. Activation of the C–Co bond of these analogs is not possible because they are lacking a ribofuranosyl moiety or a structure closely resembling it. A possible explanation for the very tight interaction between this type of analog and the enzyme is that the coenzyme binding site does not need to undergo conformational change to accommodate the more flexible alkyl chain. Type 4 cobalamins can interact with apoenzyme only at cobalamin moiety. The affinity of this type of cobalamins for enzyme is still very high. This indicates that the contribution of the interactions at adenosyl moiety to the binding of the coenzyme to the enzyme is rather small. Needless to say, the activation of the C–Co bond by enzyme is impossible in this case.

Immobilized Derivatives of Cobalamins and Their Use as Affinity Adsorbents for a Study of Enzyme–Coenzyme Interaction

Immobilized derivatives of cobalamins are useful not only for purification of cobalamin-dependent enzymes or cobalamin-binding proteins but also for a basic study of biospecific interactions between cobalamins and apoenzyme or cobalamin binders. To gain insight into the interacting positions of cobalamins with the apoprotein, several types of immobilized cobalamin derivatives were prepared newly and examined for their effectiveness as affinity adsorbents for diol dehydrase (37). Partial structures of the agarose-gel-bound cobalamins used in this study are illustrated in Figure 3. In I–IV gels, corrinoids are immobilized through a carboxyl group of the peripheral sidechains of the corrin ring. Their Coβ axial ligand is adenosyl, cytidyl, methyl, cyano, or hydroxyl group. V is a gel in which cobalamin is immobilized through an ester linkage at 5'-OH of the ribose moiety of the nucleotide ligand. In VI–IX, cobalamins are attached to agarose gel through the Coβ (upper) ligand of the cobalt ion. Affinity chromatography using these corrinoid gels as adsorbents was performed as follows: Crude extracts of *Klebsiella pneumoniae* (formerly *Aerobacter aerogenes*) ATCC 8724 containing the apoenzyme of diol dehydrase

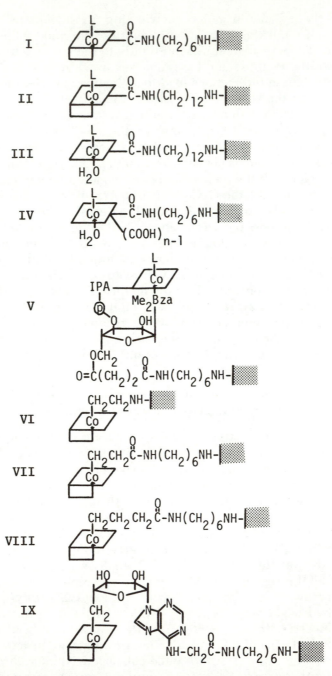

Figure 3. Partial structures of the immobilized corrinoid derivatives used in this study: L = adenosyl, cytidyl, methyl, cyano, hydroxo, etc.; IPA = 1-amino-2-propanol; Me₂Bza = 5,6-dimethylbenzimidazole

were applied to a corrinoid gel. To promote the binding, the suspension was incubated at 37°C for 1 to 2 hr with gentle stirring. The suspension was then poured into a small column and developed successively with 0.1M potassium phosphate buffer (pH 8.0) containing 2% propanediol (Effluent), 0.5M potassium phosphate buffer (pH 8.0) containing 2% propanediol (Eluate 1), 0.3M Tris–HCl buffer (pH 8.0) containing 2% propanediol (Eluate 2), 0.3M Tris–HCl buffer (pH 8.0) (Eluate 3), and finally with 0.1M acetic acid containing 2% propanediol (Eluate 4). A typical example with an adenosyl form of **II** is

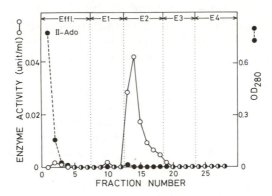

Journal of Biological Chemistry

Figure 4.   Affinity chromatography of diol dehydrase on the adenosyl form of **II** *(37). About 1 unit of enzyme was applied to 1 mL of packed corrinoid gel in 0.1 M potassium phosphate buffer (pH 8.0) containing 2% 1,2-propanediol, in a total volume of 2 mL. Affinity chromatography was carried out as described in the text. Two-milliliter fractions were collected.*

shown in Figure 4. Diol dehydrase was adsorbed completely on the gel, and was not eluted by increasing the buffer concentration. The enzyme could be eluted with Tris–HCl buffer or triethanolamine–HCl buffer (pH 8.0) containing substrate. The addition of KCl in Tris–HCl buffer depressed the elution significantly. These findings indicate that removal of K^+ from the system affected dissociation of the enzyme from the affinity adsorbent, and were consistent with the fact that K^+ or other monovalent cations of similar size are required absolutely for the binding of adenosylcobalamin or its analogs to diol dehydrase (*9, 10, 11*). Such elution, based on the dependence of the enzyme–coenzyme binding in an ionic environment, would give a new and interesting example of a selective and mild elution method, and might be applicable to some other systems.

Of the corrinoid–agarose gels mentioned above, the corrinoids immobilized through sidechains of the corrin ring were the most effective affinity adsorbents. Although the enzyme was adsorbed completely on **II** irrespective of Coβ ligands, some leakage was observed with **I**, and its degree was dependent on the Coβ ligands. With its cyano form, the leakage was more significant than that with its adenosyl form. These results suggest that a spacer of sufficient length is necessary for the firm binding, and also that the enzyme binds more tightly to an adenosyl form adsorbent. This indicates that the adenosyl moiety of the coenzyme is also important for specific interaction, and is consistent with the finding that **IX** was not an effective affinity adsorbent. That the enzyme was not adsorbed by **V** indicates that the specific interaction between enzyme and coenzyme is destroyed by attachment at the ribose moiety of the nucleotide loop. This is in agreement with the observation that a cyano–aqua form of nucleotide-lacking corrinoids immobilized via sidechains of the corrin ring with 1,12-diaminododecane as a spacer (*III*—CN) did not serve as an affinity adsorbent, and is suggestive of the important participation of the lower nucleotide ligand in the binding of cobalamins to the enzyme. Among the cobalamins immobilized through Coβ ligands, only **VIII** was a partially effective affinity adsorbent for diol dehydrase. **VI** was not effective, presumably because the distance between the agarose matrix and the cobalamin moiety is too short. **VII** was intrinsically unstable and decomposed under the conditions for affinity chromatography. The results of affinity chromatography provide important information for understanding of the enzyme–coenzyme interactions.

Concluding Remarks

It must be kept in mind that in some cases quantitative comparison of kinetic parameters for coenzyme analogs listed in the tables may be invalid, because the conditions employed by each investigator are not the same. However, the data obtained so far permit some speculation on the nature of the interactions between apoenzyme and coenzyme. It is likely that the high affinity of the enzyme for adenosylcobalamin with an apparent K_m less than $1\mu M$ would be possible as a result of integration of multiple weak interactions, such as hydrogen bond, hydrophobic bonds, ionic bonds, and van der Waals forces, at a number of different sites of the coenzyme with the enzyme protein. Thus, even if some interactions become impossible by modification of some group(s) of the coenzyme, analogs still are able to bind to the enzyme tightly. Based on the assumption that the apoprotein has a binding site for the adenosyl group in the Coβ position, it is probable that interactions between enzyme and coenzyme at the adenosyl moiety as well as at the peripheral amide sidechains of the corrin ring play essential roles in activation of the C–Co bond of the coenzyme. This assumption seems reasonable, since we have observed that adenosine is an inhibitor of the diol-dehydrase reaction (*38*). Furthermore, direct evidence for the binding site of diol–dehydrase apoenzyme for adenosyl portion of the coenzyme recently has been obtained (*39*).

Addendum

Recently, Anton et al. have proposed, from the analysis of the carbon-13 NMR spectra, that the tentatively identified cyanocobalamin-*b*-, -*e*-, and -*d*-carboxylic acid (Table II) should be cyanocobalamin-*d*-, -*b*-, and -*e*-carboxylic acid, respectively (*40*).

Literature Cited

1. Barker, H. A.; Weissbach, H.; Smyth, R. D. *Proc. Nat. Acad. Sci. USA.* **1958**, *44*, 1093.
2. Weissbach, H.; Toohey, J. I.; Barker, H. A. *Proc. Nat. Acad. Sci. USA.* **1959**, *45*, 521.
3. Abeles, R. H.; Lee, H. A. Jr. *J. Biol. Chem.* **1961**, *236*, 2347.
4. Lee, H. A. Jr.; Abeles, R. H. *J. Biol. Chem.* **1963**, *238*, 2367.
5. Toraya, T.; Shirakashi, T.; Kosuga, T.; Fukui, S. *Biochem. Biophys. Res. Commun.* **1976**, *69*, 475.
6. Poznanskaja, A. A.; Tanizawa, K.; Soda, K.; Toraya, T.; Fukui, S. *Arch. Biochem. Biophys.* **1979**, *194*, 379.
7. Toraya, T.; Uesaka, M.; Fukui, S. *Biochem. Biophys. Res. Commun.* **1973**, *52*, 350.
8. Toraya, T.; Uesaka, M.; Fukui, S. *Biochemistry* **1974**, *13*, 3895.
9. Toraya, T.; Sugimoto, Y.; Tamao, Y.; Shimizu, S.; Fukui, S. *Biochem. Biophys. Res. Commun.* **1970**, *41*, 1314.
10. Toraya, T.; Sugimoto, Y.; Tamao, Y.; Shimizu, S.; Fukui, S. *Biochemistry* **1971**, *10*, 3475.
11. Toraya, T.; Kondo, M.; Isemura, Y.; Fukui, S. *Biochemistry* **1972**, *11*, 2599.
12. Toraya, T.; Fukui, S. *Biochim. Biophys. Acta* **1972**, *284*, 536.
13. Abeles, R. H. In "The Enzymes"; Boyer, P. D., Ed.; Academic: New York, 1971; Vol. 5, p. 481.
14. Abeles, R. H. *Adv. Chem. Ser.* **1971**, *100*, 346.
15. Abeles, R. H.; Dolphin, D. *Acc. Chem. Res.* **1976**, *9*, 114.
16. Schrauzer, G. N. *Adv. Chem. Ser.* **1971**, *100*, 1.
17. Corey, E. J.; Cooper, N. J.; Green, M. L. H. *Proc. Natl. Acad. Sci. U.S.A.* **1977**, *74*, 811.
18. Tamao, Y.; Morikawa, Y.; Shimizu, S.; Fukui, S. *Biochim. Biophys. Acta* **1968**, *151*, 260.
19. Tkachuck, R. D.; Grant, M. E.; Hogenkamp, H. P. C. *Biochemistry* **1974**, *13*, 2645.
20. Stoeckli–Evans, H.; Edmond, E.; Hodgkin, D. C. *J. Chem. Soc. Perkin Trans. 2* **1972**, 605.
21. Toraya, T.; Shirakashi, T.; Fukui, S.; Hogenkamp, H. P. C. *Biochemistry* **1975**, *14*, 3949.
22. Rapp, P. *Hoppe–Seyler's Z. Physiol. Chem.* **1972**, *353*, 887.
23. Toraya, T.; Krodel, E.; Mildvan, A. S.; Abeles, R. H. *Biochemistry* **1979**, *18*, 417.
24. Tamao, Y.; Morikawa, Y.; Shimizu, S.; Fukui, S. *Bitamin* **1970**, *41*, 45.
25. Rapp, P.; Hildebrand, R. *Hoppe–Seyler's Z. Physiol. Chem.* **1972**, *353*, 1141.
26. Kato, T.; Shimizu, S.; Fukui, S. *J. Vitaminol. Kyoto* **1964**, *10*, 89.
27. Toraya, T.; Ohashi, K.; Ueno, H.; Fukui, S. *Bioinorg. Chem.* **1975**, *4*, 245.
28. Hayashi, M.; Kamikubo, T. *FEBS Lett.* **1971**, *15*, 213.
29. Kamikubo, T.; Narahara, H.; Murai, K.; Kumamoto, N. *Biochem. Z.* **1966**, *346*, 159.

30. Uchida, Y.; Hayashi, M.; Kamikubo, T. *Bitamin* **1973,** *47,* 27.
31. Friedrich, W.; Heinrich, H. C.; Königk, E.; Schulze, P. *Ann. N. Y. Acad. Sci.* **1964,** *112,* 601.
32. Hogenkamp, H. P. C.; Oikawa, T. G. *J. Biol. Chem.* **1964,** *239,* 1911.
33. Kerwar, S. S.; Smith, T. A.; Abeles, R. H. *J. Biol. Chem.* **1970,** *245,* 1169.
34. Toraya, T.; Ushio, K.; Fukui, S.; Hogenkamp, H. P. C. *J. Biol. Chem.* **1977,** *252,* 963.
35. Zagalak, B.; Pawelkiewicz, J. *Acta Biochim. Pol.* **1965,** *12,* 103.
36. Sando, G. N.; Blakley, R. L.; Hogenkamp, H. P. C.; Hoffmann, P. J. *J. Biol. Chem.* **1975,** *250,* 8774.
37. Toraya, T.; Fukui, S. *J. Biol. Chem.,* in press.
38. Yamane, T.; Shimizu, S.; Fukui, S. *Biochim. Biophys. Acta* **1965,** *110,* 616.
39. Toraya, T.; Abeles, R. H., unpublished data.
40. Anton, D. L.; Hogenkamp, H. P. C.; Walker, T. E.; Matwiyoff, N. A. *J. Amer. Chem. Soc.* **1980,** *102,* 2215.

RECEIVED May 21, 1979.

Some Mechanistic Aspects of Coenzyme B_{12}-Dependent Rearrangements

Homolytic Cleavage of Metal–Carbon Bonds

JACK HALPERN

Department of Chemistry, University of Chicago, Chicago, IL 60637

A widely accepted mechanism of coenzyme B_{12}-dependent rearrangements encompasses, as the initial step, the homolytic cleavage of the carbon–cobalt bond to generate the 5'-deoxyadenosyl radical. The thermodynamic and kinetic aspects of this and related processes involving the homolysis of transition metal–carbon bonds are discussed.

Coenzyme B_{12} (5'-deoxyadenosylcobalamin, abbreviated RCH_2—B_{12}) serves as a cofactor for a variety of enzymic reactions, a common feature of which involves the 1,2-interchange of an H atom and another group (X = OH, NH_2, $CH(NH_2)COOH$, etc.) on adjacent carbon atoms (Equation 1) (*1, 2*).

$$\underset{C_1}{\overset{H}{|}} - \underset{C_2}{\overset{X}{|}} \rightleftarrows \underset{C_1}{\overset{X}{|}} - \underset{C_2}{\overset{H}{|}} \tag{1}$$

A widely accepted mechanistic interpretation of these reactions, supported by a variety of evidence from studies on the enzymic processes as well as on model systems, is depicted by Equations 2 and 3 (*1, 2, 3*).

$$RCH_2-B_{12} \underset{\text{enzyme}}{\overset{\text{enzyme}}{\rightleftharpoons}} RCH_2\cdot + B_{12r} \tag{2}$$

0-8412-0514-0/80/33-191-165$05.00/0
© 1980 American Chemical Society

$$
\begin{array}{c}
\overset{H}{\underset{|}{-C_1}}\overset{X}{\underset{|}{-C_2}}- \xrightarrow[(-RCH_3)]{RCH_2\cdot} \left[\overset{X}{\underset{|}{-\dot{C}_1}}\overset{}{\underset{|}{-C_2}}- \right] \rightarrow \\[2ex]
\left[\overset{X}{\underset{|}{-C_1}}\overset{}{\underset{|}{-\dot{C}_2}}- \right] \xrightarrow[(-RCH_2\cdot)]{RCH_3} \overset{X}{\underset{|}{-C_1}}\overset{H}{\underset{|}{-C_2}}- \quad (3)
\end{array}
$$

This mechanism encompasses the following sequence of steps: (i) enzyme-induced homolysis of the Co–C bond to generate cob(II)alamin (i.e., vitamin B_{12r}) and a 5'-deoxyadenosyl radical (abbreviated $RCH_2\cdot$); (ii) H-atom abstraction from the substrate to generate a substrate radical and 5'-deoxyadenosine (RCH_3); (iii) rearrangement of the resulting substrate radical (through a mechanism that is not fully understood and that probably differs from substrate to substrate); and (iv) abstraction of an H atom from RCH_3 by the rearranged radical to complete the rearrangement reaction.

While the possible involvement of the cobalt complex in the substrate radical rearrangement step has been suggested (1), the evidence for this is inconclusive. At this stage it appears that the principal, if not only, role of the organometallic cofactor (i.e., of coenzyme B_{12}) in these reactions is to serve as a precursor for an organic free radical, which presumably implies facile homolytic dissociation of the Co–alkyl bond.

A troublesome feature of this mechanistic interpretation is the absence of direct supporting evidence that the Co–C bond in coenzyme B_{12} (whose dissociation energy has not yet been determined) is sufficiently weak that facile homolysis under the mild conditions of the enzymic reactions is a plausible process. In fact, alkylcobalamins, including coenzyme B_{12}, exhibit considerable thermal stability and typically do not decompose at measurable rates, in the absence of light or reagents such as O_2, until fairly elevated temperatures ($\geqslant$ 200°C for methylcobalamin) (4). Among the possible interpretations of this behavior are:

1. The above mechanistic interpretation is incorrect and Co–C bond homolysis is not involved in coenzyme B_{12}-dependent rearrangements.

2. Co–C bond weakening and homolysis is induced by interaction of the coenzyme with the enzyme (e.g., through *trans*-axial ligand substitution or conformational distortion of the corrin ring).

3. Co–C bond homolysis, (i.e., according to Equation 1) occurs spontaneously under mild conditions, but is reversible and, hence, is not reflected in net decomposition.

(In this case, decomposition should be induced by appropriate radical traps such as O$_2$. The results of numerous studies on the thermal and photochemical stabilities of alkylcobalamins as well as of model compounds are consistent with this.) (*5, 6, 7, 8*).

This chapter addresses the following themes: The estimation of the metal–alkyl bond dissociation energies of coenzyme B$_{12}$ and related compounds. The factors that influence metal–alkyl bond dissociation energies of organocobalamins and related compounds. Reactions between free radicals and metal complexes.

The third theme is of significance in this context, not only because the proposed mechanism of coenzyme B$_{12}$-dependent rearrangements (Equations 2 and 3) involves the generation of organic free radicals in the presence of a metal complex (i.e., vitamin B$_{12r}$), but also because of the potential utility of metal complexes as radical traps in studies involving kinetic approaches to the estimation of metal–alkyl bond dissociation energies.

Some Reactions of Free Radicals with Metal Complexes

Low-spin cobalt(II) complexes characteristically react with organic halides to generate free radicals through halogen abstraction processes of the type depicted by Equation 4. Such reactions, which have been identified for pentacyanocobaltate(II) (*9, 10, 11*), for *bis*(dioximato)– and Schiffs-base–cobalt(II) complexes (*12, 13, 14*) as well as for vitamin B$_{12r}$ (*15*), are among the cleanest thermal routes for generating organic free radicals in solution, particularly for the purpose of studying the reactions of free radicals with metal complexes.

$$L_5Co^{II} + R-X \rightarrow L_5Co^{III}X + R \cdot \qquad (4)$$

The overall course of reaction of pentacyanocobaltate(II) with methyl and benzyl halides is depicted by the scheme of Equations 5, 6, and 7, in which the initial halogen abstraction step is rate determining.

$$[Co(CN)_5]^{3-} + R-X \longrightarrow [X-Co(CN)_5]^{3-} + R \cdot \qquad (5)$$

$$[Co(CN)_5]^{3-} + R \cdot \xrightarrow{\text{fast}} [R-Co(CN)_5]^{3-} \qquad (6)$$

Overall reaction: $2[Co(CN)_5]^{3-} + R-X \rightarrow$
$$[X-Co(CN)_5]^{3-} + [R-Co(CN)_5]^{3-} \qquad (7)$$

For radicals containing β-hydrogen atoms such as ethyl or isopropyl, β-hydrogen abstraction by $[Co(CN)_5]^{3-}$ competes with combination of $R \cdot$ and $[Co(Cn_5)]^{3-}$ resulting in the formation of $[H—Co(CN)_5]^{3-}$ and an olefin through parallel paths as depicted below: for $R = C_2H_5$, $k_{9a}/k_{9b} \sim 4$ (11).

$$[Co(CN)_5]^{3-} + C_2H_5X \rightarrow [X—Co(CN)_3]^{3-} + C_2H_5 \cdot \qquad (8)$$

$$[Co(CN)_5]^{3-} + C_2H_5 \cdot \begin{cases} \xrightarrow{k_{9a}} [C_2H_5—Co(CN)_5]^{3-} & (9a) \\ \xrightarrow{k_{9b}} [H—Co(CN)_5]^{3-} + CH_2{=}CH_2 & (9b) \end{cases}$$

Finally, when $[Co(CN)_5]^{3-}$ is reacted with organic halides in the presence of $[H—Co(CN)_5]^{3-}$, efficient trapping of $R \cdot$ by H-atom abstraction from the metal hydride occurs (Equation 10), resulting in the catalytic cycle depicted by Equations 5, 10, and 11.

$$[Co(CN)_5]^{3-} + R—X \rightarrow [X—Co(CN)_5]^{3-} + R \cdot \qquad (5)$$

$$R \cdot + [H—Co(CN)_5]^{3-} \xrightarrow{fast} [Co(CN)_5]^{3-} + R—H \qquad (10)$$

$$\overline{[H—Co(CN)_5]^{3-} + R—X \rightarrow [X—Co(CN)_5]^{3-} + R—H \qquad (11)}$$

The reactions in which the free radicals $(R \cdot)$ are consumed in these systems (i.e., Equations 12, 13, and 14, where $L_nM \cdot = [Co(CN)_5]^{3-}$ for the cases cited), are all sufficiently fast that they compete effectively with the radical-coupling process, $2R \cdot \rightarrow R_2$, and no formation of R_2 can be detected. From such competition studies it can be deduced that Reactions 12, 13, and 14 must all proceed at rates that are close to diffusion-controlled and that the activation barriers for such reactions do not exceed a few kcal/mol.

$$L_nM \cdot + R \cdot \rightarrow L_nM—R \qquad (12)$$

$$L_nM \cdot + RCH_2CH_2 \cdot \rightarrow L_nM—H + RCH{=}CH_2 \qquad (13)$$

$$R \cdot + L_nM—H \rightarrow R—H + L_nM \cdot \qquad (14)$$

Reaction 12 is the reverse of the metal–alkyl bond dissociation process (Equation 15). Hence, the activation enthalpies $(\Delta H^\dagger)$ of such homolytic bond dissociation reactions are expected to be close to the corresponding bond-dissociation energies.

$$L_nM—R \rightarrow L_nM \cdot + R \cdot \qquad (15)$$

As elaborated below, the facile occurrence of Reactions 13 and 14 is important in this context because of the potential usefulness of these reactions as radical trapping processes in kinetic approaches to the determination of metal–alkyl bond dissociation energies.

Approaches to the Estimation of Metal–Alkyl Bond-Dissociation Energies

Few transition-metal–alkyl bond-dissociation energies are known reliably (*16*). Potential approaches to the estimation of such dissociation energies encompass the following:

1. *Thermochemical.* Application to the estimation of the enthalpy of a process such as that depicted by Equation 15 requires determination of the heats of formation of L_nM-R, $R\cdot$, and $L_nM\cdot$. The latter usually is not accessible to measurement although it is in the case of alkylcobalamins (where $L_nM\cdot$ corresponds to vitamin B_{12r}, a stable and accessible compound). Thus, thermochemical approaches, in principle, are potentially applicable to the estimation of the Co–C bond dissociation energy in coenzyme B_{12}. However, the practical difficulties are considerable and the probable accuracy of the result is questionable.

2. *Kinetic.* This approach entails determination of the activation enthalpy $(\Delta H^{\ddagger})$ of the homolytic dissociation process depicted by Equation 15 and the identification of $\Delta H^{\ddagger}$ with the corresponding bond-dissociation energy (i.e., assuming that the activation energy of the reverse process, namely the recombination of $L_nM\cdot$ and $R\cdot$, is small). As noted earlier, the latter assumption probably is valid generally. Successful application of this approach may be compromised by interference from other accompanying modes of decomposition or by complicating secondary reactions (including recombination to form L_nM-R) of the initial radical products unless the latter are scavenged efficiently by appropriate radical traps.

3. *Equilibrium.* Reactions of the type depicted by Equation 16 have been identified as synthetic routes to organocobalt compounds (*17, 18*). Cobalt–alkyl bond-dissociation energies could be deduced from the enthalpies of such reactions if the latter could be determined, for example from the temperature dependence of the corresponding equilibrium constants.

$$L_n\mathrm{Co^{II}} + \ \ \overset{\diagdown}{\underset{\diagup}{}}C{=}C\overset{\diagup}{\underset{\diagdown}{}} \ + \tfrac{1}{2}\mathrm{H_2} \rightleftharpoons L_n\mathrm{Co}-\overset{\diagdown}{\underset{\diagup}{}}C-C\overset{\diagup}{\underset{\diagdown}{}}-\mathrm{H} \qquad (16)$$

4. *Photochemical.* Determination of the threshold wavelength for the photolytic dissociation of a metal–alkyl bond yields an upper limit for the corresponding thermal bond dissociation energy (*5, 19*), but the assumption that the photochemical threshold approximates the bond dissociation energy does not appear to be warranted.

Among these alternative approaches, 1 and 2 are considered the most promising for the determination of metal–alkyl bond-dissociation energies in coenzyme B_{12} and related compounds. The application of these approaches will be elaborated below.

Transferability of Information about Metal–Alkyl Bond-Dissociation Energies

Failure to achieve a reliable determination of the Co–C bond-dissociation energy in coenzyme B_{12} reflects significant obstacles associated with this task. Accordingly, it seems appropriate to explore indirect approaches to estimating this energy, notably through extrapolating information about metal–alkyl bond-dissociation energies in related compounds.

Unfortunately, hardly any other transition-metal–alkyl bond-dissociation energies are known reliably (16). Our approach has encompassed attempts to estimate such dissociation energies for a wide variety of transition-metal–alkyl compounds, including both recognizable organocobalt B_{12} analogues as well as less directly related compounds. The objectives of these studies have been twofold: to test approaches to the estimation of metal–alkyl bond-dissociation energies on appropriate test compounds with a view to applying such approaches to alkylcobalamins; and to accumulate systematic information about metal–alkyl bond-dissociation energies for a variety of organometallic compounds with a view to identifying trends from which the Co–C bond-dissociation energy of coenzyme B_{12} might be deduced by extrapolation and/or interpolation.

Our studies have encompassed the following series of compounds ($R—ML_n$), which are arranged below in order of decreasing apparent "stabilities" of the corresponding hydrides ($H—ML_n$). The values in parentheses below the compounds are estimates of the M–H bond dissociation energies (kcal/mol) of the hydrides deduced from the temperature dependence and/or position of the equilibrium corresponding to Equation 17 (20, 21, 22, 23). It seems likely that, at least in the absence of steric factors, the corresponding metal–alkyl bond-dissociation energies (e.g., of $CH_3—ML_n$) would follow a similar trend ($DMGH_2$ = dimethylglyoxime).

$$[R—Mn(CO)_5] > [R—Co(CN)_5]^{3-} >$$
$$(\gtrsim 60) \qquad\qquad (57)$$

$$[R—Co(DMGH)_2L] > [R—Cobalamin]$$
$$(\sim 53) \qquad\qquad\qquad (\gtrsim 50)$$

$$2 \cdot ML_n + H_2 \rightleftarrows 2H—ML_n \qquad\qquad (17)$$

Our initial studies, in the context of this approach, relate to attempts to estimate metal–alkyl bond-dissociation energies in compounds of the type R—Mn(CO)$_5$ and R—Mn(CO)$_4$(PR$'_3$). Among the reasons for this choice of compounds are the following: the CH$_3$–Mn(CO)$_5$ bond-dissociation energy (~ 30 kcal/mol) is one of the few transition-metal–alkyl bond-dissociation energies to have been determined to date *(24)*; the relatively high stabilities of alkyl- and benzyl-manganese pentacarbonyl compounds (compared, for example, with the corresponding cobalamins) afford some advantages in experimental convenience; the high thermal stability and accessibility of the hydride, H—Mn(CO)$_5$, enables it to be used as a radical-trapping agent in experiments involving kinetic approaches to the estimation of metal–alkyl bond-dissociation energies in the corresponding R—Mn(CO)$_5$ compounds.

Kinetic Approaches to the Estimation of Metal–Alkyl Bond-Dissociation Energies

C$_6$H$_5$CH$_2$—Mn(CO)$_5$ reacts cleanly with H—Mn(CO)$_5$, at rates conveniently measurable in the temperature range 40°–80°C, according to the stoichiometry of Equation 18 and the first-order rate law, Equation 19, where k_{19} (2.0×10^{-5}sec^{-1} at 45°C) is independent of the concentration of H—Mn(CO)$_5$. Measurements of the temperature dependence of k_{19} yield the activation parameters, $\Delta H^\ddagger_{19} = 25$ kcal/mol and $\Delta S^\ddagger_{19} = 0$ cal/(mol deg) *(25)*.

$$C_6H_5CH_2\text{—Mn(CO)}_5 + H\text{—Mn(CO)}_5 \rightarrow C_6H_5CH_3 + Mn_2(CO)_{10} \quad (18)$$

$$\text{Rate} = k_{19}[C_6H_5CH_2\text{—Mn(CO)}_5] \quad (19)$$

These observations imply a unimolecular rate-determining reaction of C$_6$H$_5$CH$_2$Mn(CO)$_5$ to generate an intermediate that reacts rapidly with HMn(CO)$_5$ to form the observed products. Possible candidate reactions for the rate-determining step are: loss of CO to form C$_6$H$_5$CH$_2$Mn(CO)$_4$; migratory insertion rearrangement to form C$_6$H$_5$CH$_2$(CO)Mn(CO)$_4$; and homolytic Mn–C bond dissociation to form C$_6$H$_5$CH$_2\cdot$ and $\cdot$Mn(CO)$_5$. The last possibility, which is of particular interest in this context, would lead to the mechanistic sequence of Equations 20, 21, and 22.

$$C_6H_5CH_2\text{—Mn(CO)}_5 \xrightarrow{k_{19}} C_6H_5CH_2\cdot + \cdot Mn(CO)_5 \quad (20)$$

$$C_6H_5CH_2\cdot + H\text{—Mn(CO)}_5 \xrightarrow{\text{fast}} C_6H_5CH_3 + \cdot Mn(CO)_5 \quad (21)$$

$$2 \cdot Mn(CO)_5 \xrightarrow{\text{fast}} Mn_2(CO)_{10} \quad (22)$$

The identification of $\Delta H^{\ddagger}_{19}$ with the $C_6H_5CH_2-Mn(CO)_5$ bond dissociation energy, according to this interpretation, yields a value of approximately 25 kcal/mol for the latter. This is consistent with the value expected from the thermochemically determined bond-dissociation energy (~ 30 kcal/mol) of $CH_3-Mn(CO)_5$. If this interpretation is invoked, a puzzling issue is the observation that the thermal decomposition of $C_6H_5CH_2-Mn(CO)_5$, in the absence of $H-Mn(CO)_5$ (or other radical traps), is much slower than would correspond to k_{19}. This could imply that, in the absence of such radical-trapping agents, selective recombination of $C_6H_5CH_2\cdot$ and $\cdot Mn(CO)_5$ to re-form $C_6H_5CH_2-Mn(CO)_5$ is favored relative to the self-coupling reactions to form $(C_6H_5CH_2)_2$ and $Mn_2(CO)_{10}$. The reasons for such behavior are unclear and the investigation of this and related systems is continuing. The issue is pertinent directly to the behavior of coenzyme B_{12} and other alkylcobalamins that also exhibit higher apparent thermal (and photochemical) stabilities in the absence, than in the presence, of radical-trapping agents.

The direct extension of similar approaches to other systems, notably alkyl-cobalamins and related alkyl-cobalt compounds, is constrained by the instability of the corresponding hydrides, which are consequently unsuitable as radical traps. Modification of the approach to utilize other radical traps (e.g., O_2) potentially is feasible but needs further investigation in view of possible complicating features such as reaction between the trapping agent and the parent metal-alkyl and ambiguities between homolytic dissociation and other (e.g., concerted) mechanisms (26, 27).

One possible variant of the kinetic approach to the estimation of metal–alkyl bond-dissociation energies, which potentially is applicable to alkyl groups having β-hydrogen atoms, involves hydrogen abstraction, Equation 13, as the radical-quenching step. This may be a feature of some observations that we have made on the thermal decompositions of certain organobis(dimethylglyoximato)cobalt compounds ($R = CH_3$, $CH_2C_6H_5$, C_6H_5, etc. and py = pyridine) that proceed under mild conditions ($25°-80°C$) according to the stoichiometry of Equation 23 and the first-order rate law, 24 (k_{24} at $25°C = 1 \times 10^{-7}$ and 8×10^{-4} sec^{-1}; $\Delta H^{\ddagger}_{24} = 31$ and 21 kcal/mol for $R = CH_3$ and C_6H_5, respectively) (28).

$$\left[(py)(DMGH)_2Co-C \underset{H}{\overset{R}{\diagup}} CH_3 \right] \rightarrow$$

$$[(py)(DMGH)_2Co(II)] + CH_2{=}C \underset{H}{\overset{R}{\diagup}} + \tfrac{1}{2}H_2 \quad (23)$$

$$\left[\text{Rate} = k_{24}\ (py)(DMGH)_2Co-C\begin{smallmatrix}R\\ \diagup\\ \text{---}CH_3\\ \diagdown\\ H\end{smallmatrix} \right] \qquad (24)$$

A plausible mechanistic interpretation of these reactions is as follows:

$$\left[(py)(DMGH)_2Co-C\begin{smallmatrix}R\\ \diagup\\ \text{---}CH_3\\ \diagdown\\ H\end{smallmatrix} \right]$$

$$\xrightarrow{k_{24}} [(py)(DMGH)_2Co(II)] + \cdot C\begin{smallmatrix}R\\ \diagup\\ \text{---}CH_3\\ \diagdown\\ H\end{smallmatrix}$$

$$\xrightarrow{\text{fast}} [(py)(DMGH)_2Co-H] + CH_2{=}C\begin{smallmatrix}R\\ \diagup\\ \\ \diagdown\\ H\end{smallmatrix} \qquad (25)$$

$$[(py)(DMGH)_2Co-H] \xrightarrow{\text{fast}} [(py)(DMGH)_2Co(II)] + \tfrac{1}{2}H_2 \qquad (26)$$

According to this interpretation, $\Delta H^{\ddagger}{}_{24}$ may be identified with the cobalt–alkyl bond-dissociation energies in these compounds. However, at this stage other mechanistic interpretations, such as that involving a concerted olefin elimination, cannot be excluded (29). The investigation of these systems is continuing.

Equilibrium Approaches to the Estimation of Metal–Alkyl Bond-Dissociation Energies

We have found that in certain cases (e.g., when $R = C_6H_5$) the reaction depicted in Equation 23 attains a measurable equilibrium permitting the spectrophotometric determination of the equilibrium constant, K_{27}, defined by Equation 27 (28). Measurements of the temperature dependence of K_{27} yielded corresponding values of $\Delta H^{\circ}{}_{27}$ and $\Delta S^{\circ}{}_{27}$. For the case, $CH_2{=}CHR =$ styrene, the following values were determined in toluene solution: $K_{27}(25°C) = 1.3 \times 10^{-5}M^{3/2}$; $\Delta G^{\circ}{}_{27}(25°C) = 6.7$ kcal/mol; $\Delta H^{\circ}{}_{27} = 22.1$ kcal/mol; $\Delta S^{\circ}{}_{27} = 52$ cal/ (mol deg). The same values of K_{27} were obtained when the equilibrium was approached from the opposite direction, that is, starting with $[(py)(DMGH)_2Co(II)]$, $C_6H_5CH{=}CH_2$ and H_2.

$$K_{27} = \frac{[(py)(DMGH)_2Co(II)][C_6H_5CH{=}CH_2][H_2]^{1/2}}{[(py)(DMGH)_2Co-CH(CH_3)C_6H_5]} \qquad (27)$$

Using available data for the heats of formation of $C_6H_5CH=CH_2$ [$\Delta H°_f(25°C) = 35.2$ kcal/mol] (30) and of the $C_6H_5\dot{C}HCH_3$ radical [$\Delta H°_f(25°C) = 33$ kcal/mol] (31), the Co–C bond-dissociation energy of [(py)(DMGH)$_2$Co—CH(CH$_3$)C$_6$H$_5$] can be deduced to be 19.9 kcal/mol using the following thermochemical cycle:

$$\Delta H°$$

$$[(py)(DMGH)_2Co—CH(CH_3)C_6H_5] \rightleftarrows$$
$$[(py)(DMGH)_2Co(II)] + C_6H_5CH=CH_2 + \tfrac{1}{2}H_2 \qquad 22.1 \quad (28)$$

$$C_6H_5CH=CH_2 + \tfrac{1}{2}H_2 \rightleftarrows C_6H_5CHCH_3 \qquad -2.2 \quad (29)$$

$$[(py)(DMGH)_2Co—CH(CH_3)C_6H_5] \rightleftarrows$$
$$[(py)(DMGH)_2Co(II)] + C_6H_5\dot{C}HCH_3 \qquad 19.9 \quad (30)$$

This determination of the Co–C bond-dissociation energy rests entirely upon thermodynamic considerations and is independent of the mechanism of the reaction depicted in Equation 28. However, the value so determined (19.9 kcal/mol) is close to the value (21 kcal/mol) deduced above from kinetic measurements, assuming the mechanism depicted by Equations 25 and 26, and identifying $\Delta H^{\ddagger}$ for Reaction 28 with the Co–C bond-dissociation energy. This internal consistency is encouraging. We presently are attempting to extend this "equilibrium" approach to the determination of metal–alkyl bond-dissociation energies in other organocobalt compounds including cobalamins.

Finally, in the context of these results, the well recognized, apparently lower thermal (and photochemical) stability of secondary cobalt-alkyls, relative to primary ones, may reflect the greater accessibility of irreversible decomposition pathways involving olefin elimination (i.e., through schemes such as that in Reactions 25–26), in addition to some probable lowering of the metal–alkyl bond-dissociation energy.

Cobalamins

Although we are attempting to extend these approaches to the determination of Co–alkyl bond-dissociation energies in coenzyme B_{12} and other organocobalamins, no reliable estimates of such bond-dissociation energies have been reported thus far.

However, some indirect indication of the strength of alkyl–Co bonds in organo-cobalamin, relative to those in other alkyl-cobalt compounds, is provided by observations concerning the stability of benzylcobalamin. Attempts to prepare benzylcobalamin by either the B_{12s} route (Reaction 31) (32) or the B_{12r} route (Reactions 32–34) (15) have yielded spectroscopic evidence for its initial formation in solution. However, benzylcobalamin proved to be too unstable for isolation and decomposed within a few hours; bibenzyl was among the decomposition products. If this decomposition is assumed to be caused by benzyl–Co bond homolysis, the rate of decomposition (assuming $\Delta S^\ddagger \sim 0$) would correspond to an activation enthalpy, and hence a bond-dissociation energy, of approximately 15–20 kcal/mol. The usual difference of about 5–10 kcal/mol between benzyl and alkyl bond dissociation energies would, accordingly, imply a value in the range of 20–30 kcal/mol for the Co–C bond-dissociation energies of coenzyme B$_{12}$ and other primary alkylcobalamins.

$$B_{12s} + C_6H_5CH_2\text{—}Br \rightarrow C_6H_5CH_2\text{—}B_{12} + Br^- \tag{31}$$

$$B_{12r} + C_6H_5CH_2\text{—}Br \rightarrow Br\text{—}B_{12} + C_6H_5CH_2\cdot \tag{32}$$

$$B_{12r} + C_6H_5CH_2\cdot \rightarrow C_6H_5CH_2\text{—}B_{12} \tag{33}$$

$$2B_{12r} + C_6H_5CH_2\text{—}Br \rightarrow$$
$$C_6H_5CH_2\text{—}B_{12} + Br\text{—}B_{12}\left(\overset{H_2O}{\rightleftharpoons} B_{12a} + Br^-\right) \tag{34}$$

The thermal stability of benzylcobalamin is significantly lower than that of the benzyl derivatives of various other cobalt complexes for example $[C_6H_5CH_2\text{—}Co(DMGH)_2(py)]$ and $[C_6H_5CH_2\text{—}Co(CN)_5]^{3-}$, that have been invoked as vitamin B$_{12}$ models (9, 10, 12, 13).

Concluding Remarks

It is remarkable that, notwithstanding the widespread recognition of the probable role of Co–C bond homolysis in coenzyme B$_{12}$-dependent rearrangements, virtually no information has been available concerning the Co–C bond dissociation energy in coenzyme B$_{12}$ or closely related compounds.

The studies discussed in this chapter reveal that for a series of alkyl- and benzyl-transition metal compounds, including members of the series $[R\text{—}Mn(CO)_5]$ and $[R\text{—}Co(DMGH)_2(py)]$, metal–C bond-dissociation energies consistently appear to lie in the range of 20–30 kcal/mol. Comparisons of the relative stabilities of the corresponding hydrides and benzyl derivatives suggest that bonds to cobalamin probably are weaker than those to the other metal complexes encompassed by these studies. Hence, it seems likely that the Co–C bond-dissociation energy in coenzyme B$_{12}$ also lies in the same low range, that is, 20–30 kcal/mol. This is compatible with the proposed role of coenzyme B$_{12}$ as a 5'-deoxyadenosyl-radical precursor; very little additional activation would be required for homolysis of such a weak bond to occur at rates consistent with those of the enzymatic processes. It thus appears that the role of coenzyme B$_{12}$, one of the rare roles of an organometallic compound in biochemical processes, is associated with the characteristic weakness of transition-metal–alkyl bonds, making such compounds ideal free-radical precursors. At this stage, no other role for coenzyme B_{12} has been identified conclusively.

Finally, the apparent thermal stabilities of alkyl-cobalamins, as well as of some of the other transition-metal–alkyl compounds that have been examined in the course of these studies, generally are higher than would correspond to their metal–C bond-dissociation energies. The most probable explanation for this is that, in the absence of effective radical scavengers, homolytic dissociation of metal–alkyl bonds occurs reversibly because of selective recombination of the initially produced radicals and metal complexes.

Acknowledgments

The material discussed in this chapter encompasses the research and unpublished results of several co-workers, notably H. U. Blaser, R. L. Sweany, M. J. H. Russell, S. Diefenbach, G. L. Rempel, and Flora T. T. Ng, whose contributions are acknowledged gratefully. This research was supported through grants from the National Institutes of Health (AM-13339) and the National Science Foundation (CHE78-01192).

Literature Cited

1. Abeles, R. H.; Dolphin, D. *Acc. Chem. Res.* **1976,** *9,* 114.
2. Babior, B. M. *Acc. Chem. Res.* **1975,** *8,* 376.
3. Halpern, J. *Ann. N.Y. Acad. Sci.* **1974,** *239,* 2.
4. Schrauzer, G. N.; Sibert, J. W.; Windgassen, R. J. *J. Am. Chem. Soc.* **1968,** *90,* 6681.
5. Endicott, J. F.; Ferraudi, G. J. *J. Am. Chem. Soc.* **1977,** *99,* 243.
6. Pratt, J. M. *J. Chem. Soc.* **1964,** 5154.
7. Yamada, R.; Shimizu, S.; Fukui, S. *Biochim. Biophys. Acta* **1966,** *124,* 195.
8. Golding, B. T.; Kemp, T. J.; Sellers, P. J. *J. Chem. Soc., Dalton Trans.* **1977,** 1266.
9. Halpern, J.; Maher, J. P. *J. Am. Chem. Soc.* **1964,** *86,* 2311.
10. Halpern, J.; Maher, J. P. *J. Am. Chem. Soc.* **1965,** *87,* 5361.
11. Chock, P. B.; Halpern, J. *J. Am. Chem. Soc.* **1968,** *90,* 6959.
12. Schneider, P. W.; Phelan, P. F.; Halpern, J. *J. Am. Chem. Soc.* **1969,** *91,* 77.
13. Halpern, J.; Phelan, P. F. *J. Am. Chem. Soc.* **1972,** *94,* 1881.
14. Marzilli, L. G.; Marzilli, P. A.; Halpern, J. *J. Am. Chem. Soc.* **1971,** *93,* 1374.
15. Bläser, H.; Halpern, J. *J. Am. Chem. Soc.* **1980,** *102,* 1684.
16. Connor, J. A. *Top. in Curr. Chem.* **1977,** *71,* 71.
17. Schrauzer, G. N.; Windgassen, R. J. *J. Am. Chem. Soc.* **1967,** *89,* 1999.
18. Simandi, L; Szeverenyi, Z.; Budo–Zahonyi, E. *Inorg. Nucl. Chem. Lett.* **1975,** *11,* 773.
19. Mok, C. Y.; Endicott, J. F. *J. Am. Chem. Soc.* **1978,** *100,* 123.
20. Halpern, J.; Pribanic, M. *Inorg. Chem.* **1970,** *9,* 2616.
21. de Vries, B. *J. Catalysis* **1962,** *1,* 489.
22. Chao, T. H.; Espenson, J. H. *J. Am. Chem. Soc.* **1978,** *100,* 129.

23. Schrauzer, G. N.; Holland, R. J. *J. Am. Chem. Soc.* **1971**, *93*, 4060.
24. Brown, D. L. S.; Connor, J. A.; Skinner, H. A. *J. Organomet. Chem.* **1974**, *81*, 403.
25. Halpern, J.; Sweany, R. L.; Russell, M. J. H., unpublished results.
26. Fountaine, C.; Duong, K. N. V.; Nerienne, C.; Gaudemer, A.; Giannotti, C. *J. Organomet. Chem.* **1972**, *38*, 167.
27. Jensen, F. R.; Kiskis, R. C. *J. Am. Chem. Soc.* **1975**, *97*, 5825.
28. Halpern, J.; Ng, Flora, T. T.; Rempel, G. L. *J. Am. Chem. Soc.* **1979**, *101*, 7124.
29. Duong, K. N. V.; Ahond, A.; Merienne, C.; Gaudemer, A. *J. Organomet. Chem.* **1973**, *55*, 375.
30. Stull, D. R.; Westrum, E. F.; Sinke, G. C. "The Chemical Thermodynamics of Organic Compounds"; John Wiley and Sons: New York, 1969.
31. Kerr, J. A. *Chem. Rev.* **1966**, *66*, 465.
32. Tachkova, E. M.; Rudakova, I. P.; Yurkevich, A. M. *Zh. Obshch. Khim.* **1974**, *44*, 2594.

RECEIVED June 11, 1979.

Vitamin B_{12} Models with Macrocyclic Ligands

YUKITO MURAKAMI

Department of Organic Synthesis, Faculty of Engineering, Kyushu University, Fukuoka 812, Japan

Bisdehydrocorrin (BDHC) and corrinoid complexes are comparable in the electronic nature of the nuclear cobalt. Consequently, the double bonds at the periphery of the A and D rings of BDHC are not in conjugation with the interior double bonds. The steric effect is more pronounced for Co(I)(BDHC) than for vitamin B_{12s} when the reaction with alkyl halides is carried out, and is attributed to the 1,3-diaxial-type interaction between angular methyl groups placed at the C(1) and C(19) positions of BDHC and an approaching alkyl ligand. The photolysis of the methylated and ethylated Co(III)(BDHC) complexes results in the normal homolytic Co—C cleavage under anaerobic conditions. On the other hand, the Co—C bond in the isopropyl derivative undergoes heterolytic cleavage to yield the isopropyl anion and Co(III)(BDHC). The Co(BDHC) complex can be used as a catalyst for selective hydrogenation of primary alkyl halides using sodium hydroborate as the stoichiometric reducing agent.

$\mathbf{A}$lthough the cobalt corrinoids have been studied extensively in the last two decades (*1*), the significance of corrin as an equatorial ligand is not well understood. To characterize coenzyme B_{12} as an organocobalt derivative, a search for model cobalt complexes that can form a Co–C bond axial to a planar equatorial ligand has been stimulated. Studies on model systems (*2–13*), particularly on the cobaloxime derivatives (*2–7*), characterized their respective chemistry, but it is still not easy to establish a general correlation between the structure of an equatorial ligand and the properties of cobalt complex

0-8412-0514-0/80/33-191-179$05.00/0
© 1980 American Chemical Society

regarding the redox behavior of the nuclear cobalt, the reactivity of the Co(I) nucleophile, and the nature of the Co–C bond. To determine the structure–reactivity correlation, a set of model compounds should be chosen carefully so that the alteration in structure can be manipulated with minimal (ideally one) structural parameters. In this respect, we have studied the cobalt complexes of the modified corrins, 8,12-diethyl-1,2,3,7,13,17,18,19-octamethyl-*AD*-bisdehydrocorrin (BDHC) (*14*) and its tetradehydro analogue (TDHC) (*14, 15*). Co(II) (TDHC) and Co(III) (CN)$_2$ (BDHC) were first prepared by Johnson et al., and characterization of the latter complex by electronic and NMR spectroscopy as well as formation and photolysis of the methylated derivative was described briefly (*14*). Both BDHC and TDHC have additional double bonds at the peripheral positions that would cause electronic perturbations of the interior conjugation system of the corrinoid. Another important structural aspect of BDHC and TDHC is that these modified corrins possess angular methyl groups at the $C(1)$ and $C(19)$ positions, while the corrinoid has only one at $C(1)$. Thus, an axial ligand in the Co(BDHC) complex, regardless of its location at the upper or the lower side of the macrocycle, may be subjected to a steric interaction with one of the angular methyls. Consequently, the differences that might be found in properties among the BDHC, TDHC, and corrinoid complexes can be interpreted in a more straightforward manner.

Co(II) (BDHC) perchlorate was prepared by hydrogenation of the corresponding Co(II) (TDHC) perchlorate and purified by preparative thin-layer chromatography (TLC) on silica gel (*16*).

The general electronic structure and the coordination and reaction behaviors of the bisdehydrocorrin complexes are discussed here in reference to those of the corrinoid and tetradehydrocorrin complexes.

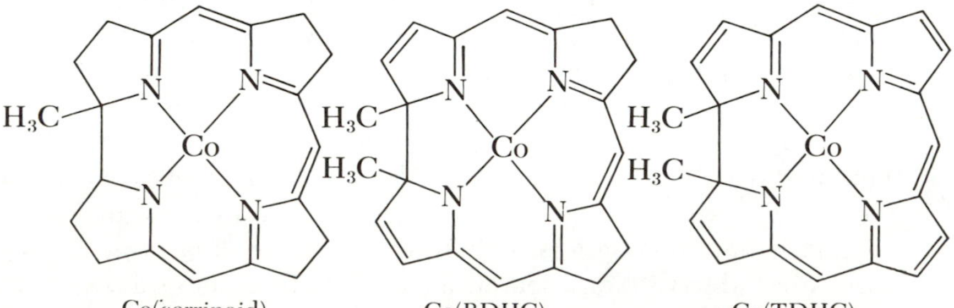

Co(corrinoid)   Co(BDHC)   Co(TDHC)

Electronic Properties of the Macrocyclic Chromophores

The π-conjugation effects in corrin, bisdehydrocorrin, and tetradehydrocorrin rings are reflected in the electronic spectra of their

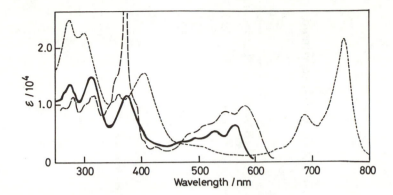

Figure 1. Electronic absorption spectra of (——) (CN)₂Co(III)-(BDHC), (– – –) (CN)₂Co(III)(cobinamide), and (- - - -) (CN)₂Co(III)-(TDHC) in water (for the former two complexes) and in methanol (for the last one) at room temperature. The spectrum for the cobinamide complex is taken from Ref. 19.

dicyanocobalt complexes (*see* Figure 1). The overall spectral features of the BDHC complex resemble those of the corrinoid complex (*14, 17*), but are far different from those of the TDHC complex. The four absorption bands of equal spacing (1.28×10^3 cm^{-1}) observed in the visible region are attributed to a π–π^* transition along with vibrational fine structure. Similar spectral features were observed for the corrinoid complex (*18, 19*). Both BDHC and corrinoid complexes show three main absorption bands in the near-UV region, but the intensity ratio between the γ-band (in 360–370-nm range) and the immediate higher energy band is reversed among these cobalt complexes. This may reflect degeneracy of π energy levels, which is characteristic of the corrinoid complex that is different from the BDHC complex. The α-, β-, and γ-bands for both corrinoid and BDHC complexes with various axial ligands (*X* and *Y*) are summarized in Table I.

The separation of π–π^* energy levels in the BDHC complexes is comparable with that in the corrinoid complexes. The extent of electronic perturbation on the π and π^* levels by the axial ligands is nearly the same for the two complexes, judging from the extent of band shifting. The red shift is increased by the axial ligands as follows: (H₂O, H₂O) < (H₂O, OH⁻) < (CN⁻, H₂O) < (OH⁻, OH⁻) < (CN⁻, Py) < (CN⁻, OH⁻) < (CN⁻, CN⁻). A molecular orbital treatment advanced for the elucidation of axial ligation effects on electronic spectra of the corrin complexes (*19*) may be applied in this case.

The diamagnetic shielding effects provided by BDHC and corrinoid macrocycles are the same, based on NMR measurements of Co(III)(*n*-C₃H₇)(BDHC) and the corresponding *n*-propylated cobalt corrinoid (*17, 21*) (Figure 2).

Table I.   Assignments of α-, β-, and γ-Bands (in nm) for XY—Co(III)(BDHC) and XY—Co(III)(corrinoid) as Measured in Water at Room Temperature[a]

X	Y	γ		β		α	
		BDHC	Corrinoid	BDHC	Corrinoid	BDHC	Corrinoid
H_2O	H_2O	352	349	474	490	499	520
H_2O	OH^-	350–360[b]	351	c	490	c	520
OH^-	OH^-	360–370[b]	d	496	505	~525	533
CN^-	H_2O	362	354	481	496	516	526
CN^-	Py	374	—	502	—	534	—
CN^-	OH^-	372	361	510	520	540	552
CN^-	CN^-	378	369	529	545	568	586

[a] The corrinoid complex is cobinamide and those coordinated with cyano group or groups are cited from Ref. 20.
[b] Shoulder.
[c] α- and β-Bands are not well separated in 480–500 nm.
[d] Split into two bands, 340 and 355 nm.

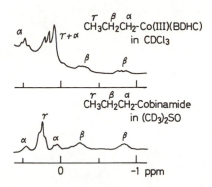

Figure 2.   NMR spectra of the Co-(III)(BDHC) complex with a n-propyl group in CDCl₃ and the corresponding cobinamide complex in (CD₃)₂SO; only the high-field region, where the proton signals due to the n-propyl group are observed, is shown here. The spectrum for the cobinamide complex is taken from Ref. 21.

Electronic Properties of the Nuclear Cobalt

The ESR spectrum of $[Co(II)(BDHC)]ClO_4$ in chloroform–benzene (2:1 v/v) at 77 K is typical for a low spin complex with d^7-configuration (Figure 3). Eight hyperfine splitting lines through interaction with ^{59}Co nucleus ($I = \frac{7}{2}$) are observed in the $g_{\parallel}$ range. Upon addition of pyridine, a triplet superhyperfine structure can be seen on the hyperfine lines of the $g_{\parallel}$ component, as shown in Figure 3. This may be caused by the interaction between the nitrogen nucleus of the axially coordinated pyridine molecule and the electron occupying the d_{z^2} orbital of the nuclear cobalt. Further spectral change was not observed with increasing pyridine concentration. Consequently, only one pyridine molecule can be coordinated with the cationic Co(II)(BDHC). The coupling constants ($A_{\parallel}{}^{Co}$ and $A_{\parallel}{}^{N}$) for the pyridine adduct of BDHC complex are comparable with those for Co(II)(cobinamide) coordinated with pyridine, as shown in Table II.

The oxygenated BDHC complexes, $(O_2)Co(II)(BDHC)$ and $(Py)-(O_2)Co(II)(BDHC)$, show a small $A_{\parallel}{}^{Co}$ value, which indicates reduced

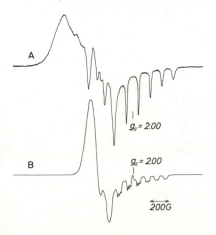

Figure 3.   X-Band ESR spectra of $[Co(II)(BDHC)]ClO_4$ in chloroform–benzene (2:1 v/v) at 77 K: (A) without pyridine; (B) a drop of pyridine added.

Table II. ESR Spin Hamiltonian Parameters for Cobalt(II) Complexes

Complex	Temperature (K)	Medium	$A_\parallel^{Co}(\times 10^{-4}\ cm^{-1})$	$A_\parallel^{N}(\times 10^{-4}\ cm^{-1})$	Reference
Co(II)(BDHC)	77	$CHCl_3$–C_6H_6(2:1 v/v)	127	—	This work
Py—Co(II)(BDHC)	77	$CHCl_3$–C_6H_6(2:1 v/v)	101	17.1	This work
O_2—Co(II)(BDHC)	77	$CHCl_3$–C_6H_6(2:1 v/v)	18	—	This work
(Py)(O_2)—Co(II)(BDHC)	77	$CHCl_3$–C_6H_6(2:1 v/v)	12	—	This work
Co(II)cobinamide	77	H_2O	135	—	(23)
Py—Co(II)(cobinamide)	77	CH_3OH	105	17.7	(23)
Vitamin B_{12r}	77	H_2O	102.7	—	(24)
O_2—Vitamin B_{12r}	77	H_2O	14.0	—	(24)

electron density at the cobalt nucleus and pronounced contribution of the resonance structure $Co(III)-O_2^-$. The oxygenated BDHC complex with pyridine coordinated to cobalt at the other axial site shows a $A_{\parallel}^{Co}$ value comparable with that for the oxygenated vitamin B_{12r} (Table II).

Cyclic voltammograms for the BDHC and TDHC complexes are shown in Figure 4; cyclic voltammetry of the latter complex was studied by Elson et al. (22). The clearly distinguishable redox pairs of reversible character are referred to as $Co(III) \rightleftharpoons Co(II)$ and $Co(II) \rightleftharpoons Co(I)$. The $Co(II)/Co(I)$ potential for the BDHC complex is different from that for the TDHC complex, but closer to that for vitamin B_{12}. The TDHC complex shows another redox pair in *N,N*-dimethylformamide (DMF) and dichloromethane at -1.31 and -1.57 V vs. SCE, respectively. This redox pair is referred to as $Co(I)(TDHC) \rightleftharpoons [Co(II)(TDHC)]^-$. $(CN)_2Co(III)(TDHC)$ shows a peak potential for two-electron reduction in the range of -0.6 V and the corresponding oxidation potential in the range of 0 V vs. SCE. The redox reaction is represented by

$$(CN)_2Co(III)(TDHC) \overset{2e}{\rightleftharpoons} Co(I)(TDHC) + 2CN^-$$

Further reduction beyond -0.6 V yields another peak potential at -1.3 V, which is referred to the following reaction.

$$Co(I)(TDHC) \overset{e}{\rightleftharpoons} [Co(II)(TDHC)]^-$$

The redox potentials for TDHC, BDHC, and corrinoid complexes are compared in Table III.

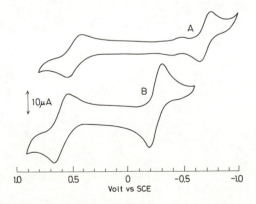

Figure 4. Cyclic voltammograms of (A) [Co(II)(BDHC)]ClO$_4$ and (B) [Co(II)(TDHC)]ClO$_4$ in methanol; scan rate, 100 mV sec^{-1}; concentrations: cobalt complexes, 5.0 × 10^{-4}M; (Bu$_4$N)ClO$_4$, 5.0 × 10^{-2}M.

Table III. Redox Potentials (V vs. SCE) for BDHC, TDHC, and Corrinoid Complexes

Complex	Medium	$Co(III)/Co(II)$	$Co(II)/Co(I)$	$Co(I)/[Co(II)L]^{-a}$	$(CN)_2Co(III) \rightarrow Co(I)$
Co(BDHC)	CH$_3$OH	+0.47	-0.71	—	—
	DMF	—	-0.55	—	—
	CH$_2$Cl$_2$	—	-0.57	—	—
Co(TDHC)	CH$_3$OH	+0.59	-0.25	—	—
	DMF	—	-0.11	-1.31	—
	CH$_2$Cl$_2$	+0.97	-0.13	-1.57	—
Vitamin B$_{12}$^{b}	H$_2$O	~+0.3	-0.742	—	—
(CN)$_2$Co(III)(TDHC)	DMF	—	—	—	-0.58
	CH$_2$Cl$_2$	—	—	—	-0.73
(CN)$_2$Co(III)(BDHC)	DMF	—	—	—	-1.38
	CH$_2$Cl$_2$	—	—	—	<-1.45
(CN)$_2$Co(III)(cobinamide)	DMF	—	—	—	-1.14
(CN)$_2$Co(III)(cobinamide)^{c}	H$_2$O	—	—	—	-1.18
(CN)$_2$Co(III)(cobalamin)^{d}	H$_2$O	—	—	—	-1.33

[a] Co(I)/[Co(II)L]$^-$ is the redox potential for Co(I)(TDHC) ⇌ [Co(II)(TDHC)]$^-$.
[b] Data from Ref. 25 and 26.
[c] Data from Ref. 27.
[d] Data from Ref. 28.

The ESR spin Hamiltonian parameters and the redox potentials for the BDHC complexes are similar in magnitude to those for the corresponding corrinoid complexes. Thus, BDHC and corrinoid complexes are similar in the electronic nature of their nuclear cobalt.

Axial Coordination Behavior

Acid-dissociation equilibria for the coordinated water in $[Co(III)(H_2O)_2(BDHC)]^{2+}$ and $[Co(III)(H_2O)(CN)(BDHC)]^+$ are accompanied by the spectral changes shown in Figure 5; the pK_a values are summarized in Table IV along with those for the corrinoid complexes (29). The pK_a values are comparable with each other between these two complex systems.

The addition of pyridine bases to $[Co(III)(H_2O)(CN)(BDHC)]^+$ in aqueous media results in a spectral change that can be attributed to substitution of the coordinated water by a pyridine base. This substitution is represented in the following equation and equilibrium.

$$[Co(III)(H_2O)(CN)(BDHC)]^+ + Py \rightleftharpoons$$
$$[Co(III)(Py)(CN)(BDHC)]^+ + H_2O$$

$$K = \frac{[Co(III)(Py)(CN)(BDHC)]}{[Co(III)(H_2O)(CN)(BDHC)][Py]}$$ $\log K = 2.17$ (with pyridine)

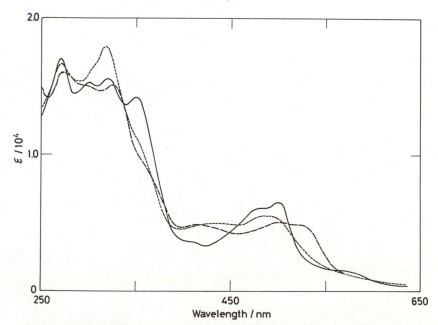

Figure 5. Electronic absorption spectra of cobalt(III) complexes in water at room temperature: (——) $[(H_2O)_2Co(III)(BDHC)]^{2+}$; (-----) $[(H_2O)(OH)Co(III)(BDHC)]^+$; (-----) $(OH)_2Co(III)(BDHC)$.

Table IV.   Acid Dissociation Constants for Coordinated Water[a]

pK_a	BDHC Complex	Cobinamide[b]
pK_{a1}	6.2	6.0
pK_{a2}	11.2	c
pK_{a3}	11.3	11.0

[a] $H_2O-Co(III)-OH_2 \xrightarrow{K_{a1}} H_2O-Co(III)-OH \xrightarrow{K_{a2}}$
$HO-Co(III)-OH; \ NC-Co(III)-OH_2 \xrightarrow{K_{a3}} NC-Co(III)$
$-OH$
[b] Data from Ref. 29.
[c] Not reported, a spectral change was observed around pH 11.

This log K value (2.17) is very close to that reported for the pyridine adduct of corrinoid complex (log K = 2.6) (30). The linear correlation between the pK_a's of pyridine bases and the log K value for the BDHC complex system is shown in Figure 6.

The coordination equilibria for pyridine bases with [Co(II)-(BDHC)]ClO$_4$ were investigated spectrophotometrically in dichloromethane and the results are summarized in Table V.

On the basis of the foregoing results and discussion, the electronic effect of BDHC on the nuclear cobalt is similar to that of the corrinoids. Consequently, the double bonds at the periphery of the A and

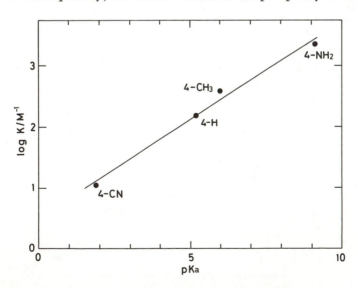

Figure 6.   Correlation between pK_a's of pyridine bases and log K's for the coordination of pyridine bases to $[(H_2O)(CN)Co(III)(BDHC)]^+$ in water at 25.0 ± 0.1°C; substituent groups on the pyridine moiety are given in the figure.

Table V.   Stability Constants for the Coordination of Pyridine Bases to [Co(II)(BDHC)]ClO₄ in Dichloromethane at 25°C

Pyridine Base	$pK_a{}^a$	$K/(L\ mol^{-1})$
Py	5.19	2.7×10^3
4-Me—Py	6.02	5.1×10^3
4-NH₂—Py	9.12	6.1×10^3
4-CN—Py	1.90	2.1×10^2
2-Me—Py	5.97	9.1
2,4,6-Me₃—Py	7.48	8.4

[a] Data from Ref. 45.

D rings of BDHC are of isolated character. On the other hand, the peripheral double bonds in the B and C rings of TDHC are in conjugation with the interior double bonds, extending the conjugation of the π-bond system. In fact, the NMR signals for the *meso*-protons (at the 5, 10, and 15 positions) of Co(III)(CN)₂(TDHC) are shifted downfield by approximately 1 ppm relative to those of Co(III)(CN)₂(BDHC) by this extended π-conjugation. The π-conjugation effect of the equatorial ligand is of primary importance in controlling the electronic nature of the nuclear cobalt, which is reflected in the redox potentials. The Co(III)/Co(II) and Co(II)/Co(I) redox potentials are listed in Table VI for the cobalt complexes of macrocyclic tetrapyrroles.

These potentials are affected by two primary factors: the size of the negative charge on the equatorial ligand; and the π^* energy level relative to the $4p$ or $3d$ (Co) orbital (the π^* energy level is affected by

Table VI.   Redox Potentials (V vs. SCE) for Various Cobalt Complexes

Complex	Medium	$Co(III)/Co(II)$	$Co(II)/Co(I)$
Co(TDHC)	CH₃OH	+0.59	−0.25
	CH₂Cl₂	+0.97	−0.13
Co(BDHC)	CH₃OH	+0.47	−0.71
	CH₂Cl₂	[a]	−0.55
Vitamin B₁₂[b]	H₂O	~+0.3	−0.742
Co(EtioP)[c]	CH₃(CH₂)₂CN	+0.30[d]	—
	DMF	—	−1.04[e]
Co(corrole)[f]	CH₂Cl₂	−0.26	—

[a] Reversible redox behavior was not observed.
[b] Data from Ref. 25 and 26.
[c] EtioP stands for etioporphyrin.
[d] Data from Ref. 31.
[e] Data from Ref. 32.
[f] Data from Ref. 33.

the extent of π conjugation in the macrocyclic ligand). Corrin and its analogues (corrin, TDHC, and BDHC) act as monoanionic ligands, while porphyrin and corrole carry -2 and -3 charges, respectively, when coordinated with cobalt; as the negative charge on the equatorial ligand increases, the complex becomes more resistant to electrolytic reduction. When the π^* level in the equatorial ligand is lowered through the π-conjugation effect, the π-back donation from $4p$ or $3d$ (Co) to π^* would occur. Such a nephelauxetic effect tends to stabilize the lower oxidation state of the nuclear cobalt, as would be the case for the TDHC complex. Further electrolytic reduction of Co(I)(TDHC) results in the formation of $[Co(II)(TDHC)]^-$; instead of reducing the nuclear cobalt, two electrons are transferred to the π^* level of TDHC and cobalt is oxidized to the divalent state (34).

To observe the B_{12} function, the equatorial ligand must have either -1 or zero charge (cobaloxime) when coordinated with cobalt so that the higher oxidation state of cobalt is not too stabilized. However, if the equatorial ligand involves a lower π^* level that can accept an electron from the metal orbital, the univalent cobalt state is stabilized to the extent that the cobalt is not reactive.

Steric Effects on the Reactivity at the Axial Site

The reaction of a Co(I) nucleophile with an appropriate alkyl donor is used most frequently for the formation of a Co–C bond, which also can be formed readily by addition of a Co(I) complex to an acetylenic compound or an electron-deficient olefin (5). The nucleophilicity of Co(I) in Co(I)(BDHC) is expected to be similar to that in the corrinoid complex, as indicated by their redox potentials. The formation of Co–C σ-bond is the attractive criterion for vitamin B_{12} models. Sodium hydroborate (NaBH$_4$) was used for the reduction of Co(III)(CN)$_2$(BDHC) in tetrahydrofuran–water (1:1 or 2:1 v/v). The univalent cobalt complex thus obtained, Co(I)(BDHC), was converted readily to an organometallic derivative in which the axial position of cobalt was alkylated on treatment with an alkyl iodide or bromide. As expected for organo-cobalt derivatives, the resulting alkylated complexes were photolabile (17).

The second-order rate constants for reactions of Co(I)(BDHC) with alkyl halides were determined spectrophotometrically at 400 nm (17). These rate constants are listed in Table VII along with those for Co(I)(corrinoid)(vitamin B_{12s}) in methanol at 25°C (35). These data indicate that the S_N2 mechanism is operative in the reaction of Co(I)(BDHC); the iodides are more reactive with the cobalt complex than the bromides, and the rate decreases with increasing bulkiness of the alkyl donor. The steric effect is more pronounced for Co(I)(BDHC) than for vitamin B_{12s}, which is confirmed by the rate ratios for

Table VII.   Second-Order Rate Constants for Reactions of Co(I)(BDHC)[a] and Vitamin B$_{12s}$[b] with Alkyl Halides[c]

Alkyl Halide	k(BDHC) ($L\ mol^{-1}\ sec^{-1}$)	k(B$_{12s}$) ($L\ mol^{-1}\ sec^{-1}$)	$\frac{k(BDHC)}{k(B_{12s})}$	Reference
CH$_3$I	3.0×10^4	3.4×10^4	0.88	(17)
CH$_3$CH$_2$I	3.1×10^2	—	—	(17)
CH$_3$CH$_2$CH$_2$I	1.1×10^2	—	—	(17)
(CH$_3$)$_2$CHI	0.71	2.3×10^2	3.1×10^{-3}	(17)
CH$_3$Br	8.5×10^2	1.6×10^3	0.53	(17)
CH$_3$CH$_2$Br	3.9	3.1×10	0.13	(17)
CH$_3$CH$_2$CH$_2$Br	1.8	1.4×10	0.13	(17)
(CH$_3$)$_2$CHBr	7×10^{-4}	1.8	4×10^{-4}	(17)
CH$_3$(CH$_2$)$_{11}$Br	3.0×10^{-1}	—	—	This work
CH$_3$CHBr(CH$_2$)$_5$CH$_3$	4×10^{-4}	—	—	This work
⬡—CH$_2$Br	2.1×10^{-2}	—	—	This work
methylcyclohexyl—Br	5×10^{-4}	—	—	This work

[a] With CH$_3$X, CH$_3$CH$_2$X, CH$_3$CH$_2$CH$_2$X, and (CH$_3$)$_2$CHX (X, Br or I), in THF–water (1:1 v/v); with the rest of alkyl halides, in THF–water (2:1 v/v).
[b] In methanol containing 0.10M NaOH (from Ref. 35).
[c] Co(I)(BDHC), at 25.0 ± 0.1°C; vitamin B$_{12s}$, at 25 ± 2°C with ethyl, n-propyl, and isopropyl halides, and at 25.0°C with methyl halides.

Co(I)(BDHC) vs. vitamin B_{12s}, $k(BDHC)/k(B_{12s})$, listed in Table VII. The significant steric effect seems to arise from the 1,3-diaxial type interaction between the angular methyl groups placed at the $C(1)$ and $C(19)$ positions of the macrocyclic ligand and an approaching alkyl group, which is bound finally to the cobalt atom. A methyl group is placed only at $C(1)$ in the corrinoid system and an alkyl halide may approach from the less hindered side of the equatorial skeleton to avoid steric interaction with the CH_3—$C(1)$.

The methylated complex of cobyric acid is in isomer equilibrium under photolytic or thermal condition to give *anti*- and *syn*-forms at a $92:8$ molar ratio (Figure 7). The *anti*- to *syn*-form ratio for the ethyl-

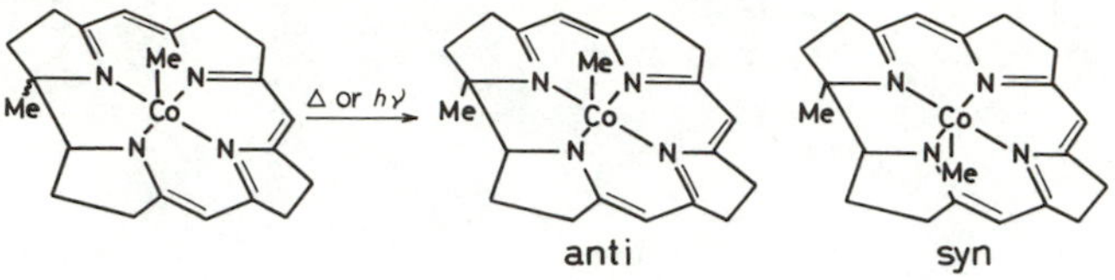

Figure 7. Equilibrium formation of anti- *and* syn-*forms of the methylated complex of cobyric acid under photolytic or thermal condition (36).*

ated complex is $99:1$ (*36*). Thus, the predominant formation of *anti*-species is attributed primarily to the steric repulsion between an axial alkyl ligand and the angular methyl of the corrinoid. When the cyanide ion and the benzimidazole group are coordinated with the corrinoid complex at the side where the angular methyl group is sticking out, these donor groups assume configurations to minimize the steric interaction with the methyl group. This configuration is confirmed by x-ray diffraction studies (*37*).

A cyclohexanemethyl group is subjected to a large steric effect, as seen in Table VII; its local structure (**1**) bears a close resemblance to that of methylmalonyl CoA (**2**).

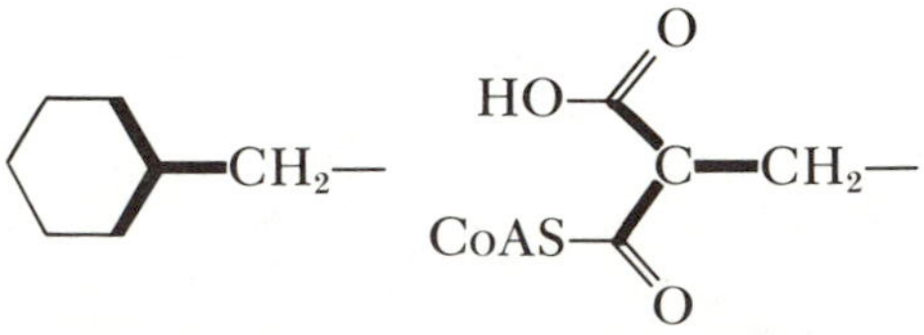

It is reasonable to assume that the S_N2 reactions listed in Table VII are controlled primarily by steric effects. The steric repulsion energies for selected reactions are estimated as follows.

1. The free energy difference between the *anti-* and *syn-* forms of the methylated cobyric acid: $RT\ln (92/8) \simeq 1.4$ kcal, $CH_3—CH_3$ repulsion energy.

2. The free energy difference between *anti-* and *syn-*forms of the ethylated cobyric acid: $RT\ln (99/1) \simeq 2.8$ kcal, $CH_3—C_2H_5$ repulsion energy.

3. The difference in activation free energy between methylation and isopropylation of Co(I)(BDHC): $RT\ln (3.0 \times 10^4/0.71) \simeq 6.4$ kcal. Taking $a, b, c,$ and d as the repulsion energies for i-C_3H_7—CH_3, i-C_3H_7—macrocyclic skeleton (BDHC), CH_3—CH_3, and CH_3—macrocyclic skeleton (BDHC), respectively, the following equation can be given: $(a + b) - (c + d) = 6.4$.

4. The difference in activation free energy between methylation and isopropylation of B_{12s}: $RT\ln (3.4 \times 10^4/2.3 \times 10^2) \simeq 3$ kcal, corresponding to b if $d \simeq 0$.

5. The difference in activation free energy between methylation and ethylation of Co(I)(BDHC): $RT\ln (3.0 \times 10^4/3.1 \times 10^2) \simeq 1.3$ kcal.

The estimated steric repulsion energies are listed in Table VIII.

Table VIII.   Steric Repulsion Energies for Alkylation of Co(I)(BDHC) with Alkyl Iodides

Alkyl Group (R)	*Repulsion Energy (kcal mol^{-1})*	
	$R—CH_3$	$R—Skeleton$
CH_3	1.4	0^a
CH_3CH_2	2.8	1.3
$(CH_3)_2CH$	5	3

[a] The steric repulsion energy between the methyl group and the macrocyclic skeleton (BDHC) is assumed to be negligible.

Photolysis

The alkylated BDHC complexes are very unstable in general and cannot be isolated in pure form from the reaction mixture. A large portion of an alkylated product decomposes during the extraction procedure with organic solvents. Thus, the decomposition behavior of the alkylated BDHC complexes was investigated as they were prepared in aqueous media, after decomposition of the excess $NaBH_4$ with acid in the presence of excess alkyl halide. An alkylated BDHC complex also can be prepared by the reaction of $[Co(II)(BDHC)]ClO_4$ with an alkyl iodide under alkaline condition (pH $\geq$ 12); the disproportionation reaction yields the Co(III)(BDHC) species and an alkylated BDHC complex.

The methylated complex, $CH_3—Co(III)(BDHC)$, undergoes aerobic photolysis by irradiation with a 200-W tungsten lamp (from a distance of 60 cm) to give $[(H_2O)_2Co(III)(BDHC)]^{2+}$. The isosbestic points are observed as shown in Figure 8; no stable intermediate was formed during the reaction: rate, $7 \times 10^{-3} \, sec^{-1}$ (half life ~100 sec) for pH 2.37–4.99. Under anaerobic conditions, the photolysis rate was reduced to have a half life of 3×10^2 min and $Co(II)(BDHC)$ was obtained. Thus, the homolytic radical cleavage takes place for the methylated complex. The acceleration of photolysis under aerobic conditions apparently is caused by the trapping of a methyl radical with oxygen (5), which inhibits the recombination of the radical with the Co(II) species. This photolysis is shown in the following reaction sequence.

$$CH_3—Co(III)(BDHC) \overset{h\nu}{\rightleftharpoons} Co(II)(BDHC) + CH_3\cdot$$
$$\phantom{CH_3—Co(III)(BDHC) \overset{h\nu}{\rightleftharpoons} Co(II)(BDHC) } \overset{O_2}{\longrightarrow} CH_3—O—O\cdot$$

Methylcobalamin was photolyzed under anaerobic conditions to yield CH_3 radicals and Co(II) species that rapidly recombined to form the original complex (38).

The cyanide ion greatly enhanced the reaction under anaerobic conditions in alkaline media (pH 9.93, carbonate buffer) (see Table IX). The reaction gave $(CN)_2Co(III)(BDHC)$ and no stable intermediate was detected spectrophotometrically since clear isosbestic points

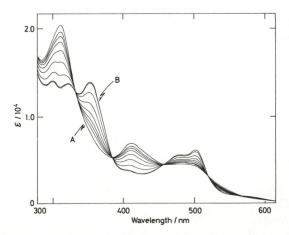

Figure 8. Spectral change for the aerobic photolysis of $[(CH_3)$-$Co(III)(BDHC)]^+$ in water containing 1.0% (v/v) methanol at pH 4.43 (acetate buffer): irradiated with a 200-W tungsten lamp from a distance of 60 cm; duration of photolysis, 0, 10, 20, 40, 85, 175, 355, 955, and 1285 sec (read from A to B).

**Table IX.   Anaerobic Photolysis
of CH$_3$—Co(III)(BDHC) in
the Presence of Cyanide Ion
at pH 9.93 (Carbonate Buffer)**

[CN$^-$](M)	Half Life (min)
0	3×10^2
1.2×10^{-4}	7.8
1.0×10^{-3}	7.3
0.8×10^{-2}	7.3

were observed. The rate enhancement by the cyanide ion was also observed under acidic conditions to give [(H$_2$O)(CN)Co(III)(BDHC)]$^+$. Because the oxidation of Co(II)(BDHC) is much slower than the photolysis of the methylated complex under the same conditions, the photolysis must yield directly [(H$_2$O)(CN)Co(III)(BDHC)] without forming Co(II)(BDHC). A plausible reaction scheme for the cyanide-promoted photolysis is given as follows.

$$CH_3—Co(III)(BDHC) \overset{h\nu}{\rightleftharpoons} [CH_3\cdot + Co(II)(BDHC)] \xrightarrow{CN^-}$$

$$CH_3\cdot + Co(II)(BDHC) \rightarrow CH_3^- + Co(III)(BDHC)$$

$$\underset{CN}{|} \qquad\qquad \underset{CN}{|}$$

The increased reduction ability of the Co(II) species by coordination with the cyanide ion seems to be responsible for the above reaction (39).

The aerobic photolysis of the ethylated BDHC complex yielded [(H$_2$O)$_2$Co(III)(BDHC)]$^{2+}$ under acidic conditions, while the anaerobic photolysis yielded Co(II)(BDHC) in the same manner as was observed for the methylated complex. On the other hand, the isopropyl derivative of Co(III)(BDHC) gave [(H$_2$O)Co(III)(BDHC)]$^{2+}$ under dark acidic conditions and photolysis conditions. The half lives for these reactions are listed in Table X.

**Table X.   Half Lives for Photolysis[a] of Alkylated
BDHC Complexes**

Alkyl Moiety	$\tau_{1/2}$(aerobic)(min)	$\tau_{1/2}$(anaerobic)(min)
CH$_3$	1.7	$\sim 3 \times 10^2$
CH$_3$CH$_2$	2.4	7
(CH$_3$)$_2$CH	0.5	0.5

[a] Irradiated with a 200-W tungsten lamp from a distance of 60 cm.

On the basis of these findings, photolysis of the methylated and ethylated complexes results in the normal homolytic Co—C cleavage under anaerobic conditions. The faster rate for anaerobic photolysis of the ethyl derivative relative to the methyl derivative would be attributed partly to olefin formation through the following reaction sequence.

$$CH_3CH_2-Co(III)(BDHC) \xrightarrow{h\nu} CH_3CH_2\cdot + Co(II)(BDHC)$$
$$\longrightarrow CH_2{=}CH_2 + \tfrac{1}{2}H_2$$

The isopropyl derivative does not seem to undergo normal homolytic cleavage of the Co—C bond since the photolysis rates are comparable under both aerobic and anaerobic conditions. Thus, the Co–C bond undergoes heterolytic cleavage to yield the isopropyl anion and Co(III)(BDHC); intrinsic heterolysis, or initial homolysis is followed by rapid electron transfer before the radical is trapped by oxygen molecule. Owing to the steric pressure of the BDHC skeleton and the angular methyl groups at the $C(1)$ and $C(19)$ positions against the bulky isopropyl group, the cobalt species and the alkyl anion diffuse out mutually after heterolysis. The mutual electronic interaction is not retained.

The Fe(II)/Fe(III) redox potential for the electron transfer from (octaethylporphinato)iron(II) to the methyl radical (40) is reported to be approximately 0 V vs. SCE (41).

$$Fe(II)(OEP) + CH_3\cdot \rightarrow Fe(III)(OEP) + CH_3{}^{-}$$
$$\xrightarrow{H^+} CH_4$$

From the electrochemical viewpoint, the metal complex system can reduce the methyl radical if its redox potential is at most 0 V vs. SCE. The Co(III)/Co(II) redox potential for the BDHC complex without any axial ligand is +0.5 V vs. SCE. Even if the nuclear cobalt(II) in the complex does not have a high reducing ability, the Co(II) complex can reduce the alkyl radical upon input of 12 kcal of energy (corresponding to the potential difference of 0.5 V) from other sources. The estimated steric energy at the transition state for the reaction between Co(I)(BDHC) and $(CH_3)_2CHI$ is 8 kcal mol^{-1}, as described above. Thus, the strain energy release upon the Co–C bond cleavage for the isopropyl derivative must be over 8 kcal mol^{-1}. On the basis of these energy estimations, it is reasonable to expect that the BDHC complex with a bulky axial ligand may undergo novel heterolysis of the Co–C bond.

Reactivity of the Cobalt–Carbon Bond

The BDHC complex with the dodecyl group at its axial site was obtained by the reaction of Co(I)(BDHC) with 1-dodecyl bromide. After decomposition of excess sodium hydroborate with acetic acid, the solution of the alkylated BDHC complex in aqueous micellar hexadecyltrimethylammonium bromide (CTAB) was irradiated with a 200-W tungsten lamp under nitrogen. The main reaction product was dodecane, which was recovered in a 50% yield. Since the methyl radical does not abstract a hydrogen atom from water (*42, 43*), dodecane is not formed appreciably through the radical mechanism. In addition, the methyl radical formed by the photolysis of methylcobaloxime does not incorporate the methyl hydrogen of the peripheral methyl groups in cobaloxime (*44*). Thus, the alkyl radical, if formed, does not seem to take up a hydrogen atom from the peripheral alkyl substituents of BDHC.

Alkylated cobaloximes yield the corresponding dimeric species of alkyl radicals by photolysis under acidic conditions. But the BDHC complex with a hexyl or benzyl group at its axial site does not yield the corresponding dimeric species by photolysis (dodecane and bibenzyl, respectively). Consequently, the hydrogenation product must be obtained through the formation of a carbanion by heterolytic cleavage of the Co–C bond, followed by its protonation.

Based on the kinetic data on the alkylation of Co(I)(BDHC) (Table VII), the BDHC complex can be used as a catalyst for selective hydrogenation of primary alkyl halides using sodium hydroborate as the stoichiometric reducing reagent (*see* Scheme 1). In fact, when 1-dodecyl bromide was used as a substrate in aqueous micellar CTAB, dodecane was obtained in 500% yield on the basis of the BDHC complex. In the competitive hydrogenation of 1-dodecyl bromide and 2-octyl bromide, the latter was not hydrogenated appreciably. 1-Bromo-3-bromomethylcyclohexane, which has both primary and secondary bromo groups, was hydrogenated to yield 1-bromo-3-methylcyclohexane and bromomethylcyclohexane at a 10 : 1 ratio. The selectivity of the primary bromo group should be noted. Hydrogenation of the bromo compounds is not generally a facile reac-

Scheme 1.

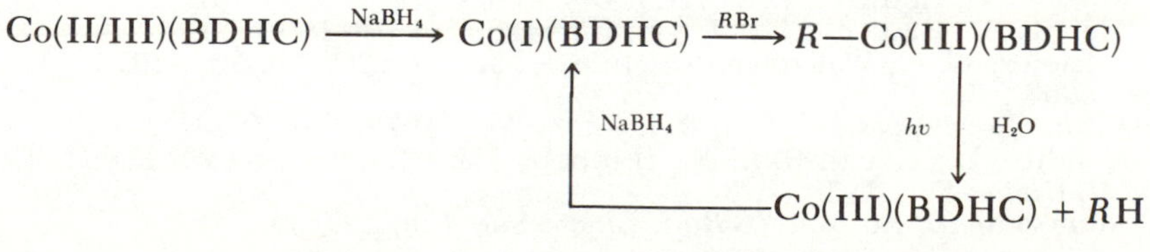

Co(II/III)(BDHC) $\xrightarrow{\text{NaBH}_4}$ Co(I)(BDHC) $\xrightarrow{R\text{Br}}$ R—Co(III)(BDHC)

NaBH$_4$ $h\nu$ | H$_2$O

Co(III)(BDHC) + RH

tion. The hydrogenation described in this chapter prevails over other methods in the following respects. (1) The hydrogenation (or reduction) can be performed in an aqueous medium. (2) The reaction can be carried out selectively with the primary alkyl halides.

The results of this investigation may provide a useful criterion for the development of vitamin B_{12} models. The cobalt complex of BDHC bears a close resemblance to vitamin B_{12} in electronic structure, but the methyl groups at the $C(1)$ and $C(19)$ positions of the former complex exercise a pronounced steric effect on the axial ligand, which gives the Co—C a reactivity quite different from that of the corrinoid complex. The novel heterolysis of the Co–C bond involved in the BDHC complex with a bulky axial ligand can be applied to the selective hydrogenation of primary alkyl halides. To simulate various isomerization reactions observed for the vitamin B_{12} system, the reaction intermediate and the cobalt species formed by the Co–C bond cleavage need to be retained in mutual proximity without diffusing so that intimate electronic interaction between the two species can be promoted.

Acknowledgment

The research I have reviewed here is the result of the experimental and intellectual efforts of my students and of collaborators whose names appeared in the references. Particularly, I am grateful to my associate Yasuhiro Aoyama, who has guided the more quantitative aspects of this work.

Literature Cited

1. Pratt, J. M. "Inorganic Chemistry of Vitamin B_{12}"; Academic: London, 1972.
2. Schrauzer, G. N. *Acc. Chem. Res.* **1968,** *1*, 97.
3. Costa, G. *Coord. Chem. Rev.* **1972,** *8*, 63.
4. Biggetto, A.; Costa, G.; Mestroni, G.; Pellizer, G.; Puxeddu, A.; Reisenhofer, E.; Stefani, L.; Tauzher, G. *Inorg. Chim. Acta Rev.* **1970,** *4*, 41.
5. Dodd, D.; Johnson, M. D. *Organomet. Chem. Rev.* **1973,** *52*, 1.
6. Schrauzer, G. N. *Angew. Chem.* **1976,** *88*, 465.
7. Esperson, J. H.; Martin, A. H. *J. Am. Chem. Soc.* **1977,** *99*, 5953.
8. Ochiai, E.; Long, K. M.; Sperati, C. R.; Busch, D. H. *J. Am. Chem. Soc.* **1969,** *91*, 3201.
9. Farmery, K.; Busch, D. H. *Inorg. Chem.* **1972,** *11*, 2901.
10. Mok, C. Y.; Endicott, J. F. *J. Am. Chem. Soc.* **1977,** *99*, 1276.
11. Mok, C. Y.; Endicott, J. F. *J. Am. Chem. Soc.* **1978,** *100*, 123.
12. Schaefer, W. P.; Waltzman, R.; Huie, B. T. *J. Am. Chem. Soc.* **1978,** *100*, 5063.
13. Elroi, H.; Meyerstein, D. *J. Am. Chem. Soc.* **1978,** *100*, 5540.
14. Dolphin, D.; Harris, R. L. N.; Huppatz, J. L.; Johnson, A. W.; Kay, I. T. *J. Chem. Soc.* **1966,** *C*, 30.
15. Murakami, Y.; Aoyama, Y. *Bull. Chem. Soc. Jpn.* **1976,** *49*, 683.

16. Murakami, Y.; Aoyama, Y.; Tokunaga, K. *Inorg. Nucl. Chem. Lett.* **1979,** *15,* 7.
17. Murakami, Y.; Aoyama, Y.; Nakanishi, S. *Chem. Lett.* **1977,** 991.
18. Firth, R. A.; Hill, H. A. O.; Mann, B. E.; Pratt, J. M.; Thorp, R. G.; Williams, R. J. P. *J. Chem. Soc.* **1968,** A, 2419.
19. Offenhartz, P. O.; Offenhartz, B. H.; Fung, M. M. *J. Am. Chem. Soc.* **1970,** *92,* 2966.
20. Firth, R. A.; Hill, H. A. O.; Pratt, J. M.; Thorp, R. G.; Williams, R. J. P. *J. Chem. Soc.* **1968,** A, 2428.
21. Brodie, J. D.; Poe, M. *Biochemistry* **1971,** *10,* 914.
22. Elson, C. M.; Hamilton, A.; Johnson, A. W. *J. Chem. Soc., Perkin Trans. 1* **1973,** 775.
23. Bayston, J. H.; Looney, F. D.; Pilbow, J. R.; Winfield, M. E. *Biochemistry* **1970,** *9,* 2164.
24. Bayston, J. H.; King, N. K.; Looney, F. D.; Winfield, M. E. *J. Am. Chem. Soc.* **1969,** *91,* 2775.
25. Lexa, D.; Saveant, J. M. *J. Am. Chem. Soc.* **1976,** *98,* 2652.
26. Lexa, D.; Saveant, J. M.; Zickler, J. *J. Am. Chem. Soc.* **1977,** *99,* 2786.
27. Hogenkamp, H. P. C.; Holmes, S. *Biochemistry* **1970,** *9,* 1886.
28. Pratt, J. M. "Inorganic Chemistry of Vitamin B₁₂"; Academic: London, 1972; Chapter 7.
29. Hayward, G. C.; Hill, H. A. O; Pratt, J. M.; Vanston, N. J.; Williams, R. J. P. *J. Chem. Soc.* **1965,** 6485.
30. Hayward, G. C.; Hill, H. A. O.; Pratt, J. M.; Williams, R. J. P. *J. Chem. Soc.* **1971,** A 196.
31. Stanienda, A.; Biebl, G. *Z. Phys. Chem. (Frankfurt am Main)* [*N. S.*] **1967,** *52,* 254.
32. Felton, R. H.; Linschitz, H. *J. Am. Chem. Soc.* **1966,** *88,* 1113.
33. Conlon, M.; Johnson, A. W.; Overend, W. R.; Rajapaksa, D.; Elson, C. M. *J. Chem. Soc., Perkin Trans. 1* **1973,** 2281.
34. Hush, N. S.; Woolsey, I. S. *J. Am. Chem. Soc.* **1972,** *94,* 4107.
35. Schrauzer, G. N.; Deutsch, E. *J. Am. Chem. Soc.* **1969,** *91,* 3341.
36. Friedrich, W.; Messerschmidt, R. *Z. Naturforsch.* **1969,** *24b,* 465.
37. Pratt, J. M. "Inorganic Chemistry of Vitamin B₁₂"; Academic: London, 1972; Chapter 6.
38. Endicott, J. F.; Ferraudi, G. J. *J. Am. Chem. Soc.* **1977,** *99,* 243.
39. Schrauzer, G. N.; Sibert, J. W.; Windgassen, R. J. *J. Am. Chem. Soc.* **1968,** *90,* 6681.
40. Castro, C. E.; Robertson, C.; Davis, H. F. *Bioorg. Chem.* **1974,** *3,* 343.
41. Davis, D. G.; Bynum, L. M. *Bioelectrochem. Bioenerg.* **1975,** *2,* 184.
42. Gilbert, B. C.; Norman, R. O. C.; Placucci, G.; Sealy, R. C. *J. Chem. Soc., Perkin Trans. 2* **1975,** 885.
43. Thomas, J. K. *J. Phys. Chem.* **1967,** *71,* 1919.
44. Golding, B. T.; Kemp, T. J.; Sellers, P. J.; Nocchi, E. *J. Chem. Soc., Dalton Trans.* **1977,** 1266.
45. Jencks, W. P.; Regenstein, J. "Handbook of Biochemistry and Molecular Biology"; Fasman, G. D., Ed.; CRC Press: Cleveland, 1976; Vol. 1.

RECEIVED May 14, 1979.

Nature of the Iron–Oxygen Bond and Control of Oxygen Affinity of the Heme by the Structure of the Globin in Hemoglobin

M. F. PERUTZ

MRC Laboratory of Molecular Biology, Cambridge CB2 2QH, England

Spectroscopic and chemical evidence speak in favor of the iron–oxygen bond being polar. X-ray analysis shows that the oxygen molecule is inclined at an angle of about 115° to the heme plane. Cooperative binding of oxygen by hemoglobin is attributable to an equilibrium between two alternative structures that differ in oxygen affinity by the equivalent of 3–3.5 kcal/mol. The author has proposed that in the low-affinity structure the globin opposes the movement of the iron atom from its pentacoordinated pyramidal geometry in the heme of deoxyhemoglobin to its hexacoordinated planar geometry in the heme of oxyhemoglobin, while in the high-affinity structure this restraint is absent. Recent evidence supporting this mechanism is described.

This chapter discusses the Fe–O bond in myoglobin and hemoglobin and the origin of the cooperativity of the reaction of hemoglobin with oxygen. The nature of the Fe–O bond has been the subject of speculation and experiment. The oxygen molecule has a spin of $S = 1$ and the ferrous iron in deoxyhemoglobin has a spin of $S = 2$, yet when the two combine to form oxyhemoglobin Pauling and Coryell found the compound to be diamagnetic; they argued that the Fe–O bond should have the resonating structure:

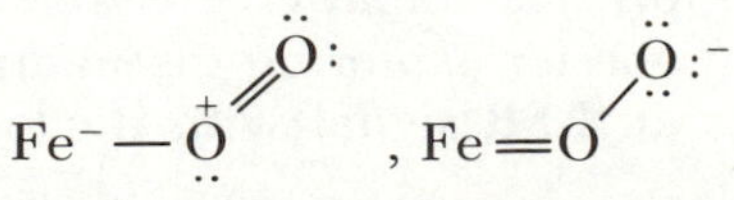

$$\text{Fe}^- \!-\! \overset{+}{\underset{..}{\text{O}}} \!\!\overset{\displaystyle \ddot{\text{O}}:}{\diagup} \quad , \quad \text{Fe} \!=\! \text{O} \!\!\overset{\displaystyle \ddot{\text{O}}:^-}{\diagup}$$

0-8412-0514-0/80/33-191-201$05.00/0
© 1980 American Chemical Society

that makes all electrons paired (*1*). In modern terms this means that the two $1\pi_g^*$ orbitals no longer have the same energy, so that their electrons pair in the single π_y^*, which has a lower energy than π_x^*, because it lies at right angles to the Fe—O—O plane. On Pauling's model, the bond between the iron and oxygen would be made by hybridization between the π orbitals of the oxygen and the d_{xz} and d_{yz} orbitals of the iron, with some net transfer of charge from the oxygen to the iron. This model was challenged by J. J. Weiss, who suggested that the bond might be ionic between a ferric ion and a superoxide ion, net charge being transferred from the iron to the oxygen $(Fe^{3+}O_2^-)$ (*2*). Experimental support for Weiss' model was first advanced by Misra and Fridovich (*3*). They showed the autoxidation of oxyhemoglobin to be a first-order reaction depending only on $[HbO_2]$; when epinephrin was added to the solution it was oxidized to adrenochrome, but this oxidation was inhibited in the simultaneous presence of superoxide dismutase and catalase, which suggests that superoxide ion is liberated on autoxidation of oxyhemoglobin. Recently, Demma and Salhany (*4*) have shown that liberation of oxygen by flash photolysis of oxyhemoglobin reduces cytochrome c, and this, too, is inhibited by superoxide dismutase and catalase. It could be argued that the superoxide ion is the result of an excited state induced by the flash, but solvent effects also speak in favor of a polar $(Fe^{3+}O_2^-)$ bond. Brinigar et al. (*5*) synthesized a heme with a covalently attached pyridyl base that allows it to combine reversibly with molecular oxygen at $-45°$. In dimethylformamide ($\epsilon = 36$) half saturation of this complex requires an oxygen pressure of only 5 torr, in 10% *N*-methylpyrrolidine–90% toluene 28 torr, and in pure toluene ($\epsilon = 2.4$), about 400 torr. The $(Fe^{3+}O_2^-)$ structure is also supported by spectroscopic evidence. The IR O—O stretching frequency in oxyhemoglobin is 1107 cm^{-1} (*6*), which is in the superoxide ion range (1150–1100 cm^{-1}), much lower than that of the oxygen molecule (1556 cm^{-1}), and higher than that of a single O—O bond (~800 cm^{-1}). X-ray fluorescence also points to the presence of unpaired electron density on the iron atom (*7*). Cobalt porphyrins and hemoglobins combine reversibly with molecular oxygen. Again solvent effects speak in favor of a polar Co–O bond. Stynes and Ibers (*8*) found that the oxygen affinity of cobalt-porphyrin complexes rises with polarity of the solvent. At $-23°C$ Co(II)protoporphyrin IX dimethylester methyl-imidazole in dimethylformamide requires an oxygen pressure of 12.6 torr for half saturation; substitution of toluene as a solvent raises that pressure to 417 torr. The unpaired electron of the cobaltous d^7 ion provides a useful probe for exploring the nature of the Co–O bond. This electron gives an ESR signal with nuclear hyperfine splitting from which the unpaired electron density can be located. In the deoxy derivatives hyperfine splitting from the cobalt is combined with that of a single nitrogen atom of the proximal histidine, which shows that the unpaired electron occupies the d_{z^2} orbital pointing towards the histidine. In the oxy derivatives, the nitrogen splitting disappears and the separation of the hyperfine lines due to cobalt is reduced to about a third. Since that separation is directly proportional to the unpaired electron density on the cobalt, it is inferred that a substantial fraction of the density has gone to the oxygen (9, 10, 11). This has been confirmed by a similar experiment in reverse, using $CoHb^{17}O_2$. ^{17}O has a nuclear spin of $I = \frac{5}{2}$, so should give rise to hyperfine splitting if the unpaired electron density is transferred from the Co to O_2. This is indeed observed, and its magnitude suggests that about 60% of the unpaired electron density is transferred (12). It could be argued that the metal–oxygen bonds might be different in the Fe and Co derivatives, but the similarity of the O—O stretching frequencies, 1107 cm^{-1} for Fe and 1106 cm^{-1} for Co, suggests that they are similar (13). Unfortunately, Weiss died before his prediction was confirmed experimentally. The $Fe^{3+}O_2^-$ model can be reconciled with diamagnetism or weak paramagnetism of the complex by postulating that the transferred d electron of the iron pairs with one of the two π^* electrons of the oxygen, and that the spin of the other π^* electron is paired with the odd d electron left behind on the iron by antiferromagnetic coupling. The diamagnetism of oxyhemoglobin has been challenged recently by Cerdonio et al., who produced evidence of a low-lying triplet state that makes it weakly paramagnetic at room temperature (14, 15). Their observations, though apparently flawless, do raise problems. For example, if oxyhemoglobin at room temperatures had a molar susceptibility per heme of $+2460 \times 10^{-6}$ cgs/mol, as they report, one would expect its NMR spectrum to exhibit hyperfine-shifted heme proton resonances, but these have not been observed.

The oxygen adducts of the picket-fence complex definitely are diamagnetic (16). The oxygen molecules are bent to the heme axis and lie in four alternative orientations. Because of that disorder it has not been possible to determine the coordinates of the terminal oxygen as accurately as those of the other atoms. In the 1-methylimidazole (1-MeIm) complex O—O = 1.16 Å and Fe—O—O = 131°, and in the 2-MeIm complex O—O = 1.22 ± 0.02 Å and Fe—O—O = 129 ± 1°. The authors state that they may have underestimated the O—O distance in the unhindered complex by as much as 0.15 Å. The geometry agrees with Pauling's prediction of a bent FeO_2 bond, and the O—O distance is close to that of 1.27 Å, predicted by him in a recent paper. It is slightly shorter than that of 1.34 Å in the superoxide anion, in agreement with the ESR results, which show that no more than two-thirds of the density of one electron is transferred from the metal to the antibonding π^* orbitals of the oxygen.

The structures of sperm whale deoxy- and oxymyoglobin have been refined to a resolution of 1.6 Å (17, 18). In deoxymyoglobin the iron atom lies at the apex of a pyramid with the four nitrogen atoms at its base. The displacement of the iron from the mean plane of the four porphyrin nitrogens is 0.4 Å and from the mean porphyrin plane, including the α carbons, 0.55 Å. These displacements are the same, within error, as in the model compound (2-MeIm) meso-tetra-phenylporphinato Fe(II) shown in Figure 1 and also the same as in deoxyhemoglobin (19). On the other hand, unlike the picket-fence hemes, the heme in oxyhemoglobin is not planar, but the iron is displaced from the plane of the four porphyrin nitrogens by 0.18 Å towards the proximal histidine. The difference may be caused by the dihedral angle between the plane of the histidine imidazole and the Fe–pyrrole bonds. In the picket-fence complex the imidazole lies at

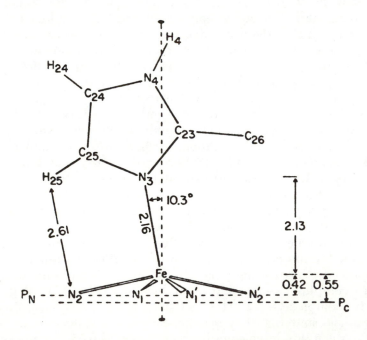

Figure 1.   Coordination of the iron in (2-methylimidazole)mesotetra-phenylporphinato Fe(II). As far as can be judged at the resolution of the deoxymyoglobin and hemoglobin Fouriers, the stereochemistry of the iron is the same in these proteins as in this model compound. P_N is the mean plane of the porphyrin nitrogens; P_C the mean plane of the porphyrin nitrogens and carbons (53).

about 45° to the N—Fe—N bonds (*16*), which minimizes van der Waals repulsion and maximizes overlap of the Fe *d* orbitals with the π orbitals of the imidazole nitrogen. In oxymyoglobin, on the other hand, the plane of the imidazole makes an angle of only 7° with the N—Fe—N bonds. In this eclipsed orientation repulsion is maximized and overlap minimized so that the proximal histidine tends to pull the iron away from the porphyrin plane. The oxygen molecule occupies a single ordered position with Fe—O—O = 115° and Fe—O = 1.8 Å. The imidazole of the proximal histidine and Fe—O—O are approximately coplanar, the oxygen being constrained to that orientation by steric hindrance of the distal histidine, valine, and phenylalanine. N_ϵ of the distal histidine is in contact with the first, iron-bound oxygen atom, but it is not clear from the x-ray data whether this is a van der Waals contact or a hydrogen bond (Figure 2). Chemical evidence speaks in favor of the former.

I now come to cooperative oxygen binding, also known as heme–heme interaction, which is exhibited by all vertebrate hemoglobins. These hemoglobins are tetrameric and have sigmoid oxygen equilibrium curves, which means that their oxygen affinity rises with increasing oxygen saturation. This cooperative behavior is attributable to a transition between two alternative structures in equilibrium, one with a low and the other with a high oxygen affinity (*20, 21, 22, 23*). They are distinguished by the internal structure of the four subunits, by the mutual arrangement of the subunits, and by the number of energy of the bonds between them. In the oxy, or R structure, the iron atoms are hexacoordinated and the structure appears to put no significant constraints on the heme that are not also present in free subunits; the oxygen affinity of this structure is only slightly higher than that of free subunits. In the deoxy, or T structure, the iron atoms are pentacoordinated and the hemes are constrained by additional bonds within and between the subunits; the oxygen affinity of that structure is lower than that of free subunits or that of the R structure by the equivalent of 1.5–3.5 kcal/mol Fe, depending on the strength of the constraining bonds.

For understanding heme–heme interaction, the two basic questions are: how does combination of ligands with the heme irons change the quaternary structure of the globin from T to R?; and conversely, how does the change from R to T lower the ligand affinity of the heme iron? I proposed that the equilibrium between the two structures is governed by the displacement of the iron atoms and the proximal histidines from the plane of the porphyrins and by the steric effect of the ligand on the distal valines in the β subunits (*24*). By the laws of action and reaction, if movement of the iron and the proximal histidine towards the porphyrin on ligand binding changes the structure from T

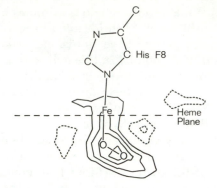

Figure 2a. Difference Fourier synthesis of oxymyoglobin, showing the electron density for the bound oxygen. The Fourier synthesis was computed with $|F_{observed}|$ − $|F_{calculated}|$ as coefficients; the calculated structure amplitudes were derived from the positions of all the atoms except the two oxygens.

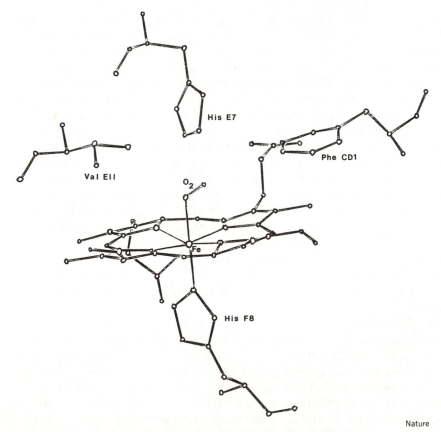

Nature

Figure 2b. Stereochemistry of the heme pocket with bound oxygen (17).

to R, then a transition from R to T must pull the iron and histidine away from the porphyrin. In that case, the T structure should exercise a tension on the heme that restrains the iron from moving into the porphyrin plane (25). The existence of such a restraint should be detectable by physical methods.

To study the influence of the quaternary structure of the globin on the state of the heme, we needed a method of changing that structure without changing the sixth ligand at the heme. There are two ways of doing this: one is to use a valency hybrid in which the hemes in either the α or the β subunits are ferric; combination of the ferrous hemes with ligand is used to change the quaternary structure and the effect on the ferric hemes is studied spectroscopically. Alternatively, addition of the allosteric effector inositolhexaphosphate (IHP) may switch the quaternary structure from R to T. In certain fish hemoglobins the transition may be accomplished by merely lowering the pH.

Gibson (26) first observed a change in the Soret (γ) band on transition from what are now known to have been deoxyhemoglobin dimers, which are equivalent to the R structure, to tetramers in the T structure; fuller descriptions of the spectral changes have been given by Brunori et al. (27) and by Perutz et al. (28). They consist of blue shifts of the Soret, visible, and near IR bands, together with the appearance of a shoulder at 590 nm flanking the peak at 556 nm. The spectral changes in mammalian and fish hemoglobins are similar. Sugita (29) showed that they are caused almost entirely by the hemes in the α subunits. Perutz et al. (30) suggested that these shifts may be attributable to an increase in Fe–N bond distances on transition from the R to the T structure, but it was not clear if this interpretation was correct.

The evidence from resonance Raman spectra is conflicting. At first such spectra of deoxyhemoglobin in the R and T structures seemed indistinguishable, but with improved techniques differences have been found. The most striking difference occurs in the low frequency region which has been explored by Nagai et al. (31) and by Desbois et al. (32). It consists of a shift of a band at 215 cm^{-1} in the T structure to 220 cm^{-1} in the R structure. Hori and Kitagawa have assigned this band to the Fe—N$_\epsilon$ stretching frequency (33). Using difference Raman spectroscopy Shelnutt et al. have found differences also in the high frequency region (34). Bands that lie at 1357, 1471, 1567, and 1605 cm^{-1} in the T structure shift to lower frequencies by between 1.3 and 2.2 cm^{-1} in the R structure (Figure 3), but no shifts occur in any of the bands that are believed to be markers for the Fe–N$_{porph}$ bond lengths. The shifts occur in bands known to be sensitive to the occupation of the π^* antibonding orbitals of the porphyrin, and correspond to a charge transfer to these orbitals on going from the T to the R structure. Since the binding of oxygen requires charge transfer from the heme to the oxygen molecule, the

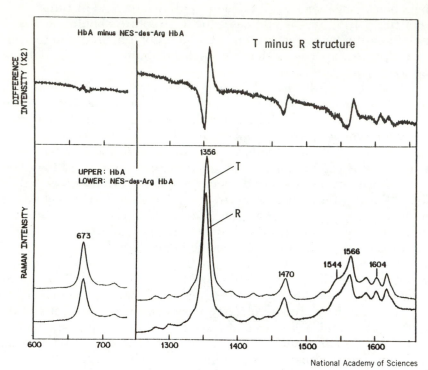

National Academy of Sciences

*Figure 3.   Resonance Raman spectra of human deoxyhemoglobin in
the T and R structures with difference spectrum, showing the shifts to
lower frequency on transition from T to R (34).*

presence of extra charge in the π^* orbitals of the heme would favor
higher oxygen affinity. At this stage it is not clear where the charge
comes from, nor how much this effect contributes to the free energy of
heme–heme interaction.

Nitrosylhemoglobin (HbNO) proved the most revealing of the
ferrous derivatives. The R → T transition gives rise to blue shifts and
reduction in intensity of the α, β, and γ bands, and the appearance of
"high spin" bands at 495, 518, and 603 nm, though the complex remains
low spin ($S = \frac{1}{2}$) (35, 36). Studies of hybrid hemoglobins show that
these spectral changes are mainly caused by the α subunits, just as in
deoxyhemoglobin. Nitrosylhemoglobin in the R structure shows a
single IR stretching frequency characteristic for hexacoordinated nitrosyl hemes; transition to the T structure causes the appearance of a
second IR band, of intensity equal to the first, characteristic of pentacoordinated hemes (37). Similarly, nitrosylhemoglobin in the R
structure shows an ESR spectrum similar to that of hexacoordinated
nitrosyl hemes, while nitrosylhemoglobin in the T structure shows a
composite of penta- and hexacoordinated nitrosyl hemes (38). Similar
results have been obtained from comparisons of the resonance Raman

spectra of penta- and hexacoordinated nitrosyl hemes with those of nitrosylhemoglobin in the R and T structures (*39, 40*) (Figure 4). Taken together, these results imply that in the R structure all four hemes are hexacoordinated, but in the T structure only the hemes in the β subunits remain hexacoordinated; those in the α subunits become pentacoordinated, the iron–histidine bond having been broken by the restraints that impede the movement of the proximal histidine towards the porphyrin. This experiment corroborates the restraint at the hemes in the T structure, but it does not determine its energy equivalent, though kinetic experiments showed that an activation energy of 17 kcal/mol was needed to break the $Fe–N_\epsilon$ bond (*35, 41, 42*). The bond breaks because occupation of the antibonding Fe d_{z^2} orbital by the unpaired NO electron has weakened it.

We now come to the class of ferric derivatives that have provided the most useful information concerning the effect of changes of quaternary structure on the state of the heme. These are the mixed-spin derivatives in which there exists a thermal equilibrium between the spin states of $S = \frac{5}{2}$ and $S = \frac{1}{2}$. The derivatives investigated include hydroxyl, azide, thiocyanate, and nitrite methemoglobin (*34, 44*). The spectral changes induced by the R → T transition in these derivatives include blue shifts of the Soret band, increases in intensity of the visible high-spin bands and of the charge-transfer bands in the near IR, and decreases in intensity of the low-spin bands. The most striking spectral changes were seen in nitrite methemoglobin of carp, which has the red color characteristic of low-spin ferric hemoglobins in the R structure and the brown color characteristic of high-spin ferric hemoglobins in the T structure.

Magnetic measurements of human hemoglobins by NMR indicated that the R → T transition caused the paramagnetic susceptibility of hydroxymethemoglobin to rise by 45% and that of thiocyanatemethemoglobin by 11%. Human azide methemoglobin cannot be converted to the T structure, but two abnormal hemoglobins form valency hybrids that allowed the effect of the R → T transition on the paramagnetic susceptibility of the α and β subunits to be measured separately by IR absorption spectroscopy. This showed that in the R structure the α hemes are less than 10% high spin, while in the T structure with IHP the high-spin fraction rises to 27%. The high-spin fraction of the β hemes in the R structure is probably about 10%; in the T structure with IHP it rises to 35%. In trout IV and carp azide methemoglobin measurements of the magnetic susceptibility and of the azide stretching frequencies showed a rise of the high-spin fraction in the tetramer from about 10% in the R structure to 50% in the T structure, which is equivalent to a free energy change of approximately 1 kcal/mol heme (Figure 5). All these measurements were done at a single temperature near 20°C. Qualitatively, these results showed

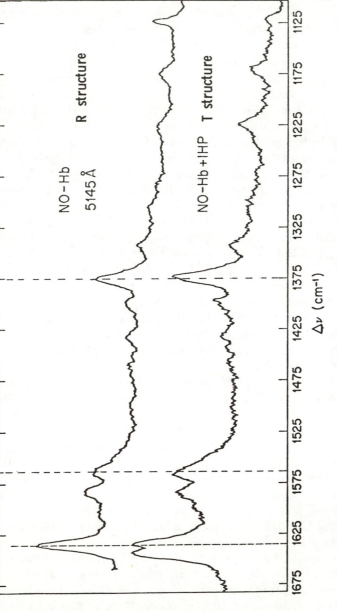

Figure 4. Resonance Raman spectra of human nitrosylhemoglobin in the T and R structures. The R structure shows a single band at 1634 cm^{-1} characteristic of hexacoordinated nitrosyl heme. The T structure shows two bands of equal intensity, one at 1634 cm^{-1} and the other at 1644 cm^{-1}, characteristic of hexa- and pentacoordinated hemes, respectively (40).

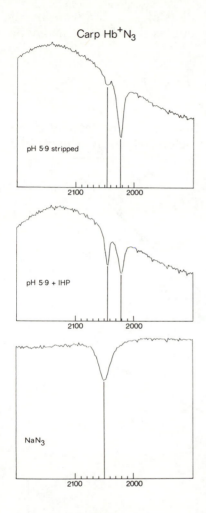

Carp Hb$^+$N$_3$

pH 5·9 stripped

2100   2000

pH 5·9 + IHP

2100   2000

NaN$_3$

2100   2000

Biochemistry

Figure 5. IR absorption spectra of carp azide methemoglobin at 20°C. In the R structure the azide stretching frequency at 2023 cm^{-1}, characteristic of low-spin azide hemes, dominates. In the T structure two peaks of equal intensity appear at 2023 and 2046 cm^{-1}, characteristic of low- and high-spin azide hemes, respectively (43).

that in mixed-spin derivatives the R → T transition causes a change to higher spin, equivalent to a stretching of the iron–nitrogen bonds (43).

Thermal spin equilibria in methemoglobin derivatives have been studied extensively (for review *see* Ref. 45), but measurements have been confined to derivatives in the R structure. Messana et al. (44) have determined the effect of the R → T transition on spin equilibria by measuring magnetic susceptibilities between 300 and 90 K with a high-resolution, superconducting magnetometer. The derivatives used were carp azide, nitrite, and thiocyanate methemoglobin. The authors expected to find a dependence upon 1/T like the theoretical curves for mixed-spin derivatives, that is, a low-spin ground state followed by a gradual transition to higher spin above some critical temperature, but the actual results were rather different.

At the lowest temperatures all the plots of χ vs. $1/T$ were linear (Figure 6). From the linear parts of the curves the effective magnetic moments, μ_e, could be derived:

$$\mu_e = \sqrt{\frac{3\chi kT}{\sqrt{N}\,\beta^2}} = 2.828\ \sqrt{\chi T}$$

where N is Avogadro's number, k, the Boltzmann constant, and β, the Bohr magneton. For all but one of the hemoglobins in the R structure, the

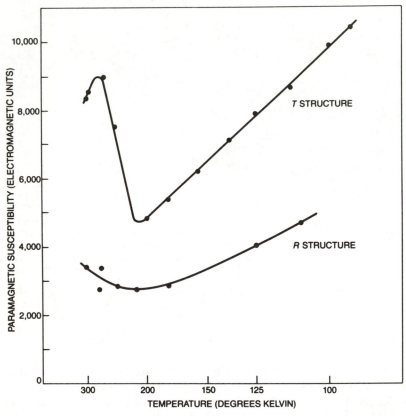

Figure 6. Direct measurement of the magnetic changes observed on switching carp azide methemoglobin from the R to the T structure. The paramagnetism of the iron atoms is higher in the T than in the R structure at all temperatures. In the R structure at low temperature all the iron atoms are low spin and the Curie Law is obeyed. At about 200 K the thermal energy approaches ΔE, the difference in energy between the $^6A_1$ and $^2T_2$ states, the iron atoms begin to oscillate between them, and the susceptibility rises with rising temperature. In the T structure at low temperatures a random mixture of high- and low-spin iron atoms is frozen in. At about 250 K the fraction of high-spin iron atoms rises sharply, only to drop again at higher temperatures for reasons that are not clear (44).

moments have values characteristic of low-spin heme complexes, but for all hemoglobins in the T structure and for the thiocyanate derivative in the R structure, their values are intermediate between low and high spin. In all instances the effective magnetic moment for the frozen hemoglobins in the T structure is larger than in the R structure. In solution, several derivatives show reverse Curie behavior characteristic of compounds in which there exists a thermal equilibrium between two spin states. This behavior persists after freezing down to temperatures of between 250 and 200 K. The magnetic moments in solution in the T structure are also larger than in the R structure; the rise just above the freezing point varies from 1.26-fold in thiocyanate to 1.84-fold in the nitrite derivative. The 1.52-fold rise in carp azide methemoglobin at room temperature corresponds to a 2.4-fold rise in magnetic susceptibility, which agrees with the 2.5-fold increase found in the closely related trout IV azide methemoglobin by NMR and also with the increase calculated from the change in the relative intensities of the high- and low-spin IR azide stretching frequencies of carp azide methemoglobin mentioned above (*43*).

Figure 7 shows a plot of spin equilibria calculated from the data in Figure 6. The two sloping lines correspond to the temperature range where a thermal spin equilibrium exists. From the difference in height between them, the free energy difference between the equilibria in the R and T structures is calculated as 0.9–1.2 kcal/mol Fe, in agreement with the value derived from the IR measurements mentioned above, and with a recently published determination of the spin equilibrium by resonance Raman spectroscopy (*46*).

So much for the physical measurements to determine the energy equivalent of the restraint that opposes transition to the low-spin state of the T structure. The first chemical evidence that restraint of the kind I had envisaged could actually diminish the ligand affinity of heme came from an experiment by Rougee and Brault (*47*). They measured the carbon monoxide affinity of deuteroheme in benzene at 25°C. That affinity dropped 200-fold on substitution of the sterically hindered 2-methylimidazole for imidazole as the distal base. Collman, Traylor, and their co-workers (*48–51*) later asked whether restraint of the kind I had envisaged could actually lead to a reduced oxygen affinity. They measured the thermodynamic constants of oxygen binding to some of the cobalt and iron picket-fence complexes with either *N*-methylimidazole or 1,2-methylimidazole as the fifth ligand. The former combines with the cobalt or iron atom without steric hindrance; the latter is restrained by close contact of the 2-methyl group with the porphyrin, so that it opposes the movement of the metal atom into the plane of the porphyrin on ligation with oxygen. This is the same effect as I imagined the globin has on the T structure of hemoglobin. The results show that the mean oxygen affinity (p_{50}) for the unhindered

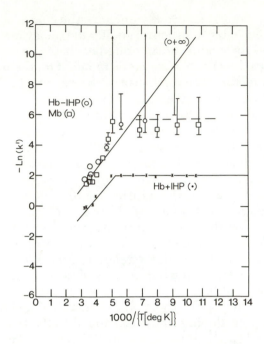

Biochemistry

Figure 7. Experimental points and theoretical curves replotted as the negative logarithm of the apparent spin equilibrium constant $K' = (8\chi T - \mu_L^2)/(\mu_H^2 - 8\chi T)$, where $-\ln K' = \Delta G/RT$. χ is the paramagnetic susceptibility, $\mu_L^2 = 4.0\,\mu_B^2$ and $\mu_H^2 = 35\,\mu_B^2$; (---) indicates the low temperature asymptote of $-\ln K'$ for azide metmyoglobin. Error bars are drawn for an arbitrarily assumed error of $\pm 3\%$ in χ for all data and tend to become very large when μ_e^2 approaches μ_L^2. Hence, the low temperature data are not useful for determining ΔG (44).

N-methyl cobalt complex is 150 torr, compared with 960 torr in the hindered 1,2-methylimidazole cobalt complex. In the iron complexes the corresponding values are 0.59 and 38 torr, corresponding to a difference in free energy of oxygen binding of 2.5 kcal/mol, comparable with the free energy of heme–heme interaction in hemoglobin. The answer to the question asked at the beginning of this paragraph is therefore in the affirmative, restraint does lead to reduced oxygen affinity.

On the basis of this, or indeed of any theory of heme–heme interaction, one would have expected the Fe—O bond to be weaker in the T than in the R structure, weaker bonds being associated normally with lower affinity. Nagai et al. (31) have tested this point by resonance Raman scattering, but found the Fe—O stretching frequency of oxyhemoglobin to be within 5 cm^{-1} the same in the two structures. They have now repeated this experiment with the oxygenated picket-fence complex, comparing the Fe–O stretching frequency in the unhindered 1-methylimidazole complex, which has an oxygen affinity comparable with that of the R structure, with that in the hindered 1,2-methylimidazole complex, which has an oxygen affinity comparable with that of the T structure. The result was the same: the Fe–O stretching frequencies were identical. They were not able to detect the Fe–N_{Im} stretching frequency. Since the Fe–O stretching frequency is 570 cm^{-1}, compared with 215 cm^{-1} for the Fe–N_{Im} bond (in deoxyhemeoglobin), the latter is clearly the weaker of the two and therefore may be the one that is stretched by the constraints of the T structure or the steric hindrance of the picket-fence complex. Therefore, this bond is the one where a major part of the free energy of heme–heme interaction is likely to be stored, but so far it has not been possible to measure its stretching frequency in any liganded hemoglobin derivative.

Conclusion

Ferrous iron in hemoglobin binds oxygen end on, with an Fe—O—O angle of 115°. The ligand pocket is constructed so it can accommodate an oxygen molecule in this orientation and constrain it within narrow limits to one azimuth. The Fe–O bond is polar with transfer of charge from the iron to the oxygen. The distances of the iron atom and the proximal histidine from the porphyrin plane play a vital part in determining the equilibrium between the T and R structures of the hemoglobin molecule. The low oxygen affinity of the T structure is associated with a restraint or tension that opposes the movement of the iron atom towards the porphyrin plane. The resulting strain is distributed between the heme and the protein in varying proportions in different derivatives. In oxyhemoglobin no significant effect of the T structure on the heme has been detected so far, which suggests that most of the strain is stored in the protein. In deoxyhemoglobin it seems to be associated with a stretching of the Fe–N_ϵ bond. In nitrosylhemoglobin it causes the hemes in the α subunits to be torn from their bonds with the proximal histidines of the globin. In derivatives where there exists a thermal spin equilibrium the restraints of the T structure shift that equilibrium towards higher spin, that is, towards the form with the longer iron–nitrogen bonds.

Model compounds designed to mimic the steric restraint of the proximal histidine in the T structure have lower oxygen affinities than do similar model compounds lacking such restraint, and the degree by which the oxygen affinity is lowered is of the same magnitude as in the T compared with the R structure, which shows that the concept of steric restraint is viable chemically.

Literature Cited

1. Pauling, L.; Coryell, C. *Proc. Natl. Acad. Sci. USA* **1936**, *22*, 210–216.
2. Weiss, J. J. *Nature* **1964**, *202*, 83–84.

3. Misra, H. P.; Fridovich, I. *J. Biol. Chem.* **1972**, *247*, 6960–6962.
4. Demma, L. S.; Salhany, J. K. *J. Biol. Chem.* **1977**, *252*, 1226–1231.
5. Brinigar, W. S.; Chang, C. K.; Geibel, J.; Traylor, T. G. *J. Am. Chem. Soc.* **1974**, *96*, 5597–5599.
6. Barlow, C. H.; Maxwell, J. C.; Wallace, W. J.; Caughey, W. S. *Biochem. Biophys. Res. Commun.* **1973**, *55*, 91–95.
7. Koster, A. S. *J. Chem. Phys.* **1975**, *63*, 3284–3286.
8. Stynes, H. C.; Ibers, J. A. *J. Am. Chem. Soc.* **1972**, *94*, 5125–5127.
9. Hoffmann, B. M.; Petering, D. H. *Proc. Natl. Acad. Sci. USA* **1970**, *67*, 637–643.
10. Chien, J. C. W.; Dickinson, L. C. *Proc. Natl. Acad. Sci. USA* **1972**, *69*, 2783–2787.
11. Yonetani, T.; Yamamoto, H.; Iizuka, T. *J. Biol. Chem.* **1974**, *249*, 2168–2173.
12. Gupta, R. K.; Mildvan, A. S.; Yonetani, T.; Strivastava, T. S. *Biochem. Biophys. Res. Commun.* **1975**, *67*, 1005–1012.
13. Maxwell, J. C.; Caughey, W. S. *Biochem. Biophys. Res. Commun.* **1974**, *60*, 1309–1314.
14. Cerdonio, M.; Congiu–Castellano, A.; Mogno, F.; Pispisa, B.; Romani, G. L.; Vitale, S. *Proc. Natl. Acad. Sci. USA* **1977**, *74*, 398–400.
15. Cerdonio, M.; Congiu–Castellano, A.; Calabresi, L.; Morante, S.; Pispisa, B.; Vitale, S. *Proc. Natl. Acad. Sci. USA* **1978**, *75*, 4916–4919.
16. Collman, J. P. *Acc. Chem. Res.* **1977**, *10*, 265–272.
17. Phillips, S. E. V. *Nature* **1978**, *273*, 247–248.
18. Phillips, S. E. V., private communication.
19. Fermi, G. *J. Mol. Biol.* **1975**, *97*, 237–256.
20. Perutz, M. F. *Br. Med. Bull.* **1976**, *32*, 195–208.
21. Perutz, M. F. *Sci. Am.* **1978**, *239*, December, 92–125.
22. Perutz, M. F. *Rev. Biochem.* **1979**, *48*, 327–386.
23. Baldwin, J. M. *Prog. Biophys. Mol. Biol.* **1975**, *29*, 225–320.
24. Perutz, M. F. *Nature* **1970**, *228*, 726–739.
25. Perutz, M. F. *Nature* **1972**, *237*, 495–499.
26. Gibson, Q. H. *Biochem. J.* **1959**, *71*, 293–303.
27. Brunori, M.; Antonini, E.; Wyman, J.; Anderson, S. R. *J. Mol. Biol.* **1968**, *34*, 357–359.
28. Perutz, M. F.; Ladner, J. E.; Simon, S. E.; Ho, C. *Biochemistry* **1974**, *13*, 2163–2173.
29. Sugita, Y. *J. Biol. Chem.* **1975**, *250*, 1251–1256.
30. Perutz, M. F.; Heidner, E. J.; Ladner, J. E.; Beetlestone, J. G.; Ho, C.; Slade, E. F. *Biochemistry* **1974**, *14*, 2187–2200.
31. Nagai, K.; Kitagawa, T. *Proc. Natl. Acad. Sci. USA* **1980**, *77*, 2033–2037.
32. Desbois, A.; Lutz, M.; Banerjee, R. *Biochemistry* **1979**, in press.
33. Hori, H.; Kitagawa, T. *J. Am. Chem. Soc.* **1980**, *102*, 3608–3613.
34. Shelnutt, J. A.; Rousseau, D. L.; Friedman, J. L.; Simon, S. R. *Proc. Natl. Acad. Sci. USA* **1979**, *76*, 4409–4413.
35. Perutz, M. F.; Kilmartin, J. V.; Nagai, K.; Szabo, A.; Simon, S. R. *Biochemistry* **1976**, *15*, 378–387.
36. Cassoly, R. *C. R. Acad. Sci. Ser. D.* **1974**, *278*, 1417–1419.
37. Maxwell, J. C.; Caughey, W. S. *Biochemistry* **1976**, *15*, 388–396.
38. Szabo, A.; Perutz, M. F. *Biochemistry* **1976**, *15*, 4427–4428.
39. Szabo, A.; Barron, L. D. *J. Am. Chem. Soc.* **1975**, *97*, 660–662.
40. Strong, J. D.; Burke, J. M.; Daly, P.; Wright, P.; Spiro, T. G. *J. Am. Chem. Soc.*, in press.
41. Salhany, J. M.; Ogawa, S.; Shulman, R. *Proc. Natl. Acad. Sci. USA* **1974**, *71*, 3359–3362.
42. Salhany, J. M.; Ogawa, S.; Shulman, R. G. *Biochemistry* **1975**, *14*, 2180–2190.
43. Perutz, M. F.; Sanders, J. K. M.; Chenery, D. H.; Noble, R. W.; Pennelly, R. R.; Fung, L. W.-M.; Ho, C.; Giannini, I.; Pörschke, D.; Winkler, H. *Biochemistry* **1978**, *17*, 3640–3652.

44. Messana, C.; Cerdonio, M.; Shenkin, P.; Noble, R. W.; Fermi, G.; Perutz, R. N.; Perutz, M. F. *Biochemistry* **1978**, *17*, 3652–3662.
45. Iizuka, T.; Yonetani, T. *Adv. Biophys.* **1970**, *1*, 157–211.
46. Scholler, D. M.; Hoffmann, B. M. *J. Am. Chem. Soc.* **1979**, *101*, 1655–1662.
47. Rougee, M.; Brault, D. *Biochemistry* **1975**, *14*, 4100–4106.
48. Collman, J. P.; Brauman, J. I.; Doxsee, K. M.; Halbert, T. R.; Hayes, S. E.; Suslick, K. S. *J. Am. Chem. Soc.* **1978**, *100*, 2761–2766.
49. Collman, J. P.; Brauman, J. I.; Doxsee, K. M.; Halbert, T. R.; Suslick, K. S. *Proc. Natl. Acad. Sci. USA* **1978**, *75*, 564–568.
50. Geibel, J.; Cannon, J.; Campbell, D.; Traylor, T. G. *J. Am. Chem. Soc.* **1978**, *100*, 3575–3585.
51. White, D. K.; Cannon, J. B.; Traylor, T. G. *J. Am. Chem. Soc.* **1979**, *101*, 2443–2454.
52. Warshel, A. *Proc. Natl. Acad. Sci. USA* **1977**, *74*, 1789–1793.
53. Hoard, J. L., private communication.

RECEIVED May 15, 1979.

The Chemical Basis of Variations in Hemoglobin Reactivity

TEDDY G. TRAYLOR, ALBIN P. BERZINIS, JOHN B. CANNON,
DWANE H. CAMPBELL, JON F. GEIBEL, TERRY MINCEY, S. TSUCHIYA,
and DABNEY K. WHITE

Department of Chemistry, University of California, San Diego, La Jolla, CA
92093

Model studies for the R- and T-states of hemoglobin indicate that molecular control over dioxygen and carbon monoxide binding is achieved principally by changes in the tension between the axial base and the heme. Electronic changes in the axial base, heme periphery, or surrounding medium substantially affect the dynamics of O_2, but not of CO, binding. Distal side steric effects are viewed as relatively unimportant in R- and T-state hemoglobin, although they may play a role in ligand binding of other heme proteins. Two new synthetic models with distal side steric effects are presented.

Hemoglobin, a tetrameric protein with one heme per chain, has the fascinating ability to bind dioxygen to one or two of these hemes and then use this binding to increase the affinity of the remaining hemes in the molecule (*1*). This classic example of cooperativity in a biological process has held the attention of researchers from physics to medicine for over a hundred years. The cooperative ligation currently is viewed as resulting from three phenomena (*2*): the binding of the first ligand; a protein conformational change; and alteration of the dynamic behavior of another heme in the molecule to improve its affinity.

Structural studies of normal and abnormal hemoglobins (*3, 4, 5*) have revealed a great deal about the protein conformational change and have suggested several possibilities for triggering this change. Such structural studies of the heme site area in these hemoglobins have also led to several suggestions for alteration of heme reactivity

0-8412-0514-0/80/33-191-219$05.00/0
© 1980 American Chemical Society

resulting from protein conformational change. The most widely held view is that of Perutz (6), following suggestions by Hoard (7), in which there are two states governed by the protein conformation: the R-state (relaxed, reactive), which has no strain, and the T-state (tense, unreactive) in which steric effects make ligation, and its accompanying change in heme geometry, more difficult. This change can be represented as the ratio of equilibria for a ligand binding to the R-state to that for binding to the T-state, K_R/K_T.

The most widely studied equilibria for hemoglobin are those involving oxygen binding. The binding is very sensitive to the concentrations of protons, NaCl, and various organic phosphates. It is also sensitive to temperature, and to the concentration of hemoglobin (which has a rather facile tetramer $\rightleftharpoons$ dimer equilibrium). In addition, the actual values reported depend upon the mathematical model used to describe the ligand binding. It is therefore not surprising that the K_R/K_T ratio varies widely, at least from 40 to 1000 (3, 4). The corresponding carbon monoxide ratio varies also, at least from 100 to 850 (8, 9, 10, 11).

Not only the equilibria, but also the kinetics of hemoglobin binding have been studied by several groups, in particular those of Gibson (9, 12, 13, 14) and Antonini (1). Two representative sets of values for human Hb A are given in Table I. These illustrate the dependence of the observed rates on buffer, pH, phosphates, and the α,β chain heterogeneity. However, despite this sensitivity, the data suggest ways in which model compounds can be used to test some of the detailed molecular theories of hemoglobin cooperativity. In particular, while

Table I.   Representative Hemoglobin Kinetic and Equilibrium Data.[a]

	Carbon Monoxide		Oxygen	
	Phosphate[b]	Borate[c]	Phosphate[b,d]	Borate[c]
$k_T{}^{on}$	1.1×10^5	5.3×10^5	$2.9, 11.8 \times 10^6$	4.2×10^6
$k_R{}^{on}$	6.0×10^6	1.24×10^7	$5.9, 5.9 \times 10^7$	5.9×10^7
$k_T{}^{off}$	0.09	0.023	183, 2480	116
$k_R{}^{off}$	0.006–0.009	0.0045	12, 21	11.5
K_T	1.2×10^6	2.3×10^7	$16, 4.8 \times 10^3$	3.6×10^4
K_R	7×10^8– 1×10^9	2.8×10^9	$4.9, 2.8 \times 10^6$	5.1×10^6
K_R/K_T	580–830	120	300, 580	135
Reference	8, 10, 11	9, 10	13	12

[a] On a per heme basis, see Ref. 3, 4, 9, and 12 for calculation details.
[b] pH 7, 0.05–0.1M phosphate at 20°C.
[c] pH 9, 0.1M borate at 20°C.
[d] α and β chains, respectively.

hemoglobin acquires increased affinity for both O_2 and CO in the change from T- to R-state, it seems to do so for different reasons. The increased O_2 affinity appears to be largely a result of a decrease in the dissociation rate. In contrast, the increased CO affinity results from a faster association rate and a slower dissociation rate, with the association rate showing the greater change. With kinetic measurements for the on and off rates for O_2 and CO in R- and T-state hemoglobin available in the literature, we had eight rates to match, and hoped to find models that would allow us to explain the molecular control of heme binding without having to use different effects for O_2 and for CO binding.

Development of the Model

A view of the hemoprotein binding site is shown in Figure 1. The primary requirement for a model system is that it reversibly bind O_2. We thought initially (15) that three features would be required: a distal-side protecting group; a method of maintaining pentacoordination; and a heme whose structure resembled the natural one. Early synthetic efforts in our own laboratory (cyclophane heme) (15), followed by more definitive results of Collman et al. (picket fence) (16) and Almog et al. (capped) (17) hemes provided models designed to incorporate the first of the three requirements given above.

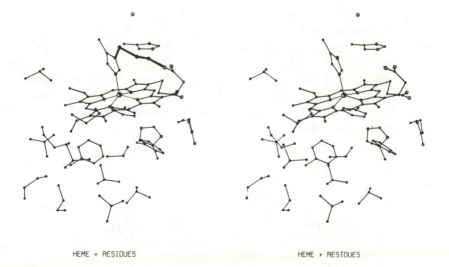

HEME + RESIDUES   HEME + RESIDUES

Figure 1. Hemoprotein binding site in a stereoscopic view. The broadened line in the left figure represents the method of covalently attaching the proximal imidazole to give a strain-free pentacoordinated heme, discussed in the text.

This requirement, having a distal-side protecting group, was important to inhibit the oxy complex from reacting with another heme to form a μ-peroxy complex, with subsequent oxidation of both hemes and loss of reversible O_2 binding. However, shortly after synthesizing the first cyclophane porphyrin, we discovered that distal-side protection was unnecessary for kinetic studies (18, 19). The flash photolysis of mixtures of heme, CO, and O_2, first pioneered with hemoproteins (14), was surprisingly easy to apply to simple hemes. This method depends on the observation that although a heme or heme protein has a higher affinity for CO, it binds O_2 more quickly. Thus a model heme without distal-side protection may be oxygenated and deoxygenated hundreds of times if each oxygenation cycle is short enough and if any heme molecule not found as an oxy complex is protected as the carbonmonoxy complex (19).

The second requirement, a method of maintaining pentacoordination, was important because addition of a second imidazole to the pentacoordinated heme–imidazole complex is very facile ($K_B{}^B > K^B$) and the presence of dibase hexacoordinated species complicates kinetic studies (20).

The three groups used three different approaches. The capped or protected porphyrin approach was used by Baldwin (17), who found that it not only allowed reversible binding of O_2, but also prevented hexacoordination. In a second approach, Collman and Reed (21) used the hindered 2-methylimidazole, which binds only once to the heme. However, Rougee and Brault (22) found that this method of maintaining pentacoordination reduced the binding of other ligands to the sixth position (opposite the 2-methylimidazole). We have since found that a sterically hindered imidazole dramatically increases the participation of tetracoordinated heme in the dynamic process of ligation, the base-elimination mechanism shown in Reaction 1 (20). This approach is therefore not useful as a general kinetic solution, although it has been very successful for structural studies (23).

$$
\begin{array}{ccccc}
\text{Im} & \text{Im} & & \text{Co} & \\
| & | & & \| & \\
-\text{Fe}- \rightleftharpoons & -\text{Fe}- \rightleftharpoons & -\text{Fe}- \rightleftharpoons & -\text{Fe}- & \quad (1)\\
| & & & | & \\
\text{CO} & & & \text{CO} &
\end{array}
$$

Our approach followed an older idea (24, 25) of attaching the base to a sidechain already on the heme to form what we call a "chelated heme" (26). Addition of the sidechain is a one-step synthesis (26); the resulting product is unstrained, and, in most cases, reacts as the pen-

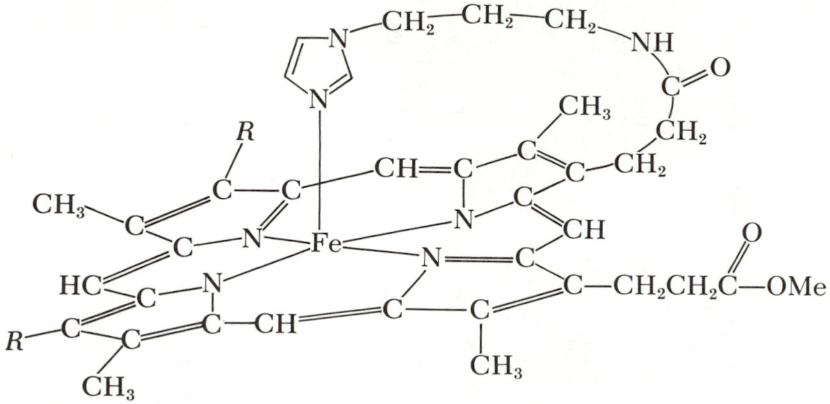

R = ethyl: chelated mesoheme
R = vinyl: chelated protoheme

tacoordinated heme–imidazole unit. Worries that the chelated heme would dimerize (27) have proved groundless for Fe(II) hemes at the concentrations used in kinetic studies ($<10^{-5}M$) (26, 28), although there is evidence that chelated Fe(III) hemes do dimerize (to give one tetra- and one hexacoordinated heme) at high concentrations in some solvents (29).

The ability of chelated hemes to mimic the kinetics and equilibria of ligation of R-state hemoglobin is seen by comparing the data for chelated protoheme with those for R-state hemoglobin (Table II). Not only are the spectroscopic and equilibrium properties of R-state hemoglobin matched by those of the chelated heme, but the association and dissociation rates are also well duplicated in aqueous cetyl-trimethylammonium bromide (CTAB). These rate constants agree with the general idea that R-state hemoglobin contains an unstrained or unhindered heme, and lead us to conclude that, in a kinetic sense, the globin serves only to protect the heme, maintain pentacoordination, and transmit the information that one heme has bound to the other hemes. Thus it seems that the bent geometry (31) of the CO in car-boxyhemoglobin (R-state) and its causative steric repulsion do not alter greatly the CO affinity. A corollary is that steric differentiation between O_2 and CO in R-state hemoglobin is also of minor importance. The situation in T-state hemoglobin is discussed below.

These results and conclusions contrast the suggestion that model heme compounds bind CO much more strongly than do hemoproteins (32). In the solid state, models appear to bind CO "irreversibly" at room temperature (18, 32), but this is in part due to the high affinity of all hemes, and may, in addition, reflect difficulties in removing the CO from the solid. In solution, models bind CO very strongly

Table II. Comparison of Spectroscopic and Kinetic Properties of Hemoglobin and Chelated Protoheme[a]

| Compound | Spectra[b] | | | Rate Constants[c] | | Equilibrium[a] Constant (M^{-1}) |
	Soret	α	β	On ($M^{-1}\,sec^{-1}$)	Off (sec^{-1})	
Hb	430 (133)		555 (12.5)			
Chelated protoheme	430 (114)		558 (13.5)			
Hb(CO)$_4$	419 (191)	569 (13.4)	540 (13.4)	6×10^6	0.006– 0.009	7×10^8– 1×10^9
Chelated protoheme–CO	420 (203)	569 (15.2)	540 (16.3)	4×10^6	0.007	6×10^8
Hb(O$_2$)$_4$	415 (125)	577 (14.6)	541 (13.8)	6×10^7	17	4×10^6
Chelated protoheme–O$_2$[e]	414 (121)	575 (14.2)	543 (16.5)	2×10^7	44	5×10^5

[a] Chelated protoheme was suspended in 2% CTAB–phosphate buffer at pH 7.3 (28).

[b] Extinction coefficients, mM, in parentheses; data for Hb from Ref. 1; for chelated protoheme from Ref. 26 and 30.

[c] See Table I for hemoglobin references; the oxygen rates are the average of the α and β chains. The chelated protoheme rates are from Ref. 26 and 30.

[d] k_{on}/k_{off}

[e] Static visible spectrum determined in 70 : 30 (v/v) DMF : H$_2$O.

$(K_B^{CO}$ (deuteroheme dimethylester, imidazole, benzene) = 4.3 × $10^8 M^{-1}$ (20, 22), K_B^{CO} (chelated protoheme, aqueous CTAB) = 7 × $10^8 M^{-1}$ (26), and K_B^{CO} (microperoxidase, water) = 1–2 × $10^9 M^{-1}$ (33)). However, these values are not significantly greater than those of R-state hemoglobin (7–30 × 10^8 (10)) or isolated α chains (2–4 × 10^8 (1)). When binding is expressed in pressure for half saturation, then the model compounds appear to have a higher affinity for CO in benzene or toluene where CO is more soluble $(P_{1/2} \simeq 0.0002$ torr) than they do in aqueous CTAB $(P_{1/2} \simeq 0.001$ torr). However, as stated above, this value in aqueous CTAB is close to that of R-state hemoglobin $(P_{1/2} = 0.0002–0.001$ torr) or α chains in aqueous solution $(P_{1/2} = 0.002–0.004$ torr).

Proposals for the Control of Ligand Binding

The salient features of the heme pocket appear to be (1) the heme is connected to the globin through the proximal imidazole, (2) the proximal imidazole may hydrogen bond to a nearby amide, (3) the vinyl groups are subject to steric removal from conjugation by rotation (with a resulting change in the electronic density of the heme), (4) a distal imidazole is close to the binding site, and (5) the distal pocket is both small and hydrophobic. Hemoglobin researchers have suggested a variety of ways in which these features might be used to control O_2 or CO affinity (1), and some of these are given in Figure 2. Features 2, 3

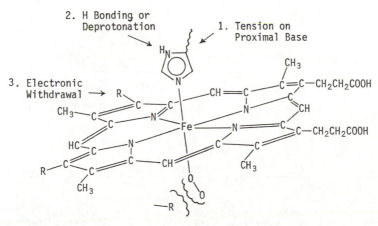

Figure 2. Proposed effects on O_2 and CO binding

and 5 can be grouped together as electronic and solvent effects, and are discussed together in the first section below. Feature 1, steric control of ligand binding (the Perutz mechanism), is discussed in the second section. Feature 4, the distal-side steric effect, is discussed in the last section.

Electronic and Solvent Effects

Our first concern was to model possible electronic effects on ligand binding by synthesis of a set of models with varying axial bases or porphyrin sidechains (Figure 3). These models are not the subject of steric strain, as shown by the fact that the compounds with $n = 3$ and $n = 4$ have similar rate constants for addition of CO (28). Neither the association rate constant of O_2 nor that of CO was altered greatly when either the proximal base, the sidechain basicity, or the solvent was changed. Therefore, we list in Figure 3 only the effect of these changes on the O_2 dissociation rate constant, k, and the CO dissociation rate constant, l. Figure 3 shows that increasing the electron withdrawal

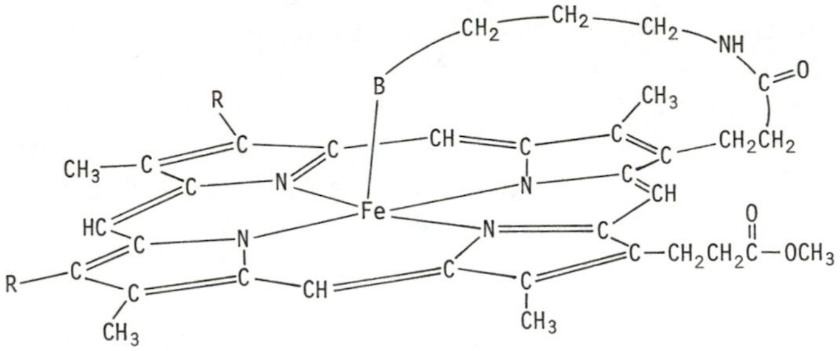

Side Chain Effect			Proximal Basicity Effect[a]			Solvent Effect		
R	l (30)	k (34)	B	l (30)	k (33)	Solvent	l (30)	k
Et	0.019	22	Im	0.06	23	2% CTAB/H_2O	0.005-0.008[b]	47 (34)
Vinyl	0.005	47	Im[c]	0.04	24	50/50 MeOH/H_2O	0.032	—
Acetyl	0.006	400	Pyr	0.034	380	95/5 toluene/MeOH	0.025	—
						90/10 toluene/CH_2Cl_2	—	1700 (26)

Figure 3. Effect of side chain, proximal basicity, and solvent on CO and O_2 dissociation rates. B = Im, R = vinyl unless otherwise noted. Solvent is aqueous detergent. Exact conditions vary but are internally consistent in any column. [a] R = ethyl. [b] Variations in kinetic method and in suspending agent. [c] The heme was prepared from $NH_2(CH_2)_4N$⬡N instead of $NH_2(CH_2)_3N$⬡N used in the other imidazole models.

retards CO dissociation and accelerates O_2 dissociation. In addition, O_2 is more sensitive (by a factor of five or greater) than is CO to these changes. This is interpreted as evidence in favor of the usual picture that there is more charge separation in the O_2 complex than there is in the CO complex (*19*). This interpretation is further strengthened by the

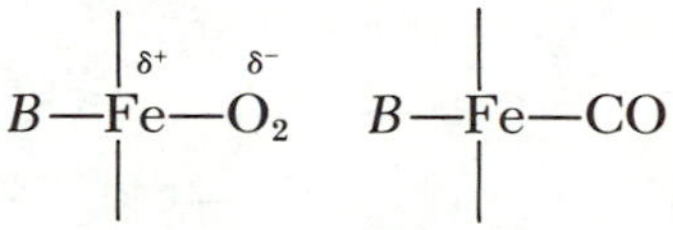

$$B-\overset{|}{\underset{|}{Fe}}-\overset{\delta^+ \quad \delta^-}{O_2} \qquad B-\overset{|}{\underset{|}{Fe}}-CO$$

effect of solvent on the off rates, which is far greater for O_2 than for CO. The main import of these results, however, is that electronic changes are probably not responsible for the difference between the T- and R-states. The hemoglobin T-to-R switch reduces the affinity for both O_2 and CO. Electron withdrawal in the models reduces O_2 affinity but increases CO affinity, and it is therefore unlikely that electronic effects are responsible for the difference between the T- and R-states.

The Proximal Tension Effect

Because protein conformational changes are changes in geometry, it has been tempting to consider a steric effect as the controlling factor in heme reactivity. This effect, as proposed by Hoard and Perutz (*6, 7*), involves a tension imposed by the protein on the proximal base, which in turn either pulls on the center of the heme as if it were a pump diaphragm or fits the naturally domed shape of the deoxyheme. Re-

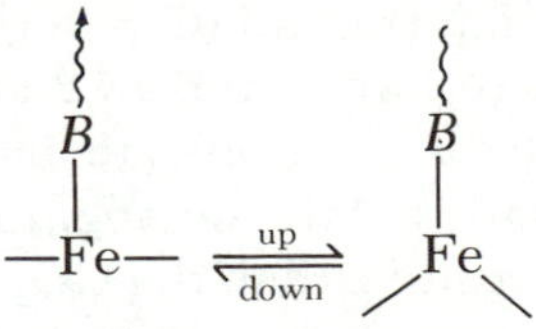

$$-\overset{B}{\underset{|}{Fe}}- \quad \underset{down}{\overset{up}{\rightleftharpoons}} \quad \overset{B}{\underset{\diagdown}{\diagup}}Fe$$

gardless of the exact mechanics of doming, the displaced iron becomes less accessible to hexacoordination. When B is let go or pushed, the heme (diaphragm) returns to its reactive (R) state.

We have modeled this diaphragm tension in three ways using chelated hemes as shown in Figure 4 (*28*): 1. Putting a large group (A) on the base itself; 2. Placing large groups (G) under the arm, thus lifting it so as to pull the base away; 3. Decreasing the arm length (n) to pull the imidazole to one side and thus strain it. Introduction of steric strain into chelated hemes by any of these three methods resulted in a change of mechanism from one of direct addition ($HB + L \rightarrow HBL$) to at least partly one of base elimination ($HB \rightarrow H \rightarrow HL \rightarrow HBL$ (*28*).

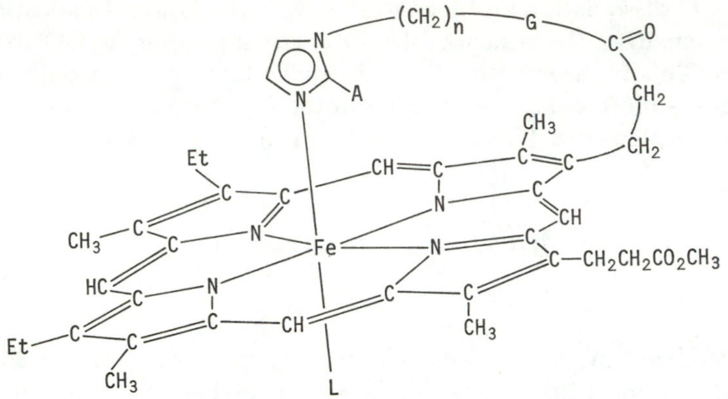

Figure 4.   Strained chelated mesoheme derivatives. See the text for the meanings of A, n, and G.

The first model introduced a methyl group in the 2 position of the axial imidazole, "pushing" the imidazole away from the heme plane. In this instance the steric strain was severe. This model oxidized very easily, even in the presence of CO. In addition, the rate of CO addition increased, rather than decreasing. Both of these characteristics arise from the easy loss of the axial base, which allows the ligation (and oxidation) to proceed through the tetracoordinated heme (28). The other two classes of models, however, contained much less strain, which resulted in some base elimination during CO addition, but did not seem to change the mechanism of oxygen addition (28). In the first of these, strain was introduced by substituting an alkyl sidechain for the hydrogen on the amide NH bond (G = NHR in Figure 4) and the O_2 on- and off-rates increased by factors of 2 and 7, respectively. The corresponding CO direct dissociation rate increased by less than a factor of 2. This model mimicked the hemoglobin T-to-R switch in that the O_2 off-rate was more sensitive than the CO rate to this type of steric strain, but the changes were quite small. Similar O_2 kinetics were obtained when the sidechain was shortened from $n = 3$ to $n = 2$ (G = oxygen in Figure 4).

The problem of base elimination in the CO kinetics was circumvented by measuring the kinetics of hemes with external 1- and 2-methylimidazole (20). The former is a strain-free system, and serves as a model for R-state hemoglobin, while the latter is strained (2-MeIm does not bind twice (21)), and serves as a model for T-state hemoglobin. The rate constants for addition of CO to the pentacoordinated heme-1-methylimidazole in CTAB buffer (Figure 5) are almost identical to those of R-state hemoglobin in Table I. The change from

1-methylimidazole to 2-methylimidazole resulted in a twelvefold increase in the off-rate and in a fifteenfold decrease in the on-rate, compared with ten- and sixtyfold changes, respectively, for hemoglobin in pH 7 phosphate. The comparison indicates that the steric strain in this model system affects the on- and off-rates almost equally, while in hemoglobin the T-to-R switch affects the on-rate more than the off-rate. Our results show that different models for T-state hemoglobin not only show different amounts of strain (reflected in the equilibrium constants) but also different effects of strain (reflected in the on- and off-rates). In view of the range of values measured for both hemoglobin and the models, it is probably premature to assign the kinetic differences between them to any specific effect.

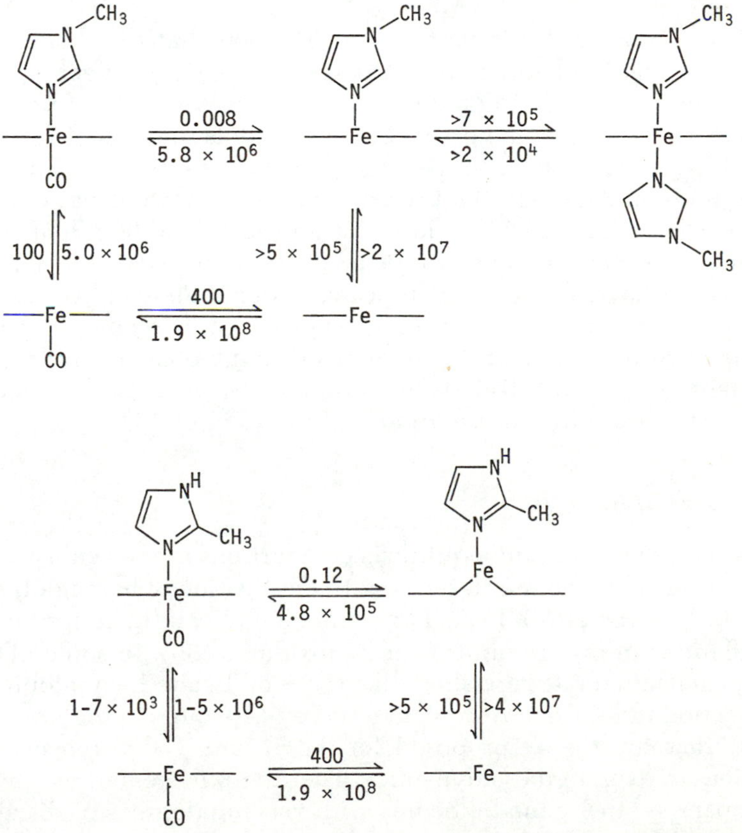

Figure 5.   Rate constants for reactions of CO with mesoheme dimethyl ester and the indicated imidazoles in 2% aqueous CTAB at pH 7.3. On-rates and in $M^{-1} sec^{-1}$, off-rates in sec^{-1}. Data is from Ref. 20. Similar steric effects on kinetic constants were observed in benzene.

Equilibria present a somewhat simpler picture. For CO in aqueous CTAB (20), just as in benzene (22), the affinity decreases by about 190 as a result of the change from 1-methylimidazole to 2-methylimidazole. The oxygen equilibria have been studied by Collman et al. (32) who find that their R-state model (Fe(pivalamidophenyl)$_3$ heme with a chelating imidazole, $P_{1/2} = 0.60$ torr) binds O_2 approximately 65 times more strongly than their T-state model (Fe(pivalamidophenyl)$_4$ heme with external 1,2-dimethylimidazole, $P_{1/2} = 38$ torr). These ratios for CO and O_2 are within a factor of 2 of the ratios reported for hemoglobin in pH 9 borate buffer and a factor of 4–8 less than the ratios reported for hemoglobin in pH 7 phosphate buffer (Table I). Thus it is apparent that proximal strain models resemble hemoglobin to similar extents for CO and O_2 binding.

This similarity leads us to the conclusion that distal-side steric effects are small in hemoglobin. Any distal-side steric effect presumably would destabilize CO more than O_2. However, since CO and O_2 binding are mimicked almost equally by models that contain no distal steric hindrance, it seems unnecessary to invoke this steric hindrance in the protein. Although the present studies do not present evidence for the effect, hemoglobin equilibria are heavily dependent on the conditions of measurement, and it might be that distal-side steric effects would have a role in CO binding under some conditions. However, at the accuracy of our present knowledge, it seems most likely that hemoglobin cooperativity operates through changes in proximal-base tension, without distal-side steric effects, polarity, or electronic effects being required to explain the T-to-R switch.

Distal-Side Steric Effects

Although kinetic and equilibria comparisons argue against important distal-side steric interference with the ligation of hemoglobin, the phenomenon remains a logical one and a good candidate for lowered CO affinities of myoglobin (1) and peroxidases (35). In some of these cases, association rates are slow, like those of T-state hemoglobin, and dissociation rates are also slow, like those of R-state hemoglobin. This means that the theory proposed for the R and T differences is not sufficient to explain these dynamics. Therefore, it seemed worthwhile to prepare synthetic model hemes with varying degrees of distal-side steric repulsion toward CO or other ligands and to study the effects of such repulsion on the dynamics of O_2 and CO ligation.

In order to have available an adjustable steric effect, we have prepared the anthracene heme cyclophane and the bicyclophane shown below (36). The severe interference of both cyclophane "pock-

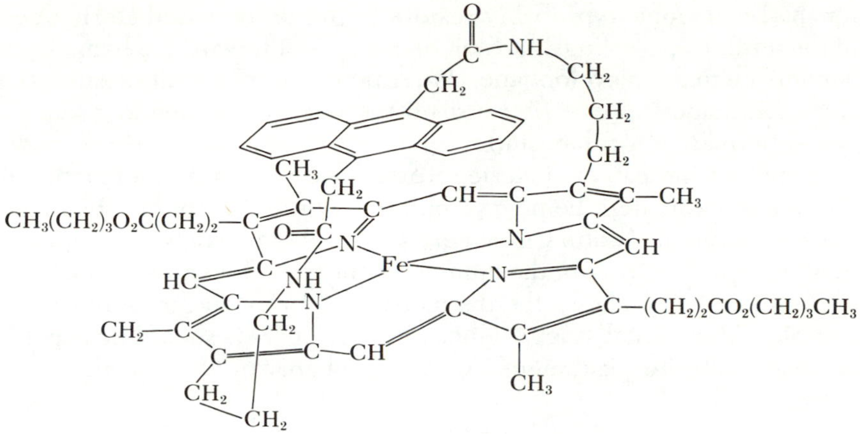

Anthracene heme cyclophane

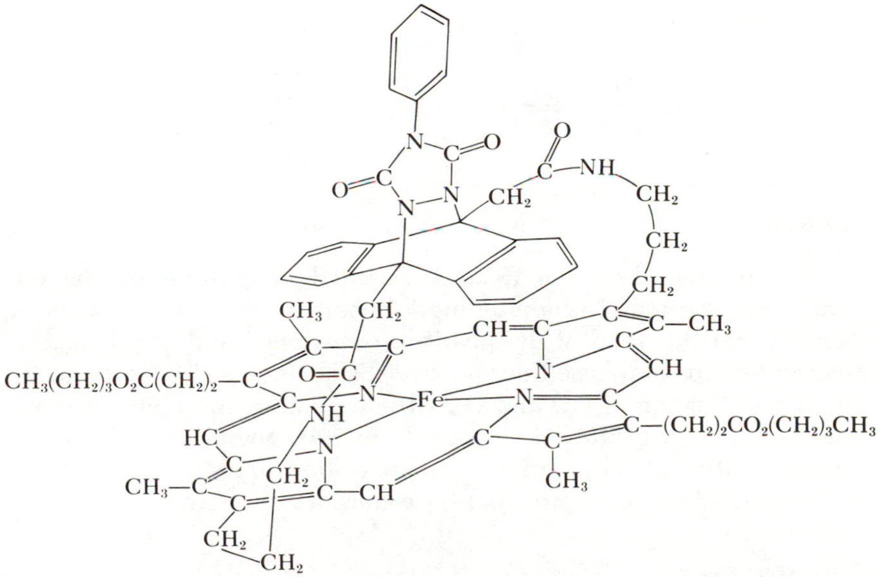

Bicyclophane heme
"Iron(II) pagoda porphyrin"

ets" with CO is seen in a comparison of the association rate constants, l', with that of an electronically similar pentacoordinated heme, chelated mesoheme. The rate constant decreases from $10^7 M^{-1} \sec^{-1}$ for chelated mesoheme to less than $10^4 M^{-1} \sec^{-1}$ for the cyclophane heme mono-1-methylimidazole complex and less than $10^3 M^{-1} \sec^{-1}$ for the bicyclophane heme mono-1-methylimidazole complex. The lower rate for the bicyclophane probably results from the increased steric effect of the anthracene "wings", which become bent toward the heme upon conversion to the bicyclophane. (We report an upper limit because this steric hindrance apparently causes the observed reaction to go by the base-elimination mechanism.)

Another indication of steric interference is seen in the reaction of the tetracoordinated heme cyclophane with CO in dry benzene. Under conditions (1 atm CO) where deuteroheme exists as an approximately equal mixture of the mono-CO and di-CO complexes (37), the cyclophane heme shows the distinct spectrum of the pure mono-CO complex. Thus distal-side residues in heme pockets should be capable of greatly altering ligation of CO, and probably of O_2 as well.

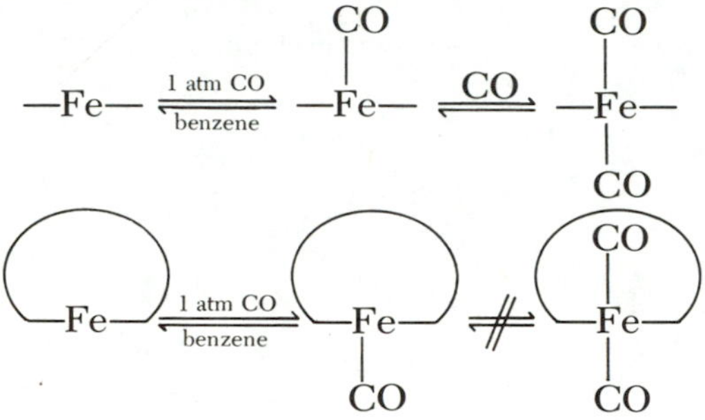

Conclusion

Comparisons between R- and T-state hemoglobins on the one hand and a variety of synthetic model compounds on the other have allowed an evaluation of the possible occurrence and importance of electronic, proximal-base tension, and distal-side steric effects on the kinetics of ligation of CO and O_2. Although all of these effects could influence the reactivities of hemoproteins, we conclude that hemoglobin reactivity and cooperativity are controlled predominantly by the presence or absence of proximal-base tension.

Acknowledgments

We are grateful to C. K. Chang for help and advice in the cyclophane syntheses, and to the National Institutes of Health for support of this research (Grant HL-13581) and for the computer facilities that were used (Grant RR-00757, Division of Research Resources).

Literature Cited

1. Antonini, E.; Brunori, M. "Hemoglobin and Myoglobin in Their Reactions with Ligands"; North Holland Publishing Co.: Amsterdam, 1971.
2. Perutz, M. F. Br. Med. Bull. 1976, 32, 195–208.

3. Baldwin, J. M. *Prog. Biophys. Molec. Biol.* **1975**, *29*, 225–320.
4. Shulman, R. G.; Hopfield, J. J.; Ogawa, S. *Q. Rev. Biophys.* **1975**, *8*, 325–420.
5. Tucker, P. W.; Phillips, S. E. V.; Perutz, M. F.; Houtchens, R.; Caughey, W. S. *Proc. Natl. Acad. Sci. USA* **1978**, *75*, 1076–1080.
6. Perutz, M. F. *Nature (London)* **1970**, *228*, 726–739.
7. Hoard, J. L. In "Hemes and Hemoproteins"; Chance, B., Estabrook, R. W., Yonetani, T., Eds.; Academic: New York, 1966; p. 9.
8. Antonini, E.; Gibson, Q. H. *Biochem. J.* **1960**, *76*, 534–538.
9. Sawicki, C. A.; Gibson, Q. H. *J. Biol. Chem.* **1976**, *251*, 1533–1542.
10. DeYoung, A.; Pennelly, R. R.; Tan–Wilson, A. L.; Noble, R. W. *J. Biol. Chem.* **1976**, *251*, 6692–6698.
11. Sharma, V. S.; Schmidt, M. R.; Ranney, H. M. *J. Biol. Chem.* **1976**, *251*, 4267–4272.
12. Sawicki, C. A.; Gibson, Q. H. *J. Biol. Chem.* **1977**, *252*, 5783–5788.
13. Ibid., 7538–7547.
14. Noble, R. W.; Gibson, Q. H.; Brunori, M.; Antonini, E.; Wyman, J. *J. Biol. Chem.* **1969**, *244*, 3905–3908.
15. Diekmann, H.; Chang, C. K.; Traylor, T. G. *J. Am. Chem. Soc.* **1971**, *93*, 4068–4070.
16. Collman, J. P.; Gagne, R. R.; Halbert, T. R.; Marchon, J.-C.; Reed, C. A. *J. Am. Chem. Soc.* **1973**, *95*, 7868–7870.
17. Almog, J.; Baldwin, J. E.; Dyer, R. L.; Peters, M. *J. Am. Chem. Soc.* **1975**, *97*, 226.
18. Chang, C. K.; Traylor, T. G. *Proc. Natl. Acad. Sci. USA* **1973**, *70*, 2647–2650.
19. Chang, C. K.; Traylor, T. G. *Proc. Natl. Acad. Sci. USA* **1975**, *72*, 1166–1170.
20. White, D. K.; Cannon, J.; Traylor, T. G. *J. Am. Chem. Soc.* **1979**, *101*, 2443–2454.
21. Collman, J. P.; Reed, C. A. *J. Am. Chem. Soc.* **1973**, *95*, 2048–2049.
22. Rougee, M.; Brault, D. *Biochemistry* **1975**, *14*, 4100–4106.
23. Jameson, G. B.; Molinaro, F. S.; Ibers, J. A.; Collman, J. P.; Brauman, J. I.; Rose, E.; Suslick, K. S. *J. Am. Chem. Soc.* **1978**, *100*, 6769–6770.
24. Losse, G.; Müller, G. *Z. Physiol. Chem.* **1962**, *327*, 205–216.
25. Warme, P. K.; Hager, L. P. *Biochemistry* **1970**, *9*, 1599–1614.
26. Traylor, T. G.; Chang, C. K.; Geibel, J.; Berzinis, A.; Mincey, T.; Cannon, J. *J. Am. Chem. Soc.* **1979**, *101*, 6716–6731.
27. Collman, J. P. *Acc. Chem. Res.* **1977**, *10*, 265–272.
28. Geibel, J.; Cannon, J.; Campbell, D.; Traylor, T. G. *J. Am. Chem. Soc.* **1978**, *100*, 3575–3585.
29. Momenteau, M.; Rougee, M.; Loock, B. *Eur. J. Biochem.* **1976**, *71*, 63–76.
30. Traylor, T. G.; Campbell, D.; Sharma, V.; Geibel, J. *J. Am. Chem. Soc.* **1979**, *101*, 5376–5383.
31. Heidner, E. J.; Ladner, R. C.; Perutz, M. F. *J. Mol. Biol.* **1976**, *104*, 707–722.
32. Collman, J. P.; Brauman, J. I.; Doxsee, K. M.; Halbert, T. R.; Suslick, K. S. *Proc. Natl. Acad. Sci. USA* **1978**, *75*, 564–568.
33. Geibel, J.; Traylor, T. G., unpublished data.
34. White, D. K.; Berzinis, A. P.; Campbell, D. H.; Traylor, T. G., in preparation.
35. Dunford, H. B.; Stillman, J. S. *Coord. Chem. Rev.* **1976**, *19*, 187–251.
36. Traylor, T. G.; Campbell, D.; Tsuchiya, S. *J. Am. Chem. Soc.* **1979**, *101*, 4748, 4749.
37. Rougee, M.; Brault, D. *Biochem. Biophys. Res. Commun.* **1973**, *55*, 1364–1369.

RECEIVED June 5, 1979.

Evidence Regarding Mechanisms for Protein Control of Heme Reactivity

BRIAN M. HOFFMAN, JAMES C. SWARTZ, and MARLENE A. STANFORD

Department of Chemistry, Northwestern University, Evanston, IL 60201

QUENTIN H. GIBSON

Department of Biochemistry, Molecular and Cell Biology, Cornell University, Ithaca, NY 14853

We present studies designed to test three different proposed mechanisms for protein control of heme reactivity. We have used measurements of the kinetics of ligand binding to manganese-substituted hemoglobin and to ferro-porphyrin model compounds as the indicator of heme reactivity, thus eliminating the problem of relating spectroscopic indicators to chemical reactivity. Two mechanisms under consideration, steric control of ligand access and peripheral electronic control through π-donor/acceptor interactions with the porphyrin, seem to be inadequate to explain either the function of hemoglobin, or, by extension, the function of other proteins. However, the third mechanism, proximal electronic control through modulation of the properties of the metal-bound imidazole by hydrogen bonding, is able to exert a powerful influence on heme reactivity. Although probably not applicable to hemoglobin, this mechanism clearly must be given careful consideration in discussions of other heme and nonheme proteins.

One of the obvious goals in the study of hemoproteins is an understanding of the mechanisms by which the physical and chemical properties of the heme group are modulated by the protein environment. The primary control of heme reactivity clearly is exercised in the "selection" of axial ligand(s) through molecular evolution. However, powerful secondary control mechanisms also exist. As the

0-8412-0514-0/80/33-191-235$05.00/0
© 1980 American Chemical Society

best-known example, cooperative ligation of hemoglobin reflects a conformational equilibrium between one protein form (T) with low ligand affinity and a second form (R) with approximately 10^2-fold higher affinity, yet the single endogenous heme–ligand is the same in both forms (1, 2, 3). This influence of conformation on reactivity is expressed correspondingly in modulation of ligand-binding kinetics: for example, the CO on-rate for the T-state is about 20–60-fold less than for the R-state (4).

The most widely known and extensively articulated explanation for the control of heme reactivity within hemoglobin is the trigger mechanism elaborated by Perutz (5) following suggestions of Williams (6) and Hoard (7). As originally formulated, the mechanism involved strain induced by tension applied by the protein to the heme. The first direct evidence for the inadequacy of this mechanism was obtained in our studies of coboglobin, the cobalt-substituted hemoglobin analogue (8, 9, 10). Since then a variety of experimental techniques has been used in largely unavailing search for evidence of heme strain (9–17). Presumably as a result of these and other studies, there has been a progressive, if underplayed, evolution of the trigger mechanism to a new one involving restraint in the formation of a liganded heme within the T-state (3). Equilibrium (18) and kinetic (19) binding studies as well as crystallographic (20) and spectroscopic measurements and theoretical calculations (21, 22) have explored the ability of this new mechanism to explain the facts of Hb ligation.

To a substantial degree, the wide dissemination of the above trigger mechanism has diverted attention from other proposals for protein control of heme reactivity. This is unfortunate, partly because of possible shortcomings of the trigger mechanism as applied to Hb, but more importantly because these other forms of control might be used by nature in other proteins (23), including those with other prosthetic groups. Proposed mechanisms range from suggestions of purely steric effects at one extreme through suggestions of purely electronic control at the other.

In this chapter we discuss some of our studies (24, 25, 26, 27) designed to test selected alternate mechanisms of protein control of heme reactivity. For illustrative purposes we discuss on the one hand a proposal involving only steric effects that are not local to the heme (5). On the other hand we discuss two different control proposals involving purely electronic effects local to the heme, one associated with protein-induced perturbations of the proximal histidine (28), the other with perturbations of the porphyrin ring (29). We use measurements of the kinetics of ligand binding to a metal-substituted hemoglobin anlogue and to ferro-porphyrin model compounds as the indicator of heme reactivity, thus eliminating the problems that arise in attempts to relate spectroscopic indicators to chemical reactivity.

Steric Control of Heme Reactivity

One early proposal for controlling the reactivity of a prosthetic group avoided the necessity of direct protein modulation of the prosthetic group properties. Perutz suggested that both the rates of ligand binding to individual hemoglobin chains and the changes in binding rate associated with the T → R transaction are in part determined by steric factors that control the access of a ligand to the heme iron (5). This mechanism, which was based on considerations of the size of the heme crevice, has the attractive feature that it is consistent with the measurements that find no evidence for heme strain.

In this section, we test the proposal through studies of the reactions of manganese-substituted hemoglobin (MnHb). The similarities of NO binding to MnHb and of CO binding to Hb extend to the finest details of the process (24, 25). These include cooperativity, photosensitivity, and a T → R transition that is observable directly upon NO photodissociation. These similarities show MnHb is a faithful analogue of Hb; thus, we may compare the rates of NO binding by Hb and MnHb. This comparison supports the conclusion that steric control of access to the metal center cannot be responsible for the changes in ligation rates caused by the T → R transformation.

Properties of MnHb.   We have previously shown that the heme iron of hemoglobin can be replaced by manganese with minimal perturbations of protein structure. For example, the structure of metmanganoglobin (met-MnHb) and that of native hemoglobin have been compared by x-ray difference Fourier techniques at 2.5-Å resolution (30). The quaternary structures are identical and the tertiary structures are similar; thus, met-MnHb retains the major structural properties of methemoglobin. This result, plus a variety of studies of function (10), suggested that the binding of NO to MnHb might show a close parallel to cooperative hemoglobin ligation.

This suggestion was supported further by studies of model compounds. Upon binding a nitrogenous base, an Mn(II) porphyrin adopts the pentacoordinate geometry imposed by the heme pocket of hemoglobin and provides a close structural analogue to a pentacoordinated Fe(II) porphyrin (31). The pentacoordinated Mn(II) porphyrin will bind reversibly NO as a sixth ligand. However, the Mn(II) + NO system is isoelectronic with Fe(II) + CO, not with Fe(II) + NO, indicating that NO binding by MnHb should be analogous to CO binding by Hb, not to NO binding. This conclusion was strengthened by x-ray studies of model compounds showing that in both Fe—CO and Mn—NO the diatomic molecule binds in a linear fashion (32, 33), and by our successful prediction that the two linkages would be comparably photolabile (34).

Binding of NO to MnHb.   MnHb binds NO in a second-order reaction. It is measured readily by stopped-flow techniques, with a typical observed association rate constant of j' (intrinsic) = $35mM^{-1}$ sec^{-1} (0.05M Bis–Tris, pH 7 (20°C)). Under similar conditions the Hb + CO reaction proceeds at a comparable (four times faster) rate. In contrast, the Hb + NO reaction is much more rapid, about 700 times faster, and is observed only with difficulty by stopped-flow techniques.

Detailed examination of the time course of NO binding to MnHb shows an accelerating or "autocatalytic" time course. Figure 1 is a plot of the apparent rate constant against the proportion of the reaction completed. It shows an increase of 30% over the course of the reaction, which is qualitatively and quantitatively similar to the effect observed in CO binding to Hb by Gibson and Roughton (35). This acceleration is the kinetic manifestation of cooperative (sigmoidal) equilibrium ligand binding (2). Furthermore, the rate of combination of MnHb with NO is appreciably reduced (approximately one-fifth) by the addition of IHP; thus, kinetically demonstrating a heterotropic linkage to the binding of organic phosphates, which is again analogous to that seen in the CO + Hb reaction.

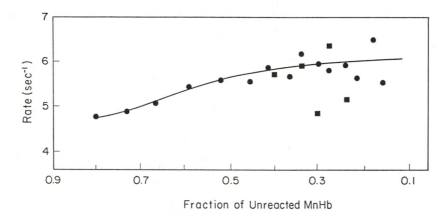

The Journal of Biological Chemistry

Figure 1. Observed rate for the reaction of MnHb (5 μM) with NO (150 μM), plotted vs. the fraction of unreacted MnHb: (●) the whole reaction followed by stopped flow; (■) the latter stages were also obtained from the slow phase of the reaction observed when the reagents were mixed together and then examined by flash photolysis; conditions: 0.05M Pi, pH 7.0, 20°C, observation at 434 nm, 1-cm path length (25).

Flash Photolysis of MnHb NO. MnHb NO is photodissociated readily as is HbCO, in sharp contrast to HbNO, which under most conditions shows a very low apparent quantum yield. Experiments performed with varying light intensities indicated that the quantum yields for dissociation are of the same order for MnHb NO and HbCO.

Thus, as we have discussed elsewhere (*34*), the isoelectronic nature of these two systems is reflected not only in the structures of the liganded metalloporphyrins, but also in the properties of their photoexcited states.

NO binding to MnHb after flash dissociation gives further evidence of cooperative ligand binding similar to that of Hb. Unlike the stopped-flow experiment, NO association after flash dissociation is markedly biphasic; in a dilute solution roughly 70% of the absorbance change occurs approximately 30 times more rapidly than in the stopped-flow experiment and the remainder at the same rate as in the stopped-flow experiment. The amount of rapid reaction is concentration dependent and thus appears to be related to the dissociation of MnHb NO into dimers. This implies that, as with Hb, unliganded dimers react more rapidly than do unliganded tetramers.

The binding of NO to MnHb NO that is dissociated partially (~15%) is substantially faster than that after full dissociation. This increased rate of binding for the last NO also demonstrates cooperative ligand binding to MnHb, analogous to results for CO and Hb, and shows that partially saturated intermediates react faster than does unliganded MnHb (Figure 2). This is in full conformance with the ideas that MnHb + NO behaves precisely like Hb + CO, that unliganded

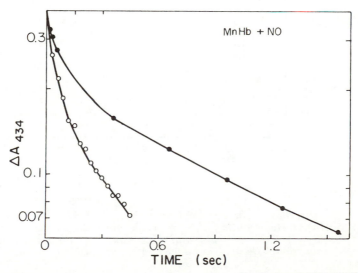

The Journal of Biological Chemistry

Figure 2.   Reaction of MnHb (15 µM, monomer) with NO (~70 µM): (●) after full flash photolysis of MnHbNO; (○) after partial photolysis of MnHbNO. The actual OD excursion upon partial flash was only ~15% that upon full flash, but excursions in the lower curve are scaled to give the same initial ΔA as after full flash. Conditions are as in Figure 1, but the path length is 2 mm (25).

MnHb is in the low affinity T-form, that NO binding causes a switch to the high affinity R-form, and that the two forms differ in NO on-rates by about 30-fold.

Conformation Measurements.   The NO binding kinetics indicate that manganese substitution does not disrupt the T → R conformational change upon ligation. This has been corroborated through a number of solution measurements that provide an index of hemoglobin conformation (24, 25), but the most satisfying verification is the direct observation of the T → R transition achieved by following absorbance changes near the ligation isosbestic point (429 nm) (25). No isosbestic point was found, but after a short lag period, there was a decrease in absorbance with a half time of about 250 μsec. This change is not associated with ligand binding because its rate is unaffected by changing the concentration of NO from 50 to 100μM (Figure 3, Curves

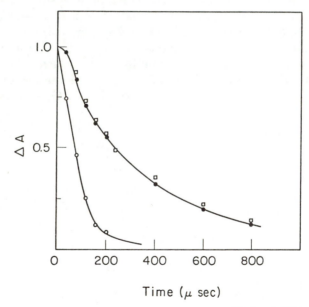

The Journal of Biological Chemistry

Figure 3.   Absorbance change (ΔA) vs. time upon flash photolysis of 10μM MnHbNO. Reaction is followed at 429 nm with: A, (●) [NO] = 50μM; and B, (□) [NO] = 100μM; and at 424 nm: C, (○) [NO] = 50μM; scale: full excursion, 0.04 in absorbance at 429 nm, 0.465 at 424 nm (25).

A and B). It does not represent the rate of removal of ligand by the flash, which (followed at 424 nm) is represented by Curve C. This has a half time of about 80 μsec, and shows no sign of the lag period at the beginning of Curves A and B. Under the conditions of Curve A, the half time for rebinding of NO in the rapid phase of the reaction (due to dimers) is 3 msec and in the slow phase, 70 msec; these times are outside the range of Figure 3. These results are compatible with the identification of the change at 429 nm as caused by the R → T transition of the unliganded Mn^{2+} compound. The initial lag in Curves A and B arises because the R → T transition does not begin until more than one molecule of NO has been removed from a tetramer; the rate of the later part should represent the rate of the R → T transition, about 3500 sec^{-1}. This is similar to, but somewhat less than, the rate for the corresponding transition in the iron protein (36).

Steric Control of Heme Reactivity(?). Examination of the MnHb + NO binding reaction shows that the likeness to the Hb + CO reaction extends to the finest details of the ligation process, including slow, cooperative, binding and a T → R switch. In contrast, Cassoly and Gibson have shown that Hb binds NO in a rapid, noncooperative manner (37). The observations of Cassoly and Gibson suggest that the steric factors do not control access of a ligand to the heme ion. The combination of their results with those for MnHb indicate this in a more immediate way.

The R-structure of manganoglobin has been shown by x-ray diffraction to correspond closely to that of hemoglobin (30), and all the considerations given above indicate that the same correspondence should obtain for the T-state. Thus, the two different types of kinetic behavior, slow and cooperative (MnHb) or rapid and noncooperative (Hb), are exhibited by a single ligand, NO, interacting with equivalent structures in the two hemoglobins. Viewed from a slightly different perspective, one that can be discussed more deeply (25), comparison of NO binding by Hb and MnHb shows that the NO can reach the vicinity of the Mn at least 700 times faster than it is bound to the T-state protein. Thus the approximately 30-fold increase of the NO binding rate in R-state MnHb cannot arise from increased access of NO to the metal site. The functional identity of MnHb and Hb means that restricted access cannot account for the quaternary structure influences on the rates of ligation by Hb, a conclusion that is anticipated in earlier studies of fluorescence quenching by O_2 (38). Rather, the influence of the T-state structure must occur after initial encounter between ligand and metal. This is in agreement with the idea that steric features inferred from the x-ray studies of the crystals do not give a sufficient guide to the accessibility of the hemes to ligands, and/or that such features are not important contributors to the ensemble of structures occurring in solution.

Electronic Control of Reactivity

Proximal Effects. One attractive mechanism for electronic control of heme reactivity has been advanced by Chevion and co-workers (28). They propose that conformational changes cause alteration in hydrogen bonding to the proton on the amino nitrogen of a metal-coordinated imidazole. The stronger the hydrogen bond formed by the proton, the more electron-rich is imidazole; the limit of full proton transfer is the imidazolate anion. Differences in the imidazole "protonation state" would alter the electron-donating power of the imino nitrogen, thus modifying the electronic structure of the metal to which it is coordinated.

Considering hemoglobin within this framework, in the low reactivity T-state the proximal histidine is thought to be in its neutral form with a proton on $N(1)$. The increased ligand affinity of the R-state is thought to occur because the proximal imidazole is partially "deprotonated" through formation of a strong hydrogen bond, and therefore is a better electron donor. Such a mechanism also has been invoked to discuss the electronic structure of cytochrome c from different organisms (23) and the crystal structures of a number of hemoproteins (39). We have now tested the relation of heme reactivity and the imidazole protonation state by direct comparison of the CO binding rate for a pentacoordinated ferrous porphyrin, Fe(P), model with neutral imidazole as fifth ligand, Fe(P)(Im), with that for the model with deprotonated imidazole as the fifth ligand, Fe(P)(Im$^-$); here, P = m-tetraphenylporphrin (TPP) or deuteroporphyrin dimethyl ester (DPD). We find that deprotonation can alter the CO binding rates to a degree in excess of the difference in binding rates to the T- and R-states of Hb. Furthermore, the studies of Mincey and Traylor (40) show that CO binding rates and affinities change in parallel. However, the difference is in the opposite sense to that which would normally be expected: deprotonation decreases the rate of CO binding (26).

Fe(Im)(CO). The Fe(TPP)(Im)(CO) complex is highly photolabile and the photoproduct rebinds CO with a pseudo first-order rate constant, $k_{obs}\alpha$[CO] (Figure 4). The product obtained immediately after the flash (~20 μsec) does not correspond to the formation of Fe(TPP)(Im), which should have a Soret maximum near 435 nm (41). Rather, the kinetic difference spectrum is the same as the static difference spectrum [Fe(TPP)(Im)$_2$] − [Fe(TPP)(Im)(CO)], obtained by direct subtraction of the absorbance spectra of the appropriate complexes, which is also presented in Figure 5A. This indicates that a second Im is bound within the flash lifetime. Therefore, the overall stoichiometry of the CO rebinding reaction as observed in the regeneration of Fe(TPP)(Im)(CO) after the photolysis flash is ligand replacement, Fe(TPP)(Im)$_2$ + CO → Fe(TPP)(Im)(CO) + Im, not the ligand addition reaction, Fe(TPP)(Im) + CO $\xrightarrow{k_s}$ Fe(TPP)(Im)(CO), which would correspond to the CO binding reaction by Hb.

This observation is expected from previous studies, such as those of Traylor (19); binding of CO to an Fe(II) porphyrin in the presence of excess base (B) usually can be described in terms of pre-equilibrium

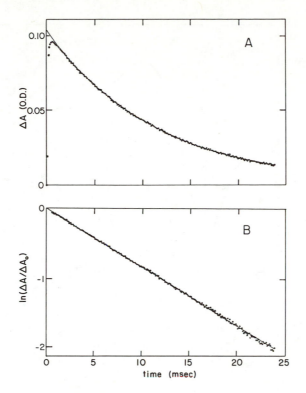

Figure 4.   (A) *Typical kinetic trace of absorbance change* (ΔA) *for CO recombination after flash photolysis of* $Fe(TPP)(Py)(CO)$: $[Fe(TPP)] = 6 \times 10^{-6} M$; $[Py] = 0.1$ M; $P_{CO} = 1$ atm; CH_2Cl_2; *experimental data:* (...) *digitized by a Biomation 610B;* (−) *nonlinear least squares 1st order fit;* $\lambda = 430$ nm; *slit width, 2 nm.* (B) *Plot of* $\ln(\Delta A/\Delta A_0)$ *for the same experiment:* (...) *experimental data;* (−) *linear least squares fit.*

Scheme 1.

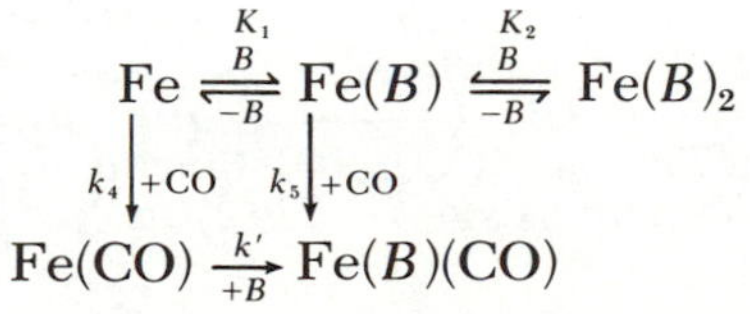

$$\text{Fe} \underset{-B}{\overset{K_1}{\underset{B}{\rightleftharpoons}}} \text{Fe}(B) \underset{-B}{\overset{K_2}{\underset{B}{\rightleftharpoons}}} \text{Fe}(B)_2$$

$$k_4 \Big\downarrow +\text{CO} \qquad k_5 \Big\downarrow +\text{CO}$$

$$\text{Fe(CO)} \xrightarrow[+B]{k'} \text{Fe}(B)(\text{CO})$$

between the $Fe(B)_n$ ($n = 0, 1, 2$) (*40, 19*), where we have suppressed the porphyrin abbreviation (*see* Scheme 1.). Here K_1 and K_2 are the measured static binding constants, and k_4 and k_5 are the second-order CO binding rates, of which k_5 is of interest; the reverse reaction, loss of CO, can be neglected on our time scale. When comparing rates for different bases, we will keep track by writing $k_5(B)$.

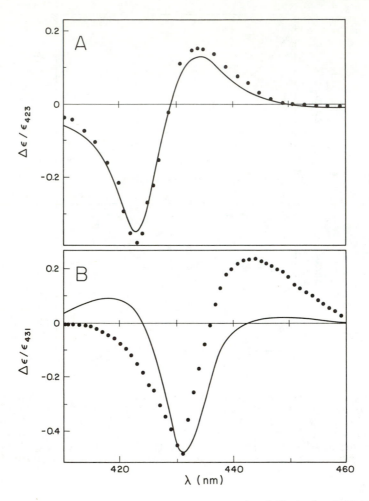

Journal of the American Chemical Society

Figure 5. (A) (———) *Static difference spectrum,* $Fe(TPP)(Im)_2$ − $Fe(TPP)(Im)(CO)$; (●) *kinetic difference spectrum after flash photolysis of* $Fe(TPP)(Im)(CO)$. (B) (———) *Static difference spectrum,* $Fe(TPP)(Im^-)_2$ − $Fe(TPP)(Im^-)(CO)$; (●) *kinetic difference spectrum after flash photolysis of* $Fe(TPP)(Im^-)(CO)$ (26).

Under the conditions of our experiments, the results of White et al. (19) show that CO addition is rate limiting, and that the ligand rebinding rate obeys an equation, which we write as Equation 1,

$$\frac{k_{obs}}{[CO]} = \frac{k_4}{\Sigma} + \frac{k_5 K_1[B]}{\Sigma} \tag{1}$$

where $\Sigma = 1 + K_1[B] + K_1K_2[B]^2$. The equilibrium constants for binding Im by Fe(TPP) are known: $K_1 = 8.8 \times 10^3 M^{-1}$ and $K_2 = 7.9 \times 10^4 M^{-1}$ (*41*). We performed a series of rate measurements at constant [CO] and varying [Im]. A plot of Equation 2 vs. [*B*] is linear, Figure 6. We obtain k_4 from the intercept and k_5 from the slope (Table I).

$$k_{obs}\Sigma/[CO] = k_4 + k_5K_1[B] \tag{2}$$

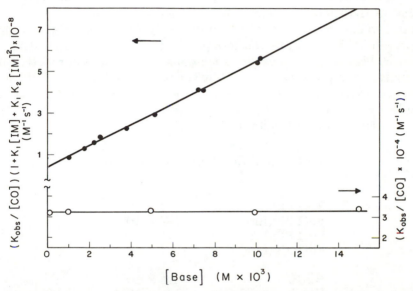

Figure 6. CO binding kinetics for Fe(TPP) as a function of nitrogenous base concentration, [B]: (●) represent $k_{obs} \times [1 + K_1[B] + K_1K_2[B]^2]/[CO]$ *(left ordinate) for B = Im, [CO] = 5 × 10⁻⁴M, [Fe(TPP)] = 6 × 10⁻⁶M; (○) represent* $k_{obs}/[CO]$ *(right ordinate) for B = Im⁻, [DC = 18-C-6] = 0.2M, [CO] = 7.5 × 10⁻³M, [Fe(TPP)] = 6 × 10⁻⁶M. The CO binding rate constants obtained from this data are listed in Table I. The data for B = Im and derived constants differ somewhat from those in Ref. 26; see footnote b, Table I.*

Fe(Im⁻)(CO). The Fe(TPP)(Im⁻)(CO) complex is also photolabile, and the photoproduct also returns to the starting materials with a rate $\alpha[CO]$. The kinetic difference spectrum upon photolyzing Fe(TPP)(Im⁻)(CO) does not correspond to the static [Fe(TPP)(Im⁻)₂] − [Fe(TPP)(Im⁻)(CO)] difference obtained by subtraction of the appropriate absorbance spectra, (Figure 5B). It also does not agree with what would be observed if the tetracoordinated Fe(TPP) were formed (*41*). Instead the red-shifted peak with maximum at 444 nm corresponds to the formation of the high-spin, pentacoordinated Fe(TPP)(Im⁻) (*40*). In addition, the isosbestic point observed at 436 nm is independent of

[Im⁻], and the rebinding of CO takes place with pseudo first-order kinetics. We conclude that the pentacoordinated Fe(TPP)(Im⁻) produced upon photolysis directly adds CO to regenerate the Fe(TPP)(Im⁻)(CO) without observable formation of Fe(TPP)(Im⁻)$_2$ on the time scale of the experiment. Thus, although under equilibrium conditions Fe(TPP)(Im⁻) can bind weakly a second imidazolate, the above observations further indicate that for purposes of the analysis of kinetic data by Equation 1, we may ignore the term in Σ that is proportional to $[B]^2$, which is equivalent to setting $K_2 = 0$. In a series of measurements at constant [CO] and with $10^2 M \leq [Im^-] < 0.2M$, the observed CO binding rate was independent of [Im⁻] (Figure 6). Examination of Equation 1 shows that such behavior validates our method of analysis and implies that the binding constant of Im⁻ and Fe(TPP), K_1, is much greater than $10^2 M^{-1}$ and that the observed pseudo first-order binding rate, k_{obs}, is just $k_5[CO]$. The second-order rate, $k_5(Im^-)$, for Fe(TPP)(Im⁻) is listed in Table I.

Table I.   Rate Constants for CO Binding

$$k^a (mol^{-1} sec^{-1})$$

Fe(TPP)	$5.0 \times 10^{7\,b}$
Fe(TPP)(Im)	$5.7 \times 10^{6\,b}$
Fe(TPP)(Im⁻)	$3.4 \times 10^{4\,c}$
Fe(TPP)(Py)	$3.8 \times 10^{6\,b}$
	$3.4 \times 10^{6\,d}$
Hb(R)	$1.1 \times 10^{7\,e}$
Hb(T)	$5 \times 10^{5\,e}$
Mb	$5 \times 10^{5\,f}$

^{a} For Fe(TPP), $k = k_4$; for Fe(TPP)(B), $k = k_5$ (Scheme I in text); k for the proteins is equivalent to k_5.
^{b} Toluene, 21°C. Ref. 26 and 27. These k_4 and k_5 values differ from those in Ref. 26, which suffered from a systematic error.
^{c} Toluene solution, ~0.2M DC-18-C-6 and KIm, 21°C. Ref. 26 and 27.
^{d} Methylene chloride, 21°C, Ref. 27. Pyridine binding constants equal to those in toluene have been assumed; *see* text.
^{e} Conditions: 0.05M sodium borate buffer, pH 9.2, 20°C, Ref. 42.
^{f} Conditions: 0.1M Pi buffer, pH 7, 20°C, Ref. 2.

Proximal Electronic Control of Heme Reactivity (?). The deprotonation of a coordinated imidazole results in a large reduction in CO binding rate constants: $k_5(Im)/k_5(Im^-) = 170$ (Table I). However, the changes are inverse to expectation: deprotonation causes the rate constant to decrease, whereas if increased electron donation by the fifth ligand was the critical factor, the rate constant should increase. It is probable that Im⁻ as an axial ligand stabilizes the pentacoordinated ferroporphyrin and that this effect far overbalances any influence of increased charge donation by Im⁻ in the transition state on the reaction path by which the hexacoordinated Fe(P)(Im⁻)(CO) is formed.

The rate constants for CO binding to T- and R-state Hb, although sensitive to solution conditions, differ only by a factor of 20–60 (Table I), and both fall within the range spanned by the models studied here: $k_5(Im^-) \ll k_5(T) < k_5(R) \approx k_5(Im)$. Thus, the Peisach mechanism, if inverted, is in principle more than adequate to explain the protein modulation of the CO binding rates, and only a relatively minor change in protonation state upon the T ↔ R conversion need be invoked. However, the kinetic difference spectra for the Fe(TPP) complex shows that the Soret band of the pentacoordinated imidazolate adduct is appreciably red-shifted from that of the imidazole adduct, and the same is expected for heme itself. In contrast, the Soret band of T-state Hb is blue-shifted from that of the R-state (*42*). Moreover, the analysis of hemoprotein crystal structures suggested stronger hydrogen bonding occurs in the R-state (*39*).

The present results as well as others make it unlikely that the Peisach mechanism applies to hemoglobin function. However, the large change in ligation rates brought about by altering the properties of the fifth heme ligand is of intrinsic interest and importance. Moreover, it enhances the plausibility of applying this mechanism to other proteins (*23*).

Peripheral Effects

An alternate mechanism for electronic control of heme reactivity involves π-donor/acceptor(*D/A*) interactions between the heme macrocycle as acceptor and an aromatic amino acid (e.g., phenylalanine, tyrosine, histidine) as donors (*29, 43, 44*). If these interactions significantly alter the electron density on the ring, they also could influence indirectly the electron density at the metal and thus its reactivity. Considering hemoglobin in particular, two aromatic residues in contact with the heme, the phenyl rings of Phe CD1 and G5, have altered orientations in the T- and R-structures (*45*). Enhanced *D/A* interactions in the R-form could account for the increased ligand affinity; the results of high precision, difference resonance Raman studies have been interpreted in this fashion (*44*). The proposal of control through electronic interactions at the heme periphery can be compared usefully with the suggestion that, on going from the T- to the R-state, changes in van der Waals contacts with the porphyrin are important in controlling reactivity (*46*).

We have examined the effects of electron donors on the CO binding rate for a pentacoordinated ferrous porphyrin (*27*). Comparison of the rates measured in toluene with those in a nonaromatic solvent indicate the maximal effect obtainable from *D/A* interactions with the phenyl ring of phenylalanine. We have also looked for effects from the addition of N,N,N',N'-tetramethyl-p-phenylenediamine (TMPD), a much stronger donor than any of those available as amino acid side-chains (47). This kinetic measure of heme reactivity shows no effects of such D/A interactions (27).

Fe(Py)(CO) and D/A Interactions. For the study of π-donor/acceptor interactions, we examined the photolysis of Fe(TPP)(Py)-(CO) and the subsequent CO rebinding. Pyridine was chosen as the base because it is convenient to handle; it cannot hydrogen bond as does Im, and does not react with other reagents. Measurements were made in toluene, and in methylene chloride as a nonaromatic solvent, as noted above. The Fe(TPP) equilibrium binding constants for pyridine are $K_1 = 1.5 \times 10^3 M^{-1}$ and $K_2 = 1.9 \times 10^4 M^{-1}$ in toluene (48). We assume the same values for pyridine binding in methylene chloride; when analyzing kinetic data by Scheme 1 and Equations 1 and 2, we thus force any solvent effects to appear as changes in rate constants. This maximizes the probability of seeing differences in kinetic constants between the aromatic and nonaromatic solvents.

The kinetic difference spectrum upon photolysis of Fe(TPP)-(Py)(CO) is almost identical to that for Fe(TPP)(Im)(CO), as is the nature of the dependence of rebinding rate of $[B]$ and $[CO]$. Consequently, the CO binding can also be described by Scheme 1 and Equation 1. Plotting $k_{obs}\Sigma/[CO]$ vs. $[Py]$ for Fe(TPP) both in toluene and in methylene chloride gives straight lines [Equation 2, Figure 7) with slopes from which k_5 is obtained. The CO binding rate constant for the pyridine adduct of Fe(TPP) in toluene, $k_5(Py)$, is approximately three times less than that for the imidazole adduct. This rate constant is not significantly affected (~10% reduced) by the change from nonaromatic to aromatic solvent (Table I). It may be noted further that the linear behavior of the observed rate vs. $[Py]$, when plotted according to Equation 2, means that pyridine does not act additionally as a π-donor that can influence the binding rate through peripheral interaction.

Two procedures were used to look for effects of an added strong donor. In one set of experiments a fixed high concentration ($[TMPD] = 1M$) was established in methylene chloride and $[Py]$ was varied ($0.02-0.79M$); the results were plotted according to Equation 2 and compared with those in the absence of donor. In the second set of experiments, the $[Py]$ was fixed and $[TMPD]$ was varied to a maximum of $1M$.

No effect of TMPD on CO binding rates was detected by either procedure. The plot of binding rates vs. $[Py]$ according to Equation 2 is unchanged by the presence of $1M$ TMPD. Taking an Fe(TPP)(Py)(CO) solution at either low ($0.02M$) or high ($0.25M$) pyridine concentration and titrating with TMPD up to about $1M$ also

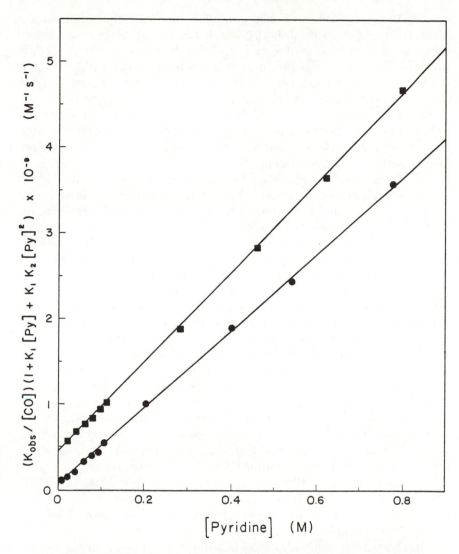

Figure 7.   CO binding kinetics for Fe(TPP) as a function of pyridine concentration in two solvents, plotted as in Figure 6: (●) *toluene;* (■) *CH₂Cl₂ data; [CO] = 0.0085M/atm in CH₂Cl₂; [CO] = 0.0075M/atm in toluene.*

produced no measurable change ($\pm 5\%$) in k_{obs}. Since the actual k_{obs} is a function of kinetic and equilibrium constants (Equations 1 and 2), these observations indicate that addition of TMPD has no measurable effect on any one of the constants k_4, $k_5(Py)$, $K_1(Py)$, $K_2(Py)$.

Peripheral *D/A* Control of Heme Reactivity (?).   The equality of CO binding rates in methylene chloride and toluene argues strongly against the importance of π-interactions between the heme and a phenyl sidechain in modulating heme reactivity. Since the change from a nonaromatic solvent to one in which the heme is surrounded by phenyl rings leaves the binding rates unchanged (to within <10%), it is unrealistic to attribute a 20- to 60-fold rate increase upon the T $\rightarrow$ R switch to a mere change in orientation of two nearby phenyl groups. This conclusion, in fact, should not be surprising. The ground state interactions between good donors/acceptors and metalloporphyrins are not particularly strong (47, 49, 50), and the phenyl group is not even a particularly good donor (47).

This argument is supported by the observations that $1M$ TMPD has no significant effect on any kinetic or equilibrium constant. At [TMPD] = $1M$, roughly 10 mol %, the encounter time for transient (or contact) D/A interactions between TMPD and the pentacoordinated porphyrin intermediate is ample to modify k_5, even if there is no appreciable D/A stabilization of a $(TMPD)(Fe(TPP)(Py))$ complex.

By considering the simplest description of an encounter, possibly with D/A interactions between the ferro-porphyrin and donor, we can estimate an upper limit to the change in k_5 in an associated donor–porphyrin pair. Since TMPD is such a strong donor, this in turn gives an upper limit to the position of the change in binding rate upon the T $\rightarrow$ R conversion that can be associated with D/A interactions. The pentacoordinated FeB in the presence of a donor will be in rapid equilibrium with the associated pair, $(D)(Fe(B))$. If the donor-associated porphyrin binds CO with a rate different from that of the free FeB, then since the CO binding is slow, the observed rate in the presence of D will be the weighted average of the rate constants for free and associated porphyrin. Carrying this treatment through (27), the failure of $1M$ TMPD to alter k_5 within the experimental error of $\pm5\%$ suggests that $Fe(TPP)(Py)$ associated with the strong donor TMPD has a k_5 that must differ from that for free $Fe(TPP)(Py)$ by a factor less than 2, far less than the 20- to 60-fold difference caused by the T $\rightarrow$ R switch (Table I).

In short, measurements of CO binding rates strongly argue against the suggestion that peripheral electronic interactions with an endogenous π-electron donor significantly alter heme reactivity in Hb. This argument applies to other proteins as well.

Conclusions

We have designed experiments to test three different mechanisms that have been proposed to explain the modulation of heme reactivity by a protein environment. Two of these, steric control of ligand access and electronic control through π-donor/acceptor interactions, seem to be insufficiently powerful to be important in explaining Hb function, and, most likely, the function of other proteins as well. However, electronic control through modulation of the properties of the metal-bound imidazole is shown to have a powerful influence on heme reactivity. Although probably not applicable to Hb, this mechanism clearly must be given careful consideration in discussions of other heme and nonheme proteins.

Acknowledgments

We acknowledge Joan S. Valentine, our collaborator on the study of proximal electronic effects, Terry E. Phillips for his technical assistance, and E. Margoliash for a preprint of work in press. This work was supported by NIH Grants 14276 (QHG) and HL 13531 (BMH) and NSF grant BMS-00478 (BMH). JCS is a NIH postdoctoral fellow (F32-HL05787-0).

Literature Cited

1. "The Porphyrins"; Dolphin, D., Ed.; Academic: New York; 1979.
2. Antonini, E.; Brunori, M. "Hemoglobin and Myoglobin in their Reactions with Ligands"; North-Holland: Amsterdam, 1971.
3. Perutz, M. F. *Br. Med. Bull.* **1976**, *32*(3), 195.
4. Gibson, Q. H. "The Porphyrins"; Dolphin, D., Ed.; Academic: New York, Vol. 5, Chapter 5, 1979.
5. Perutz, M. F. *Nature* **1970**, *228*, 726.
6. Williams, R. J. P. *Fed. Proc. Fed. Am. Soc. Exp. Biol.* **1961**, Suppl. 10, *20*, 5.
7. Hoard, J. L. In "Hemes and Hemoproteins"; Chance, B., Estabrook, R. W., Yonetani, T., Eds.; Academic: New York, 1966; p. 9.
8. Hsu, G. C.; Spilburg, C. A.; Bull, C.; Hoffman, B. M. *Proc. Natl. Acad. Sci. USA* 972, *69*(8), 2122.
9. Basolo, F.; Hoffman, B. M.; Ibers, J. A. *Acc. Chem. Res.* **1975**, *8*, 384.
10. Hoffman, B. M. "The Porphyrins"; Dolphin, D., Ed.; Academic: New York, Vol. 7, Chapter 9, 1979.
11. Shulman, R. G.; Eisenberger, P.; Kincaid, B. M.; Brown, G. S.; Teo, B. K. *Biophys. J.* **1978**, *21*, 173a.
12. Ogawa, S.; Shulman, R. G. *J. Mol. Biol.* **1972**, *70*, 315.
13. Huynh, B. H.; Papaefthymiou, G. C.; Yen, C. S.; Groves, J. L.; Wu, C. S. *J. Chem. Phys.* **1974**, *61*, 3750.
14. Scholler, D. M.; Hoffman, B. M.; Shriver, D. F. *J. Am. Chem. Soc.* **1976**, *98*, 7866.
15. Spiro, T. G.; Burke, J. M. *J. Am. Chem. Soc.* **1976**, *98*, 5482.
16. Sugita, Y. *J. Biol. Chem.* **1975**, *250*, 1251.
17. Nagai, K.; Kitagawa, T.; Morimoto, H., in press.
18. Collman, J. P.; Brauman, J. I.; Doxsee, K. M.; Halbert, T. R.; Hayes, S. E.; Suslick, K. S. *J. Am. Chem. Soc.* **1978**, *100*, 2761.
19. White, D. K.; Cannon, J. B.; Traylor, T. G. *J. Am. Chem. Soc.* **1979**, *101*, 2443.
20. Jameson, G. B.; Rodley, G. A.; Robinson, W. T.; Gagne, R. R.; Reed, C. A.; Collman, J. P. *Inorg. Chem.* **1978**, *17*, 850.
21. Gelin, B. R.; Karplus, M. *Proc. Natl. Acad. Sci. USA* **1977**, *74*(3), 801.
22. Warshel, A. *Proc. Natl. Acad. Sci. USA* **1977**, *74*(5), 1789.
23. Brautigan, D. L.; Feinberg, B. A.; Hoffman, B. M.; Margoliash, E.; Peisach, J.; Blumberg, W. E. *J. Biol. Chem.* **1977**, *252*, 574.
24. Hoffman, B. M.; Gibson, Q. H.; Bull, C.; Crépeau, R. H.; Edelstein, S. J.; Fisher, R. G.; McDonald, M. J. *Ann. N.Y. Acad. Sci.* **1975**, *244*, 174.

25. Gibson, Q. H.; Hoffman, B. M. *J. Biol. Chem.* **1979**, *254*, 4691.
26. Swartz, J. C.; Stanford, M. A.; Moy, J. N.; Hoffman, B. M. *J. Am. Chem. Soc.* **1979**, *101*, 3396.
27. Stanford, M. A.; Swartz, J. C.; Phillips, T. E.; Hoffman, B. M. *J. Am. Chem. Soc.* **1980**, *102*, 4492.
28. Chevion, M.; Salhany, J. M.; Castillo, C. L.; Peisach, J.; Blumberg, W. E. *Isr. J. Chem.* **1977**, *15*, 311.
29. Abbott, E. H.; Rafson, P. A. *J. Am. Chem. Soc.* **1974**, *96*, 7378.
30. Moffat, K.; Loe, R. S.; Hoffman, B. M. *J. Mol. Biol.* **1976**, *104*, 669.
31. Gonzales, B.; Kouba, J.; Yee, S.; Reed, C. A.; Kirner, J. F.; Scheidt, W. R. *J. Am. Chem. Soc.* **1975**, *97*, 3247.
32. Little, R. G.; Ibers, J. A. *J. Am. Chem. Soc.* **1974**, *96*, 4452.
33. Piciulo, P. L.; Rupprecht, G.; Scheidt, W. R. *J. Am. Chem. Soc.* **1974**, *96*, 5293.
34. Hoffman, B. M.; Gibson, Q. H. *Proc. Natl. Acad. Sci. USA* **1978**, *75*, 21.
35. Gibson, Q. H.; Roughton, F. J. W. *Proc. R. Soc. London Ser. B* **1957**, *146*, 206.
36. Edelstein, S. J.; Rehmar, M. J.; Olson, J. S.; Gibson, Q. H. *J. Biol. Chem.* **1970**, *245*, 4372.
37. Cassoly, R.; Gibson, Q. H. *J. Mol. Biol.* **1975**, *91*, 301.
38. Lakowicz, J. R.; Weber, G. *Biochemistry* **1973**, *12*, 4171.
39. Valentine, J. S.; Sheridan, R. P.; Allen, L. C.; Kahn, P. *Proc. Natl. Acad. Sci. USA* **1979**, *76*(3), 1009.
40. Mincey, T.; Traylor, T. G. *J. Am. Chem. Soc.* **1979**, *101*, 765.
41. Brault, D.; Rougee, M. *Biochem. Biophys. Res. Commun.* **1974**, *57*(3), 654.
42. Sawicki, C. A.; Gibson, Q. H. *J. Biol. Chem.* **1976**, *251*, 1533.
43. Shelnutt, J. A.; Rousseau, D. L.; Dethmers, J. K.; Margoliash, E. *Proc. Natl. Acad. Sci. USA* **1979**, *76*(8), 3865.
44. Shelnutt, J. A.; Rousseau, D. L.; Friedman, J. M.; Simon, S. R. *Proc. Natl. Acad. Sci. USA*, **1979**, *76*(9), 4409.
45. Ladner, R. C.; Heidner, E. J.; Perutz, M. F. *J. Mol. Biol.* **1977**, *114*, 385.
46. Hoffman, B. M. *J. Am. Chem. Soc.* **1975**, *97*, 1688.
47. Foster, R. "Organic Charge-Transfer Complexes"; Academic: New York, 1969.
48. Brault, D.; Rougee, M. *Biochemistry* **1974**, *13*(22), 4591.
49. Barry, C. D.; Hill, H. A. O.; Mann, B. E.; Sadler, P. J.; Williams, R. J. P. *J. Am. Chem. Soc.* **1973**, *95*, 4545.
50. Fulton, G. P.; LaMar, G. N. *J. Am. Chem. Soc.* **1976**, *98*, 2119.

RECEIVED June 11, 1979.

Activation of Dioxygen Using Group VIII Metal Complexes

BRIAN R. JAMES

Department of Chemistry, University of British Columbia, Vancouver, British Columbia, Canada V6T 1W5

The O_2-oxidation of inorganic and organic substrates catalyzed by Group VIII metal complexes is considered together with some analogies to enzyme systems; organometallic type and more bioinorganic type models are discussed. Net oxygen-atom transfer can occur, but rarely does this appear to involve transfer of coordinated dioxygen to a substrate coordinated at the same metal center. Other nonselective, free-radical pathways, including Haber–Weiss-catalyzed decomposition of trace hydroperoxides, usually are competitive and dominate with organic substrates, although formation of hydroperoxide or the $HO_2 \cdot$ radical via hydrogen abstraction by a metal–dioxygen moiety seems plausible. Nucleophilic substrates can release the oxidizing power of coordinated dioxygen as free peroxide or superoxide, and this (or genuine oxygen-atom transfer), if necessary coupled to a Wacker cycle, can lead to co-oxidation of substrates. Molecular H_2 can be used to provide hydride as one substrate, and the use of O_2/H_2 mixtures can result in oxidations that are formally very similar to P-450 systems. Olefin interaction with metalloporphyrins suggests possible oxidation via substrate-activation pathways.

Synthetic O_2-carriers have been known for some forty years and have maintained considerable interest because of the role of such centers in some naturally occurring oxygen-storage and -transport systems (1–8). Much of the earlier work centered around a range of formally divalent Co complexes with N- and/or O-donor ligands (3, 9),

0-8412-0514-0/80/33-191-253$06.00/0
© 1980 American Chemical Society

and indeed various forms of coordinated dioxygen moieties [monoden-
tate, bent superoxide (1), and bridging superoxide or peroxide (2)]
were first established by x-ray crystallography at Co centers.

$$O \diagdown O$$

1   2   3

Vaska's report (10) in 1963 of reversible O_2-binding by trans-
$IrCl(CO)(PPh_3)_2$ stimulated others to discover an extensive list of oxygen complexes of Group VIII metal systems with "organometallic-
like" ligands (carbonyls, tertiary phosphines, etc.); these complexes,
which were usually derived from d^8 and d^{10} platinum metal and nickel
systems, demonstrated the side-on geometry (3), the O_2 moiety being
likened either to coordinated ethylene (singlet O_2), or more usually to
peroxide when formation of the dioxygen complex is discussed under
the general classification of oxidative-addition reactions (11, 12).

Interest in dioxygen complexes of metalloporphyrins (1, 2, 3, 5, 6,
7, 8) is particularly intense because nature has evolved such systems
not only for binding and reversibly carrying O_2 (e.g., myoglobin,
hemoglobin) but also for activating O_2 via enzymic oxygenases, which
incorporate one or two atoms of O_2 to a substrate, or via oxidases that
convert both atoms of O_2 to water or hydrogen peroxide. The heme
unit (an iron-porphyrin moiety) is particularly prevalent and, for example, is found in myoglobin and hemoglobin (13, 14, 15), the
monooxygenase cytochrome P 450 (16–20), tryptophan dioxygenase
(21), and in cytochrome c oxidase—the terminal enzyme in the respiratory redox chain that reduces O_2 to water (21, 22). The enzymes
catalase and peroxidase, both containing heme centers, utilize H_2O_2
and are related to the dioxygen systems (21, 23).

It is only within the last decade that protein-free metalloporphyrin–O_2 complexes have been recognized clearly, and of interest, this again stems, at least historically, from studies on a Co(II)
system—the reaction of vitamin B_{12r} with O_2 at low temperature (24).
Co(II) systems, of course, have the advantage of forming ESR-
detectable superoxide species (Co^{III}—O_2^-). However, in this regard,
Corwin and Bruck (25) over 20 years ago almost certainly (6) had
oxygenated rather than oxidized bis(imidazole)ferrous porphyrin systems. Reaction of dioxygen with metalloporphyrins to give 1 : 1 complexes with geometry 1 or 3 has been demonstrated with Cr(II),
Mn(II), Fe(II), Co(II), and Ru(II) systems (7, 8, 26). The factors affecting the binding of dioxygen (usually "oxygen", for the sake of familiar-

ity) to transition-metal centers, whether the organometallic type systems or the more biological type porphyrin and related macrocyclic systems, are being elucidated slowly. These studies include kinetic, thermodynamic, structural, and theoretical studies, and especially a consideration of the role of the metal, ligands, and solvent. The area has been reviewed extensively in recent years (*1–8*), and other articles in this volume (*15, 27, 28*) are concerned with this topic. Studies on protein-free systems are proving extremely informative as exemplified by the elegant work of Traylor's group (*27*) and Collman's group (*29*), which is leading to a better understanding of the cooperativity phenomenon in O_2-binding by hemoglobin (*15*).

The chemical behavior of O_2 after coordination, that is the activation of O_2, is a diverse and complex subject because of the many pathways available for O_2 reactions. The search for more selective oxidations, especially using O_2, is a prime goal that has immense potential on an industrial scale (*4, 30*), and is thus an important research area. The ultimate aim would be to mimic the O_2-activation ability of enzyme systems, for example the ability of cytochrome P 450 to catalyze Reaction 1, where R is a hydrocarbon molecule.

$$RH + O_2 + NADH + H^+ \rightarrow ROH + H_2O + NAD^+ \tag{1}$$

Excellent reviews on O_2 activation by Sheldon and Kochi (*31*), and Lyons (*4*), have appeared recently, and I have used these extensively. This chapter, reflecting the author's interests, will concentrate on Group VIII metal systems, bringing in comparisons with enzymatic systems.

Oxygenation via Coordination Catalysis

The autoxidation of organic substrates catalyzed by transition-metal salts has been used widely in the petrochemical industry for many years (*32*), but the oxidations are frequently nonselective since they operate by free-radical pathways that are sometimes initiated by the transition-metal ion. An example is the chain reaction of Reactions 2, 3, 4, and 5 (*4*). Propagation is maintained via Reactions 3 and 4, and any interaction between the metal ion and O_2 would be incidental.

$$M^{n+} + RCHO \rightarrow RCO \cdot + M^{(n-1)} + H^+ \tag{2}$$

$$RCO \cdot + O_2 \rightarrow RCOO_2 \cdot \tag{3}$$

$$RCOO_2 \cdot + RCHO \rightarrow RCOO_2H + RCO \cdot \tag{4}$$

$$RCOO_2H + RCHO \rightarrow 2RCOOH \tag{5}$$

The discovery of the d^8 and d^{10} metal π-bonded dioxygen complexes (formally d^6- and d^8-peroxide systems), and the accompanying chemistry exemplified in Scheme 1, showed the possibility of attaining net oxygen-atom transfer to both inorganic and organic substrates.

Scheme 1. Chemistry typical of d^8, d^{10} Pt metal systems: M = Rh(I), Ir(I), Ni$^{(0)}$, Pd$^{(0)}$, Pt$^{(0)}$, with ancillary ligands L, typically a tertiary phosphine type ligand, CO, etc. (4, 33, 34)

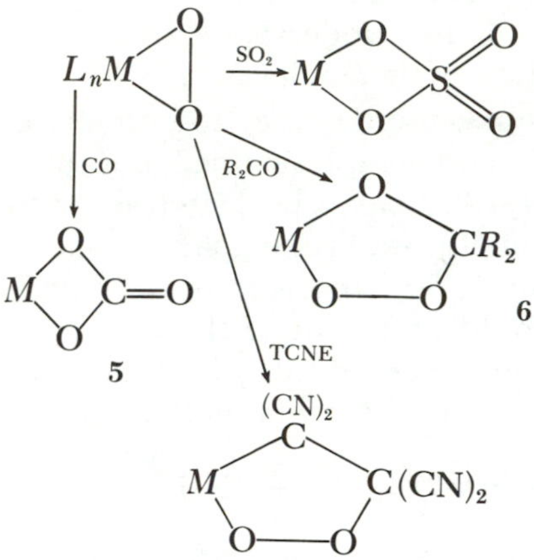

Stoichiometric oxidation of SO_2 to coordinated SO_4^{2-} (4), CO to CO_3^{2-} (5), 2NO to $2NO_2^-$, and so on, and the isolation of peroxoketones (6) and peroxometallocycles such as 7, could all be visualized as proceeding through intermediates containing both coordinated dioxygen and coordinated substrate. Indeed, the isolation of the complex IrCl(PPh$_3$)$_2$-(C$_2$H$_4$)(O$_2$) containing both π-bonded C$_2$H$_4$ and O$_2$ (35) must have resulted from attempts to selectively oxygenate ethylene. If the coordinated oxygenated substrate is replaced by further excess substrate, a catalytic oxidation cycle results (for example, Reaction 6), and catalyzed oxidations of substrates such as phosphines to phosphine oxides, and isocyanides to isocyanates, were rationalized in terms of such a mechanism (4).

$$L_2MO_2 \rightarrow M(OL)_2 \rightarrow M + 2LO \qquad (6)$$
$$\underset{+ 2L, + O_2}{\underline{\quad\qquad\qquad\qquad\qquad}}$$

However, a recent paper by Sen and Halpern (36) shows that the mechanism of direct oxygen transfer from the metal–dioxygen complex to a substrate is probably not operable in such systems. Using

PMePh$_2$ and PMe$_2$Ph as substrates, the mechanism outlined in Reaction 7 (where P = tertiary phosphine) has been substantiated for catalytic oxidation via a Pt(0) system. The phosphine substrate effects

$$PtP_3(O_2) \xrightarrow{P} PtP_4{}^{2+} + O_2{}^{2-}(\xrightarrow{H^+} HO_2{}^-)$$

$$\text{(X)}P \Big| OH^- \qquad \Big\downarrow P$$

$$O_2 \qquad OH^- + PO$$

$$H^+ + PO + PtP_4 \tag{7}$$
$$(XO)$$

displacement of the O$_2$ as peroxide by nucleophilic attack at the metal, and it is free peroxide in solution (as HO$_2{}^-$) that oxidizes the phosphine. To maintain the catalytic cycle, the Pt(II) is reduced via free phosphine and hydroxide (a two-equivalent reducing system). Only a trace amount of proton is necessary since it and the concomitant hydroxide remain in steady state concentrations. Data from this laboratory on the properties of some oxygen-containing Rh complexes also were explained best by invoking generation of free peroxide in solution (37). Superoxide can be generated similarly from metal centers that are susceptible to a one-electron oxidation (discussed later).

It is instructive to compare the chemistry outlined in Reaction 7 with that shown in Reaction 8 (where Rh$^{(I)}$ = chlororhodate(I) species) for a Rh$^{(I)} \leftrightarrows$ Rh$^{(III)}$ catalytic system (4, 38, 39, 40). The cycle shown by

$$Rh^{(I)}(CO) \xrightarrow{O_2} Rh^{(III)}(CO_3) \xrightarrow{H^+} Rh^{(III)} + CO_2 + OH^-$$

$$(XO) \quad \Big\uparrow CO \qquad \qquad \Big/ CO(X) \tag{8}$$

$$H^+ + CO_2 + Rh^{(I)} \leftarrow Rh^{(III)}(CO_2H) \xleftarrow{OH^-} Rh^{(III)}(CO)$$

the solid arrows accounts for a Rh-catalyzed conversion of CO/O$_2$ mixtures to CO$_2$ in aqueous HCl solutions. The coordinated carbonate formed (*cf.* Scheme 1) is decomposed by acid to liberate CO$_2$ and form Rh$^{(III)}$; reduction back to Rh$^{(I)}$ occurs via coordination of CO, attack by hydroxide then giving a carboxylate intermediate that decomposes via a two-electron transfer to the metal (a reductive carbonylation process). The cycle is similar in several respects to that shown in Reaction 7 with CO corresponding to P. If the Rh(I) oxidation goes via formation of a dioxygen species prior to reaction with CO, the scheme would correspond exactly to the Pt cycle. However, free peroxide formation via CO attack is unlikely and in any case free peroxide does not oxidize CO; oxygen-atom transfer to give the carbonate is preferred. If the CO (or P) substrate is replaced at a certain stage in the cycle by another oxidizable substance (X) that competes with the CO (or P), then co-oxidation is possible according to Reaction 9. Substrate X could, for

$$O_2 + X + CO \text{ (or } P) \xrightarrow{\text{catalyst}} OX + CO_2 \text{ (or } PO) \tag{9}$$

example, be an olefin and the $Rh^{(III)}/X$ reaction to give $Rh^{(I)}$, XO, and H^+ (see Reaction 8), a Wacker type oxidation, has been demonstrated for this chlororhodate(III) system (41). As in well established Pd(II)–Wacker type chemistry, reoxidation can be effected using molecular O_2 (as indicated by the dotted arrow in Reaction 8). The Pt(II) system is being studied (36).

The Wacker process, of course, gives highly selective oxidation of olefins to aldehydes or ketones (42); the function of the O_2 is to reoxidize the catalyst, and again any formation of a dioxygen complex is incidental, although such a species could be involved in the reoxidation step. Reoxidation of Cu(I) to Cu(II)/Cu(III) by O_2 appears to be involved in certain Cu-containing oxidase systems, for example, ascorbic-acid oxidase (43, 44).

The subject of co-oxidations (Reaction 9) will be returned to later.

Oxygen-Atom Transfer to Olefins

As mentioned above, catalytic oxidation of olefins via coordination catalysis with an intermediate such as L_nM (olefin) O_2 seemed an attractive possibility, and Collman's group (45) tentatively invoked such catalysis in the O_2-oxidation of cyclohexene to mainly 2-cyclohexene-1-one promoted by $IrI(CO)(PPh_3)_2$, a complex known to form a dioxygen adduct. Soon afterwards (4, 46, 47) such oxidations involving d^8 systems generally were shown to exhibit the characteristics of a radical chain process, initiated by decomposition of hydroperoxides via a Haber–Weiss mechanism, for example Reactions 10 and 11. Such oxidations catalyzed by transition-metal salts such as

$$M^+ + ROOH \rightarrow M^{2+} + RO\cdot + OH^- \tag{10}$$

$$M^{2+} + ROOH \rightarrow M^+ + RO_2\cdot + H^+ \tag{11}$$

Co(II) carboxylates were well documented, and for cyclohexene as substrate the cyclohexenyl hydroperoxide is formed in situ by attack of O_2 on the allylic radical produced by allylic hydrogen abstraction, Reaction 12 (4, 48). The products, usually those shown in Reaction 13, are formed via the metal-catalyzed decomposition of the hydroperoxide, and any O_2 coordination at the metal is incidental; there is

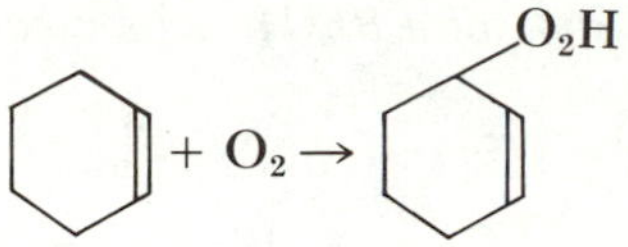

$$\text{cyclohexene} + O_2 \rightarrow \text{cyclohexenyl-O}_2\text{H} \tag{12}$$

$$\text{cyclohexene} \rightarrow \text{cyclohexanone (O)}, \text{cyclohexenol (OH)}, \text{cyclohexene oxide} \tag{13}$$

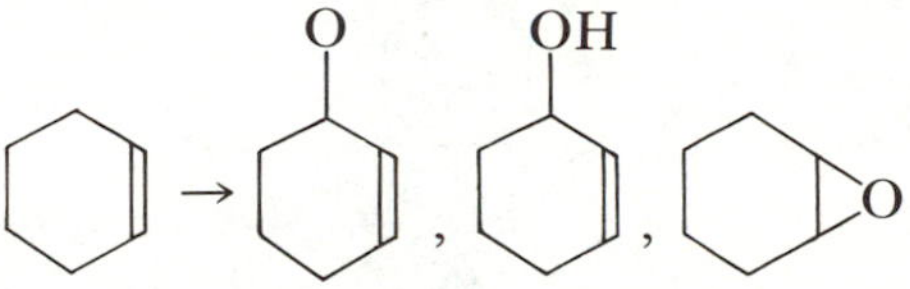

no O_2-coordination catalysis, although coordination of hydroperoxide almost certainly is involved. The decomposition of hydroperoxides catalyzed by the d^8 and d^{10} Pt-metal complexes has been studied by several groups (4, 49, 50, 51).

In cases where the olefinic substrate cannot undergo allylic hydrogen abstraction, or when the metal complex being used is inactive for hydroperoxide decomposition, radical pathways again seem dominant but alternative means of generating the initiating radical have to be proposed, for example via a dioxygen complex (Reaction 14, 52), or via oxidation of a coordinated ligand (Reaction 15, 53).

$$L_nMO_2 + \;{>}C{=}C{<}\; \rightarrow L_nMO_2{-}\overset{|}{\underset{|}{C}}{-}\overset{|}{\underset{|}{C}}\cdot \tag{14}$$

$$M(\text{acac})_3 + O_2 \rightarrow M(\text{oxide}) + CO_2 + R\cdot \tag{15}$$

We had invoked formation of hydroperoxide via the mechanism outlined in Scheme 2 for oxidation of a Rh(I) cyclooctene complex in dichloroethane, the important step being hydrogen abstraction from the substrate by the coordinated dioxygen, although it was emphasized that neither the olefin nor hydroperoxide was necessarily coordinated to the Rh (54). The whole sequence shown in Scheme 2 was based on O_2-uptake data (O_2 : Rh = 1.0) and changes in IR spectra. A catalyzed autoxidation of the solvent occurred in dimethylacetamide (DMA) solution; a Rh(II)–O_2^- species was detected by ESR and the same type of mechanism was invoked (55). There was no induction period, little inhibition by free radical inhibitors, and, interestingly, PPh_3 was not oxidized; these data tend to rule out the presence of free hydroperoxide in solution and are more consistent with catalysis via coordination, at least of the dioxygen. Olefins are unlikely to be strong enough nucleophiles to displace the dioxygen as peroxide (36). The catalytic activity of the system diminished through the formation of an inactive Rh(II) species, that was later isolated as $Rh_2Cl_6(DMA)_2^{2-}$ (56).

Scheme 2.   The O_2 oxidation of a Rh(I) cyclooctene complex (54, 55)

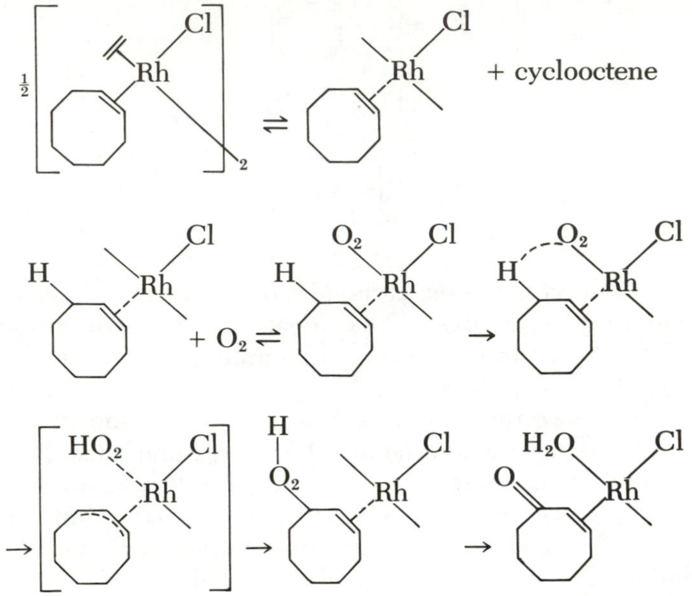

Such species would be expected to be active via Haber–Weiss type mechanisms (51), Reactions 10 and 11. Since we found that the inactive Rh(II) could be reduced by molecular H_2 back to the active Rh(I) system (51), use of H_2/O_2 mixtures for catalytic oxidation appeared realistic (discussed later).

Coordination catalysis via alkyl hydroperoxides is well documented (4, 31). Selective oxidations of olefins to epoxides (Reaction 16), using especially Group IV, V, and VI transition-metal complexes, can occur possibly via oxygen-transfer processes of the type

outlined in Equations 17 and 18 (57, 58, 59). Alternative mechanisms involve formation of oxo or peroxo metal complexes (via hydroperoxides in solution), and then subsequent oxygen transfer, for example with Mo(VI) (*see* Reaction 19) (60, 61, 62, 63). In principle, there seems no reason why similar catalysis could not occur with coordinated HO_2^- rather than HO_2R. The likely requirements of the metal complex for catalytic epoxidation of olefins by hydroperoxides have been considered (57, 64): high charge, small size, low-lying unoccupied d orbitals—in essence, good Lewis acidity to withdraw elec-

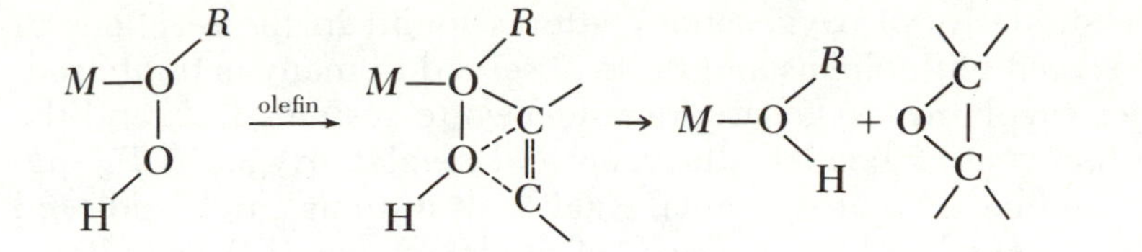

$$ \text{(17)} $$

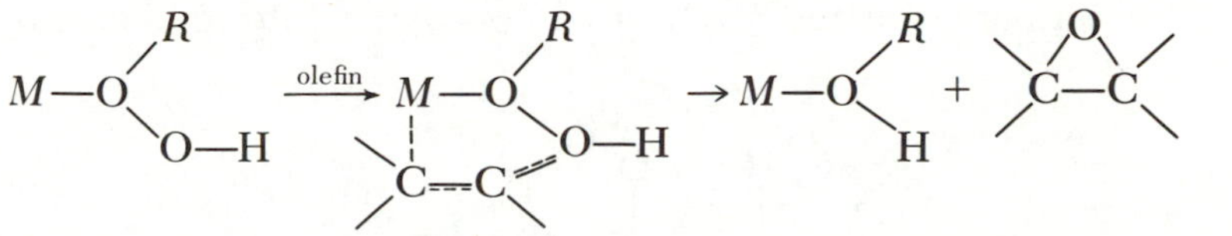

$$ \text{(18)} $$

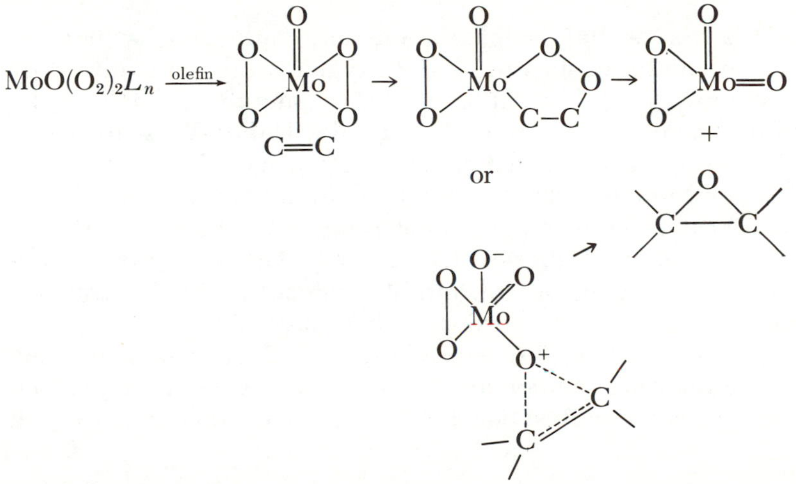

$$ MoO(O_2)_2L_n \xrightarrow{\text{olefin}} \quad \rightarrow \quad \rightarrow \quad \text{(19)} $$

or

trons from the oxygens. Participation in one-electron transfer processes also promotes homolytic cleavage of the hydroperoxide (Reactions 10 and 11), and free-radical processes result in nonselective oxidation to epoxide, ketone, and alcohol. Group VIII metal complexes have this inherent property, and it is difficult to substantiate any coordination catalysis (using hydroperoxides or oxygen) that might accompany the free-radical processes (*65*).

Several other groups have invoked initial formation of hydroperoxide by abstraction of hydrogen from organic substrates (including olefins) by a dioxygen complex. In some cases, the oxygen adducts were known to form under the reaction conditions (*67, 68*); in other cases, they have been postulated as intermediates (*69, 71*). One such report on a Pd(O) system (*67*) was later refuted by a more detailed study, which showed that the autoxidation resulted by decomposition of trace amounts of hydroperoxide impurity (*50*). (Some Co systems are considered in more detail in the next section).

   The similarity of oxygenation patterns found in the reactions of singlet oxygen with olefins and those observed in many natural product series involving oxygenase metalloenzyme systems (72), and the analogy between π-bonded dioxygen and singlet oxygen (73), suggested that the d^8 and d^{10} metal-catalyzed systems might proceed through the "ene" addition pathway of singlet oxygen (74), as in Reaction 20. The formation of the allylic hydroperoxide intermediate in

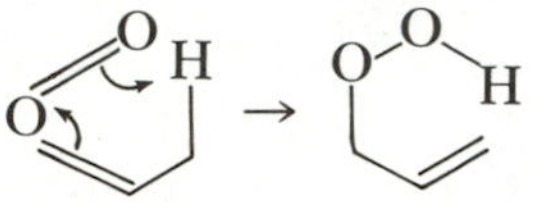

$$\tag{20}$$

Scheme 2 is similar, but cyclohexadiene and anthracene, which are known to be good acceptors of singlet oxygen, were not oxidized by our Rh system (54), and oxidation of (+)carvomenthene catalyzed by $RhCl(PPh_3)_3$ gave products more typical of a Haber–Weiss oxidation than those observed after oxidation by singlet oxygen (74). Besides our system, several other instances have been found (47, 75, 76, 77, 78), where Rh(I)-catalyzed oxidations show some differences from behavior expected from just Haber–Weiss pathways. Different product profiles, lack of an induction period, and little effect by added scavengers all suggest other pathways for forming hydroperoxides.

   Holland and Milner (79) re-examined our Rh(I) oxidation of cyclooctene in benzene solution at 74°C and demonstrated the reaction stoichiometry shown in Reaction 21. We had measured our $1:1$ O_2/Rh

$$\tfrac{1}{2}[RhCl(C_8H_{14})_2]_2 \xrightarrow{O_2} RhCl(C_8H_{14})_2O_2 \xrightarrow{\frac{1}{2}O_2}$$

$$+ H_2O + \tag{21}$$

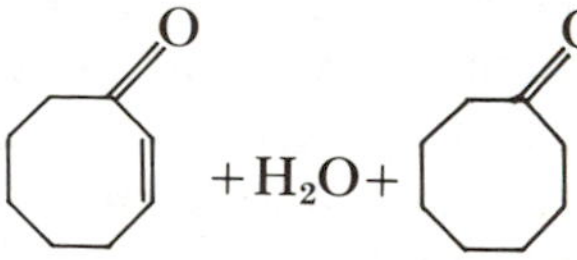

stoichiometry at 35°C, but had noted a much slower subsequent gas uptake in toluene (54). Reaction 21 was shown not to involve Wacker oxidation or radical pathways, and a rate-limiting dissociation of coordinated O_2 and insertion into allylic C–H bonds of both coordinated cyclooctene molecules was proposed (79). A related Rh(I)-catalyzed O_2 autoxidation of styrene to acetophenone and smaller amounts of benzaldehyde also appears to involve coordination catalysis (80). Other studies on oxidation of styrene (which has no allylic hydrogens) using Ir, Rh, and Pd complexes have claimed metal-centered O_2 activation but the evidence is less compelling (*81, 82, 83, 84*), and a mechanism involving radical-initiated oxidative cleavage of the double bond is indicated in some cases (*65, 66, 85*).

Read's group (*86, 87*) are studying some interesting co-oxidation reactions catalyzed particularly by $RhCl(PPh_3)_3$ in benzene at 20°C. Terminal olefins and triphenylphosphine are converted by O_2 to the corresponding methyl ketones and the phosphine oxide. Radical chain processes were not detected, and such reactions normally do not yield methyl ketone products. Net oxygen-atom transfers were considered to result via the mechanism shown in Scheme 3.

Scheme 3. The co-oxidation of triphenylphosphine and a terminal olefin at a Rh center (87)

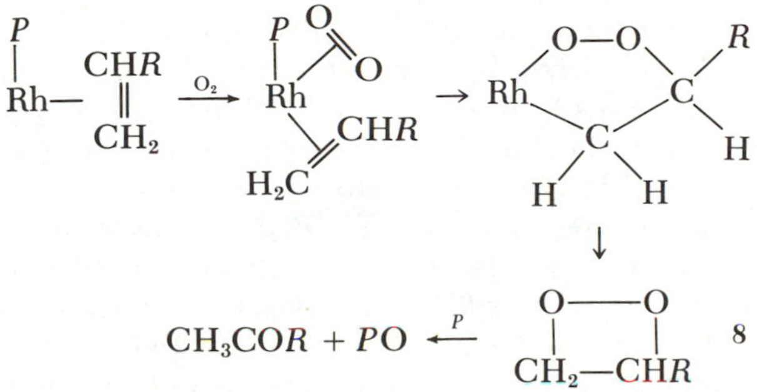

The metallocyclic intermediate (*cf.* **7** in Scheme 1) could eliminate reductively the dioxetan **8**, which on attack by PPh_3 could yield the ketone and phosphine oxide, this latter step incorporating a 1,2-hydride shift. These workers (*86*) noted the formal similarity of the overall catalytic reaction to P-450-catalyzed reactions (Reaction 1), the phosphine effectively playing the role of the hydride donor. The hydroxylation of aromatic rings effected by the monooxygenase system also is characterized by a 1, 2-hydride shift, the so called N.I.H. shift (*88*), for example Reaction 22. Epoxide intermediates have been invoked in the biological systems (*88*). No epoxides were detected in the Rh system and free epoxide was not isomerized to ketone, although isomerization via coordination at the Rh could not be ruled out (*86, 89*). The dioxetan intermediate was favored, however, since with oct-1-ene as substrate small amounts of heptanal were detected, and this could be attributed to a metal-catalyzed decomposition of this intermediate to RCHO and formaldehyde, although the latter was not detected (*87*). Several possibilities exist for the mechanism of this C–C bond cleavage in cyclic peroxo compounds (*33*).

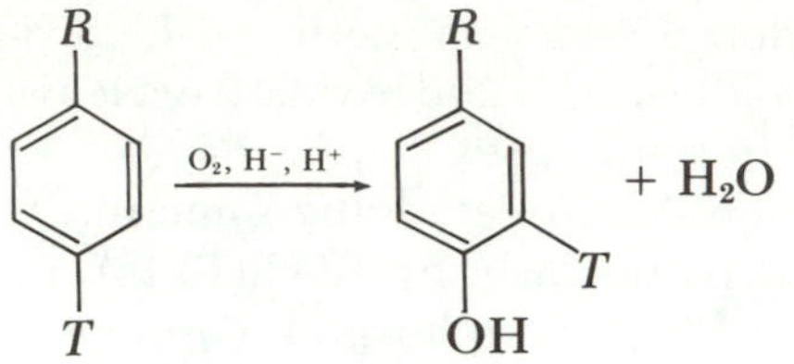

$$R \quad\quad\quad\quad R$$

$$\xrightarrow{O_2,\ H^-,\ H^+} \quad\quad + H_2O \quad\quad\quad (22)$$

$$T \quad\quad\quad OH$$

The findings of Sen and Halpern (*36, see* Reaction 7) suggest alternative pathways for the Rh-catalyzed co-oxidation, in which the phosphine is oxidized by liberated peroxide and the ketone is formed during the reduction of Rh(III) to Rh(I) (a Wacker cycle). Read and Walker (*87*) ruled out Wacker chemistry, since addition of water in their system (OH$^-$ is needed for the reduction, Reactions 7 and 8) decreased ketone yield somewhat. More convincing evidence for the absence of Wacker type chemistry comes from isotope studies using H_2O^{18}, work just published by Tang et al. (*89*).

Oxygen-atom transfer to PPh$_3$, and to olefins to give epoxide, has been accomplished via $(Ph_3P)_2PtO_2$ in the presence of electrophiles such as benzoyl chloride (*90*); transfer occurs from a peroxybenzoatobisphosphine intermediate $(Ph_3P)_2Pt(OOCOPh)Cl$ (cf. **6** in Scheme 1), which is converted to the benzoate derivative.

Mimoun and co-workers (*91*) recently have demonstrated what is essentially a four-electron oxidation by O_2 without the need for a coreducing agent such as PPh$_3$. Using mixtures of Rh and Cu salts in alcoholic solvents, sometimes with added acid, two molecules of terminal olefin could be converted to two molecules of methyl ketone per catalytic cycle. The alcohol does not act as the coreductant. The suggested mechanism involves breakdown of the peroxymetallocycle into the ketone and a Rh(III) oxo intermediate; this with acid yields the Rh(III) hydroxy species that is capable of oxidizing the second molecule of olefin via a Wacker process. The Wacker step regenerates the required acid, as in Reaction 23. Labelling studies have confirmed

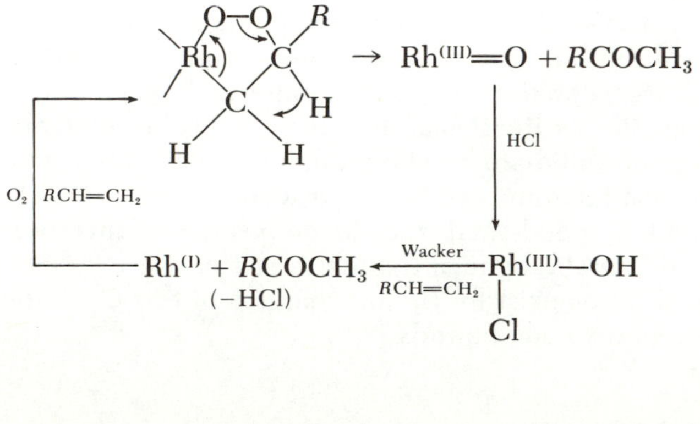

$$\xrightarrow{} Rh^{(III)}{=}O + RCOCH_3$$

$$O_2 \mid RCH{=}CH_2$$

$$Rh^{(I)} + RCOCH_3 \xleftarrow[\ RCH=CH_2\ ]{Wacker} Rh^{(III)}{-}OH \quad\quad (23)$$

$$(-HCl) \quad\quad\quad\quad\quad\quad\quad\quad Cl$$

incorporation of the coordinated O_2 of $[Rh(AsPh_3)_4O_2]^+$ ClO_4^- into a ketone product (92). The protonation of oxo species has been reported (93). Minor oxidative cleavage to aldehydes occurred in the absence of added acid, and this led to speculation that such cleavage might occur via a carbene intermediate, Reaction 24. There is precedent in the

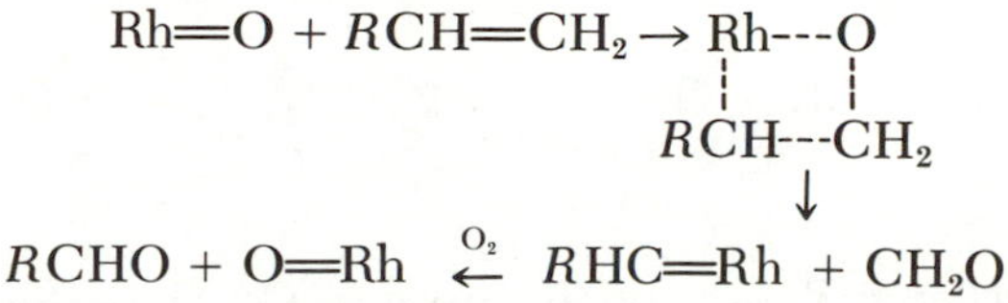

$$Rh{=}O + RCH{=}CH_2 \rightarrow \underset{RCH\text{-}\text{-}\text{-}CH_2}{Rh\text{-}\text{-}\text{-}O}$$

$$\downarrow \qquad (24)$$

$$RCHO + O{=}Rh \overset{O_2}{\underset{\leftarrow}{}} RHC{=}Rh + CH_2O$$

literature for such chemistry (94, 95, 96). Such oxo intermediates are gaining favor in P-450-catalyzed systems (18, 95), and mechanisms closely analogous to that outlined in Reaction 24 have been suggested for the enzyme system (95) (*see* Reaction 25).

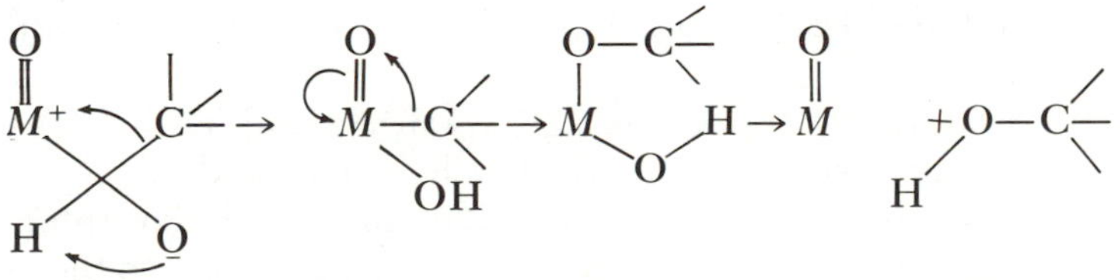

$$\text{(25)}$$

Co-oxidation of Olefins and Molecular Hydrogen

Recently, we have turned our attention to some co-oxidation reactions that are remarkably similar to a P-450 system (Reaction 1) in that hydride is used as the coreductant; the H^- and H^+ requirements of the metalloenzyme system are furnished in our systems by a metal that activates molecular H_2 via heterolytic cleavage (97), Reaction 26

$$H_2 + M \rightleftarrows MH^- + H^+ \qquad (26)$$

where $M = $ Rh,Ir. Thus, we have used H_2/O_2 mixtures to effect a net oxygen-atom addition to substrates including olefins, the other oxygen atom being reduced to water (*cf.* Reaction 1).

The work initiated from an accidental finding about a decade ago: while studying solvent effects on catalytic hydrogenation, we discovered a Rh(III)-catalyzed H_2 reduction of $(CH_3)_2SO$ to the sulfide and water (98). The suggested mechanism involved a hydridorhodium(III) intermediate containing a S-bonded sulfoxide. The reaction rates eventually decreased because of a buildup of inactive Rh(I), formed according to Reaction 27, and stabilized by the dimethylsulfide pro-

$$HRh^{(III)} \rightarrow H^+ + Rh^{(I)} \tag{27}$$

duced. In an attempt to maintain catalytic activity by reoxidation of the Rh(I), a H_2/O_2 mixture was used. This led to a surprising catalytic oxidation of the sulfoxide to the sulfone and water (98), Reaction 28.

$$(CH_3)_2SO + O_2 + H_2 \rightarrow (CH_3)_2SO_2 + H_2O \tag{28}$$
$$(H^- + H^+)$$

Neither Rh(III) nor Rh(I) species under O_2 alone performed this catalysis, implying strongly that a Rh hydride was the effective O_2 carrier. Trocha–Grimshaw and Henbest (99) also had found that isopropanol solutions of Rh(III) and Ir(III) salts carried out the same oxidation using just O_2, and such a medium is very effective for forming hydride species (100, 101). The analogy of Reaction 28 with Reaction 1 is even more marked on consideration that both catalysts are low-spin d^6 systems, (Rh(III) vs. Fe(II)) and both contain an S-donor ligand [$(CH_3)_2SO$ vs. RS^-].

The hydridorhodium(III) species $HRh(NH_3)_5{}^{2+}$ and $HRh(CN)_4$-$(H_2O)^{2-}$ are known to react with O_2 to give hydroperoxide species, Rh—OOH (102, 103, 104); the sulfoxide oxidation via such an intermediate could be accommodated by several mechanisms akin to those discussed already for olefinic substrates (cf. Reactions 17 and 18). The oxidation also could occur via liberated hydrogen peroxide, which oxidizes sulfoxide to sulfone and can be formed by treatment of Rh—OOH species with acid (104). Such a mechanism would amount to a novel catalytic conversion of H_2/O_2 mixtures to in situ H_2O_2. The part of the enzymatic cycle of P 450 that utilizes O_2 and two electrons (from a hydride source) can be bypassed by using hydrogen peroxide (16, 18).

More interesting substrates than sulfoxides are hydrocarbons, and we have been studying some olefin oxidations catalyzed by Ir systems (105); these seem to be cleaner than related Rh ones, and this may relate to the stronger metal–hydrogen bond found generally for Ir systems (106). Reaction of the 1,5-cyclooctadiene dimer [$HIrCl_2(C_8H_{12})]_2$ (107) with O_2 at 20°C has yielded a single cyclooctene-one product and an Ir species, the IR of which indicates the presence of an OH group. Oxygen-atom transfer within an IrOOH intermediate is an attractive possibility currently under study (cf. Reactions 17 and 18 with subsequent isomerization of epoxide). Reaction of the Ir(I) dimer [$IrCl(C_8H_{12})]_2$ with O_2 is reported to give a hydroxoiridium species with no oxidation of the cyclooctadiene (108), which again suggests a role for the hydride ligand in our Ir(III). Addition of excess C_8H_{12} to the $[HIrCl(C_8H_{12})]_2$ system and using O_2 or H_2/O_2 gives only a slow catalytic conversion to a mixture of products (*105*); the difficulty may be in regenerating the Ir hydride. The same Ir(III) dimer under O_2 was found inactive for the catalytic oxidation of cyclooctene added in excess, as was the cyclooctene dimer $[IrCl(C_8H_{14})_2]_2$ that we studied some years ago (*109*). However, use of a H_2/O_2 mixture has given a selective catalytic oxidation of cyclooctene to cyclooctanone and water (*105*), Reaction 29. The data appear most promising, but they are at a

$$C_8H_{14} + O_2 + H_2 \xrightarrow{\ [HIrCl_2(C_8H_{12})]_2\ } C_8H_{14}O + H_2O \qquad (29)$$

very preliminary stage, and we do have to rule out any H_2O production from the H_2/O_2 mixture by paths independent of olefin oxidation, since such hydrogenolysis of oxygen can be catalyzed at Pt metal centers (*110*). There is an accompanying catalytic hydrogenation to cyclooctane that is influenced by the partial pressure of the H_2. Such hydrogenation involves intermediate alkyls (*97*), and reaction of these with O_2 to give alkylperoxide species (e.g., $Ir—O_2R$) also could provide a pathway for the oxidation.

Molecular hydrogen can reduce cyclic peroxo complexes, for example Reaction 30 (*111*), and Sheldon and Van Doorn (*33*) had envisaged epoxide production via a similar process, Reaction 31. A mechanism based on such reactivity (with subsequent epoxide isomerization) is a possibility for our catalysis using H_2/O_2, although Reaction 31 did not occur under mild conditions (*33*). Isolation of a related peroxometallocyclic rhodium complex, $(Ph_3As)_2Rh[O_2C_2(CN)_4]^+$ (*92*), allows for a testing of Reaction 31 at a Rh center.

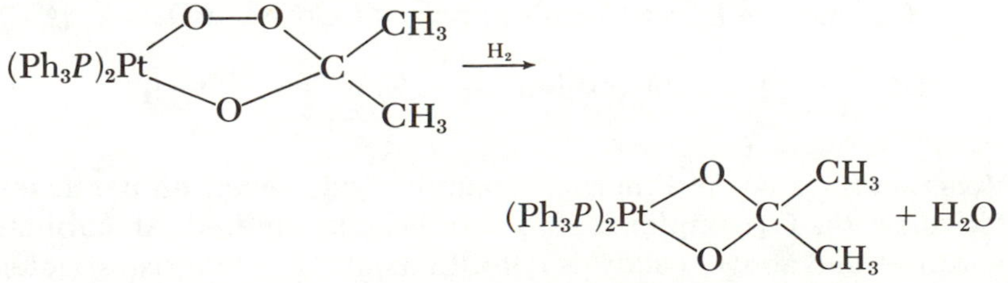

(30)

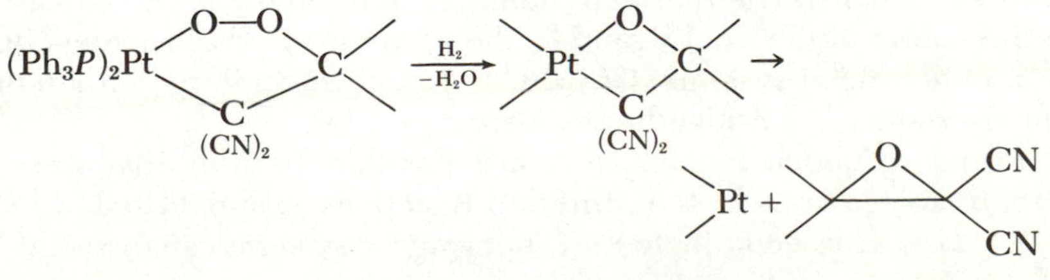

(31)

Tabushi and Koga reported the use of manganese porphyrins to catalyze the O_2-oxidation of cyclohexene to cyclohexanol and cyclohexene-ol in the presence of borohydride; these workers suggest that an equilibrium such as depicted in Reaction 32 is involved in non free-radical pathways (112).

$$Mn^{(III)}OOH \rightleftharpoons Mn^{(IV)}O \qquad (32)$$

Attempts at using CO/O_2 mixtures with Rh or Ir catalysts for selective oxygenations (cf. Reaction 9) have not proved successful (105).

Oxidation of Organic Substrates Using Metal Centers With Macrocyclic Nitrogen and Oxygen Ligands, Including Porphyrins

Transition-metal systems with macrocyclic nitrogen and oxygen donors are more analogous to biological O_2-oxidation centers than the more organometallic type systems considered thus far. Studies on the macrocyclic ligand systems follow earlier ones such as Udenfriend's systems (e.g., Fe(II)–EDTA), and the use of other transition-metal salts, sometimes with added ligands (43, 113).

Some recent systems have concentrated more on using complexes that are known to form O_2-adducts; however, in nearly all cases, 1 : 1 dioxygen–adduct formation does not result under the conditions used for the attempted catalysis. For example, the following chemistry is well established for bis(salicylaldehyde)ethylenediimine Co(II), Co(salen), and other related Schiff-base and porphyrin-ligand systems (3, 6, 9, 113), Reactions 33 and 34. Dioxygen binding is exothermic,

$$Co(salen) + L \rightleftharpoons LCo(salen) \overset{O_2, K}{=\!=\!=\!=} LCo(salen)O_2 \qquad (33)$$

$$LCo(salen)O_2 + LCo(salen) \overset{k}{\rightarrow} [LCo(salen)—O—]_2 \qquad (34)$$

and Reaction 33 is favored thermodynamically at lower temperatures ($<0°C$) where the O_2 complex is stable and can be studied. At ambient temperatures and above (catalysis conditions), bridged peroxo species are formed usually. The thermodynamic K and kinetic k values vary with the nature of the axial ligand L, the solvent, and the macrocyclic ligand. Other metal systems (Cr, Mn, Fe) behave similarly, although the final product is a bridged oxo species (6, 114).

Several oxidation mechanisms are possible besides the inner-sphere, bimolecular process outlined in Reactions 33 and 34 (5, 6, 115, 116, 117, 118). Hexacoordinate Fe(II)-porphyrin systems can dissociate a ligand and follow the inner-sphere oxidation (to a $Fe^{(III)}$—O—$Fe^{(III)}$ product), but outer-sphere oxidations involving $[L_2Fe(porp) \cdots O_2]$ seem possible. A kinetic dependence on proton concentration has been interpreted in terms of oxidation via an $HO_2 \cdot$ radical (*118*), Reaction 35. Autoxidation of hemoglobin (Reaction 36 (*119*)) and P 450 (Reac-

$$L_2Fe(porp)^{\delta +} \cdots O_2^{\delta -} \rightarrow L_2Fe(porp)^+ + HO_2 \cdot \qquad (35)$$

tion 37 (*120, 121*)) proceeds via similar formation of HO_2 through coordinated O_2. The nucleophile necessary for the HbO_2 system corresponds to the role of PPh_3 in liberating peroxide from $(Ph_3P)_2PtO_2$ (Reaction 7).

$$HbO_2 + H^+ + X^- \rightarrow HbX + HO_2 \cdot \qquad (36)$$

$$P\ 450(O_2) + H^+ \rightarrow (P\ 450)^+ + HO_2 \cdot \qquad (37)$$

Oxidation of organic substrates (flavones, phenols, indoles, thiols, ascorbic acid, etc.) using Co(salen)/O_2 systems is often interpreted in terms of initial production of radicals by H-atom (or electron) abstraction by the $Co(II)O_2$ or $Co(III)O_2Co(III)$ moieties (*122–127*) (*cf.* Scheme 2), Reaction 38. In some of the conditions used, for example in

$$Co^{(II)}—O_2\ (or\ Co^{(III)}—O_2^-) + SH \rightleftarrows Co^{(III)}—O_2H^- + S \cdot \qquad (38)$$

dimethylformamide at 20°C (*123, 128*), there will be initial partial formation of $(DMF)Co(salen)O_2$, but this oxidizes quite rapidly (*128*); nevertheless, kinetically significant amounts could be involved, or the bridged peroxide could be the catalyst. The oxidations of flavones (Reaction 39) and indoles (Reaction 40), follow in several respects those of the enzymes quercetinase (Cu) and tryptophan dioxygenase (heme, Cu), respectively (*123, 126*).

The indole oxidation has been shown to proceed via the hydroperoxide intermediate 9 (*126*), but whether this is formed via coordination catalysis, for example, as suggested in Reaction 41 for a phenol substrate ($10 \rightarrow 12,13,14$) (*124*), or via Haber–Weiss initiation, poses the same problem encountered in the organometallic type systems. A reactivity trend observed for Reaction 40 using tetraphenylporphyrin complexes (Co(II) $\gg$ Cu(II) $\gg$ Ni(II)) is reasonable in that the Co(II) system is known to give 1:1 O_2-adducts (at least, at low temperatures) but the reactivity trend also was observed for the catalyzed decomposition rate of 9. It is interesting to note that in Reac-

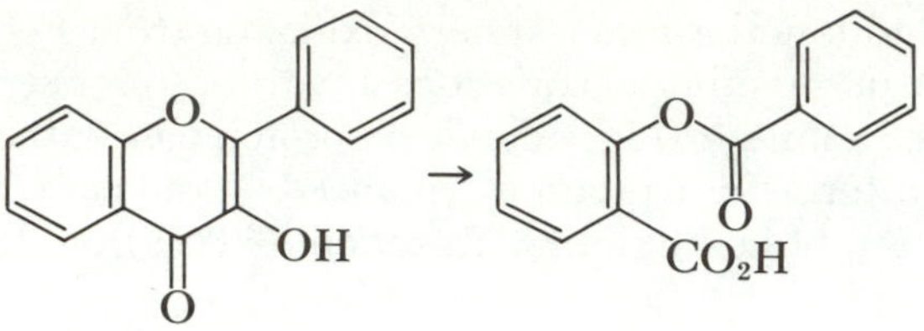

(39)

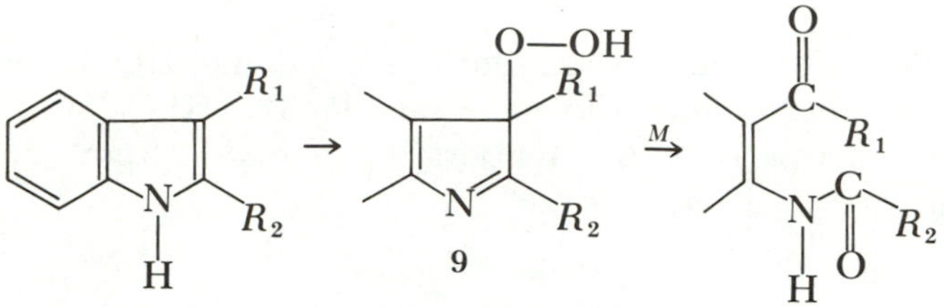

(40)

$(R_1, R_2, = H, CH_3, etc.)$

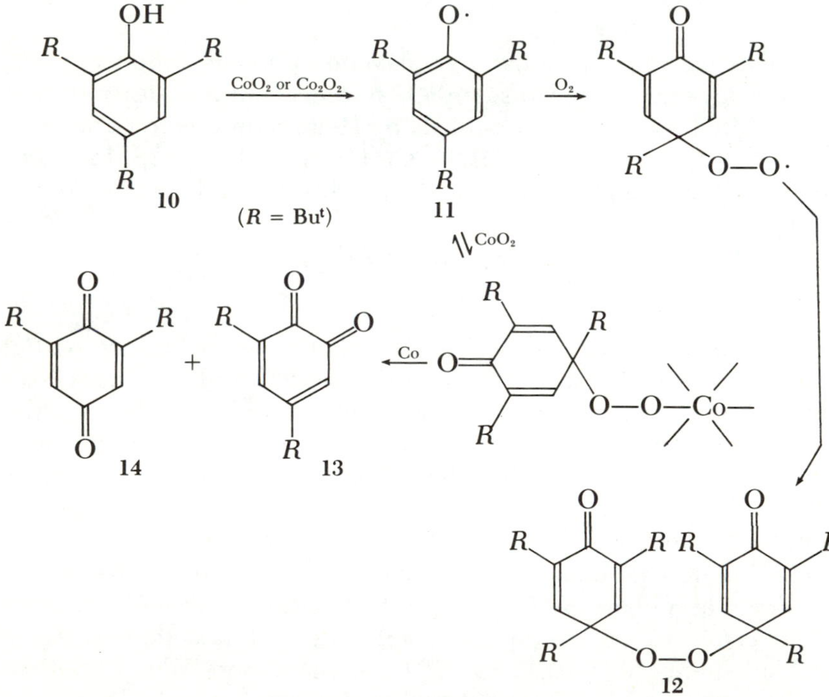

(41)

tion 39, the oxidative cleavage involves loss of a CO group; this was evolved as CO_2 resulting from a Co(salen)-catalyzed oxidation (123). However, this is not formally a co-oxidation of the type outlined in Reaction 9 since the flavone oxidation constitutes a dioxygenase system.

Metalloporphyrins catalyze the autoxidation of olefins, and with cyclohexene at least, the reaction to ketone, alcohol, and epoxide products goes via a hydroperoxide intermediate (*129, 130*). Porphyrins of Fe(II) and Co(II), the known O_2 carriers, can be used, but those of Co(III) seem most effective and no induction periods are observed then (*130*). ESR data suggest an intermediate cation radical of cyclohexene formed via interaction of the olefin with the Co(III) porphyrin; this then implies possible catalysis via olefin activation rather than O_2 activation. A Mn(II) porphyrin has been shown to complex with tetracyanoethylene with charge transfer to the substrate (*131*), and we have shown that a Ru(II) porphyrin complexes with ethylene (*8*). Metalloporphyrins remain as attractive catalysts via such substrate activation, and epoxidation of squalene with no concomitant allylic oxidation has been noted and is thought to proceed via such a mechanism (*130*). Phthalocyanine complexes also have been used to catalyze autoxidation reactions (*69*).

No reports have appeared on oxidation of substrates using metalloporphyrins under conditions at which dioxygen complexes are known to exist; at ambient temperatures, for example, the O_2 adducts of specially designed "picket fence" (*5, 29, 132*) and polymer-supported metalloporphyrins (*133*), (py)Cr(*TPP*)O_2 (*134*), and (DMF)Ru(OEP)O_2 (*8, 135*) can be formed from molecular O_2, while Ti(OEP)O_2 (*136*) and Mo(TPP)(O_2)$_2$ (*137*) can be made from peroxide addition [TPP = tetraphenylporphyrin, OEP = octaethylporphyrin]. The Ru system is ineffective for oxidation of terminal olefins at least under the mild conditions (1 atm O_2, 35°C) studied thus far; even the ubiquitous substrate triphenylphosphine is not oxidized catalytically because of formation of a relatively inert Ru(OEP)(PPh_3)$_2$ complex (*138*). The catalytic potential for O_2 activation by Ru(II) porphyrins compared with Fe(II) porphyrins seems considerable, at least in principle, in view of a more readily accessible oxidation state of IV (*139*); this could circumvent the unfavorable one-electron reduction of O_2 to superoxide (*140*). Such systems seem promising generally in terms of the multi-electron redox processes that O_2 displays (*141*).

The nature of the coordinated dioxygen within the Ru porphyrins remains uncertain. Some data (lack of an ESR signal, visible spectrum) support a Ru(III)—O_2^- formulation, while the other data (comparison of visible spectral data with the corresponding carbonyl, binding constant for O_2, and binding of ethylene) imply a Ru(IV)—O_2^{2-} formulation (*8, 135*). We have reconstituted apomyoglobin with Ru porphyrins, and characterized in solution the Ru(II) species RuMb and the oxidized met-form RuMb$^+$ (*142*) as well as the carbonyl RuMb(CO), which has been reported previously (*143*). Unlike myoglobin itself, RuMb is hexacoordinate and low-spin; treatment with O_2 leads to the met-form via an outer-sphere oxidation (*cf.* Reaction 35). Our hopes of obtaining $RuMb(O_2)$, which is at least an honorary enzyme, have faded somewhat, but reconstitution via a preformed $LRu(porp)O_2$ complex remains a possibility.

Acknowledgments

I thank sincerely the graduate students, postdoctoral fellows, and colleagues who have contributed to the work carried out in these laboratories. These comprise G. L. Rempel, F. T. T. Ng, C. Y. Chan, G. Rosenberg, G. Strukul, and M. Preece, who worked on the Rh and Ir systems, and A. W. Addison, M. Cairns, D. Dolphin, N. P. Farrell, D. R. Paulson, and S. Walker, who have been involved with the Ru porphyrin studies. Financial support through NSERC of Canada in the form of operating and negotiated development grants is acknowledged gratefully, as well as the loan of Pt metals from Johnson Matthey & Co., Ltd.

Literature Cited

1. Basolo, F.; Hoffman, B. M.; Ibers, J. A. *Acc. Chem. Res.* **1975,** *8,* 384.
2. Erskine, R. W.; Field, B. O. *Struct. Bonding (Berlin)* **1976,** *28,* 1.
3. McLendon, G.; Martell, A. E. *Coord. Chem. Rev.* **1976,** *19,* 1.
4. Lyons, J. E. In "Aspects of Homogeneous Catalysis"; Ugo, R., Ed.; Reidel: Dordrecht, 1977; Vol. 3, p. 1.
5. Collman, J. P. *Acc. Chem. Res.* **1977,** *10,* 265.
6. James, B. R. In "The Porphyrins"; Dolphin, D., Ed.; Academic: New York, 1978; Vol. V., p. 205.
7. Buchler, J. W. *Angew. Chem. Int. Ed. Eng.* **1978,** *17,* 407.
8. James, B. R.; Addison, A. W.; Cairns, M.; Dolphin, D.; Farrell, N. P.; Paulson, D. R.; and Walker, S. Proc. Int. Symp. Homogeneous Catalysis, Ist, Corpus Christi 1978; Plenum: New York, 1979.
9. Olivé, G. H.; Olivé, S. *Angew. Chem. Int. Ed. Engl.* **1974,** *13,* 29.
10. Vaska, L. *Science* **1963,** *140,* 809.
11. Halpern, J. *Acc. Chem. Res.* **1970,** *3,* 386.
12. Vaska, L. *Acc. Chem. Res.* **1977,** *10,* 265.
13. Antonini, E.; Brunori, M. "Hemoglobin and Myoglobin in Their Reactions with Ligands"; North-Holland: Amsterdam, 1971.
14. Gibson, Q. H. In "The Porphyrins"; Dolphin, D., Ed.; Academic: New York, 1978; Vol. V, p. 153.
15. Perutz, M., this volume, p. 199.
16. Chang, C. K.; Dolphin, D. In "Bioorganic Chemistry"; Van Tamelen, E. E., Ed.; Academic: New York, 1978; Vol. IV, p. 37.
17. "Conf. Proc. Cytochrome P450-Structural Aspects," *Croat. Chem. Acta* **1977,** *49* (2).
18. Griffin, B. W.; Peterson, J. A.; Estabrook, R. W. In "The Porphyrins"; Dolphin, D., Ed.; Academic: New York, 1979; Vol. 7, p. 333.
19. Groves, J. T., this volume, p. 277.
20. Hoffman, B. M., this volume, p. 235.
21. Ochiai, E. "Bioinorganic Chemistry. An Introduction"; Allyn and Bacon: Boston, 1977; Chapter 7.
22. Wilson, D. F.; Erecinska, M. In "The Porphyrins"; Dolphin, D., Ed.; Academic: New York, 1979; Vol. VII, p. 2.
23. Hewson, W. D.; Hager, L. In "The Porphyrins"; Dolphin, D., Ed.; Academic: New York, 1979; Vol. VII, p. 295.

24. Bayston, J. H.; Kelso King, N.; Looney, F. D.; Winfield, M. E. *J. Am. Chem. Soc.* **1969**, *91*, 2775.
25. Corwin, A. H.; Bruck, S. D. *J. Am. Chem. Soc.* **1958**, *80*, 4736.
26. Jones, R. D.; Summerville, D. A.; Basolo, F. *J. Am. Chem. Soc.* **1978**, *100*, 4416.
27. Traylor, T. G.; Berzinis, A.; Campbell, D.; Mincey, T.; White, D., this volume, p. 217.
28. Nakahara, A.; Mori, W., this volume, p. 341.
29. Collman, J. P.; Brauman, J. I.; Doxsee, K. M.; Halbert, T. R.; Suslick, K. S. *Proc. Natl. Acad. Sci., USA* **1978**, *75*, 564.
30. Modena, G.; Sharpless, K. B.; Costa, G.; Halpern, J.; Ishii, Y.; James, B. R.; Lyons, J. E.; Mimoun, H.; Rossi, P.; Sheldon, R. A.; Teyssie, P. In "Fundamental Research in Homogeneous Catalysis"; Tsutsui, M., Ugo, R., Eds.; Plenum: New York, 1977; p. 193.
31. Sheldon, R. A.; Kochi, J. K. *Adv. Catal.* **1976**, *25*, 272.
32. Dumas, T.; Bulani, W. "Oxidation of Petrochemicals: Chemistry and Technology"; Wiley: New York, 1974.
33. Sheldon, R. A.; Van Doorn, J. A. *J. Organomet. Chem.* **1975**, *94*, 115.
34. Horn, R. W.; Weissberger, E.; Collman, J. P. *Inorg. Chem.* **1970**, *9*, 2367.
35. Van Gaal, H.; Cuppers, G.; van der Ent, A. *J. Chem. Soc., Chem. Commun.* **1970**, 1694.
36. Sen, A.; Halpern, J. *J. Am. Chem. Soc.* **1977**, *99*, 8337.
37. Cullen, W. R.; James, B. R.; Strukul, G. *Can. J. Chem.* **1978**, *17*, 484.
38. James, B. R.; Rempel, G. L. *J. Chem. Soc. A* **1969**, 78.
39. Rosenberg, G., Ph.D. thesis, University of British Columbia, 1974.
40. Stanko, J.; Petrov, G.; Thomas, C. *J. Chem. Soc., Chem. Commun.* **1969**, 1100.
41. James, B. R.; Kastner, M. *Can. J. Chem.* **1972**, *50*, 1698.
42. Henry, P. M. "Palladium(II) Catalyzed Oxidation of Hydrocarbons"; Reidel: Dordrecht, 1980.
43. Martell, A. E.; Taqui Khan, M. M. In "Inorganic Biochemistry"; Eichhorn, G., Ed.; Elsevier: Amsterdam, 1973; Vol. 2, p. 654.
44. Bossu, F. P.; Chellappa, K. L.; Margerum, D. W. *J. Am. Chem. Soc.* **1977**, *99*, 2195.
45. Collman, J. P.; Kubota, M.; Hosking, J. W. *J. Am. Chem. Soc.* **1967**, *89*, 4809.
46. Kurkov, V. P.; Pasky, J. Z.; Lavigne, J. B. *J. Am. Chem. Soc.* **1968**, *90*, 4743.
47. Fusi, A.; Ugo, R.; Pasini, A.; Cenini, S. *J. Organomet. Chem.* **1971**, *26*, 417.
48. Chalk, A. J.; Smith, J. F. *Trans. Faraday Soc.* **1957**, *53*, 1214.
49. Booth, B. L.; Haszeldine, R. N.; Neuss, G. H. R. *J. Chem. Soc., Perkin Trans. I* **1975**, 209.
50. Sheldon, R. A. *J. Chem. Soc., Chem. Commun.* **1971**, 788.
51. Arzoumanian, H.; Blanc, A. A.; Metzger, J.; Vincent, J. E. *J. Organomet. Chem.* **1974**, *82*, 261.
52. Kamiya, Y., *J. Catal.* **1972**, *24*, 69.
53. Budnik, R. A.; Kochi, J. K. *J. Org. Chem.* **1976**, *41*, 1384.
54. James, B. R.; Ochiai, E. *Can. J. Chem.* **1971**, *49*, 975.
55. James, B. R.; Ng, F. T. T.; Ochiai, E. *Can. J. Chem.* **1972**, *50*, 590.
56. James, B. R.; Rosenberg, G. *Coord. Chem. Rev.* **1975**, *16*, 153.
57. Sheldon, R. A.; Van Doorn, J. A. *J. Catal.* **1973**, *31*, 427.
58. Su, C. C.; Reed, J. W.; Gould, E. S. *Inorg. Chem.* **1973**, *12*, 337.
59. Kaloustian, J.; Lena, L.; Metzger, J. *Tetrahedron Lett.* **1975**, 599.
60. Mimoun, H.; Sérée de Roch, I.; Sajus, L. *Tetrahedron* **1970**, *26*, 37.
61. Arakawa, H.; Moro-oka, Y.; Ozaki, A. *Bull. Chem. Soc. Jpn.* **1974**, *47*, 2958.
62. Sharpless, K. B.; Townsend, J.; Williams, D. *J. Am. Chem. Soc.* **1972**, *94*, 296.

63. Arakawa, H.; Ozaki, A. *Chem. Lett.* **1975**, 1245.
64. Gould, E. S.; Hiatt, R. R.; Irwin, K. C. *J. Am. Chem. Soc.* **1968**, *90*, 4573.
65. Lyons, J. E.; Turner, J. O. *J. Org. Chem.* **1972**, *37*, 2881.
66. Lyons, J. E. *Adv. Chem. Ser.* **1974**, *132*, 64.
67. Stern, E. W. *J. Chem. Soc., Chem. Commun.* **1970**, 736.
68. Ohkatsu, Y.; Sekiguchi, O.; Osa, T. *Bull. Chem. Soc. Jpn.* **1977**, *50*, 701.
69. Ohkatsu, Y.; Hara, T.; Osa, T. *Bull. Chem. Soc. Jpn.* **1977**, *50*, 696.
70. Buxton, G. W.; Green, J. G.; Sellers, R. M. *J. Chem. Soc., Dalton Trans.* **1976**, 2160.
71. Jamieson, R. F.; Blackburn, N. J. *J. Chem. Soc., Dalton Trans.* **1976**, 534.
72. Baldwin, J. E.; Basson, H. H.; Krauss, H. Jr. *J. Chem. Soc., Chem. Commun.* **1968**, 984.
73. Mason, R. *Nature* **1968**, *217*, 543.
74. Baldwin, J. E.; Swallow, J. C. *Angew. Chem. Int. Ed. Engl.* **1969**, *8*, 601.
75. Kaneda, K.; Itoh, T.; Fujiwara, Y.; Teranishi, S. *Bull. Chem. Soc. Jpn.* **1973**, *46*, 3810.
76. Tsyskovskii, V. K.; Fedorov, V. S.; Moskovich, Y. L. *Zh. Obshch. Khim.* **1975**, *45*, 248.
77. Rubailo, V. L.; Gararina, A. B.; Emanuel, N. *Kinet. Katal.* **1974**, *15*, 891.
78. Tyutchenkova, L. D.; Privalova, L. G.; Maizus, Z. K.; Emanuel, N. M. *Izv. Akad. Nauk, SSSR, Ser. Kkim.* **1975**, 48.
79. Holland, D.; Milner, D. J. *J. Chem. Soc., Dalton Trans.* **1975**, 2440.
80. Farrar, J.; Holland, D.; Milner, D. J. *J. Chem. Soc., Dalton Trans.* **1975**, 815.
81. Takao, K.; Fujiwara, Y.; Imanaka, T.; Teranishi, S. *Bull. Chem. Soc. Jpn.* **1970**, *43*, 1153.
82. Takao, K.; Wayaku, M.; Fujiwara, Y.; Imanaka, T.; Teranishi, S. *Bull. Chem. Soc. Jpn.* **1970**, *43*, 3898.
83. Shue, R. S. *J. Catal.* **1971**, *26*, 112.
84. Muto, S.; Kamiya, Y. *J. Catal.* **1976**, *41*, 148.
85. Takao, K.; Azuma, H.; Fujiwara, Y.; Imanaka, T.; Teranishi, S. *Bull. Chem. Soc. Jpn.* **1972**, *45*, 2003.
86. Dudley, C. W.; Read, G.; Walker, P. J. C. *J. Chem. Soc., Dalton Trans.* **1974**, 1926.
87. Read, G.; Walker, P. J. C. *J. Chem. Soc., Dalton Trans.* **1977**, 883.
88. Guroff, G.; Daly, J. W.; Jerina, D. M.; Renson, J.; Witkop, B.; Udenfriend, S. *Science* **1967**, *157*, 1524.
89. Tang, R.; Mares. F.; Neary, N.; Smith, D. E. *J. Chem. Soc., Chem. Commun.* **1979**, 274.
90. Chen, M. J. Y.; Kochi, J. K. *J. Chem. Soc., Chem. Commun.* **1977**, 204.
91. Mimoun, H.; Machirant, M. M. P.; Sérée de Roch, I. *J. Am. Chem. Soc.* **1978**, *100*, 5437.
92. Igersheim, F.; Mimoun, H. *J. Chem. Soc., Chem. Commun.* **1978**, 559.
93. Newton, W. E.; Bravard, D. C.; McDonald, J. W. *Inorg. Nucl. Chem. Lett.* **1975**, *11*, 553.
94. Kolomnikov, I. S.; Koreshkov, Y. D.; Lobeeva, T. S.; Volpin, M. E. *J. Chem. Soc., Chem. Commun.* **1970**, 1432.
95. Sharpless, K. B.; Teranishi, A. Y.; Bäckvall, J. E. *J. Am. Chem. Soc.* **1977**, *99*, 3120.
96. Chong, A. O.; Oshima, K.; Sharpless, K. B. *J. Am. Chem. Soc.* **1977**, *99*, 3420.
97. James, B. R. "Homogeneous Hydrogenation"; Wiley: New York, 1973; p. 401.
98. James, B. R.; Ng, F. T. T.; Rempel, G. L. *Can. J. Chem.* **1969**, *47*, 4521.
99. Trocha–Grimshaw, J.; Henbest, H. B. *J. Chem. Soc., Chem. Commun.* **1968**, 1035.
100. Basolo, F.; Pearson, R. G. "Mechanisms of Inorganic Reactions," 2nd ed.; Wiley: New York, 1967; p. 595.

101. Haddad, Y. M. Y.; Henbest, H. B.; Trocha–Grimshaw, J. *J. Chem. Soc., Perkin Trans. 1* **1974**, 592.
102. Thomas, K.; Osborn, J. A.; Powell, A. R.; Wilkinson, G. *J. Chem. Soc. A* **1968**, 1801.
103. Johnston, L. E.; Page, J. A. *Can. J. Chem.* **1969**, *47*, 4241.
104. Roberts, H. L.; Symes, W. R. *J. Chem. Soc. A* **1968**, 1450.
105. James, B. R.; Preece, M.; Strukul, G., unpublished data.
106. Osborn, J. A.; Jardine, F. H.; Young, J. F.; Wilkinson, G. *J. Chem. Soc. A* **1966**, 1711.
107. Robinson, S. D.; Shaw, B. L. *Tetrahedron Lett.* **1964**, 1301.
108. Bonnaire, R.; Fougeroux, P. *Compt. Rend.* **1975**, *280,C*, 767.
109. Chan, C. Y.; James, B. R. *Inorg. Nucl. Chem. Lett.* **1973**, *9*, 135.
110. Hui, B. C.; James, B. R. *Can. J. Chem.* **1974**, *52*, 348.
111. Hayward, P. J.; Blake, D. M.; Wilkinson, G.; Nyman, C. J. *J. Am. Chem. Soc.* **1970**, *92*, 5873.
112. Tabushi, I.; Koga, N., this volume, p. 291.
113. McAuley, A. *Chem. Soc. Spec. Period. Rep. Inorg. React. Mech.* **1977**, *5*, 107.
114. Ochiai, E. *Inorg. Nucl. Chem. Lett.* **1974**, *10*, 453.
115. Chin, D-H.; Del Gaudio, J.; La Mar, G. N.; Balch, A. *J. Am. Chem. Soc.* **1977**, *99*, 5486.
116. Chang, C. K.; Powell, D.; Traylor, T. G. *Croat. Chem. Acta* **1977**, *49*, 295.
117. Harcourt, R. D. *J. Inorg. Nucl. Chem.* **1977**, *39*, 243.
118. Chu, M. M. L.; Castro, C. E.; Hathoway, G. M. *Biochemistry* **1978**, *17*, 481.
119. Wallace, W. J.; Maxwell, J. C.; Caughey, W. S. *Biochem. Biophys. Res. Commun.* **1974**, *57*, 1104.
120. Sligar, S. G.; Lipscomb, J. D.; Debrunner, P. G.; Gunsalus, I. C. *Biochem. Biophys. Res. Commun.* **1974**, *61*, 290.
121. Dolphin, D.; James, B. R.; Welborn, C. *J. Mol. Catal.* **1980**, *7*, 201.
122. Abel, E. W.; Pratt, J. M.; Whelan, R.; Wilkinson, P. J. *J. Am. Chem. Soc.* **1974**, *96*, 7119.
123. Nishinaga, A.; Tojo, T.; Matsuura, T. *J. Chem. Soc., Chem. Commun.* **1974**, 896.
124. Nishinaga, A.; Watanabe, K.; Matsuura, T. *Tetrahedron Lett.* **1974**, 1291.
125. Fullerton, T. J.; Ahern, S. P. *Tetrahedron Lett.* **1976**, 139.
126. Dufour–Ricroch, M. N.; Gaudemer, A. *Tetrahedron Lett.* **1976**, 4079.
127. Nishinaga, A. *Chem. Lett.* **1975**, 273.
128. Ochiai, E. *J. Inorg. Nucl. Chem.* **1973**, *35*, 1727.
129. Paulson, D. R.; Ullman, R.; Sloane, R. B.; Closs, G. L. *J. Chem. Soc., Chem. Commun.* **1974**, 186.
130. Baccouche, M.; Ernst, J.; Fuhrhop, J-H.; Schlözer, R.; Arzoumanian, H. *J. Chem. Soc., Chem. Commun.* **1977**, 821.
131. Summerville, D. A.; Cape, T. W.; Johnson, E. D.; Basolo, F. *Inorg. Chem.* **1978**, *17*, 3297.
132. Jameson, G. B.; Molinaro, F. S.; Ibers, J. A.; Collman, J. P.; Brauman, J. I.; Rose, E.; Suslick, K. S. *J. Am. Chem. Soc.* **1978**, *100*, 6769.
133. Leal, O.; Anderson, D. L.; Bowman, R. G.; Basolo, F.; Burwell, R. L. *J. Am. Chem. Soc.* **1975**, *97*, 5125.
134. Cheung, S. K.; Grimes, C. J.; Wong, J.; Reed, C. A. *J. Am. Chem. Soc.* **1976**, *98*, 5028.
135. Farrell, N. P.; Dolphin, D. H.; James, B. R. *J. Am. Chem. Soc.* **1978**, *100*, 324.
136. Guilard, R.; Fontesse, M.; Fournari, P.; Lecomte, C.; Protas, J. *J. Chem. Soc., Chem. Commun.* **1976**, 161.
137. Chevrier, B.; Diebold, T.; Weiss, R. *Inorg. Chim. Acta* **1976**, *19*, L57.
138. Dolphin, D. H.; James, B. R.; Walker, S., unpublished data.
139. Vaska, L. *Acc. Chem. Res.* **1976**, *9*, 175.

140. Bennett, L. E. *Prog. Inorg. Chem.* **1973**, *18*, 1.
141. Collman, J. P.; Elliott, C. M.; Halbert, T. R.; Tovrog, B. S. *Proc. Natl. Acad. Sci., USA.* **1977**, *74*, 18.
142. Paulson, D. R.; Addison, A. W.; Dolphin, D.; James, B. R. *J. Biol. Chem.* **1979**, *254*, 7002.
143. Strivastava, T. S. *Biochim. Biophys. Acta* **1977**, *491*, 599.

RECEIVED May 21, 1979.

Studies of the Mechanism of Oxygen Activation and Transfer Catalyzed by Cytochrome P 450

JOHN T. GROVES, S. KRISHNAN, GUILLERMINA E. AVARIA, and THOMAS E. NEMO

Department of Chemistry, University of Michigan, Ann Arbor, MI 48109

The oxidation of cyclohexene by reconstituted and peroxide-dependent cytochrome P-450$_{LM2}$ systems has been examined. The ratio of cyclohexene oxide to cyclohexenol has been shown to vary with changes in the oxygen donor used. The hydrogen-isotope effect for allylic hydroxylation has been found to be 5 ± 0.5 and independent of oxygen donor. The changes in product ratio are attributed to the presence of the oxygen donor at the active site during oxygen transfer. The invariance of the isotope effect supports a similar transition state for the NADPH/O$_2$-dependent and peroxide-dependent paths. Treatment of tetra-o-tolylporphinatoiron(III) chloride with iodosylbenzene produces a transient intermediate that transfers oxygen to substrates and degrades the porphyrin in the absence of substrate.

T he role of cytochrome P 450$_{LM}$, the heme-containing, mixed-function oxidase of liver microsomes, in catalyzing the oxidative transformations of lipid metabolism and the oxidative detoxification of drugs and other xenobiotic substances has been studied extensively (*1, 2, 3*). It is understood now that two successive, one-electron reductions of the heme center of cytochrome P 450 bind and reduce molecular oxygen to the formal oxidation state of hydrogen peroxide. A number of lines of evidence suggest that the active oxygen species in the P-450 cycle is a ferryl-ion intermediate equivalent to FeO^{3+} formed by the heterolysis of the O–O bond of hydrogen peroxide (*4, 5, 6*). The important observation that cytochrome P 450 will catalyze the

0-8412-0514-0/80/33-191-277$05.00/0
© 1980 American Chemical Society

anaerobic oxidation of hydrocarbons in the presence of hydroperoxides (5, 7, 8) has offered persuasive evidence in favor of an
overall reaction cycle such as that outlined in Scheme 1.

Scheme 1.   Oxygen activation and transfer by cytochrome P 450

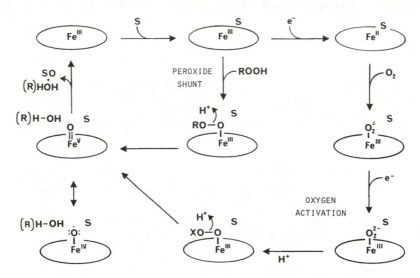

Speculation on the mechanism of oxygen insertion into unacti-
vated C–H bonds catalyzed by cytochrome P 450 has centered on the
apparent retention of configuration at the carbon and the low isotope
effects generally observed for these processes (9, 10). We and others
recently have demonstrated that with appropriately designed molecular probes very large intramolecular discrimination isotope effects can
be observed (11, 12). The oxidation of norbornane with cytochrome
P 450$_{LM2}$ gave only *exo-* and *endo*-2-norborneol with an isotope effect
of 11.5. Close examination of the mass spectra of the products revealed
that, though the hydroxylation proceeded with predominant net retention of configuration at the carbon, a substantial fraction (up to 25%) of
the oxidation had occurred with loss of stereochemistry at the oxidized
carbon. The magnitude of this isotope effect and the partial loss of
stereochemistry are similar to results observed for the oxidation of
alkanes by other transition-metal–oxo reagents such as chromate and
permanganate (13, 14).

    The free-radical mechanism proposed by Wiberg (15) for these
chemical oxidations is consistent with the observations noted above for
cytochrome P 450 and is depicted in Scheme 2 (16, 17).

    Although the peroxide shunt path to a ferryl intermediate is an
attractive hypothesis, there is very little data to support the idea that
the species obtained by this path are the same as, or similar to, that

produced by the fully reconstituted system involving NADPH, P-450 reductase, and O_2. We report here a comparison of the aerobic and peroxide-dependent oxidations of cyclohexene by cytochrome P 450$_{LM2}$ (18, 19). Product selectivities have been shown to be a function of the structure of the oxygen donor, whereas the magnitude of the hydrogen isotope effect for hydroxylation is not.

Scheme 2.

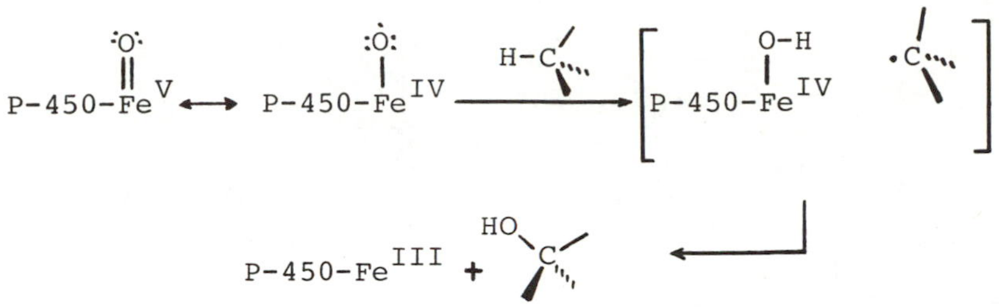

Oxidation of Cyclohexene Catalyzed by Cytochrome P 450

The oxidation of cyclohexene by a fully reconstituted cytochrome P-450$_{LM}$ system (5) gave only cyclohexene oxide (1) and cyclohexenol (2) in a ratio of 0.92 : 1. Results for the peroxide-dependent oxidation of cyclohexene in the presence of cytochrome P 450 are presented in Table I. Inspection of the data suggests that subtle differences exist between the oxidants generated from these four oxidants and lead to the observed sixfold change in the ratio of 1 and 2 in the product mixture. Also apparent is the fact that no obvious correlation exists between the ratio of the products and the nature or effectiveness of the oxidant.

Table I. Oxidation of Cyclohexene Catalyzed by Cytochrome P 450

Oxygen Donor	Cyclo-hexene Oxide (1) (nmol)	Cyclo-hexenol (2) (nmol)	$\frac{1}{2}$	Relative Effi-ciency
Cumene hydroperoxide	122	138	0.88	17.3
t-Butyl hydroperoxide	3.2	12.1	0.26	1.0
Iodobenzene diacetate	7.2	39.1	0.18	3.1
Iodosylbenzene	25.8	26.1	0.99	3.5

This variability in the product ratios with the structure of the oxygen donor requires that the oxidant produced by these reagents be different in some functional way. The simplest explanation is that the alcohol or iodobenzene fragment (X) is still at the active site when oxygen transfer takes place. Two extremes of the mechanistic spectrum that are consistent with these observations are: (a) the interaction of the substrate with the peroxidic oxygen occurring before the O–X bond is broken (3), and (b) production of a ternary cage structure (4) with X and the substrate (S) both at the active site (Scheme 3).

Scheme 3.

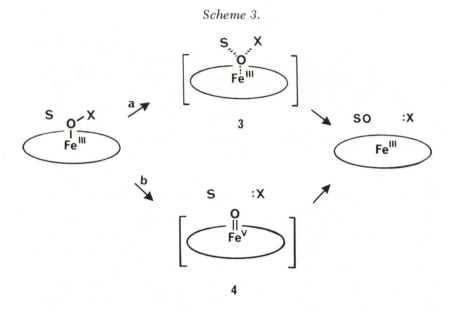

For Path (a), the structure of X would be expected to exert a profound effect on the nature of the oxidizing intermediate. In Path (b), the effect of X may be only to affect the population of various substrate-active site complexes.

Hydrogen-Isotope Effect for Hydroxylation of Cyclohexene

We have examined the hydrogen-isotope effect for cyclohexenol formation as a probe of the nature of C–H bond cleavage in the NADPH-dependent and cumene-hydroperoxide-dependent oxidation of cyclohexene.

Results for the oxidation of cyclohexene, 1,3,3-trideuterocyclohexene (5) and 3,3,6,6-tetradeuterocyclohexene (6) are given in Table II. It is apparent from the large increase in the relative amount of cyclohexene oxide upon deuteration of the allylic hydrogens that the hydroxylation path has a significant hydrogen-isotope effect (*11, 12*).

Table II. Effect of Deuteration on Cyclohexene Oxidation by Cytochrome P 450$_{LM2}$

Oxygen Donor	Substrate	Cyclo-hexene Oxide (1)	Cyclo-hexenol (2)	$\dfrac{1}{2}$
NADPH/O$_2$	cyclohexene	12.0^a	15.3^a	0.89
		43.5	48.0	0.91
	5	39.0	25.5	1.5
	6	14.2^a	2.97^a	4.7
Cumene hydroperoxide	cyclohexene	44.4	48.2	0.92
	5	38.4	25.9	1.48
	6	36.3	8.1	4.51
t-Butyl hydroperoxide	cyclohexene	1.10	3.68	0.29
	6	1.8	1.46	1.23

^{a} Absolute product yields were found to be related to the age and batch of NADPH-cytochrome P-450 reductase.

The magnitude of the isotope effect can be derived from the appropriate competitive rate expression for any two of the product ratios. The results of these calculations are given in Table III.

An independent determination of the isotope effect could be derived from the mass spectrum of the 3-trimethylsiloxycyclohexene-d_n (7) obtained from the oxidation of 5. Authentic 3-trimethylsiloxycyclohexene-d_o (8) gave prominent ions corresponding to the molecular ion (m/e = 170) and loss of ethylene by retro Diels-Alder fragmentation (m/e = 142).

Table III. Isotope Effects for Cyclohexene Hydroxylation by Cytochrome P 450$_{LM2}$

		Isotope Effect (k_H/k_D)		
			P-450–CHP System	
Rate Expression^a		NADPH/O$_2$	Cumene Hydro-peroxide	t-Butyl Hydro-peroxide
$k_H/k_D = \dfrac{[1/2]_{d_3}}{2[1/2]_{d_0} - [1/2]_{d_3}}$		4.1	4.8	—
$k_H/k_D = \dfrac{2[1/2]_{d_4} - [1/2]_{d_3}}{[1/2]_{d_3}}$		4.9	5.1	—
$k_H/k_D = \dfrac{[1/2]_{d_4}}{[1/2]_{d_0}}$		4.9	4.9	4.25

^{a} For simplicity, in the derivation of isotope effect expression, possible secondary deuterium isotope effects have been ignored.

The mass spectra of **7**, obtained from the NADPH/O$_2$/P-450 oxidation of **5**, and **8** are compared in Table IV. Three isotopic species are expected for the allylic hydroxylation of **5**. Oxidation with removal of hydrogen will produce **7**-d_3 and **7a**-d_3. Extensive controls have demonstrated that **7**-d_3 and **7a**-d_3 are indistinguishable by mass spectrometry because of rapid 1,3-migration of the siloxy group upon electron impact. Oxidation with deuterium removal leads to the production of **7**-d_2 (Scheme 4). By this analysis, the isotope effect is simply the ratio of **5**-d_3 (and **5a**-d_3) to **5**-d_2 in the oxidized sample. Deconvolution of the parent region (with appropriate correction for carbon and silicon

Table IV.   Mass Spectra of 3-Trimethylsiloxycyclohexenes Derived from the P-450$_{LM2}$-Catalyzed Oxidation of 5

7-d_n from 5	m/e	173	172	171	145	144	143
	Intensity	100	46.3	19.4	31.8	18.8	26.2
8-d_0 from cyclo-hexenol	m/e	170	169		143	142	141
	Intensity	100	37.3		6.5	45.2	4.0
		$k_H/k_D = 6.0$			$k_H/k_D = 4.5$		

Scheme 4.

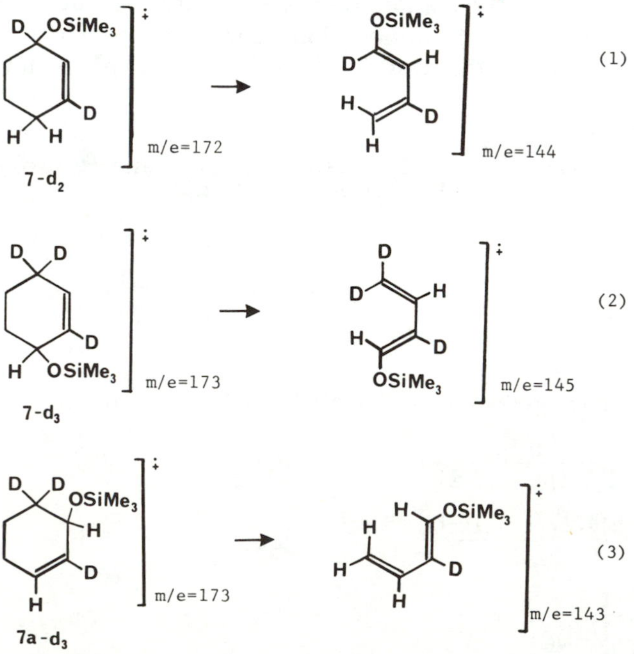

7-d_2 m/e=172   →   m/e=144   (1)

7-d_3 m/e=173   →   m/e=145   (2)

7a-d_3 m/e=173   →   m/e=143   (3)

isotopes) gave an isotope effect of 6.0 for the allylic hydroxylation of cyclohexene by the reconstituted system. Similar treatment of the retro Diels–Alder fragment gave a value of 4.5.

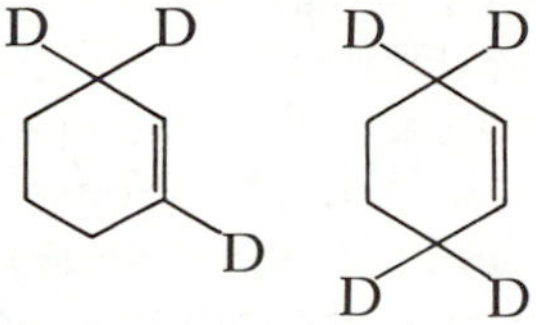

5   6

Thus, the isotope effect for the allylic oxidation of cyclohexene by cytochrome P 450 is about 5 and is the same for the reconstituted, NADPH-dependent and the peroxide-dependent paths. This similarity suggests that although product ratios may change from one oxygen donor to another, the mechanism of oxygen transfer may be invariant. Efforts to develop a clearer understanding of the relationships between the O_2-dependent and peroxide-dependent pathways for oxygen transfer catalyzed by cytochrome P 450 are currently underway.

Oxygen Transfer Catalyzed by Synthetic Metalloporphyrins

The observation that cytochrome P 450 can be driven by hydroperoxides and related oxygen donors suggests that metalloporphyrins can be made to function as oxygen-transfer catalysts in simple model systems.

In early attempts to produce an iron-oxo species (20) from typical porphyrins like chloro-$\alpha,\beta,\gamma,\delta$-tetraphenylporphinatoiron(III) [Fe(III)TPP–Cl] and chloroferriprotoporphyrin(IX)[Fe(III)PPIX–Cl], we examined the reaction of t-butyl hydroperoxide and peroxyacids with alkanes and olefins in the presence of these catalysts. With peroxyacids, decomposition of the porphyrin ring was observed, while with the t-butyl hydroperoxides, product distributions were indistinguishable from free-radical chain reactions initiated photochemically in the absence of any metals.

Therefore, it appears that the redox properties of the metalloporphyrin are required only for the initiation step in these free-radical autoxidations and that the porphyrin is not a stoichiometrically significant catalyst (21, 22, 23). The failure of these simple approaches to a reaction iron-porphine oxide or its equivalent could indicate that the ligation state of iron in the protein, presumably including an axial thiolate, is crucial to the oxygen-transfer properties of P 450. Support for this idea can be derived from the observation that cytochrome P 420, a partially denatured form of P 450, is ineffective as a catalyst for the oxidation of cyclohexene either in the fully reconstituted system or in the peroxide-dependent route. Another interpretation, however, is that the heme unit in P 450 effectively suppresses the free-radical chain reactions that dominate the chemistry in solution.

We have demonstrated recently that epoxidation and hydroxylation can be achieved with simple iron-porphine catalysts with iodosylbenzene as the oxidant (24). Cyclohexene can be oxidized with iodosylbenzene in the presence of catalytic amounts of Fe(III)TPP–Cl to give cyclohexene oxide and cyclohexenol in 55% and 15% yields, respectively. Likewise, cyclohexane is converted to cyclohexanol under these conditions. Significantly, the alcohols were not oxidized rapidly to ketones under these conditions, a selectivity shared with the enzymic hydroxylations. The distribution of products observed here, particularly the preponderance of epoxide and the lack of ketones, is distinctly different from that observed in an autoxidation reaction or in typical reactions of reagents such as chromates or permanganates (15).

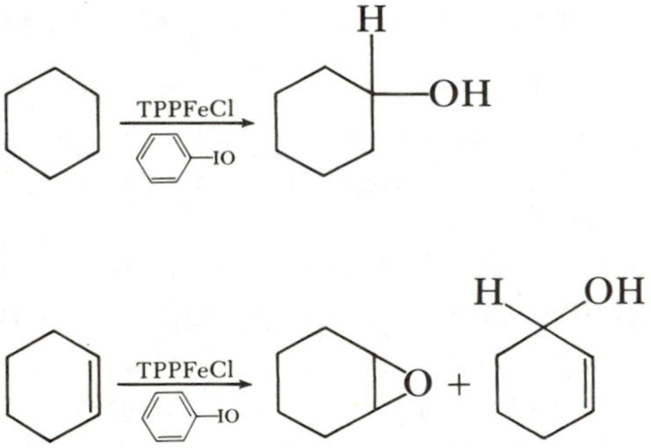

The oxidation of dioctyl Fe(III)PPIX–Cl with iodosylbenzene (9) showed that the octyl sidechains had been hydroxylated and that 60% of the hydroxylation had occurred at $C(4)$ and $C(5)$ in the middle of the chain. Molecular models indicate that these two carbon centers have the most favorable access to the center of the porphyrin ring, supporting the idea that the mechanism of this hydroxylation is an intramolecular oxygen rebound (25, 26) from iodine to iron and into the C–H bond (Scheme 5).

Scheme 5.

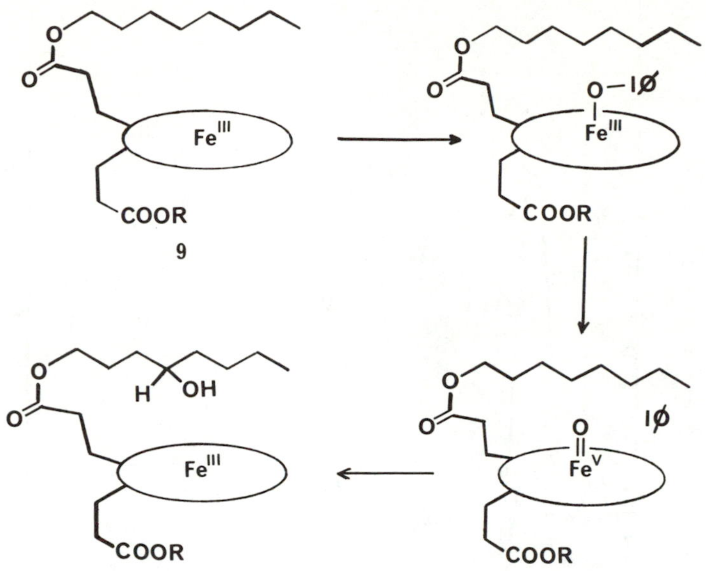

Generation of an "Active Oxygen" Metallo-Porphyrin Intermediate

Treatment of TPPFe(III)Cl or *meso*-tetra-*o*-tolylporphinato-iron(III) chloride [TTPFe(III)Cl (**10**)] with iodosylbenzene caused rapid oxidation of the porphyrin and loss of catalytic activity for hydrocarbon oxidation. Figure 1 shows changes in the visible absorption spectrum upon treatment of **10** with iodosylbenzene. These data indicate that shortly after the addition of iodosylbenzene (Scan b, Figure 1) a new porphyrin species (**11**) is formed, which then rapidly decays to oxidized porphyrin products. The kinetics of this decay process are approximately first order (Figure 2).

A mechanism for oxygen transfer catalyzed by **10** consistent with these data is given in Scheme 6. It is clear that iodosylbenzene reacts with simple iron-porphine complexes to form a reactive intermediate (**11**). This intermediate is capable of transferring oxygen to substrates such as cyclohexene and cyclohexane. In the absence of substrate, oxygen apparently is transferred to the porphyrin ring. It is not possible to assign unambiguously the structure of the reactive intermediate **11** at this time. Likely formulations are a 1:1 complex between the iron(III) porphyrin and iodosylbenzene or an iron-oxo intermediate equivalent to iron(V). The nature of this intermediate and its relevance to biological-oxygen activation and transfer are under continued study in our laboratories.

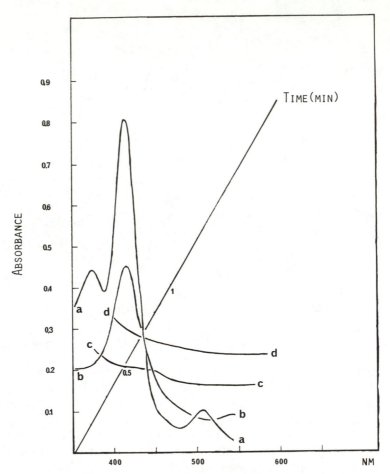

Figure 1. Visible spectrum of TTPFe(III)Cl and iodosylbenzene in methylene chloride: Scan a, time = 0; Scan b, time = 13 sec; Scan c, time = 25 sec; Scan d, time = 35 sec

Methods and Materials

Deuterated Cyclohexenes. 1,3,3-Trideuterocyclohexene (**5**) was synthesized from cyclohexanone according to the method of Fahey and Monahan (27) except that diazabicyclononane (DBN) was used instead of potassium *t*-butoxide in the final step. 3,3,6,6-Tetradeuterocyclohexene (**6**) was obtained from Merck, Sharp, and Dohme (>98%-d_4). The degree of deuteration of the labeled cyclohexenes was determined by converting them to the corresponding 1,2-dibromocyclohexanes by treatment with bromine in carbon tetrachloride. The deuterium content of these dibromocyclohexanes was determined conveniently by observing the M-bromine region of the mass spectrometry.

Aerobic Oxidation of Cyclohexene with Cytochrome P 450. In a typical aerobic cyclohexene oxidation, the reaction mixture contained 1 nmol of elec-

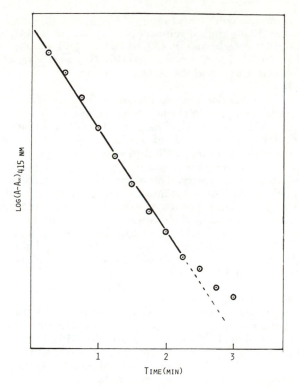

Figure 2. Reaction of TTPFe(III)Cl with iodosylbenzene in methylene chloride at 25°

Scheme 6.

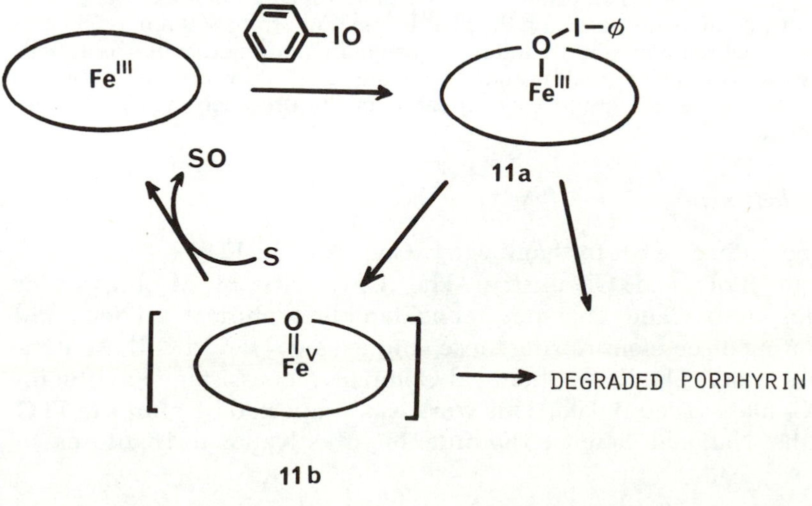

DEGRADED PORPHYRIN

trophoretically homogeneous cytochrome P 450$_{LM2}$ (5), dilauroyl glyceryl-3-phosphorylcholine (0.05 mg), cyclohexene (5 μL of a 1M solution in methanol), NADPH–cytochrome P-450 reductase (1.5 nmol), and sodium phosphate buffer, pH 7.4 (100 μmol). NADPH (2 μmol) was added in one portion (final volume 1 mL) and the reaction mixture was allowed to stand at 20°C for 15 min.

At the end of the reaction period, 100 μL of 30% sodium hydroxide was added to the reaction mixture to quench the reaction.

Quantitative Analysis of Cyclohexene Oxidation Products. Cyclohexanone or cycloheptanone (5 μL of a 10mM solution) was added as an internal standard. One milliliter of diethyl ether and 0.5 g magnesium acetate were added to the reaction mixture. Centrifugation produced a clear ether layer and an aqueous slurry. The ether layer was transferred to a vial containing a small amount of potassium carbonate. The aqueous slurry was extracted with a second 1-mL portion of ether. The combined ether extracts were placed in a 3-mL v-shaped "mini-vial." The ether layer was slowly concentrated to 200 μL in this vial by directing a nitrogen steam away from the surface of the ether with a J-tipped pipette. Aliquots of this solution were subjected to glpc and GC–MS analysis using a 4-ft, 20% Carbowax on Chrom W 60/80 or a 10-ft, 20% DEGS on Chrom W 60/80 column. Extensive controls indicated that this procedure allowed quantitative recovery of reaction products from the aqueous reaction mixture.

Peroxide-Dependent Oxidations with Cytochrome P 450. Peroxide-dependent oxidations were carried out exactly as described above except that NADPH–cytochrome P-450 reductase and NADPH were omitted. The peroxide solution (50 μL of a 20mM solution) was added to the premixed enzyme–substrate solution. The quenching solution was 100 μL of 30% sodium hydroxide saturated in sodium dithionite.

The trimethylsilyl ethers 7 and 8 were prepared by standard techniques using O,N-bis-trimethylsilylacetamide.

Reaction of TTPFe(III)Cl with Iodosylbenzene. Tetra-o-tolylporphine was synthesized according to published procedures (28). The free base was metalated with ferrous chloride in dimethylformamide according to the method of Alder (29). A 2.5-mL solution of purified tetra-o-tolylporphinatoiron(III) chloride (7.87μM) in methylene chloride was placed in a 1-cm cuvette. After scanning the visible spectrum from 360 nm to 550 nm, 5 μL of a stock solution of 0.1 g iodosylbenzene in 1 mL methanol was added, the mixture was mixed quickly, and the visible spectrum was scanned immediately. The same procedure was repeated in the presence of 0.5 mL cyclohexene.

Acknowledgments

The authors wish to thank M. J. Coon and R. E. White, Department of Biological Chemistry, The University of Michigan, for supplies of purified enzymes, considerable technical advice, and stimulating discussions during these studies; and Harold Gill, Analytical Laboratory, The Dow Chemical Company, for assistance in obtaining GC–mass spectral data. This work was supported by grants to JTG from the National Science Foundation, the National Institutes of

Health, and the Petroleum Research Fund administered by the American Chemical Society. The National Science Foundation provided funds for the purchase of a GC–mass spectrometer.

Literature Cited

1. Coon, M. J.; White, R. E. In "Metal Ion Activation of Dioxygen"; Spiro, T. G., Ed.; John Wiley and Sons: New York, 1980.
2. Groves, J. T. In "Advances in Inorganic Biochemistry"; Eichhorn, G. L., Marzilli, L. G., Eds.; Elsevier: New York, 1979; p. 119.
3. Peterson, J. A.; Ishimura, Y.; Baron, J.; Estabrook. R. W. In "Oxidases and Related Redox Systems"; King, T. E., Mason, H. S., Morrison, M., Eds.; University Press: Baltimore, 1973; p. 565.
4. Ullrich, V. *Angew. Chem. Int. Ed. Engl.* **1972**, *11*, 701.
5. Nordblom, G. D.; White, R. E.; Coon, M. J. *Arch. Biochem. Biophys.* **1976**, *175*, 524.
6. Hrycay, E. G.; Gustafsson, J.; Ingelman-Sundberg, M.; Eruster, L. *Biochem. Biophys. Res. Commun.* **1975**, *66*, 209.
7. Hrycay, E. G.; O'Brien, P. J. *Arch. Biochem. Biophys.* **1972**, *153*, 48.
8. Lichtenberger, F.; Nastainczyk, W.; Ullrich, V. *Biochem. Biophys. Res. Commun.* **1976**, *70*, 939.
9. Thompson, J. A.; Holtzman, J. L. *Drug Metab. Dispos.* **1974**, *12*, 577.
10. McMahon, R. E.; Sullivan, H. R.; Craig, J. E.; Pereira, W. E., Jr. *Arch. Biochem. Biophys.* **1969**, *132*, 575.
11. Groves, J. T.; McClusky, G. A.; White, R. E.; Coon, M. J. *Biochem. Biophys. Res. Commun.* **1978**, *81*, 154.
12. Hjelmeland, L. M.; Aronow, L.; Trudell, J. R. *Biochem. Biophys. Res. Commun.* **1977**, *76*, 541.
13. Wiberg, K. B.; Foster, G. *J. Am. Chem. Soc.* **1961**, *83*, 423.
14. Brauman, J. I.; Pandell, A. J. *J. Am. Chem. Soc.* **1970**, *92*, 329.
15. Wiberg, K. B. In "Oxidation in Organic Chemistry"; Wiberg, K. B., Ed.; Academic: New York, 1965; p. 69.
16. Groves, J. T.; McClusky, G. A. In "Clinical and Biochemical Aspects of Oxygen"; Caughey, W. S., Ed.; Academic: 1979, in press.
17. Groves, J. T. In "Metal Ion Activation of Dioxygen"; Spiro, T. G., Ed.; John Wiley and Sons: New York, 1979, in press.
18. Haugen, D. A.; van der Hoeven, T. A.; Coon, M. J. *J. Biol. Chem.* **1975**, *250*, 3567.
19. Haugen, D. A.; Coon, M. J. *J. Biol. Chem.* **1976**, *251*, 7929.
20. Sharpless, K. B.; Flood, T. C. *J. Am. Chem. Soc.* **1971**, *93*, 231.
21. Belova, V. S.; Nikonova, L. A.; Raikman, L. M.; Borukaeva, M. R. *Dolk. Akad. Nauk. SSSR* **1972**, *204*, 897.
22. Paulson, D. R.; Ullman, R.; Sloane, R. B.; Closs, G. L. *J. Chem. Soc., Chem. Commun.* **1974**, 186.
23. Baccouche, E. J.; Fuhrhop, J.-H.; Schlözer, R.; Arzoumanian, *J. Chem. Soc., Chem. Commun.* **1977**, 821.
24. Groves, J. T.; Nemo, T. E.; Myers, R. S. *J. Am. Chem. Soc.* **1979**, *101*, 1032.
25. Groves, J. T.; McClusky, G. A. *J. Am. Chem. Soc.* **1976**, *98*, 859.
26. Groves, J. T.; Van Der Puy, M. *J. Am. Chem. Soc.* **1975**, *97*, 7118.
27. Fahey, R. C.; Monahan, M. W. *J. Am. Chem. Soc.* **1970**, *92*, 2816.
28. Abraham, R. J.; Hawkes, G. H.; Hudson, M. F.; Smith, K. M. *J. Chem. Soc., Perkin Trans. 2* **1975**, 204.
29. Kobayashi, H.; Higuchi, T.; Kaizu, Y.; Osada, H.; Aaki, M. *Bull. Chem. Soc. Jpn.* **1975**, 3137.

RECEIVED June 5, 1979.

Oxygen Activation: Participation of Metalloenzymes and Related Metal Complexes

IWAO TABUSHI and NOBORU KOGA

Department of Synthetic Chemistry, Kyoto University, Yoshida, Kyoto 606, Japan

Oxygen activation by metalloenzymes is discussed as one of the most important catalytic oxidations in biological systems. Cytochrome P 450 is used as a typical example of enzymic oxygen activation, and spectroscopic approaches to possible intermediates involved in the oxidation catalysis are summarized briefly. Special attention also is directed to the participation of the unique porphyrin·iron monoxide of the "Compound I" type in P-450 catalysis. This unique and very potent oxidant may be seen in metal complexes other than iron porphyrins. From this viewpoint, a porphyrin·Mn(III)–NaBH₄–O₂ oxidation carried out by the authors is discussed in detail. This simple model system behaves similarly to the enzymic system, P-450–NADH–O₂, in that the oxygen molecule is activated by Mn(II) to form transient porphyrin–manganese–oxygen complexes, one of which is an active species for oxidizing an olefin to the corresponding epoxide; this finally gives the corresponding alcohol on further reduction with NaBH₄ and Mn(II). From spectroscopic investigations of two possible intermediates, together with the observed structure–reactivity relationship with olefinic substrates, the active oxidizing species appears to be a porphyrin·Mn monoxide of the "Compound I" type.

Interaction between molecular oxygen (in air) and organic substrates may be classified into two categories.

0-8412-0514-0/80/33-191-291$05.00/0
© 1980 American Chemical Society

INDIRECT INTERACTION. A substrate is oxidized with an oxidized form of a given enzyme (or coenzyme) to give a corresponding oxidized substrate and a reduced form of the enzyme (or coenzyme). This step is followed directly or indirectly (through the electron-transport system) by the reaction between molecular oxygen and the reduced form of the enzyme to regenerate the active oxidized form of the enzyme. In this way, the oxidation proceeds catalytically.

DIRECT INTERACTION. Through formation of an enzyme–oxygen complex, a given enzyme activates molecular oxygen to a high-potential state that oxidizes organic substrates directly. Molecular oxygen otherwise essentially is inert toward such organic substrates.

The active enzyme may be regenerated during this process or by a successive reactivation step such as NADH reduction. In this way, the oxidation also proceeds catalytically.

In this chapter, the authors discuss only direct interaction; from this viewpoint, only a few metalloenzymes can be referred to as appropriate examples in which the O_2-activation mechanism has been clarified at least in part.

Molecular oxygen may be activated by:

1. a one-electron supply to form $O_2^-\cdot$ or its equivalent, such as $HO_2\cdot$, either free or complexed with a metal ion.
2. a two-electron supply to form $O_2^=$ or its equivalent, mostly complexed with a metal ion since free HO_2^- is too weak to oxidize common organic substrates except for very electron-deficient ones.
3. an electron withdrawal to form $O_2^+\cdot$ or its equivalent.

Generally speaking, a two-electron transfer (supply or withdrawal) is more appropriate to avoid an undesirable free-radical chain reaction ("autoxidation"), which is apt to be less selective. For this reason, a two-electron supply from the central metal to O_2 will be emphasized in the following discussion in which tyrosinase and cytochrome P 450 are used as appropriate examples.

Tyrosinase. The oxidation of catechol with O_2, catalyzed by tyrosinase, was concluded by Mason (1) in 1961 not to involve any radical species; therefore, an ionic mechanism was proposed by Hamilton (2). A possible activation mechanism seems to involve the interaction between Cu(I)·protein and O_2 to give an active Cu(II)—O—OH species in which Cu(II) and OH act as electron-deficient centers, withdrawing electrons from the substrate (Figure 1).

In this metal hydroperoxide (or an equivalent such as Cu=O), molecular oxygen activates Cu to raise its oxidation potential to the required extent, but Cu also activates the oxygen in a manner quite

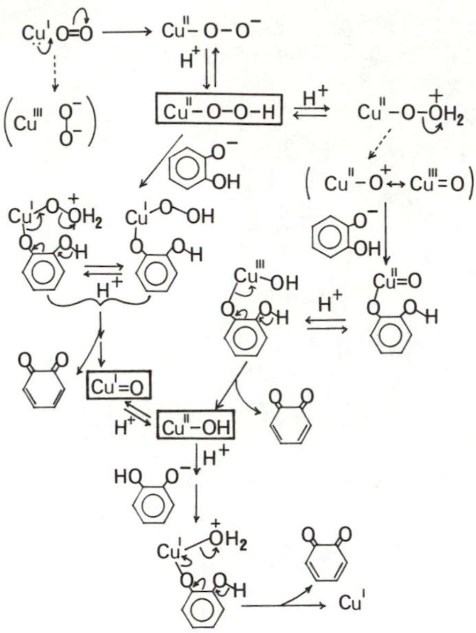

Figure 1.   Activation mechanism for tyrosinase-catalyzed oxidation of catechol with O_2

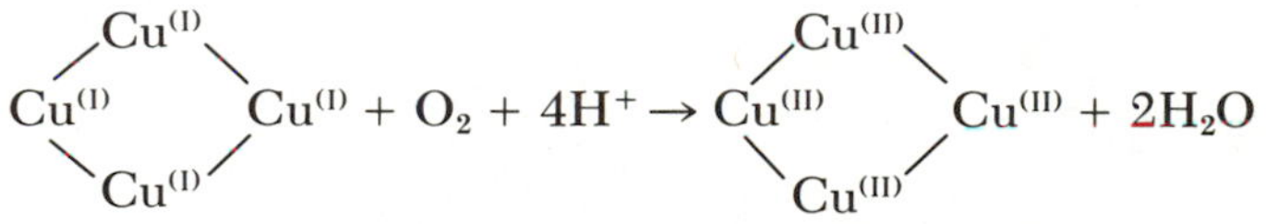

different from that of oxidases of indirect interaction such as ascorbic acid oxidase. This oxidizes a substrate via Cu(II), and the resultant Cu(I) is then reoxidized by molecular oxygen to give Cu(II).

Cytochrome P 450.   The mechanism of P-450 catalysis is outlined in Figure 2 (*3, 4*). The mechanism is based on the stoichiometry, product analysis, and labeling experiments, quenching experiments, spectral evidence for the structures of intermediates involved, and model studies (enzyme models as well as oxidant models) of P-450 catalysis.

Very significant contributions to the mechanism elucidation come from electronic, circular dichroism (CD), magnetic circular dichroism (MCD), electron paramagnetic resonance (EPR), nuclear magnetic resonance (NMR), Mössbauer, and Raman spectra of the important intermediates involved in the catalysis cycle (*5–15*). The spectral data are summarized briefly in Table I–Table VI and Figures 3 and 4. From these spectral characteristics, the oxidation–reduction states, spin states, ligands, and substrate and oxygen binding are reasonably well understood.

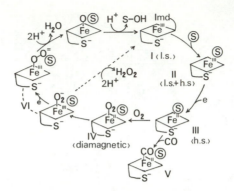

Figure 2.   Mechanism of oxygen activation by cytochrome P 450

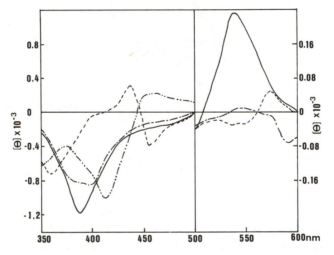

Figure 3.   Circular dichroism spectra: (-----) I; (———) II; (- - -) III;
(- - -) IV (5)

Together with these structural data, some important findings also come from chemical approaches by use of oxidants other than O_2, and by use of a deuterated substrate for the estimation of kinetic isotope effects. These results are discussed briefly below.

Formation of the porphyrin iron monoxide (of the "Compound I" type) as an active oxidizing species in the P-450 system is supported by the successful monoxidation of P 450 by an oxidizing reagent such as iodosobenzene (*see* Figure 5) (*16*). This species then effectively hydroxylates an organic substrate.

Further, oxygen and NADH, which are necessary for the P-450 catalyzed oxidation of organic substrates, also can be replaced by simple oxygenation reagents such as $NaIO_4$ or organic peroxides, as shown in Figure 6 (*17*).

Table I. Absorption Maxima in Electronic Spectra of Intermediates I–V

Intermediate	$\lambda_{max}{}^a$ (nm)				
I		417–418	535	570–571	
II	391–394		520, 540		643–647
III			490	543	
IV	355	418		544–555	
V		446–447		550–552	

[a] Data from Ref. 5–8.

Table II. Circular Dichroism Spectra of Intermediates I–V

Intermediate	Position (nm) and Sign				
I	(−)350	(−)410			
II	(−)388		(+)539		
III	(−)395		(+)547	(−)594	
IV	—				
V	(−)357	(+)441	(−)457	(−)535	(+)576

[a] Data from Ref. 4.

Table III. Electron Paramagnetic Resonance Spectra of Intermediates I–V

Intermediate	g-Value[a]				
I			2.41–2.45	2.25–2.26	1.9
II	7.9–8.1	3.7–4.0			1.7–1.8
			2.42–2.45	2.24–2.26	1.91–1.97
III	—				
IV	none				
V	—				

[a] Data from Ref. 9, 10, 11.

Table IV. Magnetic Circular Dichroism Spectra of Intermediates I–V

Intermediate	Position (nm) and Sign[a]						
I		(+)412[b] (−)430[b]			(+)527 (+)563 (−)582		
II	(−)395[b] (+)417 (−)430				(+)530 (−)556 (−)580		
III	—						
IV	(−)366[c] (+)406[c] (−)429		(+)446		(+)533 (−)560 (−)587		
V	(−)376[b] (+)413[c]			(−)456[c]		(−)586	

[a] Data from Ref. 12, 13.
[b] Strong.
[c] Very strong.

Table V.   Mössbauer Spectra of Intermediates I–V, Quadrupole Splitting and Isomer Shift

| | | Δ^a | |
Intermediate	Temperature (K)	Quadrupole Splitting ΔE_Q (mm/sec)	Isomer Shift δ (mm/sec)
I	206	2.75 ± 0.03	0.31 ± 0.02
II	210	—	—
High-spin component	—	0.78 ± 0.01	0.34 ± 0.02
Low-spin component	—	2.75 ± 0.02	0.31 ± 0.02
III	173	2.39	0.77
	213	2.36	0.76
IV	200	2.07 ± 0.04	0.27 ± 0.03
V	200	0.34	0.25

^{a} Data from Ref. 14.

Table VI.   Resonance Raman Spectra of Intermediates I–V

Intermediate	$\Delta\nu$ $(cm^{-1})^a$						
I	676	1372		1502			1635
II	675	1368	1372	1488		1570	1623
III	673	1344	1425	1466	1534	1563	1584	1601(?)
IV	—						
V		1368^b					

^{a} Data from Ref. 15; only strong absorptions are listed.
^{b} Other absorptions are not recorded.

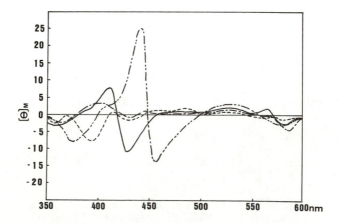

Figure 4.   Magnetic circular dichroism spectra: (——) **I**; (– – –) **II**;
(–·–·–) **III**; (– – – –) **IV** (*12, 13*)

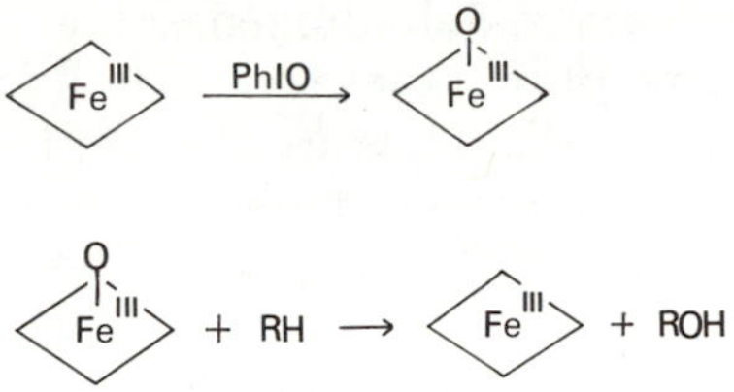

$$\text{Fe}^{III} \xrightarrow{\text{PhIO}} \text{O=Fe}^{III}$$

$$\text{O=Fe}^{III} + \text{RH} \longrightarrow \text{Fe}^{III} + \text{ROH}$$

Figure 5.   Monoxidation of P 450 using iodosobenzene

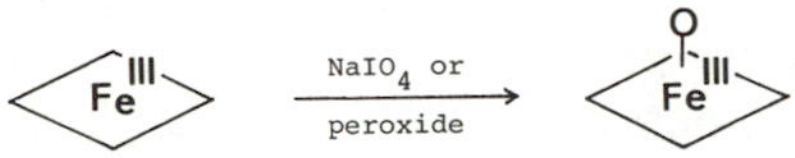

$$\text{Fe}^{III} \xrightarrow[\text{peroxide}]{\text{NaIO}_4 \text{ or}} \text{O=Fe}^{III}$$

Figure 6.   Oxidation of P 450

Cathodic reduction of molecular oxygen affords HO_2^-, which also can replace $NADH + O_2$ to activate P 450, and the oxidative demethylation of p-anisole was catalyzed effectively by the P-450–HO_2^- system (18). Addition of catalase inhibited the demethylation, demonstrating that HO_2^- was the activating species.

Significant characteristics of the porphyrin·iron monoxide are seen in the chemical reactivity. Naphthalene is converted initially to the corresponding arene oxide on treatment with P 450 (19), consistent with a molecular mechanism of oxygen transfer from an iron monoxide to the aromatic nucleus. Retention of stereochemistry in the P-450 catalyzed hydroxylation of d_1-ethylbenzene also supports the molecular mechanism. The unusually large kinetic isotope effect observed for the P-450 oxidation of dideutero 1,3-diphenylpropane, $k_H/k_D = 11$, demonstrates that C—H cleavage is involved in the rate determining step (20), probably in a very unusual environment, not incompatible with a molecular mechanism.

This potent oxidizing power of a monoxide species may not be limited to the P-450 system, but may be found in a wider range of metal catalysts for oxidations. In the later discussion, the authors demonstrate the possible participation of a porphyrin·manganese monoxide using the catalytic oxidation system, porphyrin·Mn–$NaBH_4$–O_2. Earlier discussion described the possible participation of a copper monoxide in the tyrosinase-catalyzed oxidation of catechol. However, much should be clarified before making the generalization that a metal monoxide is the active species often involved in direct oxygen activation. It is especially important to know the spectroscopic and chemical characteristics of this possibly unique and extremely potent species.

Oxygen Activation with a Porphyrin · Mn(II) Complex for Olefin Oxidation

The elucidation of the oxygen-activation mechanism in biological systems is still a most important target, although enormous efforts have been made to gain more insight into oxygenases, dehydrogenases, and hydroxylases. To understand the basic principles of these complex and highly organized systems, studies on simplified models are also important, as exemplified by the interaction of the oxygen molecule with various simple complexes of metals such as Fe (*21, 22*), Co (*23*), Mn (*24*), or Ru (*25*).

Of metalloenzymes known to activate the oxygen molecule, recently cytochrome P 450 (*26–28*) and appropriate model systems (*29, 30*) have been studied extensively in attempts to elucidate the enzyme mechanism. However, the mechanistic details of the activation have not been clarified satisfactorily.

Continuing our current studies on porphyrin·Mn complexes (*31, 32, 33*), some Mn(II) complexes have been found to activate the oxygen molecule very efficiently under certain conditions, and this chapter reports the oxidation of olefins using molecular oxygen activated by a tetraphenylporphyrin·Mn(II) complex.

Reaction Pathway and Products.   In the presence of a catalytic amount of a tetraphenylporphyrin (TPP)·Mn(III) complex (*34*) and sodium borohydride, treatment of cyclohexene with excess oxygen (air) in benzene–ethanol leads effectively to cyclohexanol and cyclohexenol. The reaction is quite different from the known "autoxidation" catalyzed by TPP·Mn(III) in the absence of NaBH$_4$ (Figure 7). The most significant characteristics of the present TPP·Mn–NaBH$_4$–O$_2$ reaction compared with the autoxidation are;

1. its striking selectivity to give cyclohexanol (Figure 7),
2. the remarkable rate acceleration (Figure 8),
3. the absence of any serious induction period (Figure 8),
4. the absence of the inhibitory effect of a typical radical inhibitor, 2,6-di-*t*-butyl-*p*-cresol, on cyclohexanol formation, in spite of a considerable inhibition of cyclohexenol formation (Figure 9).

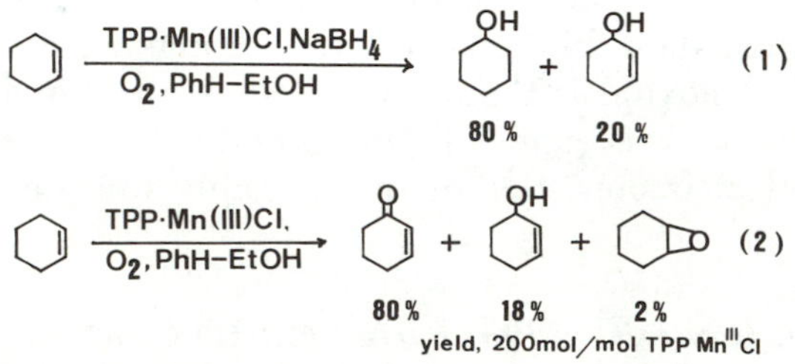

Figure 7.   TPP · Mn(III)-catalyzed autoxidation of cyclohexene: with NaBH$_4$, Reaction 1; without NaBH$_4$, Reaction 2

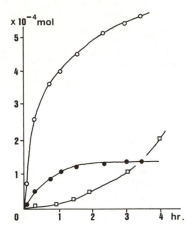

Figure 8.   TPP · Mn(III)-catalyzed autoxidation of cyclohexene: TPP · Mn(III) · Cl, 1.4 × 10⁻⁴ mol; NaBH₄, 2.6 × 10⁻³ mol; excess cyclohexene in benzene at 22 °C: (○) *with NaBH₄;* (●) *with NaBH₄;* (□) *without NaBH₄*

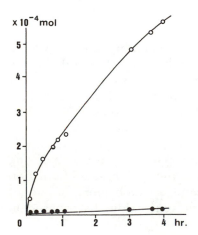

Figure 9.   Effect of radical inhibitor, *, on TPP · Mn(III)-catalyzed autoxidation of cyclohexene, conditions as described in Figure 8:* (○) *; (●)*

Thus, although a minor free-radical contribution to the product distribution is involved, the major contributing mechanism for the TPP · Mn–NaBH₄–O₂ reaction is different from that for the autoxidation; the present situation is similar to that of a P-450 oxidation in that three necessary components;

P 450   porphyrin · Fe   NADH   O₂

the present system   porphyrin · Mn   NaBH₄   O₂

take essentially the same roles (*see* Figure 10). Especially important to note is that a reducing agent is necessary to activate the metal catalyst in each system. Also noteworthy is that a metalloporphyrin acts as a

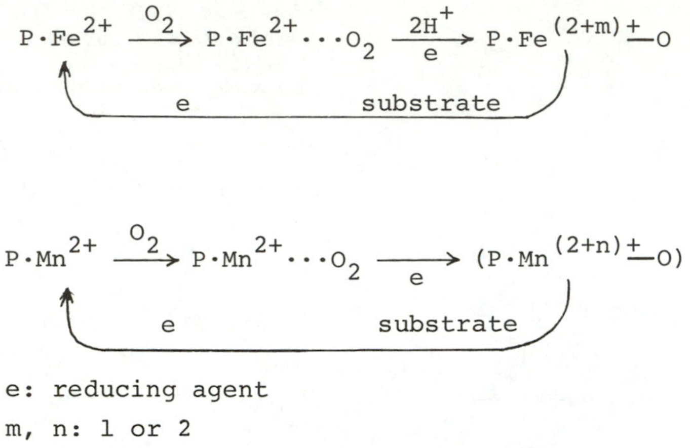

P·Fe2+
O2
P·Fe2+···O2
2H+
e
P·Fe(2+m)+—O
e
substrate
P·Mn2+
O2
P·Mn2+···O2
e
(P·Mn(2+n)+—O)
e
substrate
e: reducing agent
m, n: 1 or 2

Figure 10. Comparison of the mechanisms of P-450 autoxidation with TPP · Mn(III) oxidation

catalyst, showing a turnover number larger than unity in each system. For example, a turnover of 5 after 3.5 hr was observed for the $TPP \cdot Mn–NaBH_4–O_2$ system, and a further increase of the turnover number was observed on further addition of $NaBH_4$ (Prod/cat > 5).

Cyclohexenone, the major product of cyclohexene autoxidation was reduced to cyclohexanol and cyclohexenol in a ratio of 1 : 1.4 under the oxidation condition without oxygen, while cyclohexenol was not reduced appreciably. Thus, the reduction of cyclohexenone during the oxidation can account for only a part of the cyclohexanol formation in the $TPP \cdot Mn–NaBH_4–O_2$ reaction, and a much more important source of cyclohexanol seems to be cyclohexene oxide; this is based on the following observations (*see* Figure 11):

NaBH4
PhH-EtOH
1,2
1,4
OH
k1,2/k1,4 = 1.4
TPP·Mn(III)Cl, NaBH4
PhH-EtOH

(3)

(4)

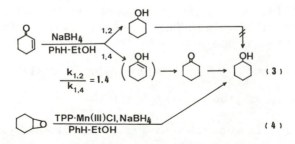

Figure 11. Cyclohexanol formation during cyclohexene oxidation

1. On replacing PhH–EtOH mixed solvent by benzene, a small amount of cyclohexene oxide was detected in the TPP·Mn–NaBH$_4$–O$_2$ reaction products even in the presence of a large excess amount of NaBH$_4$, probably because heterogeniety of the medium makes the reduction slow.

2. Cyclohexene oxide was reduced very readily under the TPP·Mn–NaBH$_4$–O$_2$ reaction conditions (or without O$_2$), giving cyclohexanol.

Thus, the reaction path to give the products observed in the TPP·Mn–NaBH$_4$–O$_2$ oxidation may be written as shown in Figure 12.

At the early stages of the reaction, contribution of the autoxidation to the product distribution is thought to be minor based on the following observations, although some autoxidation pathway may be involved to a minor extent after slow accumulation of possible initiators.

1. The observed time–conversion profile of the TPP·Mn–NaBH$_4$–O$_2$ reaction follows simple pseudo first-order kinetics (*see* Figure 13).

2. Even during the induction period of the TPP·Mn–O$_2$ autoxidation, this TPP·Mn–NaBH$_4$–O$_2$ oxidation gave a considerable amount of cyclohexenol (*see* Figure 8).

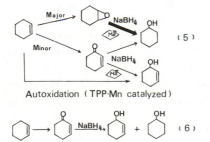

Autoxidation (TPP·Mn catalyzed)

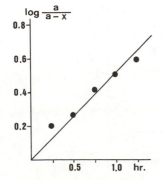

Figure 12. Direct P-450 type oxidation, Reaction 5: TPP·Mn autoxidation, Reaction 6

Figure 13. Rate of O$_2$ consumption

The latter observation is interesting, since coupled with the results of the radical inhibition, this suggests that there are two direct oxidation mechanisms as well as the autoxidation that lead to the products of the TPP·Mn–NaBH₄–O₂ reaction: a direct oxidation via a free-radical intermediate, but not through the autoxidation in which cyclohexenol formation is inhibited by 2,6-di-t-butyl-p-cresol; and another direct uninhibited oxidation that leads to cyclohexene oxide.

Based on the experimental results, a conclusion is that the present TPP·Mn–NaBH₄–O₂ reaction with cyclohexene proceeded mostly through formation of a potent oxidizing species (**X**).

Spectroscopic Characteristics of the Potent Oxidizing Species X.   Pure TPP·Mn(II), carefully prepared (*33, 35*) from TPP·Mn(III)Cl via reduction with NaBH₄ in purified benzene followed by reprecipitation on addition of n-hexane, showed a characteristic electronic spectrum when dissolved in pure benzene (Figure 14). Extremely careful purification of benzene (not only benzene for the measurements but also benzene for reaction or precipitation) is necessary for reproducibility of the spectroscopic data. Thus, freshly distilled benzene, treated with Na was redistilled in a deoxygenated vacuum line. This benzene was degassed through a "freeze and thaw" technique four times and then kept on a small amount of freshly prepared TPP·Mn(II) (TPP·Mn(II) is one of the strongest deoxygenating reagents available. Benzene was re-redistilled under Ar from this mixture just before use. Addition of a small amount of oxygen (air) to the solution of TPP·Mn(II) gave changes in the electronic spectrum, indicating the formation of at least two intermediates, **X₁** and **X₂**, before the complete conversion to the final state, TPP·Mn(III) (*see* Figure 15). The Mn(III) species no longer interacted with an olefin under the TPP·Mn–NaBH₄–O₂ reaction conditions.

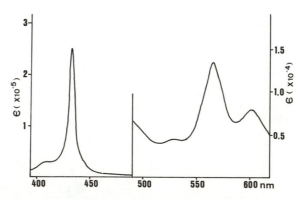

Figure 14.   Electronic spectrum of TPP · Mn(II) in benzene

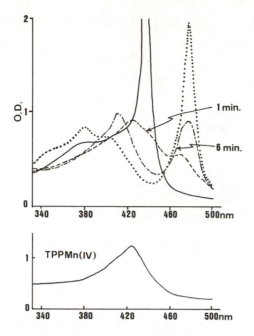

Figure 15. Electronic spectra recorded during oxidation of TPP · Mn(III): (——) TPP · Mn(II); (– – –) X₁; (– – –) X₂; (· · ·) TPP · Mn(III). After 6 min the 378-nm absorption increased uniformly. The spectrum of TPP · Mn(IV) is shown for comparison.

Another possible product, $O_2^-\cdot$, from the reaction between TPP·Mn(II) and O_2 was generated from KO_2 in benzene in the presence of 18-crown-6, but it did not affect an olefin under the present conditions. Interaction of TPP·Mn(III)Cl and KO_2 in benzene rapidly gave TPP·Mn(II) and O_2, where possible intermediates also were detected spectrophotometrically. In the presence of an olefin, an appreciable amount of the oxidation products also was obtained, indicating that the Mn(III)/$O_2^-\cdot$ reaction also produced the potent oxidizing species, X (*see* Figure 16).

The nature and the possible structures of X_1 and X_2 are indicated by their electronic spectra. The absorption maximum at 425 nm in benzene and the shape of the spectrum of X_1, the first short-lived intermediate (Figure 15), closely resemble those of the spectrum (λ_{max} at 422 nm in methylene chloride) of TPP·Mn(IV) prepared independently (*31, 32, 33, 36*), possibly indicating a charge-transfer complex (Figure 17). This assumption of a loose "side-on" complex resulting from a two-electron transfer is consistent with the previous results of the electron spin resonance (ESR) spectrum observed for an oxygen-carrying species, TPP·Mn(II) or substituted TPP·Mn(II) (*37*). The

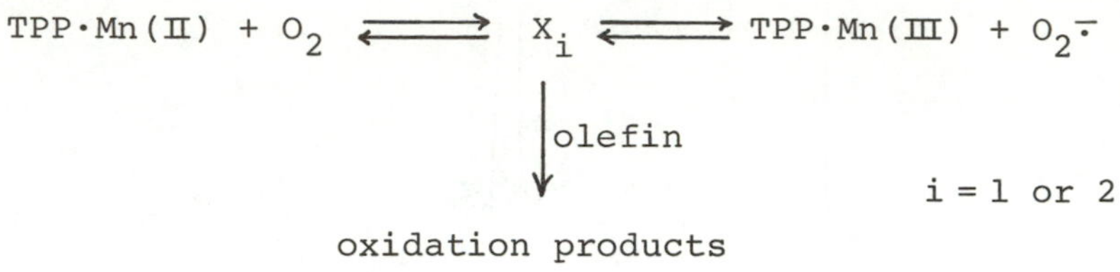

$$TPP \cdot Mn(II) + O_2 \rightleftharpoons X_i \rightleftharpoons TPP \cdot Mn(III) + O_2^-$$

↓ olefin

i = 1 or 2

oxidation products

Figure 16. Oxidation of TPP · Mn (II) with an olefin present

Figure 17. Possible structure for X₁ intermediate; lozenge denotes TPP ligand.

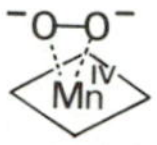

second, relatively long-lived intermediate, X_2, showed a unique electronic spectrum (Figure 15). The absorption maxima were found at 375, 410, and 477 nm in benzene, nearly identical with those of TPP·Mn(III) (375, 406, and 477 nm), although the relative intensities of the 375 and 406 (410) bands are changed drastically. These spectra suggest that X_2 is a derivative of TPP·Mn(III), possibly that shown in Figure 18.

Figure 18. Possible structure for X₂ intermediate; lozenge denotes TPP ligand.

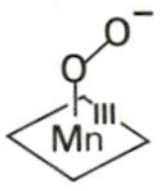

Structure–Reactivity Relationship of Olefins. The relative reactivity of a series of olefins toward the potent oxidizing species, **X**, formed by the interaction of TPP·Mn(II) with O_2, was investigated by means of a competitive reaction technique. As shown in Table VII, the relative reactivity of an olefin, as followed by gas–liquid chromatographic determination, increases on introduction of an alkyl substituent onto the olefinic carbon atom other than the reacting carbon atom. However, the introduction of an alkyl substituent onto the reacting carbon atom reduces (or compensates) the accelerative electronic effect, as seen in the comparison between cyclohexene and n-hexene. This situation becomes clearer if one compares the two dialkyl ethylenes, cyclohexene and methylenecyclohexane, where the former has a single substituent on the reacting carbon and the other has none; the observed relative reactivity is 1 : 27.2.

From the structure–reactivity relationship, a tentative conclusion may be drawn that the present oxidizing species, **X**, should be strongly electrophilic (electron deficient) and also have a strict steric require-

Table VII.   Relative Reactivities of Olefins Towards Oxidizing Species, X.

Olefins	Relative Reactivity
Cyclohexene	1.0
1-*n*-Hexene	1.4
1-Methylcyclohexene	6.1
Methylenecyclohexane	27.2

ment (very sensitive to steric hindrance). One plausible candidate for this species may be *B* derived from *A* in Figure 19, since *A* itself does not seem to have an electron-deficient oxygen.

As a conclusion, the most plausible mechanism of the present TPP·Mn–NaBH$_4$–O$_2$ oxidation of an olefin is depicted in Figure 20.

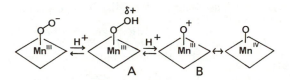

*Figure 19.   Possible formation of the oxidizing species, **X**, formed during the oxidation of TPP · Mn(II)*

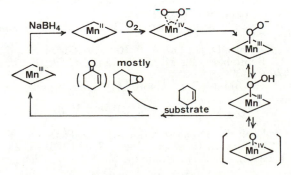

Figure 20.   TPP · Mn · NaBH$_4$–O$_2$ oxidation of an olefin

Acknowledgment

The authors wish to thank Masao Miyake for his helpful collaboration in preparation of the review part of this chapter.

Literature Cited

1. Mason, H. S.; Spencer, E.; Yamazaki, I. *Biochem. Biophys. Res. Commun.* **1961**, *4*, 236.
2. Hamilton, G. A. *Adv. Enzymol.* **1969**, *32*, 55.

3. Sato, R.; Omura, T. In "Cytochrome P-450"; Kodansha Ltd.: Tokyo, 1978.
4. Boyd, G. S.; Ullrich, V. *Biochem. Soc. Trans.* **1975**, *803*.
5. Peterson, J. A. *Arch. Biochem. Biophys.* **1971**, *144*, 678.
6. Peterson, J. A.; Ishimura, Y.; Griffin, B. W. *Arch. Biochem. Biophys.* **1972**, *149*, 197–208.
7. Tyson, C. A.; Lipscomb, J. D.; Gunsalus, I. C. *J. Biol. Chem.* **1972**, *247*, 5777.
8. Van der Hoeven, T. A.; Coon, M. J. *J. Biol. Chem.* **1974**, *249*, 6302.
9. Tsai, R.; Yu, C. A.; Gunsalus, I. C.; Peisach, J.; Blumberg, W.; Orme–Johnson, W. H.; Beinert, H. *Proc. Nat. Acad. Sci. USA.* **1970**, *66*, 1157.
10. Peisach, J.; Blumberg, W. E. *Proc. Nat. Acad. Sci. USA.* **1970**, *67*, 172.
11. Peisach, J.; Appleby, C. A.; Blumberg, W. E. *Arch. Biochem. Biophys.* **1972**, *150*, 725.
12. Dolinger, P. M.; Kielczcwski, M.; Trudell, J. R.; Barth, G.; Linder, R. E.; Bunnenberg, E.; Djerassi, C. *Proc. Nat. Acad. Sci. USA.* **1974**, *71*, 399.
13. Dawson, J. H.; Cramer, S. P. *FEBS Lett.* **1978**, *88*, 127.
14. Sharrock, M.; Derunner, P. G.; Schulz, C.; Lipscomb, J. D.; Marshall, V.; Gunsalus, I. C. *Biochim. Biophys. Acta* **1976**, *420*, 8.
15. Champion, P. M.; Gunsalus, I. C.; Wagner, G. C. *J. Am. Chem. Soc.* **1978**, *100*, 3743.
16. Lichtenberger, F.; Nastainczyk, W.; Ullrich, V. *Biochem. Biophys. Res. Commun.* **1976**, *70*, 939.
17. Hrycay, E. G.; Gustafsson, J.; Ingelman–Sundberg, M.; Ernster, L. *Biochem. Biophys. Res. Commun.* **1975**, *66*, 209.
18. Scheller, F.; Renneberg, R.; Mohr, P.; Jänig, G. R.; Ruckpaul, K. *FEBS Lett.* **1976**, *71*, 309.
19. Jerina, D. M.; Daly, J. W.; Witkop, B.; Zaltman–Nirenberg, P.; Udenfriend, S. *Biochemistry* **1970**, *9*, 147.
20. Hjelmeland, L. M.; Aronow, L.; Trudell, J. R. *Biochem. Biophys. Res. Commun.* **1977**, *76*, 541.
21. Collman, J. P. *Acc. Chem. Res.* **1977**, *10*, 265.
22. Basolo, F.; Hoffman, B. M.; Ibers, J. A. *Acc. Chem. Res.* **1975**, *8*, 334.
23. Vaska, L. *Acc. Chem. Res.* **1976**, *9*, 175.
24. Hoffman, B. M.; Szymanski, T.; Brown, T. G.; Basolo, F. *J. Am. Chem. Soc.* **1978**, *100*, 7253.
25. Farrell, N.; Dolphin, D. H.; James, B. R. *J. Am. Chem. Soc.* **1978**, *100*, 324.
26. Scheller, F.; Renneberg, R.; Mohr, P.; Jänig, G. R.; Ruckpaul, K. *FEBS Lett.* **1976**, *71*, 309.
27. Nordbrom, G. D.; White, R. E.; Coon, M. J. *Arch. Biochem. Biophys.* **1976**, *175*, 524.
28. Lichtenberger, F.; Nastainczyk, W.; Ullrich, V. *Biochem. Biophys. Res. Commun.* **1976**, *70*, 939.
29. Tang, S. C.; Koch, S.; Papaefthymiou, G. C.; Foner, S.; Frankel, R. B.; Ibers, J. A.; Holm, R. H. *J. Am. Chem. Soc.* **1976**, *98*, 2414.
30. Collman, J. P.; Sorrell, T. N.; Hoffman, B. M. *J. Am. Chem. Soc.* **1975**, *97*, 913.
31. Tabushi, I.; Kojo, S. *Tetrahedron Lett.* **1974**, *1577*.
32. Ibid., **1975**, 305.
33. Tabushi, I.; Koga, N. *Tetrahedron Lett.* **1978**, *5017*.
34. Jones, R. D.; Summerville, D. A.; Basolo, F. *J. Am. Chem. Soc.* **1978**, *100*, 4416.
35. Gonzalez, B.; Kouba, J.; Yee, S.; Reed, C. A. *J. Am. Chem.* **1975**, *97*, 3247.
36. Loach, P. A.; Calvin, M. *Biochemistry* **1963**, *2*, 361.
37. Hoffman, B. M.; Weschler, C. J.; Basolo, F. *J. Am. Chem. Soc.* **1976**, *98*, 5473.

RECEIVED May 16, 1979.

Oxygen Activation in Oxygenase Systems

Model Approach Using Iron-Porphyrin

ZEN-ICHI YOSHIDA, HIROSHI SUGIMOTO, and HISANOBU OGOSHI

Department of Synthetic Chemistry, Kyoto University Yoshida, Kyoto, 606 Japan

Fe(II)-porphyrin is shown to be the best model of the active site of L-tryptophan pyrrolase. This model complex effectively catalyzes the oxygenation of 3-substituted indoles to form products corresponding to formylkynurenine (the oxygenation product of L-tryptophan) with 40%–50% conversion of the substrate. ESR analysis of the oxygenation system indicates cooperative electron transfer (COET) occurs from the substituted indole anion to the oxygen in the ternary system: indole anion–Fe(II)-porphyrin–O_2. This type of electron transfer is a new concept that should occur in any strong donor–Fe(II)-porphyrin–acceptor system. Experimental data of the oxygenation mechanism of tryptophan and skatole are discussed in terms of COET. Oxygen activation in cytochrome P-450 and action mechanism of bleomycin, an anticancer agent, also can be explained by the COET concept.

In the oxydase reaction related to catabolism, molecular oxygen is not incorporated into the substrate. On the other hand, in the oxygenase reaction related to metabolism, the substrate incorporates molecular oxygen to yield mono- or dioxygenated products. Since independent discovery of the oxygenases tyrosinase and pyrocatechase by Mason et al. (1) and Hayaishi et al. (2), respectively, in 1955, a large number of papers concerning oxygenases have been reported (3–9). However, oxygen activation mechanisms in the oxygenase system remains obscure.

0-8412-0514-0/80/33-191-307$05.00/0
© 1980 American Chemical Society

The focus of this chapter is the reaction site of oxygenases (hemoproteins) having heme as the prosthetic group. We discuss the oxygen activation in tryptophan pyrrolase (TPO) and cytochrome P-450 based on our experimental results using iron-porphyrin complexes as the model for the active site of these enzymes.

The active site of hemoproteins is known to be heme (iron-porphyrin complex), although they exhibit the following different functions: (i) reversible oxygen binding for transport and storage (hemoglobin and myoglobin), (ii) oxygenation with molecular oxygen (monooxygenase such as cytochrome P-450, and dioxygenase such as L-tryptophan pyrrolase), (iii) oxygen reduction (cytochrome c oxidase), (iv) electron transfer (cytochrome b and c), and (v) hydrogen peroxide utilization (catalase and peroxidase).

Tryptophan Pyrrolase Model Reaction and Activation of Oxygen

Model Reaction. TPO catalyses the reaction of tryptophan $(R = CH_2CH \overset{\displaystyle NH_2}{\underset{\displaystyle CO_2H}{\big<}})$ with molecular oxygen to form formylkynurenine (FK) according to Reaction 1. The most remarkable feature of this enzyme reaction system is that the presence of a reductase system is not required. Detailed analysis of the TPO reaction indicates that the

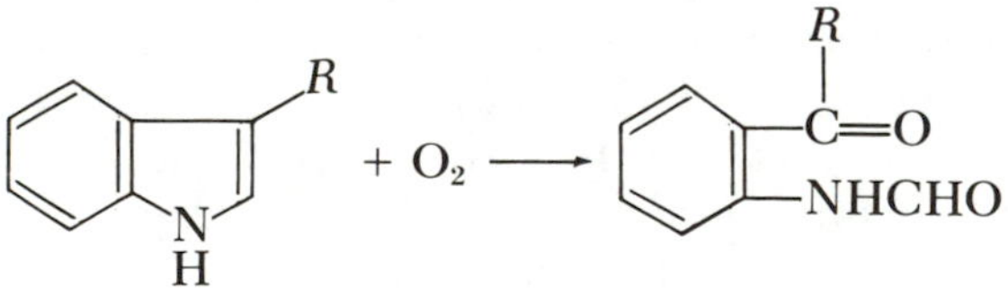

$$+ \ O_2 \ \longrightarrow \qquad\qquad (1)$$

reduced TPO (Fe(II)TPO) is the active form of this enzyme and that TPO has two kinds of binding sites that form the substrate–enzyme adduct ($EFe^{2+} \cdot S_1S_2$) with tryptophan. This adduct's interaction with molecular oxygen yields the oxygenated adduct ($EFe^{2+}S_1S_2 \cdot O_2$), which decomposes to FK and $EFe^{2+} \cdot S_1$ (*see* Figure 1). As shown in Table I, the electronic spectra of Fe(II)TPO, Fe(III)TPO, and oxygenated Fe(II)TPO (10) are very similar to those of myoglobin (11). From x-ray studies of the structure (12) of myoglobin, the heme in myoglobin (13) is considered to be surrounded by the hydrophobic peptide sidechain (Figure 2). Based on the spectral similarity between myoglobin and TPO, we deduced that the surrounding structure of the heme in TPO is similar to that in myoglobin.

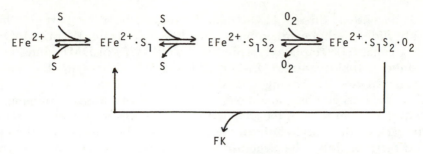

Figure 1.   *TPO reaction cycle: EFe^{2+}, reduced form of TPO; S, L-tryptophan; FK, formylkynurenine*

Table I.   Absorption Maximum of Fe(II)TPO, Fe(III)TPO, Fe(II)TPO·TRY·O_2 and Those of the Corresponding Myoglobin (nm)

	Oxidized Form			Reduced Form		Oxygenated Form		
Tryptophan pyrrolase	405	502	632	433	555	418	545	580
Myoglobin (sperm whale)	409	505	635	434	556	418	543	581

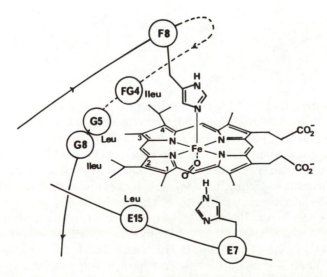

Figure 2.   *Estimated oxygen-binding structure around the heme of myoglobin*

Recently, three TPO models have been reported: *bis*(salicylidene)ethylenediaminato cobalt(II) [Co(salen)] in MeOH (*14*), cobalt(II)tetraphenylporphyrin (CoTPP) (*15*) in DMF, and manganese phthalocyanine (Mn-Pc) in DMF (*16*). These models have been reported as having the TPO-mimic function of oxygenating skatole, a tryptophan analogue, to form 2-formamidoacetophenone (FA). However, one of the most critical problems for these models is the lack of structural similarity between the TPO active site (heme) and these models. The structure of these models is not similar to the heme in TPO with respect to the central metal and/or the ligand. Furthermore, Fe(salen) does not possess the TPO-mimic function (*14*).

For designing the enzyme model, it is important to consider the functional and structural similarity between the model and the active site of enzyme. With this in mind, we chose the iron(II) complexes of octaethylporphyrin and tetraphenylporphyrin, (OEP · Fe(II) · py$_2$ and TPP · Fe(II) · py$_2$), respectively, as models of the heme and benzene as the hydrophobic environment surrounding the heme.

Skatole (2 mmol) in benzene (10 mL) is difficult to react with oxygen (1 atm) at room temperature (23°C ± 2°). However, in the presence of TPP · Fe(II) · py$_2$ or OEP · Fe(II) · py$_2$ (20 μmol, substrate/TPO model molar ratio: 100/1) oxygen absorption easily occurs and skatole is converted to 2-formamidoacetophenone (FA) as shown in Reaction 2. This reaction was monitored by the measurement of oxy-

$$\text{(2)}$$

gen absorption and TLC of the reaction mixture. The relationship between skatole conversion and reaction time is shown in Figure 3. The conversion of skatole increases with time (Figure 3). At less than 40%–50% conversion, FA is the sole product (FA yield, 100%). On the other hand, at over 50% conversion, products other than FA are formed and the amount of byproducts formed increases with time. For example, at 90% conversion of skatole, the yield of FA is 44% and the remainder (56%) of the skatole becomes 2-aminoacetophenone and polymeric materials. Both porphyrins (TPP · Fe(II) · py$_2$ and OEP · Fe(II) · py$_2$) show almost the same catalytic activity for this reaction. When *N*-ethylimidazole is added to this reaction system, the reaction of skatole with oxygen is depressed with the amount of *N*-ethylimidazole added (Figure 3). This is ascribed to the masking

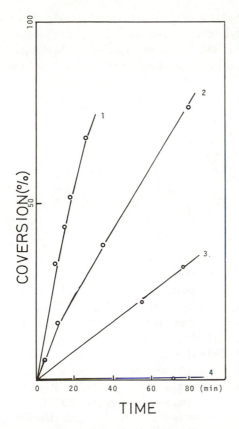

Figure 3. Relation between skatole conversion and reaction time; skatole (2 mmol), TPP·Fe(II)·py₂ (20 μmol); O₂ (1 atm in benzene, 23°C ± 2°): 1, 2, 3-dimethylindole TPPFe(II)·py₂; 2, skatole/TPPFe-(II)py₂; 3, skatole (skatole–N-ethyl-imidazole 1:0.23); 4, skatole (ska-tole–N-ethylimidazole 1:5.5)

effect of the active coordination site by the stronger ligand, indicating the importance of the coordination of skatole to Fe(II)-porphyrin. Although oxygen is known to coordinate the Fe(II) of iron-porphyrin in aprotic solvents (*17*), the electronic and NMR spectra of the skatole–Fe(II)TPP and skatole–Fe(III)TPP systems indicate that σ-coordination of skatole NH toward iron is difficult to accomplish. Therefore, the π-interaction of skatole with the iron of Fe(II) · TPP · O₂ and the σ-coordination of skatole N⁻ (produced through deprotonation by iron-porphyrin) with the iron of Fe(II) · TPP · O₂ should occur for the skatole → FA reaction; the equilibrium concentration of the latter complex should be small. When 2,3-dimethylindole, which is a stronger electron donor than skatole, is used instead of skatole, the more rapid oxygenation reaction (Figure 3) occurs, producing 100% 2-acetamidoacetophenone at 40% conversion of the substrate.

Although tryptophan does not dissolve in benzene, 3-indole-propionic acid ester (*R'*=CH₃ ∼ C₆H₁₃) is soluble in benzene, and converts smoothly to the corresponding 2'-formamidophenyl-

4-ketopropionic ester (*see* Reaction 3). The reaction proceeds more effectively with increasing alkyl chain length (R'). Therefore, judging from the oxygenation activity, the high product selectivity, the high product yield, and the structural similarity to TPO active

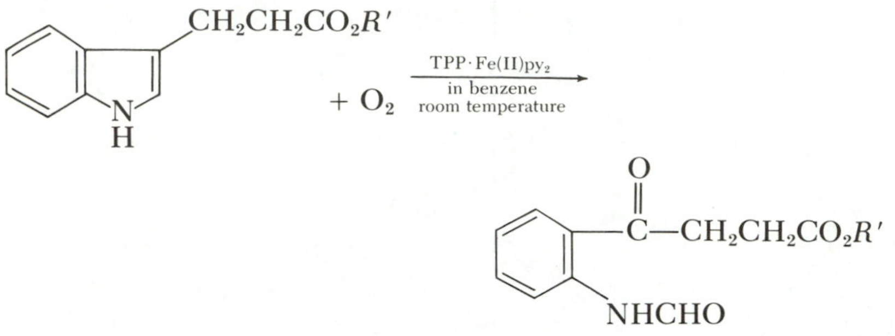

$$
\text{(indole-3-}CH_2CH_2CO_2R') + O_2 \xrightarrow[\substack{\text{in benzene} \\ \text{room temperature}}]{TPP \cdot Fe(II)py_2} \quad (3)
$$

$$
\text{(phenyl)}-\overset{\overset{\displaystyle O}{\|}}{C}-CH_2CH_2CO_2R'
$$
$$
\text{NHCHO}
$$

site, the Fe(II)-porphyrin system is considered to be the best model of the TPO active site. As for the substrate specificity of Fe(II)-porphyrin-catalyzed oxygenation, 3-alkyl- and 2,3-dialkylindoles undergo the TPO type reaction. However, indole, 2-methylindole, 1,2-dimethylindole, and 1,3-dimethylindole do not react with oxygen under the above conditions. For indole derivatives to be oxygenated catalytically by Fe(II)-porphyrin (heme), they must have a 3-alkyl group and a N—H group. This substrate specificity is different from those in photosensitized oxygenation by singlet oxygen (*18, 19, 20, 21*) and radical autoxidation of indole derivatives (*22*), suggesting that the active oxygen species produced from 3-alkylindole–Fe(II)-porphyrin–O_2 system is different than the singlet oxygen and radical oxygen species from autoxidation.

Superoxide ion ($\cdot O_2^-$) might be produced as the active oxygen from the Fe(II)-porphyrin and molecular oxygen system (binary system). However, superoxide (18-crown-6-complexed KO_2) in benzene does not react with skatole. If this reaction occurs to give FA, the reaction is stoichiometric and not catalytic. The substrate specificity mentioned above also excludes the possibility of oxygen activation via the π-coordinated skatole–Fe(II)-porphyrin–O_2 complex (**I**). So the oxygen activation in this reaction system seems to occur via a skatole-anion–coordinated-Fe(II)-porphyrin–O_2 complex (**II**).

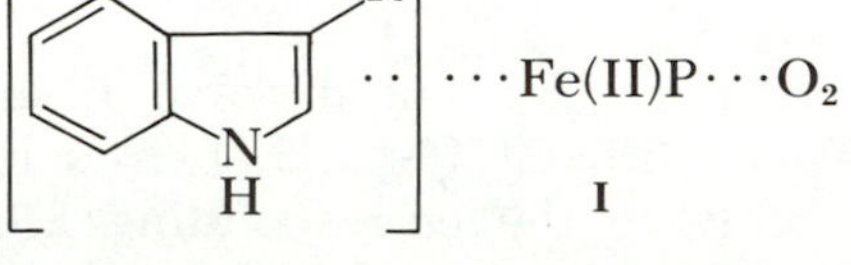

$$
\left[\text{(3-R-indole, N-H)} \right] \cdots \cdots Fe(II)P \cdots O_2
$$
$$
\textbf{I}
$$

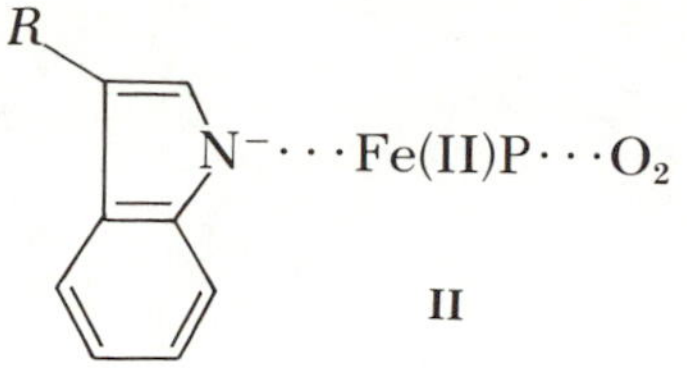

R
N⁻····Fe(II)P···O₂

II

Oxygen-Activation Mechanism. To gain insight into the oxygen-activation mechanism of this model system, ESR investigation of the following systems was carried out in toluene: 1. $TPP \cdot Fe(II) \cdot py_2$ (10 μmol)–skatole (1 mmol); 2. $TPP \cdot Fe(II)$-py_2–skatole–air; 3. $TPP \cdot Fe(III) \cdot OAc$–skatole; and 4. skatole–air system. Using air (nitrogen-diluted oxygen) instead of oxygen, we can control the reaction rate to get the ESR spectra for Systems 1–4 at different times. The ESR measurement was carried out at $-196°C$ (in liquid N_2) after allowing the solution to stand at $23°C \pm 2°$ for a given period of time. For Systems 1, 3, and 4, no ESR signal was detected for one day immediately after preparation. In the absence of molecular oxygen, electron transfer does not occur between skatole (or skatole anion) and the Fe(III)-porphyrin.

Immediately upon contact with air, the high-spin complex ($g \sim 6.1$) attributable to $Fe(III) \cdot TPP \cdot py$ (pentacoordinated complex) is observed together with the free radicals whose g values are 1.99–2.04 and the low-spin Fe(III) complex (*see* Figure 4). The formation of a pair of low-spin complexes attributable to the hexacoordinated $Fe(III) \cdot TPP$ complexes ($g = 2.66, 2.19, 1.80$ and $g = 2.31, 1.93$) occurs with time. Of those, one should be the $Fe(III) \cdot TPP \cdot py_2$ and another the product-coordinated Fe(III)–TPP complex. The appearance of ESR absorptions attributable to Fe(III) complexes (high- and low-spin) indicates that a part of the Fe(II)–TPP complex is converted to $Fe(III) \cdot TPP$ by reaction with molecular oxygen.

Absorptions other than those assigned to the Fe(III) complex are observed. The absorption at $g = 1.99$ is assigned to the superoxide ion ($\cdot O_2^-$) (23). There is a very broad band centered at $g = 2.04$ that has hyperfine splittings. Each hyperfine constant is very large (~ 20 G) compared with those for the pyrryl π-radicals (24); a large hyperfine constant is characteristic of the σ-radical. Therefore, this radical can be identified as the skatole σ-radical (**III**).

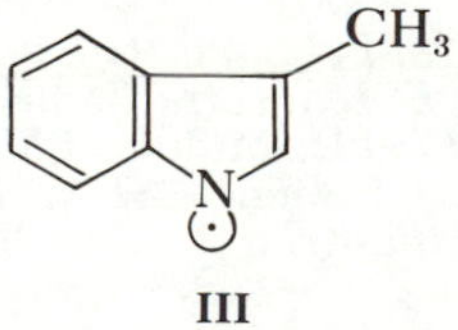

CH₃   CH₃

N   N
·

III 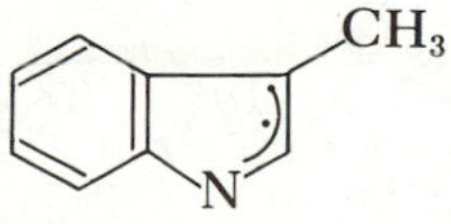   IV

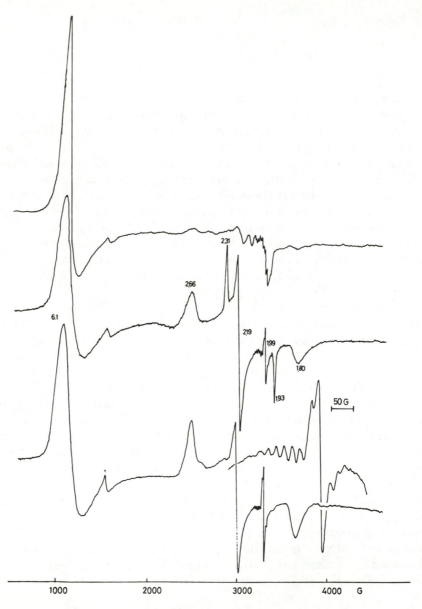

*Figure 4.   ESR spectra of TPPFe(II)py₂ (10 μmol)–skatole (1 mmol),
air system at −196°C. Top: ESR spectra immediately after contact of
the solution with air at 23°C ± 2°. Middle: ESR spectra after one-day
contact with air at 23°C ± 2°. Bottom: ESR spectra after two-day contact with air at 23°C ± 2°.*

Other than $\cdot O_2^-$ and the skatole σ-radical (**III**), a radical exists having a g-value of 2.002 with unclear hyperfine splitting. This radical may be the π-radical (**IV**) from the skatole anion.

The σ-radical (**III**) and $\cdot O_2^-$ are considered to be formed by electron transfer from the skatole anion to oxygen in the ternary complex (**II**), which is composed of a strong electron donor and a weak electron acceptor through Fe(II)-porphyrin (*see* Reaction 4). In such a complex (D–Fe(II)P–A type complex, D = electron donor, A = electron acceptor), electron transfer from the donor to the acceptor should occur more easily than the direct electron transfer in D–A complex, because in the former the cooperative interaction of the three components should decrease the energy barrier of the electron transfer.

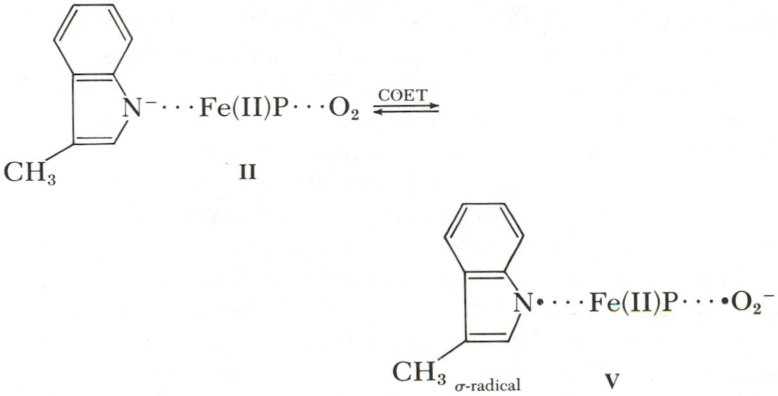

(4)

The observed ESR spectra strongly support the interpretation that no electron transfer occurs between a skatole anion and Fe(III)–porphyrin in the binary system in toluene at room temperature. On the other hand, under the same conditions electron transfer from skatole anion to oxygen occurs in the ternary complex (**II**). This is an activation mechanism of molecular oxygen in the Fe(II)-porphyrin-catalyzed oxygenation of skatole. We have termed this type of electron transfer the cooperative (or concerted) electron transfer (COET) (*see* Reaction 5). When the electron accepting power (electron affinity) of the accep-

$$D\cdots Fe(II)P\cdots A \xrightarrow{\text{COET}} D^+\cdots\cdots Fe(II)P\cdots A^- \qquad (5)$$

tor is less than that of molecular oxygen ($EA = 0.43$ eV), the electron donating power (ionization potential) of the donor must be as large as that of skatole anion for COET to occur.

The MINDO/3 ionization potentials (I_p) for skatole and the skatole anion are shown in Table II. This table indicates that by the deprotonation of the skatole NH, the HOMO (π) ionization potential becomes

Table II. MINDO/3 Ionization Potential for Skatole and Skatole Anion

Species	Ionization Potential (eV)	
	HOMO (π)	7.665
	N-lone pair (p_z)	9.428
	HOMO (π)	2.291
	N-lone pair (p_y)	3.101

significantly smaller (7.665 eV → 2.291 eV). Also, the N-lone pair (p_y) ionization potential (3.101 eV) of the skatole anion is close to the HOMO (π) ionization potential (2.291 eV). Therefore, the skatole anion is considered to be strong electron donor. From the ionization potentials of HOMO (π) and N-lone pair (p_y) in noninteracted states, the generation of the π-radical of the skatole anion should be easier than the generation of the σ-radical (III). The observed exclusive formation of σ-radical (III) in the ternary system might be caused by the effect of coordination with the Fe(II) of the heme. The planar ligand, porphyrin, should make the axial coordination of both the donor and the acceptor possible and also should be useful in the stabilization of this system.

From the arguments mentioned above, the Fe(II)-porphyrin-catalyzed oxygenation of skatole (and related compounds) most likely proceeds as shown in Scheme 1, which involves the π-radical (IV) intermediate. Process 1 is the equilibrium formation of the skatole-anion-coordinated Fe(II)-porphyrin (VI), which occurs with Fe(II) and solvent (B, π-base) assistance. When the electron-relay system II is formed by oxygen coordination, very rapid COET should take place to give the coordinated skatole σ-radical and $\cdot O_2^-$. Judging from the HOMO (π) and N-lone pair (p_y) levels, this σ-radical can interact with skatole anion to give the π-radical (IV). The equilibrium of Reaction 6 should be controlled by the stabilization of the skatole anion and the π-radical (IV). For example, the methyl groups at the 2 and 3 positions do not stabilize the skatole anion. On the other hand, the 3-methyl group can stabilize the π-radical (IV), shifting the equilibrium to the right. This π-radical (IV) should react easily with the $\cdot O_2^-$ of VII to give the final product VIII via the most probable endo-peroxide intermediate. The reproduced skatole-anion-coordinated Fe(II)-porphyrin (VI) by Process 3 is again reacted in Process 2. The Fe(II)-porphyrin-catalyzed oxygenation of skatole should proceed in this manner. This mechanism could be applicable to the TPO reaction

Scheme 1.

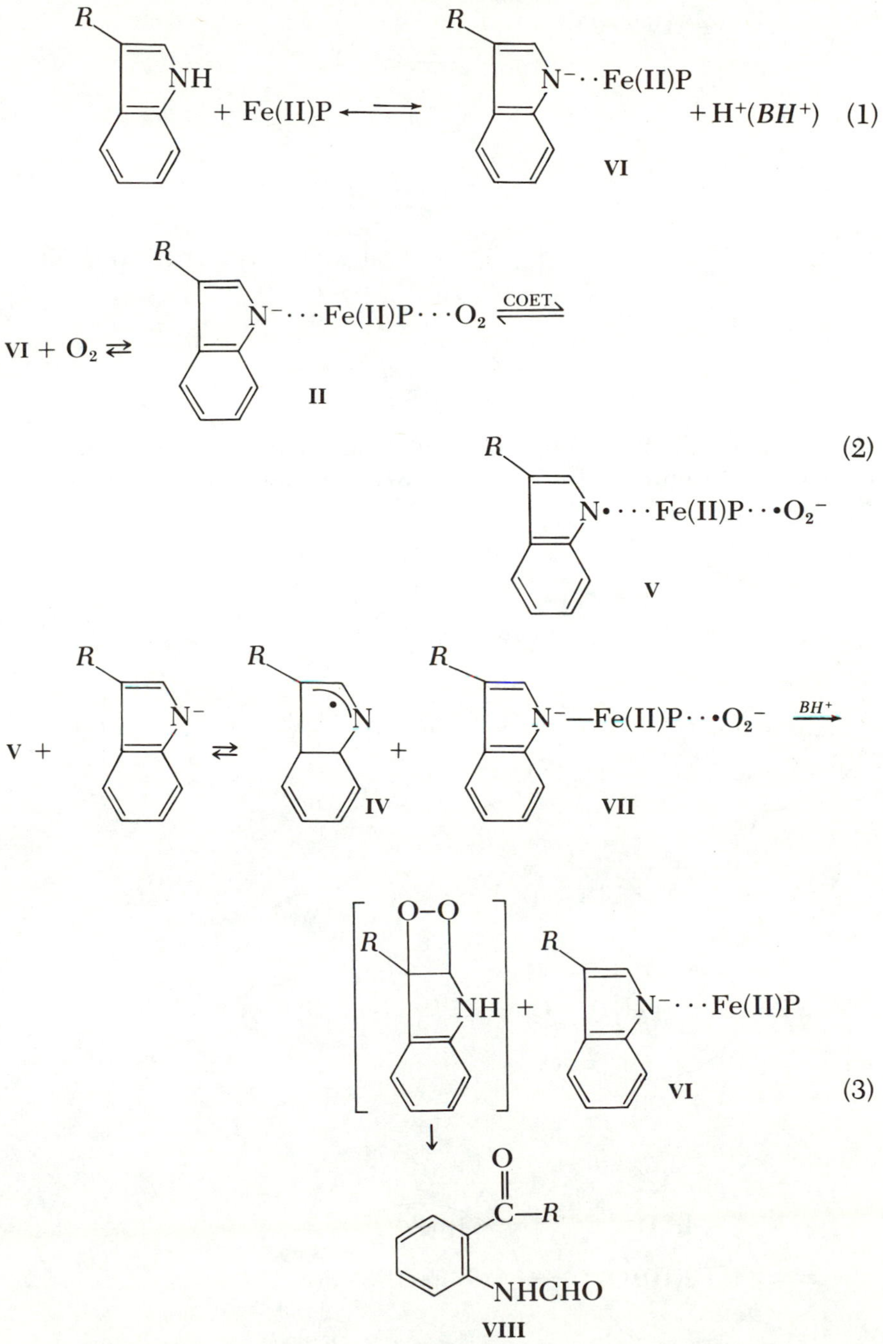
R
NH
+ Fe(II)P
N⁻· ·Fe(II)P
+H⁺(BH⁺)
(1)
VI
VI + O₂ ⇄
N⁻· · ·Fe(II)P· · ·O₂
COET
II
(2)
N• · · ·Fe(II)P· · •O₂⁻
V
V +
N⁻
N
+
N⁻—Fe(II)P· · •O₂⁻
BH⁺
IV
VII
O–O
NH
+
N⁻· · ·Fe(II)P
VI
(3)
O
C—R
NHCHO
VIII

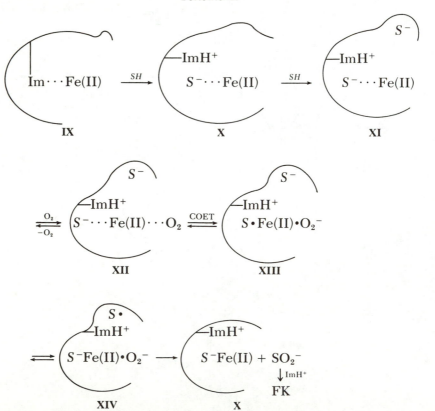

$$\text{V}$$

$$(6)$$

shown in Scheme 2, taking into account the characteristics of the TPO reaction and Figure 1. In Scheme 2, *SH* represents L-tryptophan and **IX** represents TPO.

Scheme 2.

First, the *SH* is incorporated into the activity control site to form **X** (corresponding to **VI**); a second incorporation of *SH* should occur at the binding site (recognition). The TPO reaction should proceed by the subsequent reactions (**XI** → **XII** → **XIII** → **XIV** → **X**) and repetitions of these reactions. We would like to comment on the effect of Fe(III) in a *D*–Fe(III)P–*A* system. In this complex, COET does not take place, but homolytic fission between donor (*D*) and Fe(III) does. For example, although Ar^-–Fe(III)OEP in benzene is stable to prolonged heating in the absence of molecular oxygen, in the presence of O_2, this complex reacts rapidly with oxygen to form Ar—Ar(biaryl) and Fe(II)OEP–O_2, which finally becomes the μ-oxo dimer.

$$Ar^- \cdots \cdots Fe(III)P \cdots O_2 \rightarrow Ar\cdot + Fe(II)OEP \cdots O_2 \qquad (7)$$
$$\downarrow \qquad\qquad \downarrow$$
$$Ar\text{---}Ar \quad Fe(III)OEP \cdots O_2^-$$
$$\downarrow$$
$$[Fe(III)OEP]_2O$$

μ-oxo dimer

Similar behavior also is observed for the RS^-–Fe(III)OEP complex, which forms RS—SR through contact with molecular oxygen. Such a reaction is expected for the skatole-anion–Fe(III)-porphyrin–O_2 system to form Fe(II)-porphyrin, which is useful in generating the active catalyst for the oxygenation of skatole. This was proved using Fe(III)TPP · OAc as the catalyst for oxygenation of skatole in benzene.

Oxygen Activation in Cytochrome P-450

Cytochrome P-450 found in bacteria and lever microsomes is a kind of hemoprotein that activates molecular oxygen to catalyze the hydroxylation of organic compounds in drug metabolism. For example, camphor is incorporated into cytochrome P-450 (forming P-450 cam) and then oxygenated to form 5-exo-hydroxycamphor as shown in Reaction 8 (25). The substrate specificity of this enzyme is not strict. How-

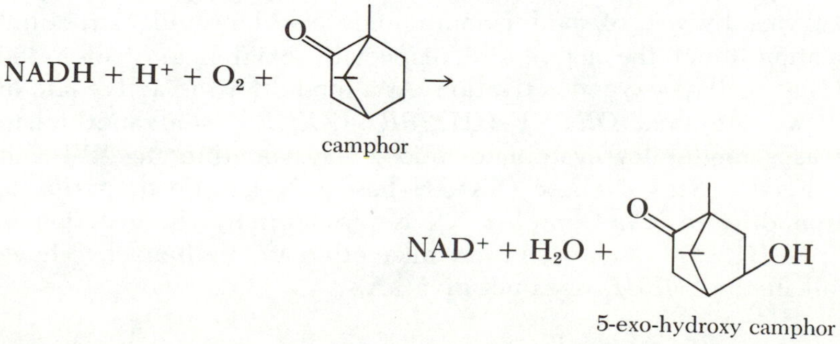

$$NADH + H^+ + O_2 +$$

camphor

(8)

$$NAD^+ + H_2O +$$

5-exo-hydroxy camphor

ever, regiospecific hydroxylation takes place. The reaction cycle of cytochrome P-450 cam, according to recent work (8, 9, 25), is shown in Figure 5, although the Fe–O bonding structure in **XIX** is still not resolved.

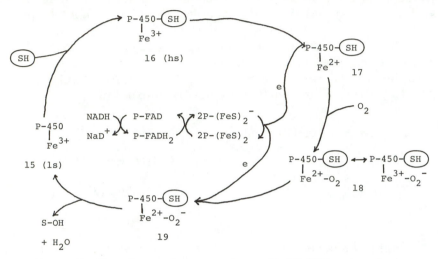

Figure 5.   Reaction cycle in cytochrome P-450 (SH = camphor)

As shown in Figure 5, the membrane enzyme P-450–Fe^{3+} (**XV**) is a low-spin (hexacoordinated) complex; through the incorporation of camphor as the substrate (SH), the spin state of P-450–$Fe^{3+} \cdot SH$ (**XVI**) becomes high-spin. Its one-electron reduction with reductase systems gives the reduced form, P-450–$Fe^{2+} \cdot SH$ (**XVII**), which is easily oxygenated with molecular oxygen to form P-450 · $Fe^{2+} \cdot O_2 \cdot SH$ (**XVIII**). Further one-electron reduction of **XVIII** should yield P-450–Fe^{2+}–O_2^-–SH (**XIX**), which should give the hydroxylated camphor (S—OH) and **XV**.

Since the x-ray structure of P-450 cam has not been determined, recent investigations have focussed on elucidating the structure of the active site of P-450. Simulation of the prothetic protoheme moiety in this enzyme by use of model compounds could provide structural information about the active site (especially axial ligand) of P-450 (26–32) as well as oxygen activation. As a model for the active site of P-450, we prepared OEP · Fe(III) · SR^- (**XX**, RS^- = alkanethiolate anion as a model for cysteinate anion) corresponding to **XVI** and OEP · Fe(III) · SR^- · N-base (**XXI**, N-base = N–Et · Im or pyridine) corresponding to **XV**. Complex **XX** is prepared by the reaction of OEP · Fe(III) · ClO_4 with sodium alkanethiolate in benzene, however alkanethiol (RSH) does not give **XX**.

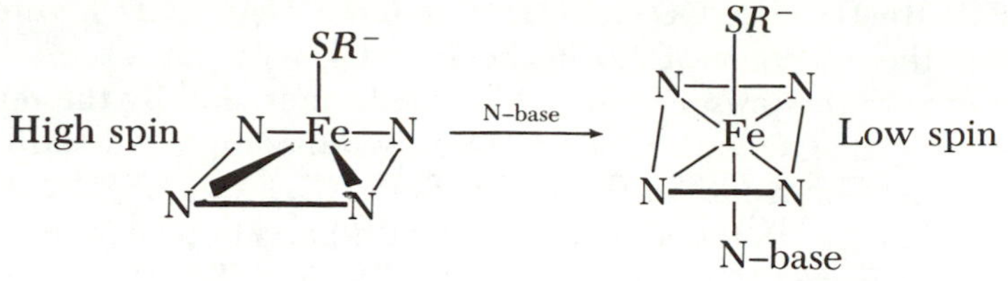

High spin   N—Fe—N $\xrightarrow{\text{N-base}}$ Low spin   (9)

SR^- ... SR^-

XX   XXI

The ESR parameters for **XX** and **XXI** are similar to those of the corresponding P-450, **XVI** and **XV**, respectively, as shown in Table III. The paramagnetic ^{1}H NMR chemical shifts for the pentacoordinated OEP·Fe(III)·L are shown in Table IV. The chemical shift of the mesoproton indicates that ^-SR ($R = t$-Bu) ligand, which is different than the other ligands, is a strong electron donor and that electron migration from ^-SR to the porphyrin ring is fairly large. Similar characteristics also have been observed for OEP·Ru(II)CO·SR. As models for the reduced P-450 carbonyl adduct, OEP·Fe(II)·SR·CO (**XXII**)

Table III.   Comparison of ESR Parameters of P-450 cam with Model Complexes at − 196°C

Complex	Spin State	g-values	Reference
P-450 cam	5/2	8, 4, 1.8	(8)
P-450 cam	1/2	2.45, 2.26, 1.91	(8)
TPPFe·SC$_6$H$_5$	5/2	8.6, 3.4, —	(26)
TPPFe·SC$_6$H$_5$ + N—MeIm	1/2	2.39, 2.26, 1.93	(26)
OEPFe·SC$_6$H$_5$	5/2	7.2, 4.7, 1.9	(27)
OEPFe·SC$_6$H$_5$	5/2	8.72, 7.22, 5.31, 2.20	this work
OEPFe·SC$_6$H$_5$ + α-picoline	1/2	2.46, 2.32, 1.91	this work
OEPFe·SC(CH$_3$)$_3$	5/2	9.42, 7.25, 4.08, 3.18, 2.43	this work
OEPFe·SC(CH$_3$)$_3$ + α-picoline	1/2	2.37, 2.26, 1.94	this work

Table IV. ^{1}HNMR Chemical Shifts (ppm from TMS) of OEPFe·L at 21°C

L	Solvent	α-CH$_2$	β-CH$_3$	meso-H
Cl	CDCl$_3$	−44.51, −40.87	−6.91	57.40
OMe	CDCl$_3$	−44.70, −41.02	−6.96	56.51
OC(CH$_3$)$_3$	CDCl$_3$	−44.81, −41.02	−7.07	56.83
SC$_6$H$_5$	CDCl$_3$	−44.50, −40.83	−6.76	57.01
SC(CH$_3$)$_3$	C$_6$D$_6$	−35.84, −33.95	−5.62	46.40
SC(CH$_3$)$_3$	CDCl$_3$	−35.82	−5.58	—

and OEP · Ru(II) · SR · CO (**XXIII**) ($R = CH_2CH_2CO_2CH_3$) were prepared by the reaction of OEP · Fe(III) · Cl with an excess of alkanethiolate in the presence of CO for the former, and by the reaction of OEP · Ru(II) · CO with dibenzo-18-crown-6-complexed sodium alkanethiolate in benzene for the latter. Both **XXII** and **XXIII** exhibit a split Soret band with the longer wavelength band appearing near 450 nm. These spectral data indicate that **XXII** and **XXIII** are feasible models for the reduced cytochrome P-450-carbonyl adduct. To discuss the oxygen-activation mechanism in the P-450 system, it is necessary to carry out the model reaction using our P-450 model. However, because of difficulties developing a suitable reduction system for the model reaction, we have not determined the mode of reaction for the **XVIII** → **XIX** process. As another approach to the oxygen-activation problem in P-450, we studied the COET process (Reaction 10), using our P-450 model.

$$R—S^- \cdots Fe(II)P \cdots O_2 \xrightarrow{\text{COET}} RS \cdot \cdots Fe(II)P \cdots \cdot O_2^- \qquad (10)$$

Contact of the RS^-–Fe(II) · OEP solution with molecular oxygen immediately produces $RS—SR$, although sodium alkanethiolate (RSNa) under same condition does not rapidly give $RS—SR$. This indicates that the effective electron transfer through COET takes place in the RS^-–Fe(II)OEP–O_2 system. To get spectroscopic evidence for COET, we have examined the electron transfer from RS^- (strong donor) to carbonyl (weak acceptor, $EA_{CO} = -1.75$ eV, cf., $EA_{O_2} = 0.43$ eV, $EA_{I_2} = 2.6$ eV, $EA_{N_2} = -2$ eV) in the stable RS^-–Fe(II) · OEP–CO complex (**XXII**) and RS^-–Ru(II)OEP–CO complex (**XXIII**), instead of in the extremely unstable RS^-–Fe(II)–OEP–O_2 complex. The observed carbonyl stretching vibrations of Ru(II)OEP · CO · L and Fe(II)OEP · CO · L are shown in Table V.

Table V. Carbonyl Stretching Vibrations in Two Stable Complexes

Complex	Solvent	L	ν_{co} (cm^{-1})
Ru(II)OEP·CO·L	Ether	PPh$_3$	1966
Ru(II)OEP·CO·L	Ether	none	1942
Ru(II)OEP·CO·L	THF	none	1938
Ru(II)OEP·CO·L	Pyridine	none	1931
Ru(II)OEP·CO·L	Benzene	none	1946
Ru(II)OEP·CO·L	Benzene	^-SR	1917
Ru(II)OEP·CO·L	DMF	none	1917
Ru(II)OEP·CO·L	DMF	^-SR	1889
Fe(II)OEP·CO·L	Benzene	none	1975
Fe(II)OEP·CO·L	Benzene	^-SR	1950
Fe(II)OEP·CO·L	DMF	none	1948
Fe(II)OEP·CO·L	DMF	^-SR	1923

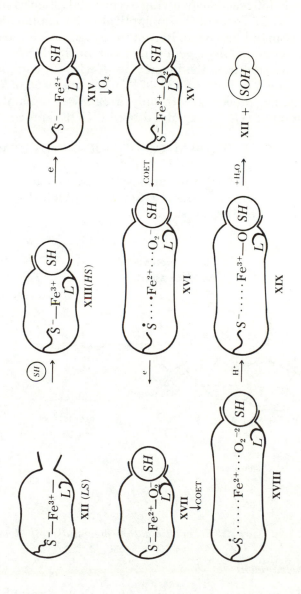

Scheme 3.

From this table, it is evident that alkanethiolate anion reduces the wavenumber of the carbonyl stretching by 28–29 cm^{-1} for the Ru(II) complex, and by 25 cm^{-1} for the Fe(II) complex, which clearly indicates COET charge transfer. Therefore, one can describe the cytochrome P-450 reaction involving oxygen activation as shown in Scheme 3. The COET concept is also applicable for explanation of the action mechanism of bleomycin. Recent investigations indicate that the anticancer action of bleomycin originates in the DNA breakdown by a $\cdot O_2^-$ generated from the bleomycin–Fe(II)–O$_2$ complex whose structure is shown in Figure 6 (33–37). In the presence of a SH compound, this reaction proceeds with only a catalytic amount of Fe(II). The generation of $\cdot O_2^-$ is explained by the COET process:

$$(RS^--Fe(II)\cdot BLM-O_2 \xrightarrow{\text{COET}} RS\cdot-Fe(II)BLM-\cdot O_2^-).$$

Thus, COET is important as the oxygen-activation process not only in heme enzymes but also in other biological systems such as the bleomycin–Fe(II)–O$_2$ system.

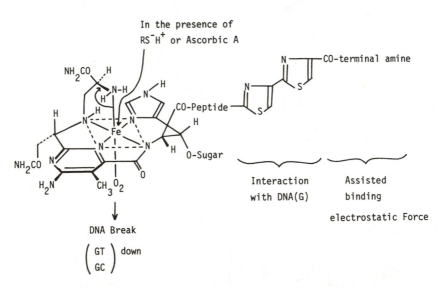

Figure 6.   Schematic action mechanism of bleomycin

Acknowledgment

This research was supported in part by the grant from the Ministry of Education, Japan.

Literature Cited

1. Mason, H. S.; Fowlks, W. L.; Peters, E. *J. Am. Chem. Soc.* **1955**, *77*, 2914.
2. Hayaishi, O.; Katagiri, M.; Rothberg, S. *J. Am. Chem. Soc.* **1955**, *77*, 5450.
3. "Oxygenases"; Hayaishi, O., Ed.; Academic: New York, 1962; Chapter 1.
4. Mason, H. S. *Ann. Rev. Biochem.* **1965**, *34*, 595.
5. Hayaishi, O. *Ann. Rev. Biochem.* **1969**, *38*, 21.
6. "Oxygenases"; Hayaishi, O., Nozaki, M., Eds.; Tokyo University Press: Tokyo, 1973; Chapter 1.
7. "Molecular Mechanisms of Oxygen Activation"; Hayaishi, O., Ed.; Academic: New York, 1974; Chapter 1.
8. Gunsalus, I. C.; Pederson, T. C.; Sligar, S. G. *Ann. Rev. Biochem.* **1975**, *44*, 377.
9. "Chemistry of Hemoproteins"; Watari, H., Ogoshi, H., Iizuka, T., Ed.; Kagakudojin: Kyoto, 1978; Chap. 6.
10. Ishimura, Y.; Nozaki, M.; Hayaishi, O.; Tamura, M.; Yamazaki, I. *J. Biol. Chem.* **1967**, *242*, 2574.
11. Hardmann, K. O.; Eylar, E. H. K. O.; Gurd., F. R. N. *J. Biol. Chem.* **1966**, *241*, 432.
12. Love, W. E.; Klock, P. A.; Lattman, E. E.; Padlan, E. A.; Ward, K. B., Jr.; Hendrickson, W. A. *Cold Spring Harbor Symp. Quant. Biol.* **1971**, *36*, 349.
13. Ogoshi, H.; Kawabe, K.; Yoshida, Z.; Imai, K.; Tyuma, I., unpublished data.
14. Nishinaga, A. *Chem. Lett.* **1975**, 273.
15. Dufour-Ricroch, M. N.; Gaudemer, A. *Tetrahedron Lett.* **1976**, 4079.
16. Uchida, K.; Soma, M.; Naito, S.; Onishi, T.; Tamaru, K. *Chem. Lett.* **1978**, 471.
17. Sugimoto, H., Ph.D. Thesis, Kyoto University, 1979.
18. Evans, N. A. *Aust. J. Chem.* **1971**, *24*, 1971.
19. Saito, I.; Imuta, M.; Matsuura, T. *Chem. Lett.* **1972**, 1173.
20. Saito, I.; Imuta, M.; Matsuura, T. *Synthesis* **1976**, 255.
21. Saito, I.; Matsuura, T.; Nakagawa, M.; Hino, T. *Acc. Chem. Res.* **1977**, *10*, 346.
22. Sundberg, R. J. "The Chemistry of Indoles"; Academic: New York, **1970**; Chap. 5.
23. Peover, M. E.; White, B. S. *Electrochim. Acta.* **1966**, *11*, 1061.
24. Samuni, A.; Neta, P. *J. Phys. Chem.* **1973**, *77*, 1629.
25. Cooper, D. Y.; Rosenthal, O.; Snyder, R.; Witmar, C. "Cytochrome P-450 and b_5"; Plenum: New York, 1975; Section II, Part 2.
26. Collman, J. P.; Sorell, T. N.; Hoffman, B. M. *J. Am. Chem. Soc.* **1975**, *97*, 913.
27. Koch, S.; Tang, S. C.; Holm, R. H.; Frankel, R. B.; Ibers, J. A. *J. Am. Chem. Soc.* **1975**, *97*, 916.
28. Ogoshi, H.; Sugimoto, H.; Yoshida, Z. *Tetrahedron Lett.* **1975**, 2285.
29. Dawson, J. H.; Trudell, J. R.; Barth, G.; Linder, R. E.; Bunnenberg, E.; Djerassi, C.; Gouterman, M.; Connell, C. R.; Sayer, P. *J. Am. Chem. Soc.* **1977**, *99*, 641.
30. Dawson, J. H.; Holm, R. H.; Trudell, J. R.; Barth, G.; Linder, R. E.; Bunnerberg, E.; Djerassi, C.; Tang, S. C. *J. Am. Chem. Soc.* **1976**, *98*, 3708–3709.
31. Chang, C. K.; Dolphin, D. *J. Am. Chem. Soc.* **1976**, *98*, 1607.
32. Ogoshi, H.; Sugimoto, H.; Yoshida, Z. *Bull. Chem. Soc. Jpn.* **1978**, *51*, 2369.
33. Sausville, E. A.; Peisach, J.; Horwitz, S. B. *Biochem. Biophys. Res. Commun.* **1976**, *73*, 814.
34. Sugiura, Y. *J. Antibiot.* **1978**, *31*, 1206, 1310.

35. Chien, M.; Grollman, A. P.; Horwitz, S. B. *Biochemistry* **1977**, *16*, 3641.
36. Kasai, H.; Nagasawa, H.; Takita, T.; Umezawa, H. *J. Antibiot.* **1978**, *31*, 1316.
37. D'Andrea, A. D.; Haseltine, W. A. *Proc. Nat. Acad. U.S.A.* **1978**, *75*, 3608.

RECEIVED May 15, 1979.

Intermediate-Spin States of Iron Porphyrins

HIROSHI KOBAYASHI and YOUKOH KAIZU

Department of Chemistry, Tokyo Institute of Technology, Ookayama, Meguro-ku, Tokyo 152, Japan

KEN EGUCHI

Department of Chemistry, University of California, Davis, CA 95616

The observed magnetic properties of the unligated Fe(II) porphyrins were reproduced by theoretical calculations taking into account all possible configuration interactions and spin–orbit coupling interactions. The calculated results were in good agreement with the observations only when it was assumed that the axial ligand field is so weak that the d_{z^2} orbital is close to the d_π and d_{xy} orbitals. The low-lying states are $^3A_2$, ^3E, ^5E, $^5B_2$ and $^5A_1$. The ground state has two sublevels with eigenvectors that are quantum mechanical admixtures of triplet states ($^3A_2$, ^3E) and quintet states ($^5A_1$). The Mössbauer quadrupole splitting should be attributed to the less temperature-dependent bonding-orbital contribution, however, the d-orbital contribution is almost constant for the ground sublevels and thus a change in the Boltzmann distribution between the sublevels yields no appreciable change in the quadrupole splitting.

The stereochemistry and functions of all iron porphyrin-containing proteins can be attributed to the varied electronic structure of iron for the oxidation and spin states that are stable in physiological environments. Theoretical descriptions of the electronic structure of iron should be, in principle, applicable to the understanding of structure–function relationships in hemeproteins.

Hydroxides and azides of ferrimyoglobin and ferrihemoglobin exhibit the intermediate values of magnetic moment between $S = \frac{1}{2}$

0-8412-0514-0/80/33-191-327$05.00/0
© 1980 American Chemical Society

and $S = \frac{5}{2}$. George, Beetlestone, and Griffith interpreted that it is near the crossover point between $S = \frac{1}{2}$ and $S = \frac{5}{2}$ and is a thermal mixture of these two spin states (1). Later, more evidence of the thermal equilibria was accumulated with ferric hemeproteins (2–6). Griffith, however, pointed out the possible existence of an intermediate-spin ground state in some ferric porphyrins (7). Ogoshi et al. (8) found an intermediate value of magnetic moment between those expected for $S = \frac{1}{2}$ and $S = \frac{5}{2}$ with their prepared octaethylporphinatoiron(III) perchlorate [OEPFeClO$_4$] and interpreted it in terms of thermal equilibrium between $S = \frac{1}{2}$ and $S = \frac{5}{2}$. Dolphin et al. (9), however, concluded the $S = \frac{3}{2}$ ground state of OEPFeClO$_4$ and the absence of any thermally induced spin crossover on the basis of the temperature-independent Mössbauer quadrupole splitting being much larger than those found for $S = \frac{1}{2}$ and $S = \frac{5}{2}$ Fe(III) porphyrins.

The magnetic susceptibility measurements and Mössbauer spectra of Fe(II) phthalocyanine [PcFe(II)] provided evidence of the ferrous intermediate-spin state (10, 11, 12). The Fe(II) ion is incorporated into the central hole of a virtually pure square-planar structure (13). Stillman and Thomson (14) assigned the ground state of Fe(II) phthalocyanine to a $^3A_{1g}$ state among possible intermediate-spin states $^3A_{1g}$, $^3B_{2g}$, ^3E_g on the basis of group-theoretical arguments based on the sign of the MCD observed in the charge-transfer band lower in energy than the lowest ring (π, π^*) band. The existence of the intermediate-spin state of ferrous porphyrin had been predicted (15). However, the recent isolation (16) and x-ray structural determination (17) of unligated tetraphenylporphinatoiron(II) [TPPFe(II)] are convincing evidence for the triplet ground state, since the complex is in a square-planar structure with a short Fe–N bond distance (1.972 Å), which argues strongly against occupation of the σ-antibonding $d_{x^2-y^2}$ orbital. Mössbauer spectra also indicate that unligated Fe(II) porphyrins are in a spin state other than the well characterized low-spin and high-spin states (17, 18). Other direct evidence of the intermediate-spin state was given by NMR studies (19). By the dominant dipolar shifts observed for the phenyl protons in TPPFe(II), an axial magnetic anisotropy with $\mu_\perp = 4.9\mu_B$ and $\mu_\parallel = 3.2\mu_B$ was concluded. The room-temperature susceptibility measurements of PcFe(II), TPPFe(II), and OEPFe(II) exhibit the effective magnetic moments of $3.7\mu_B$ (10, 12), $4.4\mu_B$(17), $4.7\mu_B$(18), respectively, while tetra($\alpha,\alpha,\alpha,\alpha$-orthopivalamide) phenylporphinatoiron(II) ["picket-fence" Fe(II) porphin] and octa-methyltetrabenzporphinatoiron(II) [OTBFe(II)] show the magnetic moments of $5.0\mu_B$ (19) and $5.9\mu_B$, respectively (20). The ground state of PcFe(II), TPPFe(II), and OEPFe(II) have been assigned to the intermediate-spin ground state, and those of the "picket-fence" Fe(II) porphin and OTBFe(II) to the high spin state. The spin-only values, however, are $2.8\mu_B$ and $4.9\mu_B$ for $S = 1$ and $S = 2$, respectively. Thus the observed values for the effective magnetic moment such as 4.4–$4.7\mu_B$ cannot be attributed necessarily to the $S = 1$ ground state. The spin–orbit coupling within iron gives rise to sublevels and admixtures of the ground state and the low-lying excited states with different spin states. The structure of the ground state is not described as simply in terms of a single intermediate-spin state. The observed temperature dependence of magnetic susceptibility is well repro-duced only by a Boltzmann distribution between sizable numbers of the low-lying sublevels that are different in the contributions of orbital angular moments and spin angular moments.

The intermediate-spin state has been suggested both for ferrous (*21*) and ferric hemeproteins (*22, 23*). Direct participation of the intermediate-spin iron porphyrins in biological processes might be less probable; however, any reasonable theoretical account of the var-ied electronic structure of iron in hemeproteins should explain the fact that the so-called intermediate-spin iron porphyrin can exist only when the axial ligand field is extremely weak.

Calculations of Ligand-Field Energy

An axially symmetrical σ-donor ligand is placed on the z axis di-recting its lone pair to the origin of the coordinate. The energy of the d orbitals of the iron ion fixed at the coordinate origin is given by the electrostatic energy term E_i and the antibonding energy term X as follows:

$$\epsilon(z^2) = X + E_0 \quad \epsilon(yz) = \epsilon(zx) = E_1 \quad \epsilon(x^2 - y^2) = \epsilon(xy) = E_2$$

Among the d orbitals, the d_{z^2} orbital is the only orbital capable of over-lapping with the σ-donor orbital on the z axis and thus is given the character of an antibonding orbital by the delocalization of ligand σ-donor electrons. Here the electrostatic energy terms correspond to the classical crystal-field energy terms. E_0, E_1, and E_2 are not necessar-ily small but much greater than X. However, the electrostatic energy terms are less angular-dependent and thus are assumed as $E_0 = E_1 = E_2 = E$.

The ligand-field splitting of the d orbitals in an octahedral ligand field is described in terms of E's and X (*24, 25*).

$$E(x^2 - y^2) = E(z^2) = 3X + 3E_0 + 3E_2 = 3X + 6E$$

$$E(xy) = E(yz) = E(zx) = 4E_1 + 2E_2 = 6E$$

$$\Delta = 3X + [3E_0 - 4E_1 + E_2] = 3X$$

This indicates that the ligand-field splitting, Δ, is attributable mainly to the delocalization effect of ligand σ-donor electrons. Similarly, the energy of the d orbitals in a tetragonal planar ligand field is given by the same parameters.

$$E(x^2 - y^2) = 3X + 4E \quad E(z^2) = X + 4E \quad E(xy) = E(yz) = E(zx) = 4E$$

However, it should be noted that the a_{1g} linear combination of σ-donor orbitals interacts not only with the d_{z^2} orbital but also with the $4s$ orbital. This results in a hybridization of the d_{z^2} and s orbitals and grants an extra stabilization to the lower component orbital. The extra stabilization energy is given approximately by

$$-\frac{\{H(d, \sigma) - S(d, \sigma)H(d, d)\}^2\{H(s, \sigma) - S(s, \sigma)H(d, d)\}^2}{\{H(d, d) - H(\sigma, \sigma)\}^2\{H(s, s) - H(d, d)\}}$$

where $H(d, d)$, $H(s, s)$, $H(d, \sigma)$, and $H(s, \sigma)$ are Hamiltonian matrices and $S(d, \sigma)$ and $S(s, \sigma)$ are group-overlap integrals. The energy stabilization of the component orbital highly contributed to by the d_{z^2} orbital (z^2 orbital) is possible for a positive value of $\{H(s, s) - H(d, d)\}$. When the gap $\{H(s, s) - H(d, d)\}$ is increased, the stabilization is reduced. However $S(s, \sigma)$ usually is greater than $S(d, \sigma)$ and therefore $\{H(s, \sigma) - S(s, \sigma)H(d, d)\}^2$ becomes an appreciable value. Thus, the extra stabilization of the z^2 orbital is quite sizable in a planar complex even for a higher value of the d–s gap.

In conclusion, the energy levels of the d-orbitals in a metalporphyrin are parametrized including the π-antibonding effect as follows:

$$E(x^2 - y^2) = \Delta$$

$$E(z^2) = \Delta - \epsilon_\sigma$$

$$E(yz) = E(zx) = \epsilon_\pi$$

$$E(xy) = 0,$$

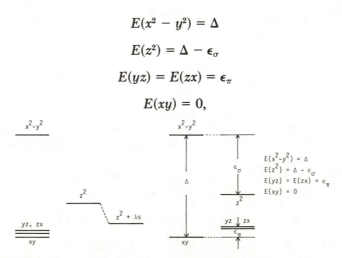

Figure 1.   The energy level of d orbitals in a metal porphyrin

where $\Delta = 3X$, the extra stabilization of the z^2 orbital in a tetragonal field is described by an increase in ϵ_σ and, thus, $\epsilon_\sigma > 2X$ in a tetragonal planar ligand field. Figure 1 shows the schematic energy level of the d-orbitals.

The d^6 and d^5 Systems of Iron Porphyrins

The possible lowest configurations of the d^6 and d^5 systems are illustrated in Figures 2 and 3. The relative energies of the lowest configurations are given approximately in terms of the energies of

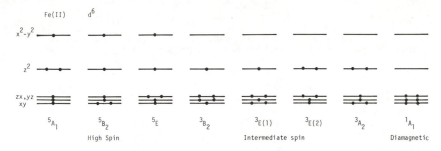

Figure 2. *The lowest configurations of an Fe(II) d^6 ion in a porphyrin*

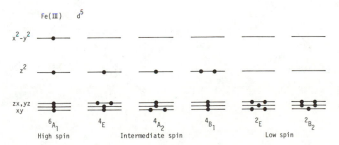

Figure 3. *The lowest configurations of an Fe(III) d^5 ion in a porphyrin*

orbital-energy promotion and electronic-pairing promotion (Table I). Assuming an empirical relationship of Racah's electrostatic interaction parameters, $C = 4B$ (26), and a reasonable value for Fe(II) porphyrin, $\Delta = 25B$ (15, 27), a Tanabe–Sugano diagram is drawn as shown in Figure 4, which indicates that the ground state is different from diamagnetic $^1A_1$, through the high-spin state, $^5B_2$, ^5E, and then $^5A_1$ when the axial ligand field ($\Delta - \epsilon_\sigma$) is decreased. The zero level in Figure 4 indicates $^5B_2$ and ^5E for $\epsilon_\pi = 0$. For a sizable π-type interaction, however, $^5B_2$ and ^5E split. Occupation of the $x^2 - y^2$ orbital displaces the iron out of the porphyrin plane, which reduces the orbital-energy promotion by ϵ (Table 1). Even if it happens, however, the

Table I. The Promotion Energies of the Lowest Configurations of the d^6-System

	Orbital-Energy Promotion[a]	Electronic-Pairing Promotion ($C = 4B$)
$^5A_1$	$3\Delta - \epsilon - 2\epsilon_\sigma$	0
$^5E, {}^5B_2$	$2\Delta - \epsilon - \epsilon_\sigma$	0
$^3B_2$	$\Delta - \epsilon_\sigma$	$33B$
$^3E(1)$	$\Delta - \epsilon_\sigma$	$27B$
$^3E(2)$	$2\Delta - 2\epsilon_\sigma$	$41B$
$^3A_2$	$2\Delta - 2\epsilon_\sigma$	$26B$
$^1A_1$	0	$37B$

[a] ϵ denotes an extra stabilization attributable to a displacement of iron from the porphyrin plane, which occurs with occupation of the $x^2 - y^2$ orbital.

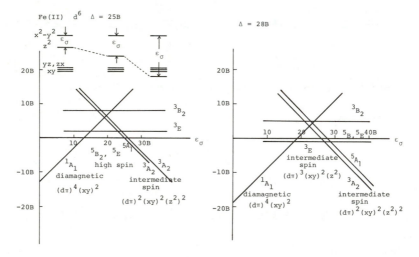

Figure 4.   The lowest states of an Fe(II) d^6 ion in a porphyrin as a function of the axial ligand field ($\Delta - \epsilon_\sigma$)

variable region of ϵ_σ for the high-spin states $^5B_2$, ^5E, and $^5A_1$ extends only slightly. For a higher value of $\Delta(28B$, for example), the intermediate-spin states ^3E and $^3A_2$ can be the ground state, depending upon the value of ϵ_σ (Figure 4). In this situation, however, $^5B_2$, ^5E, and $^5A_1$ cannot be lower than the intermediate-spin states for other ϵ_σ's unless the displacement mechanism of iron works drastically.

Since Fe(II) porphyrins have been isolated as high-spin species and low-spin (diamagnetic) species (*17, 19*), depending upon the axial ligands, Δ should be around $25B$. For $\Delta = 25B$, however, the $^5A_1$ state comes below the $^3A_2$ state. The spin–orbit coupling interaction be-

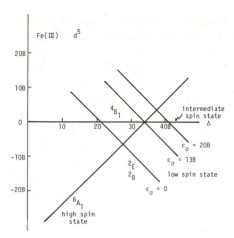

Figure 5. The lowest states of an Fe(III) d^5 ion in a porphyrin as a function of the ligand-field splitting parameter Δ

tween the $^3A_2$ and $^5A_1$ states results in admixture states, which are assigned to the so-called intermediate-spin state. This quantum mechanical admixture is discussed later.

A Tanabe–Sugano diagram is drawn also for the d^5 system in Fe(III) porphyrins. The ground state of the d^5 system is constantly $^6A_1$, regardless of the ϵ_σ values for even reasonable values of Δ in Fe(II) porphyrin. For $\Delta = 30B$, the ground state is the low-spin state ^2E or $^2B_2$ for a stronger axial ligand field ($\Delta - \epsilon_\sigma > 25B$). In fact, the low-spin Fe(III) porphyrin exists for the strong axial ligand field (28, 29). In this situation, however, the intermediate-spin state $^4B_1$ is constantly higher than the $^6A_1$ state, regardless of the axial ligand field. For a value $\Delta > 34B$, the intermediate-spin state can be the ground state of the complex. However, the low-spin states ^2E and $^2B_2$ are lower than the $^4B_1$ state, if the axial ligand field is stronger ($\Delta - \epsilon_\sigma > \Delta - 13B$). For a weak axial ligand field ($\Delta - \epsilon_\sigma < \Delta - 13B$), the intermediate-spin state is possible only for a higher value of the ligand–field splitting parameter, $\Delta > 34B$ (Figure 5). This higher value of Δ, however, may be expected only for a tripositive ion coordinated with a small number of ligands unless the coordination binding is strong, such as in diamagnetic Co(III) and its $4d$ and $5d$ analogues. Dolphin, et al. (9) claimed that the ground state of OEPFeClO$_4$ is the ferric intermediate-spin state.

Electronic Functions of Fe(II) Ion in Fe(II) Porphyrins

The nature of an Fe(II) ion incorporated in a porphyrin ring critically depends upon the spin state, which is controlled by the axial ligands. When the axial ligand field $(\Delta - \epsilon_\sigma)$ is decreased, the ground state of the complex is changed from $^1A_1$ to $^5B_2$. Since occupation of the $x^2 - y^2$ orbital occurs in $^5B_2$ state, the iron is shifted from the porphyrin plane. In the high-spin state, the z^2 orbital is occupied also and, thus, the binding of the axial ligands is destabilized. When destabilization is introduced in the binding of one of the axial ligands of an $^1A_1$ complex to an extent to bring out the spin change ($^1A_1 \rightarrow {}^5B_2$), the occupation of the z^2 orbital destabilizes or dissociates the binding of the other axial ligand. On the other hand, when Fe(II) in a $^5B_2$ state is shifted to near the crossover point by coordination of an axial ligand, the Fe(II) ion increases its affinity toward another axial ligand and changes spin state. Finally, the Fe(II) ion is in the porphyrin plane. The process involving spin-state change results in rather catastrophic changes of stereochemistry and functions. However, the process is not forbidden by the selection rule, since mixing by spin–orbit coupling within iron is enhanced appreciably near the crossover point.

Table II.   The Ionization Potentials of $d\pi$-Orbitals

State	Configuration	Energy[a] (eV)	
$^5A_1$	$(xy)(d\pi)^2(z^2)^2(x^2 - y^2)$	$\phi + 8B$	11.1
$^5B_2$	$(xy)^2(d\pi)^2(z^2)(x^2 - y^2)$	$\phi + 12B$	11.5
^5E	$(xy)(d\pi)^3(z^2)(x^2 - y^2)$	$\phi - 28B$	7.5
$^3B_2$	$(xy)(d\pi)^4(z^2)$	$\phi - 21B$	8.2
$^3E(1)$	$(xy)^2(d\pi)^3(z^2)$	$\phi - 21B$	8.2
$^3E(2)$	$(xy)(d\pi)^3(z^2)^2$	$\phi - 27B$	7.6
$^3A_2$	$(xy)^2(d\pi)^2(z^2)^2$	$\phi + B$	10.4
$^1A_1$	$(xy)^2(d\pi)^4$	$\phi - 10B$	9.3

[a] The energy was obtained by the method of Griffith (*30*); $\phi = 10.3\ ev$, $B = 0.1$ ev.

Table II shows the lowest ionization potentials of the $d\pi$ orbitals evaluated by the method proposed by Griffith (*30*). Fe(II) in a $^5B_2$ state is less $d\pi$ donating than Fe(II) in an $^1A_1$ state. Thus, the outer-sphere electron transfer through porphyrin is less possible for Fe(II) in the $^5B_2$ and $^5A_1$ states than for Fe(II) in the $^1A_1$ state. Fe(II) in the $^3A_2$ state is less $d\pi$ donating among the intermediate-spin states. Figure 6 shows the absorption and MCD spectra of TPPFe(Im)$_2$ and TPPFe(Py)$_2$ (Im, imidazole; Py, pyridine). The absorption spectra are very similar for both complexes; however, the sign of the MCD spectra in the Soret region is reversed. In the Soret region, the "metal to porphyrin" charge-transfer transition and the real $B(\pi, \pi^*)$ transition form configuration-interaction admixtures. The configuration interaction becomes sizable only when the $d\pi$ electrons migrate into the porphy-

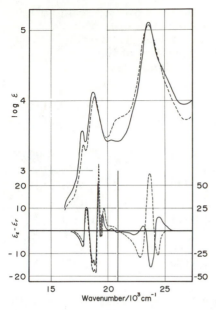

Figure 6. Absorption and MCD spectra: (———) TPPFe(Im)$_2$ in dioxane containing excess imidazole; (- - - - -) TPPFe(II)(Py)$_2$ in pyridine. MCD was measured under the magnetic field of 10,000 Gauss.

rin antibonding π^* ($4e_g$) orbital in the ground state and grants the absorption intensity of the virtual but allowed transition ($4e_g \rightarrow 2b_{1u}$) to the charge-transfer transition. TPPFe(Im)$_2$ exhibits the component transition with higher CT character in lower wavenumbers than the component transition with higher (π, π^*) character. On the other hand, the ordering is reversed in TPPFe(Py)$_2$. However, the energy of the "metal to porphyrin" charge-transfer configuration ($d\pi \rightarrow 2b_{1u}$) is estimated roughly as 3.0 ev (24000 cm^{-1}) for both complexes.

Table III shows the orbital energies of the highest filled (π) and the lowest vacant (π^*) MO's, the electrostatic stabilizations of "metal $d\pi$ to porphyrin (π^*)" and "porphyrin (π) to metal $d\pi$" charge-transfer configurations, and the charge-transfer transition energy as a function of the iron $d\pi$ ionization potential [$I_p(d\pi)$] or $d\pi$ electron affinity [$E_a(d\pi)$]. The calculation well reproduces the energy and oscillator strength of the lowest Q and B excitations. Assuming that the $d\pi$ ionization potential of the $^1A_1$ state is 9.3 ev, the calculated energy $I_p(d\pi)$ -6.3286 ev of the charge-transfer configuration ($d\pi \rightarrow 2b_{1u}$) is about 3 ev.

Figure 7 shows the electron-donating powers of the $d\pi$ and z^2 orbitals by the negative ionization potentials. The lowest spin (diamagnetic) $^1A_1$ state is more $d\pi$-donating than the high-spin states $^5B_2$ and $^5A_1$, while the $^3A_2$ and $^5A_1$ states are more z^2-donating than the $^5B_2$ state and are attractive for electron acceptors such as O$_2$. The $^5B_2$ and ^5E states are less z^2-donating but they increase in donating power as the axial ligand field ($\Delta - \epsilon_\sigma$) increases.

Table III. The Highest Filled (π) and the Lowest Vacant (π^*) Molecular Orbitals, the Electrostatic Energies of "Metal $d\pi$ to Porphyrin π^*" and "Porphyrin π to Metal $d\pi$" Charge Transfers and the Energies of the Charge-Transfer Transitions

	Orbital Energy[a] (ev)	Electrostatic Energy of Charge Transfer (ev)	Charge-Transfer Energy[b] (ev)
$3b_{2u}$	-0.6735	-5.1864	$I_p - 5.8600$
$2b_{1u}$	-2.4982	-3.8304	$I_p - 6.3286$
$4e_g$	-3.8621	-4.3843	$I_p - 8.2464$
$3a_{2u}$	-7.7134	-4.6805	$3.0329 - E_a$
$1a_{1u}$	-8.1737	-4.3911	$3.7826 - E_a$
$3e_g$	-9.7535	-4.2342	$5.5193 - E_a$
$2b_{2u}$	-9.7587	-4.3556	$5.4031 - E_a$
$2a_{2u}$	-9.7851	-4.5174	$5.2677 - E_a$

	Transition Energy (ev)	(nm)	Oscillator Strength
Q^c	2.0641	600.68	0.03230×2
B	3.1389	394.99	0.87818×2

[a] Pariser–Parr–Pople SCMO calculations assuming electroneutrality of the central metal ion.

[b] I_p and E_a denote $d\pi$ ionization potential and $d\pi$ electron affinity of the central iron, respectively.

[c] Q and B denote the lowest (π, π^*) excited states.

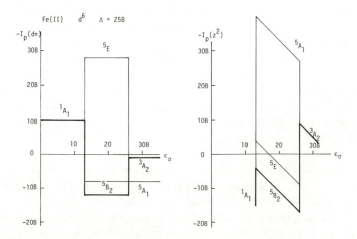

Figure 7. Electron-donating powers of $d\pi$ and z^2 orbitals as a function of the spin states

The Intermediate-Spin States of Fe(II) Porphyrin

One of the candidates for the unligated porphinatoiron(II) ground state is the $^3A_2$ state. Since the energies of the xy, $d\pi$, and z^2 orbitals are very close, the ^3E, ^5E, $^5B_2$ and $^5A_1$ states are also low-lying states as illustrated in Figure 8. In this situation, the configuration interactions

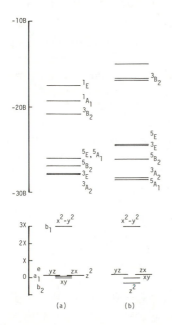

Figure 8. Calculated low-lying states of an Fe(II) d^6 ion in the unligated Fe(II) porphyrin: (a) $E(x^2 - y^2) = 3X = 26.6B$, $E(d\pi) = 0.1X$, $E(z^2) = 0.1X$, $E(xy) = 0$, $\delta = 0.4$ B, and $B = 800$ cm^{-1}; (b) $E(x^2 - y^2) = 3X = 24.5B$, $E(d\pi) = 0.2X$, $E(z^2) = -0.3X$, $E(xy) = 0$, $\delta = 0.4$ B, and $B = 800$ cm^{-1}

and spin–orbit coupling interactions play an essential role in the quantum-mechanical mixing of the low-lying states. Assuming $E(x^2 - y^2) = 3X = 26.6B$, $E(d\pi) = E(z^2) = 0.1X$, $E(xy) = 0$, the spin–orbit coupling constant $\zeta = 0.4B$, and $B = 800$ cm^{-1}, the configuration interactions were diagonalized taking into account all possible configurations up to 125, and then the spin–orbit coupling interactions were diagonalized, taking into account the low-lying spin–orbitals up to 40. From the eigenvectors thus obtained, g values were evaluated. For the calculations of magnetic susceptibility as a function of temperature, a Boltzmann distribution over all the states under 2000 cm^{-1} was taken into consideration. The calculated effective magnetic moment was $4.58\mu_B$ and anisotropic susceptibility was $\chi_\perp = 1.14 \times 10^{-2}$ erg/gauss^2 and $\chi_\parallel = 0.358 \times 10^{-2}$ erg/gauss^2 at 298 K. The calculated values were in good agreement with the values observed on TPPFe(II): $\mu_{eff} = 4.4\mu_B$ (17) and $\chi_\perp = 1 \times 10^{-2}$ erg/gauss^2 and $\chi_\parallel = 0.4 \times 10^{-2}$ erg/gauss^2 (19). The ground state has two sublevels with a gap of 83.72 cm^{-1}. Their eigenvectors are:

$-0.7307\ T(yz, zx;\ 0;\ 0) + 0.2975\ T(z^2,\ \pi;\ -1;\ 1) +0.2975\ T(z^2,\ \pi;\ 1;$
$-1) - 0.1792\ Q(x^2 - y^2,\ yz,\ zx,\ xy,\ 0;\ 0)$

and

$-0.6567\ \{T(yz, zx;\ 0;\ 1), T(yz, zx;\ 0;\ -1)\} +0.5729\ \{T(z^2,\ yz;\ y;\ 0),$
$T(z^2, zx;\ x;\ 0)\} \pm 0.2704\ \{T(z^2, zx;\ x;\ 0), T(z^2, yz;\ y;\ 0)\} \pm 0.2356\ \{T(yz,$
$zx;\ 0;\ -1), T(yz, zx;\ 0;\ 1)\} \mp 0.1791\ \{T(xy, zx;\ x;\ 0), T(xy, yz;\ y;\ 0)\}$
$-0.1546\ \{Q(x^2 - y^2,\ yz,\ zx,\ xy;\ 0;\ 1), Q(x^2 - y^2,\ yz,\ zx,\ xy;\ 0;\ -1)\}$
$-0.1246\ \{Q(x^2 - y^2,\ z^2,\ xy,\ yz;\ y;\ 0), Q(x^2 - y^2,\ z^2,\ xy,\ zx;\ x;\ 0)\}$

where T denotes a triplet state (orbitals occupied by unpaired holes; orbital angular momentum; spin momentum) and Q denotes a quintet state (orbitals occupied by unpaired holes; orbital angular momentum; spin momentum). The eigenvectors show the lowest states are the admixtures of low-lying triplets ($^3A_2$ and ^3E) and quintets ($^5A_1$). The high contribution of the $^3A_2$ state to the lowest ground state sublevels should be noted. We have tried calculations for a variety of energy parameters. The conclusions were not so different, if the orbital energies of the xy, $d\pi$, and z^2 orbitals were assumed to be very close. In the tetragonal ligand field of the unligated Fe(II) porphyrin, the hybridization of d_z2 and s orbitals significantly stabilizes the z^2 component orbital. This effect enhances the contribution of $^5A_1$ to the ground state. However, that Fe(II) ion in the unligated Fe(II) porphyrins is in the porphyrin plane indicates a higher contribution of $^3A_2$ than $^5A_1$.

The calculation assuming that $E(x^2 - y^2) = 24.5B$, $E(d\pi) = 0.2X$, $E(z^2) = -0.3X$, $E(xy) = 0$, $\zeta = 0.4B$, and $B = 800$ cm^{-1} gave $\mu_{\text{eff}} = 4.51\mu_B$, $\chi_\perp = 1.02 \times 10^{-2}$ erg/gauss^2 and $\chi_\parallel = 0.518 \times 10^{-2}$ erg/gauss^2 at 298 K. The eigenvectors of the lowest two sublevels with a gap of 63.75 cm^{-1} are as follows:

$0.6278\ T(yz, zx;\ 0;\ 0) + 0.7633\ Q(x^2 - y^2,\ yz,\ zx,\ xy;\ 0;\ 0)$

and

$-0.6091\{T(yz, zx;\ 0;\ 1), T(yz, zx;\ 0;\ -1)\} -0.7809\{Q(x^2 - y^2,\ yz,\ zx,\ xy$
$0;\ 1), Q(x^2 - y^2,\ yz,\ zx,\ xy;\ 0;\ -1)\}$

The ground state is contributed to highly by the quintet states ($^5A_1$). Even with the contributions of the $^5A_1$ and $^3A_2$ states, the calculated magnetic moment is still $4.5\mu_B$, which is coincident with the value for TPPFe(II) at room temperature (17). Figure 9 shows examples of calculations of magnetic susceptibility as a function of temperature. The temperature-dependent magnetic susceptibility of PcFe(II) (31) was well reproduced as a function of temperature. Here only two sublevels of the ground state are highly populated.

The quadrupole splitting observed in the Mössbauer spectra should be attributed to the less temperature-dependent, bonding-orbital contribution; however, the d-orbital contribution is almost con-

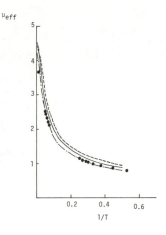

μ_eff

1/T

Figure 9. Calculated and observed magnetic susceptibilities of the unligated Fe(II) porphyrin as a function of temperature. Calculated: (———) $E(x^2 - y^2) = 3X = 26.6B$, $E(d\pi) = E(z^2) = 0.1X$, $E(xy) = 0$, $\zeta = 0.4B$, and $B = 800\ cm^{-1}$; (- - - - -) $E(x^2 - y^2) = 3X = 24.5B$, $E(d\pi) = 0.2X$, $E(z^2) = -0.3X$, $E(xy) = 0$, $\zeta = 0.4B$, and $B = 800\ cm^{-1}$; (- · — · -) $E(x^2 - y^2) = 3X = 25.6B$, $E(d\pi) = -0.1X$, $E(z^2) = -0.3X$, $E(xy) = 0$, $\zeta = 0.4B$, and $B = 800\ cm^{-1}$; (· · · · ·) the magnetic susceptibilities of Fe(II) phthalocyanine observed as a function of temperature (31).

stant for the lowest sublevels and thus a change in Boltzmann distribution upon temperature elevation between the admixture sublevels mainly comprised of the $^3A_2$, ^3E, and $^5A_1$ states results in no appreciable change of the quadrupole splitting.

Literature Cited

1. George, P.; Beetlestone, J.; Griffith, J. S. *Rev. Mod. Phys.* **1964**, *36*, 441.
2. Iizuka, T.; Kotani, M. *Biochim. Biophys. Acta* **1968**, *154*, 417.
3. Iizuka, T.; Kotani, M.; Yonetani, T. *Biochim. Biophys. Acta* **1968**, *167*, 257.
4. Iizuka, T.; Kotani, M. *Biochim. Biophys. Acta* **1969**, *181*, 275.
5. Iizuka, T.; Kotani, M. *Biochim. Biophys. Acta* **1969**, *194*, 351.
6. Iizuka, T.; Yonetani, T. *Adv. Biophys.* **1970**, *1*, 155.
7. Griffith, J. S. "The Theory of Transition–Metal Ions"; Cambridge University Press: Cambridge, 1961; p. 369.
8. Ogoshi, H.; Watanabe, E.; Yoshida, Z. *Chem. Lett.* **1973**, 989.
9. Dolphin, D. H.; Sams, J. R.; Tsin, T. B. *Inorg. Chem.* **1977**, *16*, 711.
10. Dale, B. W.; Williams, R. J. P.; Johnson, C. E.; Thorp, T. L. *J. Chem. Phys.* **1968**, *49*, 3441.
11. Dale, B. W.; Williams, R. J. P.; Edwards, P. R.; Johnson, C. E. *J. Chem. Phys.* **1968**, *49*, 3445.
12. Barraclough, C. G.; Martin, R. L.; Mitra, S.; Sherwood, R. C. *J. Chem. Phys.* **1970**, *53*, 1643.
13. Kirner, J. F.; Dow, W.; Scheidt, W. R. *Inorg. Chem.* **1976**, *15*, 1685.
14. Stillman, M. J.; Thomson, A. J. *J. Chem. Soc., Faraday Trans. 2* **1974**, *70*, 790.
15. Kobayashi, H.; Yanagawa, Y. *Bull. Chem. Soc. Jpn.* **1972**, *45*, 450.
16. Collman, J. P.; Reed, C. A. *J. Am. Chem. Soc.* **1973**, *95*, 2048.
17. Collman, J. P.; Hoard, J. L.; Kim, N.; Lang, G.; Reed, C. A. *J. Am. Chem. Soc.* **1975**, *97*, 2676.
18. Dolphin, D.; Sama, J. R.; Tsin, T. B.; Wong, K. L. *J. Am. Chem. Soc.* **1976**, *98*, 6970.
19. Collman, J. P.; Gagne, R. R.; Lang, G.; Halbert, T. R.; Reed, C. A.; Robinson, W. T. *J. Am. Chem. Soc.* **1975**, *97*, 1427.
20. Sams, J. R.; Tsin, T. B. *Chem. Phys. Lett.* **1974**, *25*, 599.
21. Maxwell, J. C.; Caughey, W. S. *Biochemistry* **1976**, *15*, 388.
22. Maltempo, M. M. *J. Chem. Phys.* **1974**, *61*, 2540.

23. Maltempo, M. M. *Q. Rev. Biophys.* **1976,** 9, 181.
24. Yamatera, H. *Bull. Chem. Soc. Jpn.* **1958,** 31, 95.
25. Schäffer, C. E.; Jørgensen, C. K. *Mol. Phys.* **1965,** 9, 401.
26. Tanabe, Y.; Sugano, S. *J. Phys. Soc. Jpn.* **1954,** 9, 766.
27. Zerner, M.; Gouterman, M.; Kobayashi, H. *Theor. Chim. Acta* **1966,** 6, 363.
28. Collins, D. M.; Countryman, R.; Hoard, J. L. *J. Am. Chem. Soc.* **1972,** 94, 2066.
29. Radonovich, L. J.; Bloom, A.; Hoard, J. L. *J. Am. Chem. Soc.* **1972,** 94, 2073.
30. Griffith, J. S. "Theory of Transition–Metal Ions"; Cambridge University Press: Cambridge, 1961; p. 101.
31. Dale, B. W.; Williams, R. J. P.; Johnson, C. E.; Thorp, T. L. *J. Chem. Phys.* **1968,** 49, 3441.

RECEIVED May 14, 1979.

The Active Site of *Sepioteuthis lessoniana* Hemocyanin

AKITSUGU NAKAHARA, WASUKE MORI, SHINNICHIRO SUZUKI, and MASAZO KIMURA

Institute of Chemistry, College of General Education, Osaka University, Toyonaka, Osaka 560, Japan

Addition of a large quantity of ethyleneglycol to a blue solution of Sepioteuthis lessoniana *hemocyanin produces the purple form of hemocyanin. The purple hemocyanin is considered to be an equilibrium mixture between approximately 60% of a slightly deformed species (A) and 40% of a seriously deformed species (B). The species A exhibits a Raman peak for the $\tilde{\nu}_{o-o}$ at around 750 cm^{-1}, being ESR-inactive and responsible for the purple color, whereas the species B has nothing to do with the origin of the purple color, exhibiting an ESR signal corresponding to approximately 20% of the total copper (A + B). Based on the superhyperfine structure of the ESR spectrum of ESR-active Cu(II), three or four of the coordinating ligands contain nitrogen.*

The hemocyanins are known as respiratory proteins for molluscs and arthropods, binding one oxygen molecule per the two copper atoms that constitute an active site unit (*1,2,3*). Since deoxyhemocyanins are colorless and diamagnetic, the coppers in the deoxy form have been considered to be Cu(I) (*4, 5*). On the other hand, oxyhemocyanins are deep blue, although they exhibit no ESR signals (*4*). For this reason, some investigators have assumed two antiferromagnetically coupled Cu(II)'s (*5*) and others, Cu(I) and Cu(III) (*3*), for the coppers in the oxy form. The binding mode of the oxygen molecule has been the object of argument, but recent resonance Raman studies on *Cancer magister* and *Busycon canaliculatum* hemocyanins, especially those using isotopically labeled oxygen, have disclosed that the Cu(II)'s are bridged by the peroxide dianion as illustrated by **1** (*6, 7, 8*).

0-8412-0514-0/80/33-191-341$05.00/0
© 1980 American Chemical Society

$$Cu^{(II)} \diagdown \overset{O^{(2-)}}{\underset{O}{\diagup}} \diagup Cu^{(II)}$$

1

However, some other structures also have been proposed in opposition to 1. Lontie and Vanquickenborne, for example, suggested the structure represented by 2 for the active site of *Helix pomatia* oxyhemocyanin (3).

$$Cu^{(I)}—L—Cu^{(III)}—O^{(2-)} \diagdown O$$

2

Various approaches have been carried out by many investigators to shed light on the structure and properties of the active site of hemocyanins. Mason and his co-workers investigated methemocyanin obtained by treating *C. magister* hemocyanin with hydrogen peroxide (9). They observed that the resulting products of H_2O_2 oxidation consist of 80%–95% of a diamagnetic species, a small amount of ESR-detectable Cu(II), and unreacted native hemocyanin. The active site of the diamagnetic species was assumed to contain a spin-coupled binuclear Cu(II) cluster whose conformation is so flexible that it easily produces magnetically isolated Cu(II) upon treatment with N_3^- (10). The uncoupling of the binuclear Cu(II) pairs in the diamagnetic species was interpreted in terms of a relaxation of superexchange through one or more bridging ligands such as tyrosine, histidine, serine, or cystein.

Solomon and his collaborators prepared half methemocyanin by employing *B. canaliculatum* hemocyanin, and spectroscopically investigated several derivatives of type Cu(I) $\cdots\cdots$ Cu(II)L, where L refers to CN^-, NO_2^-, CH_3COO^- or N_3^- (11). In the case of N_3^-, they observed that Cu(II) and Cu(I) are bridged by N_3^- in the half methemocyanin (12). They also derived met-apohemocyanin by removing Cu(I) from the half-met form (13). A comparison of the met-apo form and the half-met form suggested that the presence of Cu(I) plays an active part in binding the azide ion.

We also have been working for several years with *Sepioteuthis lessoniana* hemocyanin to clarify the structure of the environment of the coppers. Spectroscopic investigations led us to conclude that the two Cu(II) ions are each in a nonequivalent coordination geometry of very low symmetry (14).

Recently, we also found that the blue solution of oxyhemocyanin turns red–purple upon addition of a large quantity of ethyleneglycol. Characterization of the purple hemocyanin should contribute to the understanding of the structural aspects of native hemocyanin.

Experimental

Materials.   Hemocyanin was purified from the hemolymph of the squid, *Sepioteuthis lessoniana*, according to the method of Omura et al. (*15*). The dark-blue pellet of pure hemocyanin was redissolved in 0.05M Tris–HCl buffer (pH 7.5) and kept refrigerated at 0°–2°C except when the spectra were recorded. All the chemical reagents used for the experiments were of the best grade commercially available.

Measurements.   The absorption, circular dichroism (CD), and ESR spectra were recorded with a Hitachi 323 spectrophotometer, a JASCO MOE–1 magnetic circular dichroism spectrometer, and a JEOL JES–FE3X ESR spectrometer, respectively. The measurements of the spectra were carried out at 15°–20°C except for the ESR spectra, which were recorded at 77 K. The resonance Raman spectra were recorded at 10°C with 488.0-nm excitation (Ar⁺ laser) on a JEOL JRS–400D–002 spectrophotometer. The concentration of copper was determined to be 4.9mM by atomic absorption measured by a Nippon Jarrell–Ash AA–1 spectrometer.

Formation of the Purple Hemocyanin

Addition of ethyleneglycol (80 v/v% of the total volume) to a solution of *S. lessoniana* oxyhemocyanin in 0.05M Tris–HCl buffer (pH 7.5) produces a red–purple solution that is stable at 0°–2°C for a few days. The absorption maximum at 580 nm of oxyhemocyanin shifts toward shorter wavelengths with the increasing ethyleneglycol concentration, and reaches 534 nm when the concentration of ethyleneglycol becomes 80 v/v% of the total volume as represented in Figure 1. Inspection of the absorption spectra in the ultraviolet region in Figure 1 reveals that the ratio of the intensity at 280 nm to that at 347 nm also varies from 4.4 for the native oxyhemocyanin to 6.3 for the purple hemocyanin. The variation in the electronic spectra in the visible and UV region is considered to arise not only from a conformational change of the protein molecule, but also from a structural alteration around the two copper ions.

The CD spectrum of the purple hemocyanin (Figure 2) also exhibits marked difference from that of the native hemocyanin in that the shoulder at around 490 nm of the latter is a maximum in the former. Further, a decrease of the negative peaks at 565 nm and 600 nm (*14*), and of the positive peak at 700 nm is observed for the purple form compared with the native form. Thus, the information obtained through the CD study of the purple hemocyanin supports the belief that a structural deformation occurs in the environment of the copper ions with the addition of ethyleneglycol. The formation of the purple hemocyanin also could be observed with the addition of diols other than ethyleneglycol to a solution of oxyhemocyanin; the efficiency of the purple-hemocyanin formation was found to increase with increasing length of the alkyl chain, $HOCH_2CH_2OH < HOCH_2CH(OH)CH_3$

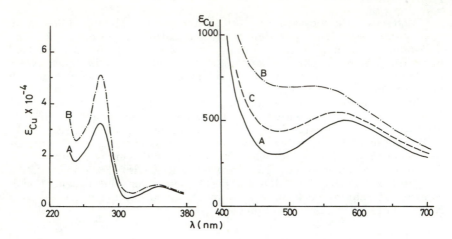

Figure 1. Absorption spectra of oxyhemocyanin (Curve A), purple hemocyanin (Curve B), and regenerated oxyhemocyanin (Curve C).

(Curve A) S. lessoniana oxyhemocyanin in 0.05M Tris–HCl buffer (pH 7.5); (Curve B) purple hemocyanin, 80 v/v% ethyleneglycol content, the period between the preparation and the measurement was 24 hr at 0°–2°C; (Curve C) the sample (regenerated hemocyanin) was obtained by adding Tris–HCl buffer (pH 7.5) to the purple hemocyanin until the final concentration of ethyleneglycol became 67 v/v%, and the period between the preparation and the measurement was 48 hr at 0°–2°C.

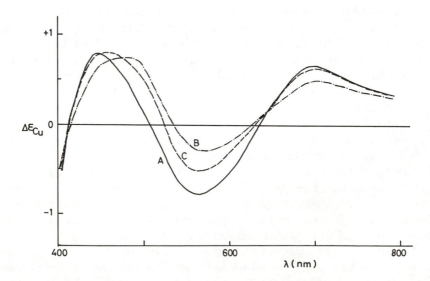

Figure 2. CD spectra of oxyhemocyanin (Curve A), purple hemocyanin (Curve B), and regenerated oxyhemocyanin (Curve C)

$\simeq$ HOCH$_2$CH$_2$CH$_2$OH $<$ HOCH$_2$CH(OH)CH$_2$CH$_3$. However, the purple state produced by diols other than ethyleneglycol was confirmed to be considerably less stable. On the other hand, addition of glycerol or polyalcohols such as 1,2,3,4,5-pentanepentanol did not produce the purple state but gave a turbid solution or precipitate from the blue solution.

Regeneration of the native oxyhemocyanin from the purple state can be attained (though not completely) upon addition of a large quantity of the buffer solution. Figures 1 (Curve C) and 2 (Curve C) show the absorption and the CD spectrum, respectively, recorded after diluting the solution of the purple hemocyanin by 0.05M Tris–HCl buffer solution until the final concentration of ethyleneglycol became 67 v/v% of the total volume. A comparison of the respective spectrum with the corresponding spectrum for the purple hemocyanin [Figures 1 (Curve B) and 2 (Curve B)] indicates that the purple form tends to revert to the native form upon dilution with the buffer solution. Accordingly, the structural deformation occurring in the purple form is not believed to be a drastic one. Addition of ethyleneglycol to the colorless deoxyhemocyanin caused no apparent variation, and the solution turned purple only on exposure to oxygen.

This finding suggested that oxygen is indispensable to the formation of the purple hemocyanin, and prompted us to carry out a Raman study on the purple hemocyanin to explore the binding mode of oxygen.

A Raman Study on the Purple Hemocyanin

Many investigators reported that the two copper ions are bridged by the peroxide dianion in oxyhemocyanin (6, 7, 8). We also observed a similar O—O stretching vibration at around 750 cm^{-1} in the resonance Raman spectrum of *S. lessoniana* oxyhemocyanin, as illustrated in Figure 3 (Curve A). The purple hemocyanin also exhibits the charac-

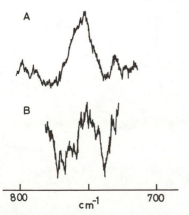

800   cm^{-1} 700

Figure 3. Resonance Raman spectra of oxyhemocyanin (Curve A) and purple hemocyanin (Curve B)

teristic peak for the peroxide dianion at the same wavenumber, though the intensity of the band is estimated to be somewhat weaker compared with that of the native oxyhemocyanin [Figure 3 (Curve B)]. This may suggest that there is no crucial difference between the binding mode of the oxygen molecule in the native and the purple hemocyanin.

ESR Studies on the Purple Hemocyanin

Oxyhemocyanin exhibits no ESR signal because of the spin-coupling of the two Cu(II) ions at the active site. The purple hemocyanin exhibits the ESR signal characteristic of the Cu(II) complexes, as represented in Figure 4 (Curve B). The amounts of ESR-active Cu(II) were calculated (based on the intensity of Cu(II)–EDTA complex) to be 13% for the measurement after 30 min and 22% after 3 hr. The pattern of the ESR spectrum of the purple hemocyanin suggests that there is no magnetic dipole–dipole interaction between the two coppers at the same active site. In other words, the ESR-active Cu(II) of the purple hemocyanin is believed to exist independently of the rest of Cu(II) present at the active site. Further, the ESR signal of the purple hemocyanin involves seven to nine superhyperfine lines attributable to nitrogens, suggesting that three or four nitrogenous ligands are coordinated with Cu(II).

As already described, the intensity of the resonance Raman peak for $\bar{\nu}_{o-o}$ of the purple hemocyanin is somewhat weaker than that of the native oxyhemocyanin. Taking this into account as well as that the amount of ESR-active Cu(II) is nearly 20%, it may be assumed that the

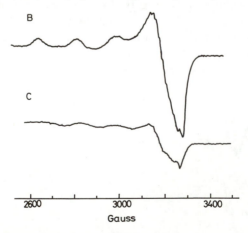

Figure 4. ESR spectra of purple hemocyanin (Curve B) and regenerated hemocyanin (Curve C): Curve B: $g_{\parallel} = 2.261$, $g_{\perp} = 2.059$, $A_{\parallel} = 173.8$ G, $A_N = 14.0$ G

purple hemocyanin is an equilibrium mixture between approximately 60% of a slightly deformed species, which gives the O—O stretching vibration at 750 cm^{-1} and approximately 40% of a rather seriously deformed species, of which one of the two coppers is ESR-detectable. As to the latter species, it remains unknown whether or not the copper binds oxygen. At least, the ESR-active Cu(II) may have no oxygen in the coordination site.

Regeneration of the native oxyhemocyanin from the purple form also was examined by use of ESR measurement, and the essential reversibility was affirmed through the spectral variation as illustrated in Figure 4 (Curve C).

Treatment of the Purple Hemocyanin with Thiocyanate

Thiocyanate ion is known to react with oxyhemocyanin to remove oxygen from the coordination sphere of the copper ions (*16, 17*). We observed that an addition of NCS$^-$ to the purple hemocyanin produces a little more reddish and transparent solution, whose absorption spectrum apparently differs from that of the purple hemocyanin, as shown in Figure 5. A drastic decrease of the absorption band at around 340 nm in Figure 5 indicates that the coordinated oxygen ligand is removed with the addition of NCS$^-$. The variation in CD spectrum accompanied by the addition of NCS$^-$ is so remarkable that the CD band centering at 430–490 nm of the purple hemocyanin almost disappears, as seen in Figure 6.

Contrary to these drastic variations, the ESR spectrum recorded after the addition of NCS$^-$ exhibits no essential difference from that of the purple hemocyanin. Addition of a reducing reagent such as $S_2O_4^{2-}$ to the reddish solution obtained by treating the purple hemocyanin with NCS$^-$ yields a colorless solution. These facts suggest that the color of the purple hemocyanin does not originate from the species containing ESR-active Cu(II), but from the species that exhibits the O—O stretching vibration at 750 cm^{-1} in the Raman spectrum. The oxidation state of the copper in the latter species must be Cu(II), and probably is kept as Cu(II) even after losing an oxygen ligand through treatment with NCS$^-$; Cu(II) is considered to be reduced to Cu(I) upon addition of $S_2O_4^{2-}$, resulting in the colorless solution as described above.

If one adds NCS$^-$ to the normal oxyhemocyanin solution before preparing the purple form, the solution becomes colorless because of oxygen loss. Addition of a large quantity of ethyleneglycol (80 v/v% of the total volume) to the resulting colorless solution does not produce the purple state. However, the final solution exhibits an ESR signal whose pattern and intensity is almost the same as those of the purple hemocyanin. This finding also coincides with the consideration that

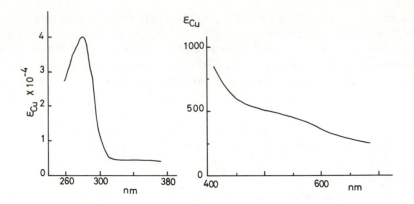

Figure 5.   Absorption spectrum of the red solution produced by treating the purple hemocyanin with NCS⁻ ion

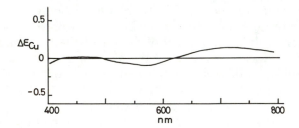

Figure 6.   CD spectrum of the red solution produced by treating the purple hemocyanin with NCS⁻ ion

the ESR-active species has nothing to do with the color of the purple hemocyanin. Considering the fact that the addition of NCS⁻ to the purple hemocyanin has no essential effect on the ESR spectrum, the species related to the color also is considered to have nothing to do with the ESR-detectable Cu(II).

Characterization of the Purple Hemocyanin

The general feature of the active site of the purple hemocyanin can be elucidated as follows. The purple hemocyanin is an equilibrium mixture of approximately 60% of a slightly deformed species (abbreviated as A) and approximately 40% of a rather seriously deformed species (abbreviated as B). The species A exhibits the Raman peak for the O—O stretching vibration at around 750 cm⁻¹, being ESR-inactive and responsible for the purple color. Though the species turns red upon the addition of NCS⁻, the oxidation state of the two coppers at the active site probably is kept as Cu(II) in an ESR-inactive structure. To satisfy this requirement, imidazole cannot be the bridging ligand since the spin–spin interaction through imidazole would not be strong enough (*18*) to hold the species *A* ESR-inactive (even after losing oxygen) upon the addition of NCS⁻. Further, the distance between the two Cu(II) ions in the peroxide-bridged structure of oxyhemocyanin (0.35–0.5 nm (*8*)) may be too short to allow the imidazole bridge (~0.6 nm (*19*)). Thus, a phenolate oxygen of tyrosine, an alcoholate oxygen of serine or threonine, a thiolate sulfur of cystein, or a carboxylate oxygen of aspartic or glutamic acid residue in the peptide chain might be most possible as the bridging atom between the two Cu(II) ions that display strong spin–spin interaction.

The species *B* has nothing to do with the origin of the purple color, but exhibits ESR signal corresponding to approximately 20% of the total copper contained in the purple hemocyanin (*A* + *B*). The ligands coordinated around the ESR-active Cu(II) are believed from the superhyperfine structure to be three or four nitrogens, which are presumably those of imidazole nitrogens of histidine residues. The other copper at the same active site may be Cu(I), considering the fact that the pattern of the ESR spectrum is characteristic of mononuclear Cu(II) complexes. The above consideration can be summarized and depicted as in Figure 7.

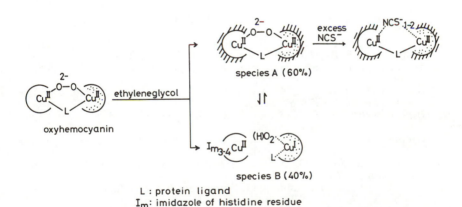

Figure 7. Structural aspects of purple hemocyanin. Coppers in the circles marked with dots illustrate different coordination geometry from those in the black circles, and shaded parts indicate the slightly deformed active site of oxyhemocyanin.

Although the information obtained through the investigation of the purple hemocyanin does not necessarily clarify the structure–function relationship in the environment of the copper ions in native hemocyanin, it would more or less help our understanding of the active site of this important copper protein.

Acknowledgments

We should like to thank Yukio Sugiura of Kyoto University for measurements of ESR spectra, and Teizo Kitagawa of Institute for Protein Research for measurements of resonance Raman spectra and for valuable discussions. The suggestions by Osamu Yamauchi of the Faculty of Pharmaceutical Sciences of Osaka University also are acknowledged. This work was supported by a grant from the Ministry of Education of Japan, to which the authors' thanks are given.

Literature Cited

1. Lontie, R.; Witters, R. In "Inorganic Biochemistry"; Eichhorn, G. L., Ed.; Elsevier: Amsterdam, 1973; p. 344.
2. Lontie, R.; Vanquickenborne, L. In "Metal Ions in Biological Systems"; Sigel, H., Ed.; Marcel Dekker: New York, 1974; Vol. 3, p. 183.
3. Lontie, R. In "Structure and Function of Haemocyanin"; Bannister, J. V., Ed.; Springer–Verlag: Berlin and Heidelberg, 1977; p. 150.
4. Nakamura, T.; Mason, H. S. *Biochem. Biophys. Res. Commun.* **1960**, *3*, 297.
5. Moss, T. H.; Gould, D. C.; Ehrenberg, A.; Loehr, J. S.; Mason, H. S. *Biochemistry* **1973**, *12*, 2444.
6. Loehr, J. S.; Freedman, T. B.; Loehr, T. M. *Biochem. Biophys. Res. Commun.* **1975**, *56*, 510.
7. Freedman, T. B.; Loehr, J. S.; Loehr, T. M. *J. Am. Chem. Soc.* **1976**, *98*, 2809.
8. Thamann, T. J.; Loehr, J. S.; Loehr, T. M. *J. Am. Chem. Soc.* **1977**, *99*, 4187.
9. Makino, N.; van der Deen, H.; McMahill, P.; Gould, D. C.; Simo, C.; Mason, H. S. *Biochem. Biophys. Acta* **1978**, *532*, 315.
10. McMahill, P. E.; Mason, H. S. *Biochem. Biophys. Res. Commun.* **1978**, *84*, 749.
11. Himmelwright, R. S.; Eickman, N. C.; Solomon, E. I. *Biochem. Biophys. Res. Commun.* **1978**, *81*, 237.
12. Himmelwright, R. S.; Eickman, N. C.; Solomon, E. I. *Biochem. Biophys. Res. Commun.* **1978**, *81*, 243.
13. Himmelwright, R. S.; Eickman, N. C.; Solomon, E. I. *Biochem. Biophys. Res. Commun.* **1978**, *84*, 300.
14. Mori, W.; Yamauchi, O.; Nakao, Y.; Nakahara, A. *Biochem. Biophys. Res. Commun.* **1975**, *66*, 725.
15. Omura, T.; Fujita, T.; Yamada, F.; Yamamoto, S. *J. Biochem.* **1961**, *50*, 400.
16. Rombauts, W. A. In "Physiology and Biochemistry of Haemocyanin"; Ghiretti, E., Ed.; Academic: London and New York, 1968; p. 75.
17. Lee, I. Y. Y.; Li, N. C.; Douglas, K. T., In "Structure and Function of Haemocyanin," Bannister, J. V., Ed.; Springer–Verlag: Berlin and Heidelberg, 1977; p. 231.
18. Inoue, M.; Kubo, M. *Coord. Chem. Rev.* **1976**, *21*, 1.
19. O'Young, C. L.; Dewan, J. C.; Lilienthal, H. R.; Lippard, S. J. *J. Am. Chem. Soc.* **1978**, *100*, 7291.

RECEIVED May 15, 1979.

Chemical Approaches to Nitrogen Fixation

WILLIAM E. NEWTON

Charles F. Kettering Research Laboratory, Yellow Springs, OH 45387

The chemistry of both aqueous- and nonaqueous-based systems that reduce N_2 (and some of its analogues) is presented together with various hypotheses that attempt to explain N_2 fixation by the enzyme nitrogenase. These systems are examined critically in light of the present knowledge of the enzymic mechanism and their inherent disadvantages relative to the requirements of an ambient temperature and pressure system for N_2 reduction are described. The concept of a common Mo cofactor is discussed with particular emphasis on the latest developments in the chemistry and composition of the Fe–Mo cofactor of nitrogenase. The evidence both for and against the synthetic mixed Fe–Mo–S cluster complexes being analogues of this cofactor is presented.

The need for augmenting the world supply of fixed nitrogen is well established (1, 2, 3). Approaches to this problem have taken chemical and biological directions. The biological approach involves the use of N_2-fixing microorganisms in symbiotic or free-living states to supply the fixed N_2 required for plant growth. This approach works well in, for example, the legumes, where nature has provided the mechanism for the biological interaction. Unfortunately, major food crops such as the cereal grains (rice, wheat, and corn), root and tuber crops, and sugar crops do not at present harbor symbiotic partners to aid in natural N_2 fixation. Although N_2-fixing bacteria often are found in the rhizosphere of some crop plants in greater than random numbers, apparently very little benefit to the plant is derived from such an "association". For crop productivity to approach levels both commercially acceptable and sufficient to meet human needs, extensive augmentation of this biologically fixed nitrogen input by a chemically synthesized fixed-nitrogen source is required.

0-8412-0514-0/80/33-191-351$06.25/0
© 1980 American Chemical Society

The biological and present day chemical processes for N_2 fixation are extremely dissimilar. The biological systems use small scale (microbial) reaction systems, work at ambient temperatures and atmospheric pressure, add nitrogen as required by soil or plant status, and use sunlight and/or photosynthate to provide a renewable energy and reducing source for N_2 fixation. In contrast, Haber–Bosch plants are enormous entities, requiring approximately 100 million dollars of capital outlay for plant construction and producing up to about 1500 metric tons of NH_3 per day. Further, nitrogen and hydrogen gases are used as feedstock in a high temperature, high pressure process. The latter reactant is produced mainly from natural gas and is a principal determinant of the price of nitrogenous fertilizer. This requirement of hydrogen from fossil fuel may be the greatest drawback of the Haber–Bosch process. The biological process is favored inherently over the chemical one as an environmentally sound, fossil-fuel-conserving system. As biological N_2 fixation is only associated in a quantitatively significant manner with a limited number of food crops, many workers currently are attempting to extend the scope of biological N_2 fixation by various means to agriculturally important crops. However, this in vivo biological approach has, as a complementary path, a technological approach, which is the main emphasis of this chapter.

The technological approach to N_2 fixation can be divided somewhat artificially into two categories. The first involves purely chemical approaches to the reduction of N_2 and takes the form of determining the basic requirements for the binding and subsequent reduction of N_2 to NH_3. The second category is concerned with simulating the composition and/or reactivity of the active site of nitrogenase by various complexes or systems. Both categories use transition metals as their bases because the Mo and Fe atoms present in nitrogenase are indicated strongly as the key moieties in the binding and reduction of N_2 and because this class of elements has been found to have the necessary properties for interaction with N_2 and similar isoelectronic molecules, such as CO and NO^+.

Chemical N_2-Reducing Systems

Molecular nitrogen is a very stable entity. It has a dissociation energy of 225 kcal/mol and an ionization potential of 15.6 eV. These properties indicate that at ambient temperature it is very difficult either to cleave the molecule or to oxidize it using common oxidants, such as halogens or oxygen. Reduction requires the addition of electrons to the lowest unoccupied molecular orbital, which is at -7 eV. The orbitals are too high in energy for reaction with all reductants except the very strongest, that is, very electropositive metals such as Li. However, these metals also react with water or oxygen under the same conditions, making interactions such as these very unlikely in the aqueous environment of the natural N_2-fixing system. Even so, systems with similar chemical basis are under study in nonaqueous conditions and have achieved some success in N_2 reduction. A transition-metal ion can, however, attack N_2 using a concerted donor–acceptor interaction. Certain metal ions in certain electronic states have the correct configuration and orientation to accept electrons at -15.6 eV from N_2 while simultaneously donating electrons to N_2, partially filling its empty acceptor orbitals at -7 eV. These conditions are not easy to satisfy and this capability depends not only on the metal, but on the other ligands in the complex and their spatial relationship to the N_2 molecule as well. This synergic donor–acceptor property forms the basis of all metal–dinitrogen chemistry.

At one time, no direct relationship could be established between the N_2-reducing systems and the isolable metal–N_2 complexes. Apparently, the intermediates in the former systems, which are likely to be the metal–N_2 complexes, were too unstable toward further N_2 reduction or loss of N_2 to be isolated, while the metal–N_2 complexes, by the fact that they could be isolated and characterized, appeared stable toward reduction to NH_3. Neither system could accomplish the three main tasks necessary for a successful catalytic N_2-fixing system, which are: (i) absorption of N_2; (ii) reduction to the appropriate level, preferably the N_2H_4 or NH_3 level; and (iii) regeneration of the fixing species with concomitant loss of the reduced nitrogen product. Much progress has been made in bridging this gap, although an efficient, ambient-temperature and -pressure system is still not available. This section is limited to a review and appraisal of those systems that achieve, at least, Steps (i) and (ii) and deals separately with the aprotic and aqueous systems.

Aprotic Systems.   The first report of an active abiological N_2-reducing system was published in 1964 (*4*). Using a reaction mixture composed of a transition-metal halide and an alkyl of an electropositive element (phenyllithium or ethylmagnesium bromide) in an aprotic solvent (ether, benzene) at ambient temperature and under 100–150 atm pressure of N_2, yields of NH_3 approaching one mol/mol of transition metal were obtained after hydrolysis of the reaction product. Presumably, organo-transition-metal compounds are produced initially, which decompose to the more reactive, reduced-metal species responsible for reaction with N_2. The reduced nitrogen-containing reaction products behave like nitride complexes and give ammonia and/or hydrazine on solvolysis.

GROUP IV SYSTEMS.   The most successful system at that time (1964) was based on titanocene dichloride [$(C_5H_5)_2TiCl_2$] with ethylmagnesium bromide (*5*), although the later use of lithium naphthalenide and $TiCl_4$ in a 15:1 molar ratio gave 1.7 mol NH_3 per mol $TiCl_4$ (6). The $[(C_5H_5)_2TiCl_2]$–ethylmagnesium-bromide system shows an induction period, with the rate of reaction dependent on N_2 pressure at pressures below 1 atm (7). Although 100–150 atm N_2 often were used, the pN_2 for half-maximum reaction rate is about 0.5 atm. The system is inhibited partly by strongly solvating solvents, CO, acetylene, and olefins (8) and variably by H_2 (9). Oxygen in small amounts does not effect the uptake of N_2, and fixation directly from air occurs with some closely allied systems (10), which also reduce "dinitrogen analogs" similarly to their reduction by nitrogenase (11). In these systems, Ti(II) is believed to be the active species (12). Although electron paramagnetic resonance (EPR) studies suggest that the principal species is diamagnetic, they are not unequivocal. Despite much work and speculation, the mechanisms of the reactions remain unclear.

These systems are not catalytic in the true sense because solvolysis, with resultant destruction of the active species, is needed to liberate NH_3. However, by controlled solvolysis followed by removal of the NH_3, a further cycle of reduction, N_2 absorption, and solvolysis often can be made. Titanium retains activity through about five such cycles in the tetra-isopropoxytitanium–sodium-naphthalenide system in ether using propan-2-ol for solvolysis (10). By using a nonprotic Lewis acid, aluminum tribromide, the catalytic effect of Ti is demonstrated. When N_2 (100 atm pressure) is treated with a mixture of titanium tetrachloride, metallic aluminum, and aluminum tribromide at 130°C as much as 284 mol of NH_3 per mol of $TiCl_4$ is obtained after hydrolysis. This, then, is a system for the catalytic nitriding of Al (13). A similar system operating electrochemically yields 6.1 mol NH_3 per g · atom Ti in 11 days (14).

Titanium-based systems also produce nitrogen-containing organic compounds. Thus, $(C_5H_5)_2TiX_2$ (X = Cl,Ph) with a fivefold excess of phenyllithium in ether solution under N_2 at room temperature gives, after hydrolysis, 0.15 mol of aniline per g · atom Ti plus 0.65 mol of NH_3 (15). Using $TiCl_4$–Na–naphthalene, direct formation of naphthylamines occurs after hydrolysis (16). An alternative synthesis uses $bis(\eta^5$-cyclopentadienyl)titanium dichloride and Mg metal in THF under N_2 (17). The resultant mixture reacts with ketones to form secondary amines. If CO_2 is introduced rather than a ketone, then an isocyanate probably is produced (18).

Since this time, more detailed studies indicate that metal–N_2 complexes indeed are formed in these reactions. Examples have been isolated by careful adjustment of conditions. That these reactions proceed to near-stoichiometric production of NH_3 and that the bis-η^5-cyclopentadienyl titanium core (and its pentamethyl-substituted analogue) limits the coordination possibilities at the metal held promise of relatively easy determination of both the structure of the complexes formed in this system and their reaction pathways. However, this expectation has not held true because a number of products are isolated from these reactions with closely related, but varying, properties. These complexes all assume an intense blue color in solution ($\lambda_{max} \sim 600$ nm). When methyl Grignard reagents are used in aprotic solvents (ethers, toluene) with (η^5-$C_5H_5)_2$TiCl at $-70°C$, a complex formulated as $[(\eta^5 - C_5H_5)_2Ti]_2N_2$ is produced (*see* Figure 1a) (*19*). A complex of similar stoichiometry has been claimed when titanocene, $[(C_5H_5)_2Ti]_2$ (produced from (η^5-$C_5H_5)_2$Ti$(CH_3)_2$ in a hydrogenation sequence), is exposed to N_2 at $-80°C$ in toluene (*20*). However, significant differences are observed for these two compounds, most noticeably the absence in the latter species of $\nu(N_2)$ at 1280 cm^{-1} observed in the former complex. A third related, but still unsubstantiated, complex, $[(\eta^5$-$C_5H_5)_2Ti]_2(N_2)_2$, is reported to form directly from titanocene (*21*).

These studies are further complicated by the problems associated with the nature of "$(C_5H_5)_2$Ti" itself. It is reported to undergo a rearrangement in solution to give a hydride-bridged species (*20, 22*), which, in contrast with $[(C_5H_5)_2Ti]_2$, is inert to N_2 (Equation 1). Other

$$[(C_5H_5)_2Ti]_2 \rightarrow [(\eta^5\text{-}C_5H_5)Ti(C_5H_4)H]_2$$
$$\Updownarrow N_2 \qquad\qquad N_2 \downarrow \qquad\qquad\qquad\qquad (1)$$
$$[(\eta^5\text{-}C_5H_5)_2Ti]_2N_2 \quad \text{No reaction}$$

evidence suggests that this N_2-inert species contains a fulvalene ligand rather than σ-C_5H_4 rings (*23, 24*). Further insight into the possible nature of titanocene and its reaction(s) with N_2 has resulted from an x-ray crystallographic study (*25*) of one such compound, (η^5-$C_5H_5)_2$-Ti-μ-$(\eta^1,\eta^5$-$C_5H_4)$Ti(η^5-$C_5H_5)$. This species reacts reversibly with N_2 to give several products, one of which is dark-blue $[(\eta^5$-$C_5H_5)_3(C_5H_4)$-Ti$_2]_2(N_2)$ (*26*), that is, one N_2 per four Ti atoms, which has no observable $\nu(N_2)$ (*see* Figure 2c). The relationships among these N_2-binding Ti complexes is not clear, although the lability of the cyclopentadienyl-ring protons is obviously a complicating factor.

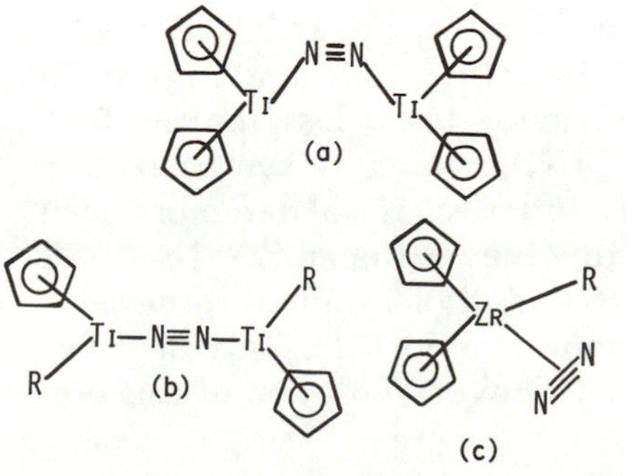

Figure 1. Arrangement of ligands in the complexes: (a)$[(\eta^5C_5H_5)_2Ti]_2$-(N_2) *as suggested in Ref. 19; (b)*$[(\eta^5\text{-}C_5H_5)TiR]_2(N_2)$ *(R = C_2H_5, C_6H_5, $CH_2C_6H_5$) as proposed in Ref. 39; and (c)* $(\eta^5\text{-}C_5H_5)_2$ Zr$[CH(SiMe_3)_2]$-(N_2) *from data in Ref. 38.*

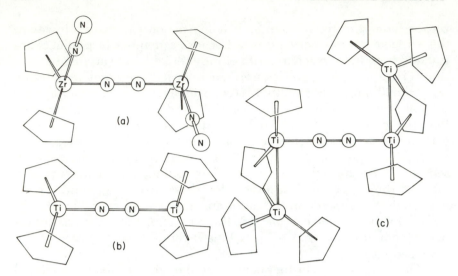

Figure 2. Structures of some Group IV metal–N$_2$ complexes: (a) [η^5-C$_5$Me$_5$)$_2$Zr(N$_2$)]$_2$(N$_2$) as determined in Ref. 30; (b) [(η^5-C$_5$Me$_5$)$_2$Ti]$_2$(N$_2$) as determined in Ref. 31, compare with the suggested structure in Figure 1a; (c) probable structure of [(η^5-C$_5$H$_5$)$_2$Ti-μ-(η^1,η^5-C$_5$H$_4$)Ti(η^5-C$_5$H$_5$)]$_2$-(N$_2$) based on the known structure of (η^5-C$_5$H$_5$)$_3$(η^1,η^5-C$_5$H$_4$)Ti$_2$, THF as in Ref. 25 and 26.

Methyl substitution of the cyclopentadienyl rings gives mononuclear (η^5-C$_5$Me$_5$)$_2$Ti as the suggested form of the titanocene analogue (27). This species is less susceptible, although not immune, to reaction at the rings and reacts to give two distinct complexes with N$_2$. Under N$_2$ in toluene, Equation 2 occurs (28) with no evidence for (η^5-

$$2(\eta^5\text{-C}_5\text{Me}_5)_2\text{Ti} \underset{>0°C}{\overset{0°C}{\rightleftharpoons}} [(\eta^5\text{-C}_5\text{Me}_5)_2\text{Ti}]_2(\text{N}_2)$$

$$[(\eta^5\text{-C}_5\text{Me}_5)_2\text{Ti(N}_2)]_2(\text{N}_2)$$

(2)

C$_5$Me$_5$)$_2$Ti(N$_2$) (27, 29). The dimeric mono-N$_2$ complex has been examined by x-ray crystallography, which shows a linear Ti—N≡N—Ti fragment with an N–N bond length of 1.16 Å (30) (*see* Figure 2b). Although [(η^5-C$_5$H$_5$)$_2$Ti(N$_2$)]$_2$(N$_2$) is unstable to N$_2$ loss above −50°C, its Zr analogue, [(η^5-C$_5$Me$_5$)$_2$Zr(N$_2$)]$_2$(N$_2$), which is synthesized by Na–Hg reduction of (η^5-C$_5$Me$_5$)$_2$ZrCl$_2$ under N$_2$, is isolated more easily (28) (*see* Figure 2a). Its structure involves a linear Zr—N≡N—Zr bridge with N–N bond length of 1.18 Å and one end-on terminal N$_2$ ligand (N—N = 1.115 Å) on each zirconium atom (31). Only the terminal N$_2$ ligands exchange with ^{15}N$_2$ or CO over a solution of this compound (28, 32).

The complex, $[(\eta^5\text{-}C_5R_5')_2Ti]_2(N_2)(R' = Me,H)$, apparently cannot be solvolyzed to produce N_2H_4 or NH_3 directly. But, at $-80°C$ in the presence of excess naphthalenide under N_2, it does give NH_3 (22). A product of the same stoichiometry, $[(\eta^5\text{-}C_5H_5)_2Ti]_2(N_2)$, produced by the MeMgI route, is reported to liberate either N_2 and N_2H_4 (0.5 mol/mol complex) or N_2 and NH_3 (0.6 mol/mol complex) with HCl at $-60°C$ depending on the solvent (19). Thus, not only is the observation of $\nu(N_2)$ in this complex different from the product formed from $[(\eta^5\text{-}C_5H_5)_2Ti]_2$ directly but its reactivity is appreciably different too. This complex also is proposed to react further with Grignard reagents in analogous fashion to $[(\eta^5\text{-}C_5H_5)_2Ti(i\text{-}C_3H_7)]_2(N_2)$ (This will be discussed later.). $[(\eta^5\text{-}C_5H_5)_3(\eta',\eta^5\text{-}C_5H_4)Ti_2]_2(N_2)$ produces 1.4 mol NH_3/mol complex after treatment with potassium naphthalenide followed by hydrolysis (26). Another complex, $[(\eta^5\text{-}C_5H_5)_2Ti(NH)]_2H$, also absorbs N_2, which is reduced stoichiometrically to NH_3 with potassium naphthalenide, followed by HCl (33).

$[(\eta^5\text{-}C_5H_5)_2Zr(N_2)]_2(N_2)$ also reacts directly with HCl at $-80°C$ in toluene to liberate two N_2 molecules and produce 0.86 mol N_2H_4 per mol complex (32), according to Equations 3 and 4. Its Ti analogue gives similar yields of N_2H_4 (30). The structural characterization of this dinuclear Zr complex shows no significant structural differences between the terminal and bridging N_2 ligands compared with numerous other N_2 complexes. As only very few of the others produce N_2H_4 or NH_3, the structural result itself gives no indication of the requirements for N_2 reduction. Isotopic labeling studies have indicated a symmetrical intermediate because the bridging and terminal N_2 ligands are scrambled in the N_2H_4 product (32). So, reduction of all three N_2 molecules equally is possible although only one N_2 per complex actually is reduced. This product distribution has been explained by Equations 3 and 4. No N_2H_4 is formed with $[(\eta^5\text{-}C_5H_5)_2Zr(CO)]_2(N_2)$, which also indicates the essential involvement of the terminal N_2 ligands and supports the concept of a symmetrical mononuclear intermediate.

$$[(\eta^5\text{-}C_5Me_5)_2Zr(N_2)]_2(N_2) + 2HCl \rightarrow$$
$$(\eta^5\text{-}C_5Me_5)_2Zr(N_2H)_2 + (\eta^5\text{-}C_5Me_5)_2ZrCl_2 + N_2 \quad (3)$$

$$(\eta^5\text{-}C_5Me_5)_2Zr(N_2H)_2 + 2HCl \rightarrow (\eta^5\text{-}C_5Me_5)_2ZrCl_2 + N_2 + N_2H_4 \quad (4)$$

This area of research is confused further by the isolation of a related series of Ti(III)–N_2 compounds from reaction sequences similar to those above. When $(i\text{-}C_3H_7)MgCl$ instead of MeMgI is used with $(\eta^5\text{-}C_5H_5)_2TiCl_n$ $(n = 1,2)$ under N_2 in ether at $-80°C$, another dark-blue complex, $[(\eta^5\text{-}C_5H_5)_2Ti(i\text{-}C_3H_7)]_2(N_2)$, forms (24). This species, which gives only N_2 on acid solvolysis, reacts further with $(i\text{-}C_3H_7)$-MgCl to give successively $[(\eta^5\text{-}C_5H_5)_2Ti]_2(N_2MgCl)$ and $(\eta^5\text{-}C_5H_5)_2$-

Ti[N(MgCl)$_2$]. The former species gives N$_2$H$_4$ (0.8 mol/mol complex) with methanolic HCl at $-60°C$ and the latter produces NH$_3$ on hydrolysis (35). A similar series of experiments using an excess of EtMgCl and (η^5-C$_5$H$_5$)$_2$TiCl$_2$ in 1,2-dimethoxyethane at 20°C, followed by exposure to N$_2$ and then addition of HCl gives NH$_3$ (0.7 mol/mol) (22). Similar blue–black N$_2$ complexes [(η^5-C$_5$H$_5$)$_2$Ti(R)]$_2$(N$_2$) also are formed directly from the Ti(III) species, (η^5-C$_5$H$_5$)$_2$Ti(R) (R = C$_2$H$_5$, substituted phenyl and benzyl), and N$_2$ in toluene at $-78°C$ (36, 37). They show no ν(N$_2$) and thus are assumed to be centrosymmetric (see Figure 1b). N$_2$ only is released on reaction with HCl. A related, but monomeric Zr(III) compound, (η^5-C$_5$H$_5$)$_2$Zr{CH(SiMe$_3$)$_2$}(N$_2$), gives 0.2 mol N$_2$H$_4$ per mol on solvolysis (see Figure 1c) (38). In contrast, the closely related Ti(III) complex, [(η^5-C$_5$H$_5$)$_2$Ti(CH$_2$SiMe$_3$)], does not pick up N$_2$ even at $-125°C$ (39).

Studies of reduction of N$_2$ in the dimeric Ti(III) compounds indicate that additional reductant is required. Reaction with four equivalents of sodium naphthalenide (NaNp) cleaves one C$_5$H$_5$ group per Ti to give {[(η^5-C$_5$H$_5$)Ti(R)]$_2$(N$_2$)}$^{2-}$, which, with HCl at $-78°C$, gives N$_2$H$_4$ (0.9 mol/mol complex) (see Equation 5a). This dianionic product on warming cleaves to two mono-anions (Equation 5b), which give

$$[(\eta^5\text{-C}_5\text{H}_5)_2\text{Ti}(R)]_2(\text{N}_2) \xrightarrow[-78°C]{4\text{NaNp}}$$

$$\{[(\eta^5\text{-C}_5\text{H}_5)\text{Ti}(R)]_2(\text{N}_2)\}\text{Na}_2 + 4\text{Np} + 2\text{Na}(\text{C}_5\text{H}_5) \quad (5a)$$

$$\{[(\eta^5\text{-C}_5\text{H}_5)\text{Ti}(R)]_2(\text{N}_2)\}^{2-} \xrightarrow{25°C} 2[(\eta^5\text{-C}_5\text{H}_5)\text{Ti}(R)\text{N}]^- \quad (5b)$$

NH$_3$ (0.6 mol/Ti) and N$_2$H$_4$ (0.15 mol/Ti) with HCl (40). With Grignard reagents (39), (η^5-C$_5$H$_5$)$_2$TiR forms intermediates similar to those proposed for the naphthalenide reactions, including the loss of one C$_5$H$_5$ group and the retention of the R group. This mechanism contrasts strongly with that proposed earlier (35) where the R group was lost preferentially (compare Equations 6 and 7). Intermediates A give

$$[(\eta^5\text{-C}_5\text{H}_5)\text{Ti}(R)]_2(\text{N}_2) \xrightarrow{R'\text{MgCl}} [(\eta^5\text{-C}_5\text{H}_5)\text{Ti}(R)]_2[\text{N}_2(\text{MgCl})_2]$$

$$A$$

$$[(\eta^5\text{-C}_5\text{H}_5)_2\text{Ti}]_2(\text{N}_2) \xrightarrow{R'\text{MgCl}} [(\eta^5\text{-C}_5\text{H}_5)_2\text{Ti}]_2(\text{N}_2\text{MgCl})$$

$$2(\eta^5\text{-C}_5\text{H}_5)_2\text{Ti}[\text{N}(\text{MgCl})_2] \longleftarrow \qquad R'\text{MgCl} \qquad (7)$$

$$B$$

$$2(\eta^5\text{-C}_5\text{H}_5)\text{Ti}(R)(\text{NMgCl}) \longleftarrow \qquad (6)$$

N$_2$H$_4$ on protonation and Intermediates B give NH$_3$. The many suggestions proposed for both the reaction intermediates and mechanism(s) still are difficult to reconcile because of the lability of these systems and their sensitivity to water and oxygen. They are, without doubt, based on dinuclear complexes.

The less well-defined Ti systems, which introduced this section and are based on the very potent reductant, Mg, have been reinvestigated (*41, 42*). The black product is of the composition [TiNMg$_2$-Cl$_2$,THF]. No ν(N$_2$) is observed. A nitride is believed to be bound to Ti in the complex and is converted quantitatively into NH$_3$ on hydrolysis (*41, 42*) or into isocyanate with CO$_2$ (*18*). Similar chemistry is observed with the VCl$_3$–Mg system (*42*).

These more recent studies support the earlier concept that Ti must be reduced, at least formally, to or below the oxidation state (II) before N$_2$ in the complex can be reduced. The formally Ti(III) complexes, [(η^5-C$_5$H$_5$)$_2$TiR]$_2$N$_2$, have only one electron per Ti atom and so have the possibility of producing diazene (N$_2$H$_2$) from N$_2$ and thus, by disproportionation, some N$_2$H$_4$ and/or NH$_3$. This reaction apparently does not occur unless additional reducing equivalents are added (*40*). Only when two electrons per Ti or Zr atom are present are substantial yields of N$_2$H$_4$ or NH$_3$ formed.

GROUP VI SYSTEMS.   The second major group of N$_2$-reducing reactions involves the mononuclear tertiary phosphine complexes of Mo and W. In contrast to the Group IV systems, the isolation of well-defined metal–N$_2$ complexes preceded the discovery of the reactions that produce NH$_3$ and/or N$_2$H$_4$ from their coordinated N$_2$ ligands. Initially, with excess of halogen acids, the reaction of *trans-M*(N$_2$)$_2$(dppe)$_2$ (*M* = Mo,W;dppe = 1,2-*bis*(diphenylphosphino)ethane) (*43, 44, 45*) (*see* Figure 3a) did not proceed past the hydrazido(2-) stage (Equation 8) (*46, 47, 48*) (*see* Figure

$$trans\text{-}Mo(N_2)_2(dppe)_2 + 2HBr \rightarrow Mo(N_2H_2)Br_2(dppe)_2 + N_2 \quad (8)$$

3c). Similar products are obtained with sulfuric acid (*49*) and tetrafluoroboric acid (*50*). However, if a mixture of Mo(N$_2$)$_2$(dppe)$_2$ and Mo(N$_2$H$_2$)Br$_2$(dppe)$_2$ is heated in *N*-methylpyrolidone with aqueous HBr, NH$_3$ is formed (0.48 mol/Mo) (*51*). Experiments involving Mo(N$_2$)$_2$(dppe)$_2$ and Mo(N$_2$H$_2$)Br$_2$(dppe)$_2$ alone under similar conditions led to the suggestion that the Mo(N$_2$H$_2$) species was an intermediate in the formation of NH$_3$. When the chelating diphosphine is replaced by PMe$_2$Ph or PPh$_2$Me to form *cis-* and *trans-M*(N$_2$)$_2$(PR_3)$_4$ (*M* = Mo,W), respectively (*48, 52*) (*see* Figure 3b), high yields of NH$_3$ can be obtained. For *M* = W, NH$_3$ yields approach 2 mol per mol of complex on treatment with sulfuric acid in MeOH or more simply (and slowly) by irradiation or heating under reflux in MeOH alone (Equation 9) (*53, 54*). The analogous Mo complexes produce up to 0.7 mol

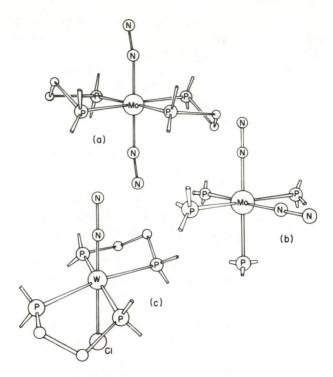

*Figure 3. Structures of some Group VI metal–N_2 and hydrazido($2-$)
complexes: (a) trans-$Mo(N_2)_2(dppe)_2$ as determined in Ref. 44; (b) suggested stereochemistry of cis-$Mo(N_2)_2(PPhMe_2)_4$ from data in Ref. 48
and 52; (c) $[W(N_2H_2)Cl(dppe)_2]^+$ as determined in Ref. 47.*

$$\textit{cis-}W(N_2)_2(PMe_2Ph)_4 \xrightarrow{\text{MeOH}}$$
$$N_2 + 2NH_3 + W(VI)\text{oxides} + 4PMe_2Ph \quad (9)$$

NH_3 with H_2SO_4–MeOH and none with MeOH alone (54, 55). This
lower yield of NH_3 from the Mo complexes has been suggested to be
caused by disproportionation of the $Mo(N_2H_2)$ moiety produced, possibly by release of N_2H_2, to give 0.67 mol of N_2 and 0.67 mol NH_3 (54).
Similar treatment of $M(N_2H_2)X_2(PMe_2Ph)_3$ with H_2SO_4–MeOH provides similar yields of NH_3 to their precursor *bis*-N_2 complexes (56).
These data, together with the isolation of complexes of other intermediate stages of reduced N_2, such as $M(N_2H)$ and $M(N_2H_3)$, by reaction
of these same complexes with halogen acids, indicate a possible stepwise sequence for the reduction and protonation of N_2 (Equation 10). A
mechanism for the reduction of N_2 on a single Mo atom in nitrogenase
has been proposed based on just such a sequence (54).

$$M(N_2)_2 \xrightarrow{H^+} M(NNH) + N_2 \xrightarrow{H^+} M(NNH_2) \xrightarrow{H^+} M(NNH_3) \rightarrow$$
$$M\equiv N + NH_3 \xrightarrow{3H^+} M(VI) + NH_3 \quad (10)$$

The key points from these experiments are that the more easily replaceable monophosphine ligands are required for the reduction of N_2, which is favored by the presence of oxo-anions. Thus, as the reaction proceeds and electron density passes from metal to N_2, the π-acceptor phosphines are replaced successively by π-donor oxo species. This change in ligand encourages further release of metal electron density onto the bound, partially reduced N_2, which results in its protonation. This resulting effective increase in the oxidation state of the metal then causes further substitution of the softer phosphines by the harder oxo-anions. These mutually enhancing effects result ultimately in complete loss of all phosphine ligands and the production of NH_3.

These same complexes also undergo C–N bond formation on reaction with alkyl or acyl halides (Equation 11) (57, 58, 59, 60). The

$$M(N_2)_2(dppe)_2 + RX \rightarrow M(N_2R)X(dppe)_2 + N_2 \qquad (11)$$

reactions with alkyl halides require irradiation for M = W (58, 59). A variety of similar products are obtained with α,ω-dibromoalkanes (60) and by the acid-catalyzed condensation reaction shown in Equation 12

$$trans\text{-}[Mo(NNH_2)F(dppe)_2]BF_4 + R_2C{=}O \rightarrow$$
$$[Mo(NNCR_2)F(dppe)_2]BF_4 + H_2O \quad (12)$$

(61). Similar reactions with the analogous monophosphines are unknown as yet, except for the reaction of $M(N_2)_2(PMe_2Ph)_4$ (M = W,Mo) with CH_3Br, which gives only $MBr_4(Me_2Ph)_2$ and N_2 (62). All the reactions of $M(N_2)_2(dppe)_2$ are believed to occur by initial loss of N_2 (on irradiation for M = W) with the attack by RX being radical in nature (63). These alkyldiazenido complexes can be degraded with release of fixed nitrogen. If $Mo(N_2Bu)Br(dppe)_2$ is treated in benzene–MeOH for 10 hr at 100°C under N_2 with sodium methoxide, 0.3 mol NH_3 and 0.3 mol of n-butylamine per Mo are produced (64). In contrast, the organohydrazido complexes, for example, $[W(N_2Me_2)Br(dppe)_2]Br$, give only amines on base, acid, or $LiAlH_4$ treatment (62).

OTHER SYSTEMS. Known reactions of salts of various metals with N_2 in the presence of a powerful reductant (4) form the basis for a more recent investigation of N_2 reduction using Mg (65). Using $CrCl_2$ and Mg in THF, a species of composition $[Cr_2N_2Mg_4Cl_4(THF)_5]$ is formed, which hydrolyzes to 1.2 mol NH_3 and 0.5 mol N_2H_4 per dinuclear unit. Insight into the possible structure of this unit comes from addition of bis(1,2-diphenylphosphino)ethane to the system when $[(dppe)_2Cr{-}N_2{-}Cr(dppe)_2]$ is suggested to be formed (66). A similar dinuclear system is postulated to be formed in the $FeCl_3$–Mg–N_2 system. In this case, $[(THF)_xMgCl_3Fe{-}N_2{-}FeCl_3Mg(THF)_x]$ produces only a small amount of hydrazine on hydrolysis (65, 67). Increased reactivity is observed with the ferric-chloride–triphenylphosphine–Grignard-reagent system. The dark-red ether solution produced absorbs one N_2 per two Fe atoms below $-40°C$. On solvolysis at $-40°C$ with hydrogen chloride, 10% of the N_2 ligand is released as N_2H_4, the remainder as N_2 gas. Hydrazine is obtained only in the presence of triphenylphosphine (68). A closely related system using ferric chloride with phenyllithium and no triphenylphosphine, on similar treatment, gives N_2H_4 (0.3 mol/mol $FeCl_3$) and NH_3 (0.15 mol/mol $FeCl_3$) (69).

N_2 is proposed to be fixed by a system containing $MoCl_4(dppe)$–Mg–alkyl-bromide/THF (65). Up to 106 molecules of N_2 per Mo atom are taken up in some unknown way to liberate N_2H_2 on hydrolysis. Here, the Mo complex appears to act as a catalyst in producing a Mg–N_2 compound, as the amount of N_2 fixed depends only on the Mg and alkyl bromide content. A similar effect may occur in the $MoCl_5$–Na—Hg–Mg^{2+}–MeOH system where up to 3.6 mol N_2H_4 per Mo atom are produced (70).

Another approach to C–N bond formation is provided by Equation 13, where the phenyl group attacks nucleophilically at the *endo*-nitrogen atom of the N_2 ligand. This product type can react further (Equation 14) and, for example, the dimethylhydrazine product can be displaced by N_2 in a cyclic process (73, 74). Such reaction systems may have significant industrial importance in the future for producing organo-nitrogen compounds using N_2 as a feedstock instead of NH_3.

$$(\eta^5\text{-}C_5H_5)Mn(CO)_2(N_2) + PhLi \rightarrow Li[(\eta^5\text{-}C_5H_5)Mn(CO)_2(N_2Ph)] \quad (13)$$

$$Li[(\eta^5\text{-}C_5H_5)Mn(CO)_2(N_2Me)] + [Me_3O]BF_4 \rightarrow$$
$$(\eta^5\text{-}C_5H_5)Mn(CO)_2[N(Me)NMe] \quad (14)$$

COMPARISON OF THE GROUP IV AND GROUP VI SYSTEMS.   Production of NH_3 in both these systems has certain features in common, although the Group IV metal systems are binuclear and the Group VI systems mononuclear. In both cases, there are two N_2 ligands in the coordination sphere of each metal but only one N_2 molecule in each complex is reduced. In both cases, the reduction of N_2 is triggered by mineral acids whose protons and counterions play key roles. Protonation of N_2 occurs as the softer ligands (phosphines and N_2), which stabilize the lower oxidation states of the metal, are replaced successively by the harder coordinating acid counterions, which tend to favor the metal's higher oxidation states. For the Mo and W complexes, the chelating diphosphines are not replaced easily, which makes the transfer of all six metal valence electrons to N_2 more difficult and, thus, decreases the tendency to form NH_3 compared with the simple phosphines. In the binuclear zirconocene system, only four electrons are available and so only hydrazine can be formed. Again, electron flow from metal to N_2 is encouraged by loss of N_2 and coordination of the appropriate anion.

While these NH_3- and N_2H_4-forming reactions are conducted under mild conditions, the reducing power of Na ($E° = -2.7V$) or Mg ($E° = -2.4V$) metals is built already into these reactions during the preparation of the metal–N_2 complexes. In addition, the $M(N_2)_2(PR_3)_4$ complexes have the disadvantage of being degraded completely to oxides during NH_3 formation. This degradation would make the likelihood of a catalytic process based on these systems less favorable compared with the zirconocene system, where the product of acid degradation is the starting material for preparation of the metal–N_2 complex. It remains to be seen if significantly milder reductants can effect the synthesis of these or similar species. In any case, it is very unlikely that the valence state Mo(O) could be reached in nitrogenase, even if all the energy of hydrolysis of ATP is transduced into reducing potential. However, transfer of electrons from Fe–S clusters to N_2, mediated by Mo, is a distinct possibility. The replacement of phosphine ligands by oxo- and other acid anions in $M(N_2)_2P_4$ complexes will result in a complex of a more reducing nature. This fact, plus the observation of N_2 reduction in conjunction with this ligand change, suggests a role for oxygen donors in all N_2-reducing systems including nitrogenase (49, 55). Recent x-ray absorption spectroscopic studies (71) of nitrogenase reveal three or four S atoms bound to Mo in nitrogenase (72). Thus, an interesting possibility arises that involves ligand substitution on Mo in nitrogenase. This change could substitute an O-donor for an S-donor and would alter the redox potential of the metal and effect electron transfer to a bound N_2. This transformation presents yet another possibility for the role of ATP in biological N_2 reduction.

Aqueous Systems. Many strong reducing agents in the presence of derivatives of transition metals have been reported to produce minute amounts of NH_3 from N_2 in water. In these cases, the low metal concentration often employed creates the possibility that contaminating species may be vitally important. Spurious results occur easily because the Nessler test for NH_3 is not specific, the system may scavenge traces of NH_3 or oxides of nitrogen (which are subsequently reduced) from the N_2 gas, and because nitrogen-containing additions to the reaction mixture may be degraded to NH_3. However, aqueous systems are known that have been proved to reduce N_2 to either N_2H_4 or NH_3 as the major product. Only the better substantiated and more visible of these will be discussed here.

HYDRAZINE-PRODUCING SYSTEMS. The essential catalyst to produce N_2H_4 from N_2 in protonic media is a reduced Mo or V salt in the presence of a substantial proportion of Mg ions and a reductant

(75). Typically, N_2 is reduced by an aqueous or aqueous-alcohol solution of sodium molybdate or oxotrichloromolybdenum(V) mixed with titanium(III) chloride at pH 10–14. The mole ratio of Mg to Ti for optimum reduction is 1:2. Hydrazine is produced at room temperature and atmospheric pressure but, at 95°C and 120 atm N_2, yields of N_2H_4 reach 88 mol per Mo atom. At the higher temperatures, some systems produce NH_3 also. Both the reductant Ti(III) and the Mo compound are required under these conditions, but the former can be replaced by V(II) or Cr(II). The system is poisoned by CO, and Mo(III) is proposed as the active entity. The mixture is heterogeneous with Mg ions keeping the Ti(III) ions apart in the hydroxide gel so that their oxidation to give H_2 is retarded. Hydroxide-bonded polynuclear entities are proposed that funnel the reducing capacity to the N_2. This system's efficiency, as measured by the rate at which reduced nitrogen species are produced per Mo atom, is about 1% of the natural system. Titanium(III) and Mg ions alone can produce N_2H_4 if the temperature is raised to above 140°C (76).

Vanadium(II), of similar electronic configuration to Mo(III), can take the place of both Mo and Ti in this system. At alkaline pH and at 100 atm N_2, it rapidly produces N_2H_4 (0.22 mol/g · atom V) (75). Carbon monoxide is said not to be inhibitory (77). Kinetic results suggest that a four-electron reaction occurs via a tetramer of V ions as in Equation 15 with each V^{2+} giving up one electron (78). The direct reduction

$$[4V^{2+}] + N_2 \xrightarrow{\text{H}_2\text{O}} [4V^{3+}] + N_2H_4 \qquad (15)$$

to N_2H_4 bypasses the two-electron reduction stage of N_2H_2, which is usually thought of as an energy-intensive hurdle. However, N_2H_2 can be stabilized on transition-metal centers (79) and a N_2H_2-level species has been implicated as an intermediate in biological N_2 reduction (80). At low temperatures, N_2H_4 is formed exclusively. However, at room temperature or higher, NH_3 is formed and H_2 evolved also (75). These products are thought to arise from a second reaction of N_2H_4 with another V(II) tetramer. The rate of N_2 reduction is faster with V than with Mo and both systems reduce acetylene to ethylene and ethane (81).

One independent reappraisal of this system confirms these observations but suggests, however, that N_2H_2 is the main product of N_2 reduction (Equation 16). Hydrazine (up to 0.15 mol/V atom) is suggested to result from N_2H_2 disproportionation (Equation 18) and NH_3 from further N_2H_4 reduction (82, 83). The presence of N_2H_2 is inferred from the observed reduction of allyl alcohol to 1-propanol and the dependence of the yield of N_2H_4 on the N_2 pressure. Further, a mononuclear unit is proposed as undergoing a two-electron oxidation to V(IV) (Equation 16), based on the quantitative conversion of acetylene to ethylene and the detection of VO^{2+} (82). Further, CO is described as an inhibitor. Obviously, combining Equations 16, 17, and 18, we have the same stoichiometry (Equation 19) as proposed previously (Equation 15) (78).

$$V^{2+} + N_2 \rightarrow V{\overset{N}{\underset{N}{\|}}} \xrightarrow{H_2O} VO^{2+} + N_2H_2 \tag{16}$$

$$VO^{2+} + V^{2+} \rightarrow 2V^{3+} \tag{17}$$

$$2N_2H_2 \rightarrow N_2H_4 + N_2 \tag{18}$$

$$4V^{2+} + N_2 \xrightarrow{H_2O} 4V^{3+} + N_2H_4 \tag{19}$$

Our studies of this system use the more strongly oxidizing nitrogenase substrates, acetylene and nitrous oxide, in attempts to distinguish between these two proposed mechanisms (84). Yields of N_2H_4 (0.2 mol/V(II)) similar to those reported previously are found. Under none of the conditions tried did the yield of ethylene per V(II) exceed 50%, strongly suggesting the one-electron V(II)–V(III) couple. However, the more powerful oxidant, nitrous oxide, produces N_2 in 100% yield at low V(II) concentrations, dropping to about 50% yield at higher concentrations. These results indicate that the two-electron V(II)–V(IV) couple is operating. Our results are consistent with Equations 20 and 21, with X being a two-electron acceptor. With $X = N_2O$, it

$$V^{2+} + X \xrightarrow{H_2O} VO^{2+} + XH_2 \tag{20}$$

$$V^{2+} + VO^{2+} \rightarrow 2V^{3+} \tag{21}$$

is such a powerful oxidant that, at low V(II) concentrations, it competes very successfully with VO^{2+} for V^{2+}, that is, Equation 20 only operates. At higher V(II) concentrations with $X = N_2O$ and at all concentrations with $X = C_2H_2$, the substrate does not compete successfully with VO^{2+} for V^{2+} and both equations operate giving an overall stoichiometry of two V^{2+} used per C_2H_2. We would, thus, expect the even less potent oxidant N_2 to suffer, at best, the same fate as C_2H_2.

NH₃-PRODUCING SYSTEMS. A homogenous aqueous-alcoholic system, composed of V(II) complexes of catechol or its derivatives, also is active in N_2 reduction. Only ligands with two o-hydroxy groups are active. This system is very sensitive to pH variations. In this case, however, NH_3 is the sole product with no N_2H_4 observed (85, 86). Hydrazine also is reduced rapidly in the system. Simultaneous H_2 evolution is observed with N_2 reduction and the suggestion has been made that both this system and nitrogenase use a mechanism involving two of the "four-electron centers" to produce one H_2 for every N_2 reduced. The first step occurs as in the hydroxide-gel system (*see* Equation 15). The second four-electron step corresponds to Equation 22. It also reduces acetylene in a specific *cis* manner to ethylene and is inhibited by CO.

$$4V^{2+} + N_2H_4 \xrightarrow{H_2O} 2NH_3 + 2V^{3+} + 2V^{2+} \xrightarrow{H_2O} H_2 + 2V^{3+} \qquad (22)$$

The second protonic, NH_3-producing system results from attempts to produce chemical models based on the knowledge that nitrogenase contains Fe, Mo, sulfide, and thiol groups. Organic thiols, sodium molybdate, and ferrous sulfate in the presence of a reducing agent, for example, $Na_2S_2O_4$ or $NaBH_4$, are reported to reduce nitrogenase substrates efficiently (87). The so-called "molybdothiol" system gives trace amounts of NH_3 from N_2 at 2000 psi pressure, for example, 3–5 μmol of NH_3 from about 5 mmol Na_2MoO_4, 2.5 mmol thioglycerol, 0.1 mmol $FeSO_4 \cdot 5H_2O$ and 0.25 g $NaBH_4$ in 50 mL of borate buffer (pH 9.6). No NH_3 is obtained in the absence of Mo. Yields up to approximately 0.04 mol NH_3/Mo atom are reported using a Mo–cysteine complex under 1 atm of N_2. Specific stimulation of activity by ATP is claimed (88), while others suggest that any acid, for example, sulfuric acid, produces the same effect, particularly as no significant ATP hydrolysis occurs (89, 90). These so-called "molybdothiol" systems are suggested to catalyze two-electron reductions only. For example, N_2 is bound side-on and reduced to N_2H_2, which disproportionates to N_2H_4 and N_2. Ammonia is then produced by a two-electron reduction of N_2H_4 (Equations 23, 24, and 25) (88). Although no intermediates have been isolated, the active principal in these reactions is thought to be a Mo(IV) species. However, recent electrochemical studies of this system indicate Mo(III) as the redox state active in substrate reduction (*91*).

$$Mo(IV) + N_2 \rightarrow Mo\overset{N}{\underset{N}{\overset{\|}{\big\langle}}} \xrightarrow{+2H^+} Mo + N_2H_2 \qquad (23)$$

$$3N_2H_2 \rightarrow 2N_2 + H_2 + N_2H_4 \qquad (24)$$

$$N_2H_4 \xrightarrow[2H^+]{2e} 2NH_3 \qquad (25)$$

These reaction mixtures still are not well understood. For example, the initial reports (87) state: (i) that while the thiol group was important, the organic residue to which it was attached had only minor effects; and (ii) that ethylene was the major product of acetylene reduction. A recent reinvestigation (92) shows: (i) that methyl groups substituted at the carbon atom adjacent to the thiol on the cysteinato ligand have a profound effect on reactivity; and (ii) that under most conditions, butadiene (C_4H_6) is the major product of acetylene reduction.

A more successful system, utilizing the Mo(IV) species, $[MoO(CN)_4(H_2O)]^{2-}$ with sodium borohydride, gives up to 0.32 mol NH_3/Mo at 10°C in 6 hr (93). Treatment of $[MoO(CN)_4(H_2O)]^{2-}$ with mild acid produces a species that can reduce N_2 to NH_3 on its own (0.07 mol/Mo) in 48 hr at 75°C (94). In contrast, although the cysteinato–Mo(V) complex dissociates (in alkaline medium) to oxomolybdenum(IV) species, it is incapable of N_2 reduction alone (95). The most successful system of this type so far produced makes use of sodium borohydride with MoO_4^{2-}–insulin (6 : 1) mixtures (96). Under optimal conditions, which includes very low (2 mmol/mL) Mo concentration, this system is reported to produce 65 mol NH_3 per Mo atom in 30 min under 1 atm N_2, in the presence of ATP, at 23°C. At molybdate concentrations of greater than 100 mmol/mL, the NH_3 yields drop to 1.1 mol/g · atom Mo atom in 6 hr under otherwise comparable conditions.

These Mo-based cyanide and insulin systems are suggested to operate similarly to the "molybdothiol" systems, via the formation of a coordinatively unsaturated oxo-molybdenum species as catalyst, which binds N_2 "side-on" and reduces it by two electrons to N_2H_2 (*see* Equations 23, 24, 25). The intermediacy of N_2H_2 in the formation of NH_3 is supported by the reduction of fumarate to succinate by these systems when operating under N_2. Succinate is not produced when other substrates, for example, acetylene, are being reduced.

Many alternative substrates of nitrogenase have been shown to be reduced by these Mo-based systems giving the same products as does nitrogenase and to this extent, they simulate the enzyme. However, our recent use of cyclopropene (97), which is reduced to cyclopropane and propene (1 : 1) by nitrogenase (98), with the cysteinato-Mo–borohydride system indicates that this model is lacking severely in that it produces only cyclopropane. Similar shortcomings were found with both the V^{2+}–Mg^{2+}–OH^- gels and the V^{2+}–catechol system (97).

A photochemical N_2-reducing system, based on TiO_2 in the rutile form containing approximately 0.4% water and impregnated with 0.2% Fe_2O_3, is known (99). This material (0.2 g) gives up to 6 μmol of NH_3, plus some N_2H_4, at 40°C under 1 atm N_2 in 3 hr on irradiation with a 360-watt mercury arc lamp.

These data, together with the much lower turnover numbers for the thiol and cyanide models compared with nitrogenase, are indicative of the required further development of these systems in order for them to become important N_2-reduction methods. These aqueous systems are extremely difficult to characterize in detail and the various mechanisms, often contradictory from different research groups, do not have the soundest of bases. However, they do show unequivocally that N_2 can be reduced in aqueous environments to give yields, in some cases substantial, of N_2H_4 and/or NH_3.

Conclusions. These chemical N_2-reducing systems are very important in providing insight into the binding and the activation of N_2 toward reduction and about the induction of internal redox reactions. The first step in enzymic N_2 reduction is undoubtedly the binding of N_2 to a transition-metal site on the enzyme. There is nothing in the known enzymology that precludes either changes in metal coordination sphere or protonation of bound N_2 (or both) as initiators of the redox process in line with effects observed in the model systems. In fact, a change in the donor-atom set might be part of the often quoted conformational change that may accompany the ATP reaction in nitrogenase.

However, all of the chemical systems now available have serious inherent disadvantages that limit and/or negate their utility for the reduction of N_2 under truly mild conditions. Thus, although protonation of $W(N_2)_2(PR_3)_4$ and $[Zr(\pi\text{-}C_5Me_5)_2]_2(N_2)_3$ to yield NH_3 and N_2H_4, respectively, appear to be "mild" reductions of N_2, it must be remembered that the N_2 complexes were formed in a reaction involving extremely powerful reductants. The "nitriding" reactions also use these powerful reductants, which would be precluded from any true catalytic cycle (and any viable commercial, ambient-conditions process) because of their incompatibility with protic media. The V(II) systems produce NH_3 in protic media but their use is limited because it has not yet proved possible to regenerate V(II) under the extremely basic conditions necessary for efficient N_2 reduction. It would seem, therefore, that inorganic systems for the reduction of N_2 in their present form almost certainly will not provide a useful means for NH_3 production in the near future. Nitrogenase, however, already is known to produce NH_3 from N_2 under ambient conditions. This enzyme, therefore, offers a working basis upon which a practical N_2-fixation process could be devised and developed concomitantly with the purely chemical systems.

Nitrogenase Models

In contrast to studying the reactivity of N_2, approaches based on knowledge of the chemical and physical properties of nitrogenase are also underway. Initially, through the original concept of the "Mo cofactor" (100, 101), which suggested that the modeling of the site of any one molybdoenzyme will impact on all the others, information gleaned from, for instance, xanthine oxidase, could be applied to a

nitrogenase model to define the parameters of the model more precisely. On this basis, a number of proposals were made for the composition of the active site and the mechanism of substrate activation by
nitrogenase. These include activation and reduction by oxidative addition (*87, 102, 103, 104, 105*) and a coupled proton–electron transfer
scheme (*106, 107*). In addition to the studies described above, complexes involving acetylenes oxidatively added to both Mo(II) and
Mo(IV) have been isolated (Equations 26 and 27). Protonation of the

$$Mo(CO)_2(R_2dtc)_2 + C_2H_2 \rightarrow Mo(C_2H_2)(CO)(R_2dtc)_2 + CO \quad (26)$$

$$OMo(R_2dtc)_2 + C_2R_2 \rightarrow OMo(C_2R_2)(R_2dtc)_2 \quad (27)$$

resulting acetylene complex produces olefins although not always in
good yield (*104, 105*). For each of the above complexes, there is convincing evidence for side-on binding of the substrate (*108*). Extrapolation of these findings to nitrogenase invokes side-on binding of substrate followed by protonation and dissociation of product. The
coupled proton–electron transfer mechanism (*106*) is based on the fact
that in its higher oxidation states the ligands coordinated to Mo will be
deprotonated, while these same ligands may be protonated in the
lower oxidation states, with other features of the coordination sphere
remaining unchanged. While the oxidation state of Mo in nitrogenase
is unknown at present, it is nevertheless likely that this state changes
during turnover. Thus, as electrons transfer from Mo to substrate, its
oxidation state increases, causing the protons on the coordinated
ligands to become acidic and transfer as well. Reactivation involves
reduction of Mo and concomitant reprotonation of the coordinated
ligands.

Which, if any, of these proposals is involved in nitrogenase action
is, at present, a matter of conjecture. Many of the proposals overlap to
some extent and it is possible that each has grasped some part of the
truth.

Fe–Mo Cofactor and its Models. It is now clear that the original
proposal of a universal Mo cofactor is in error. Recent work describes
the isolation of a small Mo-containing entity from the Mo–Fe protein
of *A. vinelandii* (*109*). It can reactivate the crude extracts of the mutant
A. vinelandii UW 45, which itself is incapable of N_2 reduction. This
cofactor contains Mo, nonheme Fe, and sulfide in a 1 : 8 : 6 atomic ratio.
It is quite stable in certain organic solvents but is very susceptible to
degradation by dioxygen. The discovery of Fe associated with this
cofactor contrasts with its absence from nitrate reductase from fungi.
These data and other reconstitution studies using extracts from organisms that produce demolybdo samples of nitrate reductase, nitrogenase, and sulfite oxidase show that two Mo–containing cofactors

exist (*110, 111*). One is called Mo-co and occurs in all known molybdo-enzymes except nitrogenase, which has its own cofactor, named FeMo-co. Thus, apparently subtle changes in the environment of Mo can alter drastically its catalytic capability and substrate selectivity.

A combination of Mössbauer and EPR spectroscopy (*112, 113, 114*), and also Mo x-ray absorption spectroscopy (*72*) indicate that the Mo and certain Fe environments in the Mo–Fe protein from *A. vinelandii, C. pasteurianum,* and FeMo-co from *A. vinelandii* are quite probably identical. FeMo-co also was shown to be the magnetic center of the enzyme responsible for generating the g = 4.3, 3.65, 2.01 EPR signal in the presence of sodium dithionite. FeMo-co remains intact upon removal from nitrogenase as judged by its yield and activity in reconstituting *A. vinelandii* UW 45 extracts in addition to its spectroscopic properties. Further, the circumstantial evidence that locates Mo at the active site of nitrogenase (*115*) suggests that FeMo-co might possess at least some of the chemical and catalytic properties of the enzyme. This notion has received some support from the reduction of acetylene to ethylene by FeMo-co using sodium borohydride in pH 9.6 borate buffer (*116*) in analogous fashion to the cysteinato-Mo systems discussed above (*87*) but with an activity two orders of magnitude higher. However, acetylene reduction is rather facile, certainly much easier to effect than N_2 reduction, and is obviously not definitive for nitrogenase activity. In fact, FeMo-co does not reduce N_2 to NH_3 under these conditions. Also, O_2-damaged cofactor is just as active in this acetylene reduction assay, although it has no activity in the biological UW 45 reconstitution assay. A very pertinent fact is that FeMo-co no longer has any biological reconstitution activity after the borohydride treatment.

Research in this area has taken two directions. The first is to determine the physical and chemical properties of FeMo-co itself and the second is the preparation of mixed Mo–Fe–S compounds in attempts to synthesize FeMo-co.

Properties of FeMo-co. The composition and the nature of FeMo-co is still not settled. Currently, it is believed to be of relatively small size with 7 ± 1 Fe atoms per Mo atom together with a number of sulfides (*109, 114, 117*). The reported value is 6 sulfides per Mo atom (*109*), but as thiomolybdates lose their sulfides less easily than Fe–S systems and because such species are produced on degradation of the Mo–Fe protein (*118*), the actual number easily could be different. No other ligands have been identified definitively. Although one report (*109*) suggests the presence of peptides, other investigations reveal none (*117, 119*). No substrate-reducing activity by FeMo-co, which allows it to retain its biological reconstitution activity, has been demonstrated.

Three redox-active states of FeMo-co, two of which are more oxidized than that produced by sodium dithionite, are observed by EPR spectroscopy on dye oxidation (*117*), in exact parallel with those observed for this chromophore in the intact Mo–Fe protein of nitrogenase (*113, 114, 120*). Thus, the redox state of FeMo-co more reduced than this EPR-active, dithionite-reduced state should correspond to the substrate-reducing, EPR-silent form of the Mo–Fe protein in the intact, fixing nitrogenase system (*121, 122, 123*). As the two more-oxidized states can be achieved, efforts are being made to generate this more-reduced state. Using the highly reducing, photoactivated 5-deazaflavin–EDTA system under CO, loss of this EPR signal occurs, which implies that the highly reduced state has been produced (*112*). However, our work shows that although the EPR signal is lost completely in the presence of 5-deazaflavin and EDTA under CO, similar experiments under argon or with no 5-deazaflavin present also did not exhibit an EPR signal (*117*). These and other results show that EDTA alone interacts with the EPR chromophore in a reaction not presently understood. It cannot, however, be a destructive perturbation as FeMo-co treated in this way is fully active in the UW 45 reconstitution assay for biological N_2-fixing activity. These results show that there is no direct correlation between the EPR spectrum and activity and that the substrate-reducing state of FeMo-co still remains elusive.

Mo–Fe–S Compounds. The guiding force and main criterion for the synthesis of Mo–Fe–S compounds to model FeMo-co is its Mo x-ray absorption spectrum (*72, 124*), particularly the analysis of the extended x-ray absorption fine structure (EXAFS) region of FeMo-co and Mo–Fe proteins. These studies are invaluable in giving the first insight into the environment of Mo in nitrogenase and showing that there are no obvious differences between the Mo site in clostridial and azotobacter Mo–Fe proteins and in FeMo-co. The Mo K-edge data eliminate the possibility of $Mo{=}O$ groups and indicate the presence of extensive S ligation. The EXAFS analysis is compatible with an environment around Mo that contains 3–4 bound S atoms at about 2.36 Å, 2–3 Fe atoms at a distance of about 2.72 Å, and 1–2 S atoms at a distance of about 2.5 Å. These, then, are the structural features with which synthetic models must be compatible.

Several reports currently are available for the preparation of Mo–Fe–S clusters; most take advantage of the techniques developed for the successful synthesis of the chemical models for the ferredoxins (125). A similar spontaneous assembly occurs when MoS_4^{2-} and $Fe(SR)_3$ are mixed in MeOH in the presence of excess thiol. In this instance, however, a more complex product results consisting of two $MoFe_3S_4(SR)_3$ cube-like structures bridged via the two Mo atoms by a sulfide and two thiolate groups, $[\{(EtSFe)_3S_4Mo\}_2S(SEt)_2]^{3-}$ (*124*). A similar product has been obtained independently by a similar method but with three bridging thiolate groups, that is, $[\{(RSFe)_3S_4Mo\}_2$-$(SR)_3]^{3-}$ $(R = Ph, Et, SCH_2CH_2OH)$ (126, 127). All structures have three Mo—Fe distances of about 2.73 Å and three Mo—S distances of about 2.37 Å to the sulfide ligands in the cubes. A distance of about 2.55 Å from Mo to the three bridging thiolates was found in the latter structures; disorder problems prevented the analogous distance from being determined in the former compound (see Figure 4a). These parameters, then, are very close to those determined for the Mo environments in the Mo–Fe proteins and FeMo-co. The x-ray absorption spectrum of $[\{(EtSFe)_3S_4Mo\}_2S(SEt)_2]^{3-}$ has been analyzed to give 3.7 S atoms of one type at 2.34 Å and 2.2 S atoms of a second type at 2.55 Å from Mo with a Mo—Fe distance of 2.76 Å (124). A comparison of the EXAFS analyses for this dicubane cluster and the Mo–Fe protein shows that the longer distance to the second type of S does not occur in the protein. However, a comparison of only the sum of the Mo–Fe and shorter Mo–S waves for both species is extremely similar. This analysis suggests that the $[(RSFe)_3S_4Mo]$ cluster might occur at the Mo site in nitrogenase (124). A preliminary report of just such a single cluster is now available (128). A fourth type of mixed Mo–Fe–S species is

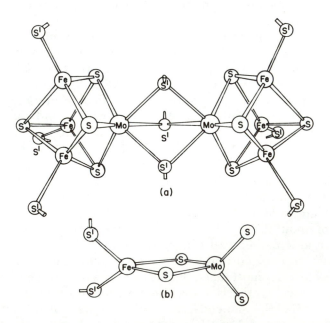

Figure 4.   Mixed Mo–Fe–S complexes: (a) structure of $[\{(RSFe)_3S_4Mo\}_2$-$(SR)_3]^{3-}$ as determined in Ref. 126 for $R = Ph$; (b) structure of $[(PhS)_2$-$FeS_2MoS_2]^{2-}$ as described in Ref. 129. In both cases, S' represents the thiolate sulfur atom.

$[(PhS)_2FeS_2MoS_2]^{2-}$, which again has a Mo environment similar to that in nitrogenase (*129*). The Mo—Fe distance is 2.75 Å with Mo—S distances of 2.25 Å and 2.15 Å in this noncubane complex (*see* Figure 4b).

Clusters of these types cannot constitute FeMo-co completely because the atomic ratio of Mo to Fe is not duplicated. Also, although the EXAFS analyses are similar, their EPR spectra are not, which highlights the problems associated with reliance on a single physical technique. A certain amount of ligand rearrangement obviously is needed if substrate is to contact Mo in the $[Fe_3S_4Mo]$ species. One attractive feature of $[(PhS)_2FeMoS_4]^{2-}$ particularly is the presence of a coordinatively unsaturated Mo atom. It may be that a $[Fe_3MoS_4]$ cluster is present in FeMo-co (but this is not certain) and joined in some unknown way to a Fe_4S_4 cluster. This arrangement would satisfy the Fe-to-Mo ratio of 7 (± 1) : 1 that is reported presently. Further, if the bridging between the $[Fe_4S_4]$ and $[Fe_3S_4Mo]$ clusters did not involve Mo, it could interact with substrate more easily. There is no doubt that the efforts underway world-wide will succeed in synthesizing many other Mo–Fe–S clusters and related compounds. Now that the proposed active site of nitrogenase is being exposed and studied, it should be just a matter of time before it is identified conclusively and produced synthetically rather than biologically.

Conclusions

The breakthroughs in Mo-cofactor research and N_2 chemistry outlined in this chapter may give rise to the availability of the key catalytic component for an ambient temperature and pressure, N_2-reducing system. These efforts indicate that the Mo-containing prosthetic group of nitrogenase is of sufficiently small size to be accessible by current synthetic techniques and that early transition metals, particularly when organized correctly, can activate N_2 very efficiently towards protonation and, thus, toward NH_3 and/or N_2H_4 formation. Therefore, as our fossil-fuel supplies continue to become increasingly scarce, it behooves us to look to the future and use our combined inventiveness to safeguard generations to come from the projected scenes of global famine. Biologists and chemists could be much more effective in this arena by working together rather than in isolation when both research areas will benefit from a mutual synergism.

Acknowledgments

I should like to thank my colleagues at the Charles F. Kettering Laboratory for the many sound and ongoing discussions. This chapter constitutes Contribution No. 662 from the Charles F. Kettering Research Laboratory.

Literature Cited

1. "Crop Productivity-Research Imperatives,"Brown, A. W. A.; Brierly, T. C.; Gibbs, M.; San Pietro, A., Eds.; Michigan State Agricultural Experiment Station: E. Lansing, 1975.
2. "World Food and Nutrition Study: The Potential Contribution of Research"; National Academy of Sciences: Washington, D.C., 1977.
3. Wittwer, S. H. "Genetic Engineering for Nitrogen Fixation," Hollaender, A., Ed.; Plenum: New York, 1977; p. 515.
4. Vol'pin, M. E.; Shur, V. B. *Dokl. Akad. Nauk SSSR* **1964**, *156*, 1102.
5. Vol'pin, M. E.; Shur, V. B.; Ilatovskaya, M. A. *Izv. Akad. Nauk SSSR, Ser. Khim.*, **1964**, *19*, 1728.
6. Henrici–Olivé, G.; Olivé, S. *Agnew. Chem. Int. Ed. Engl.* **1967**, *6*, 873.
7. Maskill, R.; Pratt, J. M. *J. Chem. Soc. (A)* **1968**, 1914.
8. Vol'pin, M. E.; Shur, V. B. *Nature* **1960**, *209*, 1236.
9. Vol'pin, M. E.; Ilatovskaya, M. A.; Larikov, E. I.; Khidekel, M. L.; Shvetsov, Yu. A.; Shur, V. B. *Dokl. Acad. Nauk SSSR* **1965**, *164*, 331.
10. van Tamelen, E. E.; Boche, G.; Greeley, R. *J. Am. Chem. Soc.* **1968**, *90*, 1677.
11. van Tamelen, E. E.; Rudler, H.; Bjorklund, C. *J. Am. Chem. Soc.* **1971**, *93*, 3526.
12. van Tamelen, E. E.; Boche, G.; Ela, S. W.; Fechter, R. B. *J. Am. Chem. Soc.* **1967**, *89*, 5707.
13. Vol'pin, M. E.; Ilatovskaya, M. A.; Kosyakova, L. V.; Shur, V. B. *Chem. Commun.* **1978**, 1074.
14. van Tamelen, E. E.; Seeley, D. A. *J. Am. Chem. Soc.* **1969**, *91*, 5194.
15. Vol'pin, M. E.; Shur, V. B.; Kudryavtsev, R. V.; Prodayko, L. A. *Chem. Commun.* **1968**, 1038.
16. van Tamelen, E. E.; Rudler, H. *J. Am. Chem. Soc.* **1970**, *92*, 5253.
17. Vol'pin, M. E.; Belyi, A. A.; Shur, V. B.; Katkov, N. A.; Nekaeva, I. M.; Kudryavtsev, R. V. *Chem. Commun.* **1971**, 246.
18. Sobota, P.; Jezowska–Trzebiatowska, B.; Janas, Z. *J. Organomet. Chem.* **1976**, *118*, 253.
19. Borodko, Yu. G.; Ivleva, I. N.; Kachapina, L. M.; Salienko, S. I.; Shilova, A. K.; Shilov, A. E. *J.C.S. Chem. Commun.* **1972**, 1178.
20. Marvich, R. H.; Brintzinger, H. H. *J. Am. Chem. Soc.* **1971**, *93*, 2046.
21. van Tamelen, E. E.; Fechter, R. B.; Schneller, S. W.; Boche, G.; Greeley, R. H.; Akermark, B. *J. Am. Chem. Soc.* **1969**, *91*, 1551.
22. Bercaw, J. E.; Marvich, R. H.; Bell, L. G.; Brintzinger, H. H. *J. Am. Chem. Soc.* **1972**, *94*, 1219.
23. Davison, A.; Wreford, S. S. *J. Am. Chem. Soc.* **1974**, *96*, 3017.
24. Brintzinger, H. H.; Bercaw, J. E. *J. Am. Chem. Soc.* **1970**, *92*, 6182.
25. Pez, G. P. *J. Am. Chem. Soc.* **1976**, *98*, 8072.
26. Pez, G. P.; Kwan, S. C. *J. Am. Chem. Soc.* **1976**, *98*, 8079.
27. Bercaw, J. E. *J. Am. Chem. Soc.* **1974**, *96*, 5087.
28. Manriquez, J. M.; McAlister, D. R.; Rosenberg, E.; Shiller, A. M.; Williamson, K. L.; Chan, S. L.; Bercaw, J. E. *J. Am. Chem. Soc.* **1978**, *100*, 3078.
29. Bercaw, J. E.; Rosenberg, E.; Roberts, J. D. *J. Am. Chem. Soc.* **1974**, *96*, 612.
30. Sanner, R. D.; Duggan, D. M.; McKenzie, T. C.; Marsh, R. E.; Bercaw, J. E. *J. Am. Chem. Soc.* **1976**, *98*, 8358.
31. Sanner, R. D.; Manriquez, J. M.; Marsh, R. E.; Bercaw, J. E. *J. Am. Chem. Soc.* **1976**, *98*, 8351.
32. Manriquez, J. M.; Sanner, R. D.; Marsh, R. E.; Bercaw, J. E. *J. Am. Chem. Soc.* **1976**, *98*, 3042.
33. Armor, J. N. *Inorg. Chem.* **1978**, *17*, 213.
34. Shilov, A. E.; Shilova, A. K.; Kvashina, E. F.; Vorontsova, T. A. *Chem. Commun.* **1971**, 1590.

35. Borodko, Yu. G.; Ivleva, I. N.; Kachapina, L. M.; Kvashina, E. F.; Shilova, A. K.; Shilov, A. E. *J.C.S. Chem. Commun.* **1973**, 169.
36. Teuben, J. H.; deLiefde Meijer, H. J. *Recl. Trav. Chem. Pays-Bas* **1971**, *90*, 360.
37. Teuben, J. H. *J. Organomet. Chem.* **1973**, *57*, 159.
38. Gynane, M. J. S.; Jeffery, J.; Lappert, M. F. *J.C.S. Chem. Commun.* **1978**, 34.
39. van der Weij, F. W.; Scholtens, H.; Teuben, J. H. *J. Organomet. Chem.* **1977**, *127*, 299.
40. van der Weij, F. W.; Teuben, J. H. *J. Organomet. Chem.* **1976**, *120*, 223.
41. Yamamoto, A.; Ookawa, M.; Ikeda, S. *Chem. Commun.* **1969**, 841.
42. Yamamoto, A.; Go, S.; Ookawa, M.; Takahashi, M.; Ikeda, S.; Keii, T. *Bull. Chem. Soc. Jpn.* **1972**, *45*, 3110.
43. Hidai, M.; Tominari, K.; Uchida, Y.; Misono, A. *Chem. Commun.* **1969**, 1392.
44. Uchida, T.; Uchida, Y.; Hidai, M.; Kodama, T. *Bull. Chem. Soc. Jpn.* **1971**, *44*, 2883.
45. Hidai, M.; Tominari, K.; Uchida, Y. *J. Am. Chem. Soc.* **1972**, *94*, 110.
46. Chatt, J.; Heath, G. A.; Richards, R. L. *J.C.S. Chem. Commun.* **1972**, 1010.
47. Heath, G. A.; Mason, R.; Thomas, K. M. *J. Am. Chem. Soc.* **1974**, *96*, 259.
48. Chatt, J.; Heath, G. A.; Richards, R. L. *J. Chem. Soc. Dalton Trans.* **1974**, 2074.
49. Chatt, J.; Pearman, A. J.; Richards, R. L. *J. Chem. Soc. Dalton Trans.* **1977**, 1852.
50. Hidai, M.; Kodama, T.; Sato, M.; Harakawa, M.; Uchida, Y. *Inorg. Chem.* **1976**, *15*, 2694.
51. Brulet, C. R.; van Tamelen, E. E. *J. Am. Chem. Soc.* **1975**, *97*, 911.
52. George, T. A.; Seibold, C. D. *Inorg. Chem.* **1973**, *12*, 2544.
53. Chatt, J.; Pearman, A. J.; Richards, R. L. *Nature* **1975**, *253*, 39.
54. Chatt, J.; Pearman, A. J.; Richards, R. L. *J. Chem. Soc. Dalton Trans.* **1977**, 1852.
55. Chatt, J.; Pearman, A. J.; Richards, R. L. *Nature* **1976**, *259*, 204.
56. Chatt, J.; Pearman, A. J.; Richards, R. L. *J. Organomet. Chem.* **1975**, *101*, C45.
57. Chatt, J.; Heath, G. A.; Leigh, G. J. *J.C.S. Chem. Commun.* **1972**, 444.
58. Day, V. W.; George, T. A.; Iske, S. D. A. *J. Am. Chem. Soc.* **1975**, *97*, 4127.
59. Diamantis, A. A.; Chatt, J.; Leigh, G. J.; Heath, G. A. *J. Organomet. Chem.* **1975**, *84*, C11.
60. Chatt, J.; Diamantis, A. A.; Heath, G. A.; Hooper, N. E.; Leigh, G. J. *J. Chem. Soc. Dalton Trans.* **1977**, 688.
61. Hidai, M.; Mizobe, Y.; Uchida, Y. *J. Am. Chem. Soc.* **1976**, *98*, 7824.
62. Bevan, P. C.; Chatt, J.; Leigh, G. J.; Leelamani, E. G. *J. Organomet. Chem.* **1977**, *139*, C59.
63. Chatt, J.; Head, R. A.; Leigh, G. J.; Pickett, C. J. *J. Chem. Soc. Chem. Commun.* **1977**, 299.
64. Busby, D. C.; George, T. A. *Inorg. Chim. Acta* **1978**, *29*, L273.
65. Sobota, P.; Jezowska–Trzebiatowska, B. *Coord. Chem. Rev.* **1978**, *26*, 71.
66. Sobota, P.; Jezowska–Trzebiatowska, B. *J. Organomet. Chem.* **1977**, *131*, 341.
67. Jezowska–Trzebiatowska, B.; Sobota, P. *J. Organomet. Chem.* **1972**, *48*, 339.
68. Borodko, Yu. G.; Broitman, M. O.; Kachapina, L. M.; Shilov, A. E.; Ukhin, L. Yu. *Chem. Commun.* **1971**, 1185.
69. Chubar, B.; Shilov, A. E.; Shilova, A. K. *Kinet. Katal.* **1975**, *16*, 1079.
70. Didenko, L. P.; Ovcharenko, A. G.; Shilov, A. E.; Shilova, A. K. *Kinet. Katal.* **1977**, *18*, 1078.
71. Cramer, S. P.; Hodgson, K. O.; Stiefel, E. I.; Newton, W. E. *J. Am. Chem. Soc.* **1978**, *100*, 2748.

72. Cramer, S. P.; Gillum, W. O.; Hodgson, K. O.; Mortenson, L. E.; Stiefel, E. I.; Chisnell, J. R.; Brill, W. J.; Shah, V. K. *J. Am. Chem. Soc.* **1978**, *100*, 3814.
73. Sellman, D.; Weiss, W. *Angew. Chem. Int. Ed. Engl.* **1977**, *16*, 880.
74. Sellman, D.; Weiss, W. *Angew. Chem. Int. Ed. Engl.* **1978**, *17*, 269.
75. Shilov, A.; Denisov, N.; Efimov, O.; Shuvalov, N.; Shuvalova, N.; Shilova, A. *Nature* **1971**, *231*, 460.
76. Kobeleva, S. I.; Denisov, N. T. *Kinet. Katal.* **1977**, *18*, 794.
77. Denisov, N. T.; Rudshtein, E. I.; Shuvalova, N. I.; Shilova, A. K.; Shilov, A. E. *Dokl. Akad. Nauk SSSR.* **1972**, *202*, 623.
78. Denisov, N. T.; Shuvalova, N. I.; Shilov, A. E. *Kinet. Katal.* **1973**, *14*, 1325.
79. Sellman, D.; Jodden, K. *Angew. Chem. Int. Ed. Engl.* **1977**, *16*, 464.
80. Newton, W. E.; Bulen, W. A.; Hadfield, K. L.; Stiefel, E. I.; Watt, G. D. "Recent Developments in Nitrogen Fixation", Newton, W. E., Postgate, J. R., Rodriguez–Barrueco, C., Eds.; Academic: London, 1977; p. 119.
81. Nikonova, L. A.; Efimov, O. N.; Ovcharenko, A. G.; Shilov, A. E. *Kinet. Katal.* **1972**, *13*, 249.
82. Zones, S. I.; Vickrey, T. M.; Palmer, J. G.; Schrauzer, G. N. *J. Am. Chem. Soc.* **1976**, *98*, 7289.
83. Zones, S. I.; Palmer, M. R.; Palmer, J. G.; Doemeny, J. M.; Schrauzer, G. N. *J. Am. Chem. Soc.* **1978**, *100*, 2113.
84. Newton, W. E., unpublished data.
85. Nikonova, L. A.; Ovcharenko, A. G.; Efimov, O. N.; Avilov, V. A., Shilov, A. E. *Kinet. Katal.* **1972**, *13*, 1602.
86. Nikonova, L. A.; Isaeva, S. A.; Pershikova, N. I.; Shilov, A. E. *J. Mol. Catal.* **1975/76**, *1*, 367.
87. Schrauzer, G. N. *Angew. Chem. Int. Ed. Engl.* **1975**, *14*, 514.
88. Schrauzer, G. N.; Kiefer, G. W.; Tano, K.; Doemeny, P. A. *J. Am. Chem. Soc.* **1974**, *96*, 641.
89. Vorontsova, T. A.; Shilov, A. E. *Kinet. Katal.* **1973**, *14*, 1326.
90. Khrushch, A. P.; Shilov, A. E.; Vorontsova, T. A. *J. Am. Chem. Soc.* **1974**, *96*, 4987.
91. Ledwith, D. A.; Schultz, F. A. *J. Am. Chem. Soc.* **1975**, *97*, 6591.
92. Corbin, J. L.; Pariyadath, N.; Stiefel, E. I. *J. Am. Chem. Soc.* **1976**, *98*, 7862.
93. Schrauzer, G. N.; Robinson, P. R.; Moorehead, E. L.; Vickrey, T. M. *J. Am. Chem. Soc.* **1976**, *98*, 2815.
94. Moorehead, E. L.; Robinson, P. R.; Vickrey, T. M.; Schrauzer, G. N. *J. Am. Chem. Soc.* **1976**, *98*, 6555.
95. Robinson, P. R.; Moorehead, E. L.; Weathers, B. J.; Ufkes, E. A.; Vickrey T. M.; Schrauzer, G. N. *J. Am. Chem. Soc.* **1977**, *99*, 3657.
96. Weathers, B. J.; Grate, J. H.; Strampach, N. A.; Schrauzer, G. N. *J. Am. Chem. Soc.* **1979**, *101*, 925.
97. McKenna, C. E.; Newton, W. E. unpublished data.
98. McKenna, C. E.; McKenna, M.-C.; Higa, M. T. *J. Am. Chem. Soc.* **1976**, *98*, 4657.
99. Schrauzer, G. N.; Guth, T. D. *J. Am. Chem. Soc.* **1977**, *99*, 7189.
100. Nason, A.; Lee, K.-Y.; Pan, S.-S.; Ketchum, P. A.; Lamberti, A.; DeVries J. *Proc. Nat. Acad. Sci. USA* **1971**, *68*, 3242.
101. Lee, K.-Y.; Pan, S.-S.; Erickson, R. H.; Nason, A. *J. Biol. Chem.* **1974**, *249*, 3941.
102. Schneider, P. W.; Bravard, D. C.; McDonald, J. W.; Newton, W. E. *J. Am. Chem. Soc.* **1972**, *94*, 8640.
103. McDonald, J. W.; Corbin, J. L.; Newton, W. E. *J. Am. Chem. Soc.* **1975**, *97*, 1970.
104. McDonald, J. W.; Newton, W. E.; Creedy, C. T. C.; Corbin, J. L. *J. Organomet. Chem.* **1975**, *92*, C25.

105. Maatta, E. A.; Wentworth, R. A. D.; Newton, W. E.; McDonald, J. W.; Watt, G. D. *J. Am. Chem. Soc.* **1978**, *100*, 1320.
106. Stiefel, E. I. *Proc. Nat. Acad. Sci. USA* **1973**, *70*, 788.
107. Stiefel, E. I.; Newton, W. E.; Watt, G. D.; Hadfield, K. L.; Bulen, W. A. In "Bioinorganic Chemistry-II," Raymond, K. N., Ed.; *Adv. Chem. Ser.* **1977**, *162*, 353.
108. Ricard, L.; Weiss, R.; Newton, W. E.; Chen, G. J.-J.; McDonald, J. W. *J. Am. Chem. Soc.* **1978**, *100*, 1318.
109. Shah, V. K.; Brill, W. J. *Proc. Nat. Acad. Sci. USA* **1977**, *74*, 3249.
110. Pienkos, P. T.; Shah, V. K.; Brill, W. J. *Proc. Nat. Acad. Sci. USA* **1977**, *74*, 5468.
111. Johnson, J. L.; Jones, H. P.; Rajagopalan, K. V. *J. Biol. Chem.* **1977**, *252*, 4994.
112. Rawlings, J.; Shah, V. K.; Chisnell, J. R.; Brill, W. J.; Zimmermann, R.; Münck, E.; Orme–Johnson, W. H. *J. Biol. Chem.* **1978**, *253*, 1001.
113. Zimmerman, R.; Münck, E.; Brill, W. J.; Shah, V. K.; Henzl, M. T.; Rawlings, J.; Orme–Johnson, W. H. *Biochim. Biophys. Acta* **1978**, *537*, 185.
114. Huynh, B. H.; Münck, E.; Orme–Johnson, W. H. *Biochim. Biophys. Acta* **1979**, *576*, 192.
115. Smith, B. E. *J. Less-Common Met.* **1977**, *54*, 465.
116. Shah, V. K.; Chisnell, J. R.; Brill, W. J. *Biochem. Biophys. Res. Commun.* **1978**, *81*, 232.
117. Newton, W. E.; Burgess, B. K.; Stiefel, E. I. "Molybdenum Chemistry of Biological Significance"; Newton, W. E., Otsuka, S., Eds.; Plenum: New York, 1980; p. 191.
118. Zumft, W. G. *Eur. J. Biochem.* **1978**, *91*, 345.
119. Smith, B. E. "Molybdenum Chemistry of Biological Significance"; Newton, W. E., Otsuka, S., Eds.; Plenum: New York, 1980; p. 179.
120. Watt, G. D.; Burns, A.; Lough, S. "Nitrogen Fixation"; Newton, W. E., Orme–Johnson, W. H., Eds.; Univ. Park: Baltimore, 1980; p. 159.
121. Orme–Johnson, W. H.; Hamilton, W. D.; Ljones, T.; Tso, M.-Y. W.; Burris, R. H.; Shah, V. K.; Brill, W. J. *Proc. Nat. Acad. Sci. USA* **1972**, *69*, 3142.
122. Mortenson, L. E.; Zumft, W. G.; Palmer, G. *Biochim. Biophys. Acta* **1973**, *292*, 422.
123. Smith, B. E.; Lowe, D. J.; Bray, R. C. *Biochem. J.* **1973**, *135*, 331.
124. Wolff, T. E.; Berg, J. M.; Hodgson, K. O.; Frankel, R. B.; Holm, R. H. *J. Am. Chem. Soc.* **1979**, *101*, 4140.
125. Averill, B. A.; Herskovitz, T.; Holm, R. H.; Ibers, J. A. *J. Am. Chem. Soc.* **1973**, *95*, 3523.
126. Christou, G.; Garner, C. D.; Mabbs, F. E.; King, T. J. *J.C.S. Chem. Commun.* **1978**, 740.
127. Christou, G.; Garner, C. D.; Mabbs, F. E.; Drew, M. G. B. *J.C.S. Chem. Commun.* **1979**, 91.
128. Otsuka, S.; Kamata, M. "Molybdenum Chemistry of Biological Significance"; Newton, W. E., Otsuka, S., Eds.; Plenum: New York, 1980, p. 229.
129. Coucouvanis, D.; Simhon, E. D.; Swenson, D.; Baenziger, N. C. *J.C.S. Chem. Commun.*, **1979**, 361.

RECEIVED May 25, 1979.

A Possible Mimic of the Nitrogenase Reaction

JOSEPH CHATT

A.R.C. Unit of Nitrogen Fixation, University of Sussex, Brighton, BN1 9RQ, United Kingdom ·

The degradation of dinitrogen in complexes of the type $[M(N_2)_2(mp)_4]$ (M = Mo or W, mp = monotertiary phosphine) to produce ammonia by treatment with acid has been investigated as possibly mimicking the reduction of dinitrogen at the molybdenum site in nitrogenase. It appears to proceed by protonation of one ligating dinitrogen molecule through the successive stages, $M^{(0)} \dddot{=} N \equiv N$ → $M \dddot{=} N \equiv NH$ → $M \equiv N \dddot{=} NH_2$ → $M \equiv N{-}NH_3{}^+$ → $M \equiv N + NH_3$ → $M \dddot{=} NH$ → $M{-}NH_2$ → $M^{(VI)} + NH_3$. The reducing electrons are supplied by oxidation of the center metal from oxidation number zero to six. Yields of up to 90%, based on the electrons available to effect the reduction, are obtained from some tungsten complexes that appear to provide the better mimic of the nitrogenase reaction at the molecular level. The reaction is stoichiometric, not catalytic.

The purpose of this chapter is to summarize the attempts of my research group to mimic the chemistry of the biological fixation of nitrogen in a stable chemical system amenable to a detailed study of the chemical mechanism. The biochemistry of nitrogenase is developing rapidly and is reasonably well understood at the descriptive level, but it has not given more than a hint of the chemical mechanism (*1, 2*).

The Nitrogenase Reaction

The fundamental nitrogenase reaction is the reduction of molecular nitrogen to ammonia and it takes place in certain very primitive microorganisms. The enzyme nitrogenase that catalyzes the reduction

0-8412-0514-0/80/33-191-379$05.00/0
© 1980 American Chemical Society

has been isolated in a very pure state from a few bacteria. It varies slightly according to its source and generally consists of two metallo-proteins. The larger is a molybdo-iron-protein of molecular weight around 220,000, and its essential metal content is ideally two atoms of molybdenum and around 24–34 of iron; the higher figure is probably the more correct. The smaller protein is an iron protein of molecular weight around 67,000 and contains four iron atoms. In both of these, the iron is associated with about an equal number of sulfide ions. The Fe-protein acts as a specific electron carrier, which in the presence of the monomagnesium salt of ATP transfers electrons from the reducing source to the MoFe-protein where the reduction of dinitrogen to ammonia is believed to occur.

The microorganisms that fix nitrogen cannot do so in the absence of molybdenum, but only minute traces are necessary for fixation to occur. This was shown in 1930 (3) and since then a molybdenum atom in the enzyme has been considered to be involved directly in some way with the dinitrogen molecule during the fixation process. Nevertheless, there is no direct evidence that the dinitrogen is reduced on a molybdenum or a dimolybdenum site, but accumulating circumstantial evidence points strongly to molybdenum being involved directly (4). The iron atoms are present in the form of iron–sulfur clusters, mainly of Fe_4S_4 type, doubtless they store and transmit the reducing electrons. The molybdenum atoms may possibly occur in place of one iron atom from each of two such clusters (5, 6).

The Basis of the Mimic

In the absence of more definite knowledge of the chemistry of the biological reaction, we needed a working hypothesis on which to base our model; at this stage of development, the model must be largely intuitive. For this reason, our model is not unique and there are others just as firmly supported by their protagonists (7). However, we can claim that our mimicking reaction is unique in giving intermediates capable of isolation and characterization, and that the mechanism to which they point makes good chemical sense.

In determining our model, we assumed that the reduction occurred at a single molybdenum atom. Our reason for choosing the single molybdenum atom site was that the MoFe-protein from *Klebsiella pneumoniae*, which is the main nitrogen-fixing organism studied in my laboratory, is often deficient in molybdenum, containing 1.3 to just under 2 g atoms per mole. Its activity is roughly proportional to its molybdenum content and not to the square, as would be expected if both atoms were involved and the molybdenum atom sites were populated randomly.

The second necessity was to find molybdenum complexes that would interact with molecular nitrogen. When we started this work none was known, but a considerable number of dinitrogen complexes of other transition metals had been prepared. Most of these contained tertiary phosphines, often together with hydride ion, as coligands. Also, the dinitrogen usually had been taken up from the gas phase during the preparation of those complexes; therefore, we chose to use tertiary phosphines as ligands for our model. Although they occur in no biological context, they appeared most likely to provide molybdenum complexes capable of binding molecular nitrogen. It was the chemical mechanism of electron transfer from molybdenum to the dinitrogen molecule and its protonation to ammonia, rather than the biochemical mechanism of feeding electrons into the molybdenum atom, that concerned us in designing our model. We also chose to study analogous tungsten chemistry because it is similar to that of molybdenum, but with the advantage that tungsten complexes are often less labile. They are easier to manipulate chemically and in general, we made our first studies with tungsten then followed through with analogous molybdenum reactions.

Dinitrogen Complexes of Molybdenum and Tungsten that Give Ammonia in Acid

After some searching, suitable mononuclear tungsten and molybdenum complexes containing dinitrogen were discovered. These were of the general type $[M(N_2)_2(L)_2]$ (M = W or Mo; L = two monotertiary phosphines or one ditertiary phosphine) (7). The monophosphine complexes give ammonia in good yields on protonation by acid at room temperature in methanol. They were prepared by the reduction of molybdenum- or tungsten-chloro-complexes by metallic reducing agents in dry tetrahydrofuran (THF) in the presence of the appropriate monophosphine under dinitrogen at atmospheric pressure and temperature (Reactions 1 and 2). Of the reducing agents tried, which included sodium amalgam, we found Grignard magnesium activated by iodine to be the best. It is essential that the starting metal complexes be prepared freshly and pure; slight oxidation or hydrolysis often spoils the reaction completely.

$$[MoCl_3(THF)_3] \xrightarrow[\text{THF, Mg, N}_2]{\text{PMe}_2\text{Ph}} cis\text{-}[Mo(N_2)_2(PMe_2Ph)_4] \qquad (1)$$

$$[WCl_4(PMe_2Ph)_2] \xrightarrow[\text{THF, Mg, N}_2]{\text{PMe}_2\text{Ph}} cis\text{-}[W(N_2)_2(PMe_2Ph)_4] \qquad (2)$$

The products of Reactions 1 and 2 are obtained in yields of 40–60%. The corresponding PMePh$_2$ complexes that have *trans-*

configurations are prepared similarly, the tungsten ones from [$WCl_4(PMePh_2)_2$]. However, the yields are capriciously inconsistent and often small. The *bis*-dinitrogen complexes thus prepared are well defined crystalline substances of the general formula *cis*- or *trans*-[$M(N_2)_2(mp)_4$] (M = W or Mo; mp = monophosphine = PMe_2Ph or $PMePh_2$). In every case, their isolation involves the use of methanol, which we believe is needed to hydrolyze some magnesium adduct (8).

The Reaction to Give Ammonia

In methanol, on addition of a little concentrated sulfuric acid at room temperature, the complexes [$M(N_2)_2(mp)_4$] degrade rapidly and completely to give ideally two molecules of ammonia together with a 0–5% yield of hydrazine (Reaction 3) (9).

$$[M(N_2)_2(mp)_4] + 6H^+ \rightarrow 2NH_3 + N_2 + M^{(VI)} \text{ products} + 4mp \quad (3)$$

When M = W and mp = PMe_2Ph, the yield of ammonia based on the reducing power of the tungsten (0) complex (6e) is 90% with 5% hydrazine and 5% dihydrogen. It seems remarkable that 95% of the electrons available for the reduction can be passed smoothly into one dinitrogen molecule that, in the presence of protons, is converted to ammonia with a small quantity of hydrazine. When the metal is molybdenum and mp = $PMePh_2$, the yield of ammonia is much lower, 35%, with only a slight trace of hydrazine. The low yield is obtained probably because molybdenum complexes are much more labile and the lower oxidation states not so strongly reducing as their tungsten analogues. Thus, during the reaction the molybdenum complex probably falls apart before the dinitrogen is reduced completely, perhaps at an N_2H_2 stage because of the stoichiometry: $3N_2H_2 = 2NH_3$ (33.3%) + $2N_2$ (*10*). However, the intermediate could not be free diazene $HN{=}NH$ because it would produce hydrazine as a sole or major nitrogen hydride product, whereas the molybdenum reaction gives only a minute trace of hydrazine. The quantity of hydrazine from the complexes, [$Mo(N_2)_2(mp)_4$], was not increased by change of solvent, but that obtained from their tungsten analogues is very solvent-dependent, the less protic solvents giving more hydrazine (*9, 12*). It also depends on the temperature (*9*) and the coligands (*13*). Indeed, [$W(NNH_2)$(quinolin-8-olato)($PMe_2Ph)_3$], where the dinitrogen already has reached the second stage of reduction (see below), gives hydrazine (0.39 mol) and no ammonia on treatment with methanol–sulfuric acid (*13*). Nevertheless, its molybdenum analogue gives 0.55 mol of ammonia and no hydrazine under the same conditions.

We are concerned with the reactions in protic solvents, particularly methanol. The fact that the molybdenum complexes give a 35%

yield of ammonia as opposed to 90% from the tungsten might suggest that the conversion of dinitrogen to ammonia is not a mimic of the reaction on nitrogenase that uses molybdenum and cannot use tungsten. However, this argument is specious. Molybdenum, in the enzyme where it is held more rigidly by an enveloping protein as a polydentate ligand, may well behave as tungsten does in an environment of monodentate ligands, the normal lability of the ligands on molybdenum being countered by the rigidity of the protein. It also may be argued that the reduction of dinitrogen in the tungsten complex involves a change of six in the oxidation state of the tungsten. Surely molybdenum cannot undergo such changes of oxidation state in the enzyme, especially in the presence of water. However, in the enzyme it is not necessary that the molybdenum undergoes such changes. The reducing electrons are believed to be stored in the iron–sulfur cluster systems and with some 34 iron atoms there is ample storage for six electrons. The only problem is the transfer of those electrons to the reducing site, but that is not part of our model. If our reaction in the tungsten complex mimics the chemistry of the reduction of dinitrogen on molybdenum in nitrogenase, it is only because all six reducing electrons are already stored in the tungsten atom. In nitrogenase, it seems more likely that the molybdenum atom in some intermediate oxidation state, perhaps even three or four, interacts with dinitrogen, which is then reduced by protic attack, and the electrons transfer one at a time from the iron–sulfur storage system through the molybdenum atom into the dinitrogen molecule.

It may also be argued against this that there is little evidence that dinitrogen forms complexes with transition metals other than in a closed-shell or almost closed-shell state. However, Shilov's (14) production of hydrazine and/or ammonia by vanadium(II) reduction of dinitrogen is indicative of an interaction with vanadium(II) that is in a d^3-state. Recently, Liebelt and Dehnicke's $Cl_4Mo–N{\equiv}N–MoCl_4$, prepared from $MoCl_5$ and IN_3, if substantiated, indicates that dinitrogen might interact with molybdenum(IV) that is in a d^2-state (15). However, in our "model" system the dinitrogen is found attached to molybdenum (0) or tungsten (0), which must be the case, if the system has to contain enough electrons to effect reduction of dinitrogen to ammonia. It is solely the acid attack at dinitrogen in our complexes $[M(N_2)_2(mp)_4]$ and the transfer of electrons from the metal to the dinitrogen molecule with its degradation to ammonia that mimics that part of the nitrogenase reaction.

Generally, when dinitrogen complexes are treated with acid the strongly reduced metal is protonated, leading to oxidation of the metal and formation of a hydrido complex. This is accompanied usually by loss of dinitrogen and often by subsequent protic attack on the hydride to evolve dihydrogen (7). Of the stable mononuclear dinitrogen complexes, only our tungsten and molybdenum complexes give good yields of ammonia on treatment with acids at room temperature in a protic solvent.

Reaction Mechanism

The course of our reaction depends on the phosphine, the acid, and the solvent. By varying these, it has been possible to isolate products containing the dinitrogen ligand in two stages of reduction, N_2H or N_2H_2.

The first stage of protonation and reduction is best studied in complexes based on the stable $M(dppe)_2$ (dppe = $Ph_2PCH_2CH_2PPh_2$) framework. Such complexes are all obtained from $trans$-$[M(N_2)_2$-$(dppe)_2]$ prepared in the same way as their monophosphine analogues, in 40–60% yields (7).

The first step may be the protonation of the terminal nitrogen atom on one molecule of ligating dinitrogen. This would stimulate electron withdrawal from the metal atom into that dinitrogen molecule, effectively oxidizing the metal. This would weaken the attachment of the second molecule and aid its displacement by the acid anion. Alternatively, the anion might displace one dinitrogen molecule so that all of the negative charge from the metal center would have to be accommodated by the other, which in turn would stimulate its protonation.

A quantitative study of the dissociation of dinitrogen from the complexes $[M(N_2)_2(dppe)_2]$ in THF shows that it occurs too slowly to be the initiating step (16) and it must be the protonation of one of the dinitrogen ligands that triggers the sequence of reactions leading to ammonia in the monophosphine complexes.

Such protonation without ligand displacement has been found in the complex $trans$-$[Mo(NCPr^n)(N_2)(dppe)_2]$, showing that under the right conditions of ligation the terminal nitrogen atom is sufficiently basic for protic attack (Reaction 4) (17).

$$trans\text{-}[Mo(NCPr^n)(N_2)(dppe)_2] + 2H^+ \rightarrow$$
$$trans\text{-}[Mo(NCPr^n)(N\text{---}NH_2)(dppe)_2]^{2+} \quad (4)$$

It is interesting that Reaction 4 occurs easily and is reversible. A weak base such as ethylamine or potassium carbonate is sufficient to remove the protons and regenerate the original complex. The N_2H_2 stage represents a definite plateau of stability; the protonation stops at the N_2H_2 stage at room temperature. This is also true of the protonation of the complexes $trans$-$[M(N_2)_2(dppe)_2]$ that occurs with ligand loss (Reaction 5) (9). The product of Reaction 5 can be deprotonated com-

$$trans\text{-}[M(N_2)_2(dppe)_2] \xrightarrow[\text{excess}]{\text{HBr}} [MBr(\equiv\!N\!\cdots\!NH_2)(dppe)_2]Br + N_2 \quad (5)$$

pletely only in the presence of a ligand capable of removing the bromide ion from the metal. In the absence of such ligands, weak alkali such as ethylamine will remove only one proton to give an iminonitrosyl or diazenido complex as shown in Reaction 6 (20). The diazenido complexes cannot be prepared pure by the action of one mole of acid on *trans*-$[M(N_2)_2(dppe)_2]$, but such reaction gives a mixture of the N_2H and N_2H_2 products together with the unchanged complex; the pure N_2H product is prepared best by Reaction 6. It is reasonably stable when M = W, but much less so when M = Mo. The N_2H complexes react immediately with acid to regenerate the N_2H_2 complexes and undoubtedly represent the product of the first stage of protic attack on ligating dinitrogen in the above tungsten and molybdenum complexes.

$$trans\text{-}[M\mathrm{Br}(\equiv\!\!\!\!=N\cdots NH_2)(dppe)_2]\mathrm{Br} \xrightarrow{\mathrm{NEt_3}}$$
$$trans\text{-}[M\mathrm{Br}(\equiv\!\!\!\equiv N\equiv\!\!\!=NH)(dppe)_2] + \mathrm{NHEt_3Br} \quad (6)$$

Protonation of the complexes *trans*-$[M(N_2)_2(dppe)_2]$ by methanol–sulfuric acid also stops at the N_2H_2 stage and stable complexes $[M(N_2H_2)(HSO_4)(dppe)_2]HSO_4$ are produced. To go beyond the N_2H_2 stage it appears to be necessary to displace the diphosphine and higher temperatures are needed, but monophosphines are displaced so easily from their complexes that the reaction with acid continues to give ammonia at room temperature as outlined in Reaction 3. In the methanol–sulfuric acid reaction with $[M(N_2)_2(mp)_4]$ type complexes, it is difficult to isolate any intermediates. Nevertheless, the monophosphine complex *cis*-$[M(N_2)_2(PMe_2Ph)_4]$, by careful treatment with just 2 mol of sulfuric acid in THF followed by addition of pentane after 30 min, precipitates a product of composition $W(N_2H_2)(PMe_2Ph)_2(HSO_4)_2$ (Reaction 7a), shown to contain the $N\!-\!NH_2$ group by its IR spectrum. This material decomposes in any solvent, whether or not it is protic, and methanol degrades it immediately to give nitrogen hydrides quantitatively (Reaction 7b) (10).

$$cis\text{-}[W(N_2)_2(PMe_2Ph)_4] + 2H_2SO_4$$
$$\xrightarrow{} W(N\!-\!NH_2)(PMe_2Ph)_2(HSO_4)_2\!\downarrow \quad (7a)$$
$$\xrightarrow{\mathrm{MeOH}} NH_3(95\%) + N_2H_4(5\%) \quad (7b)$$

In principle, the nitrogen hydride intermediates formed during the protonation of the complexes $[M(N_2)_2(mp)_4]$ by methanol–sulfuric acid can be followed by ^{15}N NMR spectroscopy but the method is not yet sufficiently sensitive to show them all, even with 95% $^{15}N_2$. It shows the original ligating N_2 being consumed and the $N\!-\!NH_2$ stage coming up, followed by the formation of ammonium salt (11). The ^{15}N NMR study emphasizes again that the $N\!-\!NH_2$ stage is the most persistent.

By using halogen acids, various phosphines, and solvent media, it has been possible to isolate complexes containing dinitrogen in two stages of protonation, N_2H and N_2H_2. A few examples of the complexes are given in Table I. Only the diphosphine gave examples of complexes containing the first stage of reduction with sufficient stability to be isolated; they are probably (N_2H), but possibly $H(N_2)$ complexes. Many examples containing dinitrogen at the N_2H_2 stage have been isolated with both di- and monophosphines. Also, a number of their structures have been determined by x-rays. These all have an essentially linear $M—N—N$ system and bond distances indicative of strong conjugation to the metal (Table II). These include an example, $[WCl_3(NNH_2)(PMe_2Ph)_2]$, where even the hydrogen atoms are resolved and the $W—N—NH_2$ system is planar (*see* Table II). Despite the high level of conjugation, protonation proceeds further under the right conditions. It appears to occur most readily at the terminal nitrogen atom, from which point the reaction towards ammonia proceeds rapidly and spontaneously as in Reaction 7.

Table I. Some Examples of Complexes of Molecular Nitrogen in the Initial Stages of Reduction Towards Ammonia or Hydrazine

Atomic Grouping   *Typical Examples*[a]

(1) $M—N{\equiv}N$ $[M(N_2)_2(dppe)_2]$, $[M(N_2)_2(PR_3)_4]$
(2) $M—N{=}NH$ $[MF(N_2H)(dppe)_2]$ cf. $[MX(NO)(dppe)_2]$
(3) $M{=}N—NH_2$ $[MX(N_2H_2)(dppe)_2]^+$, $[MX_2(N_2H_2)(PR_3)_3]$
                                  $[MX(N_2H_2)(py)(PR_3)_3]^+$

$$\qquad\quad H$$
$$\qquad\quad /$$
(4) $M{=\!=}N{\cdots}NH_2$ $WCl_3H(N_2H_2)(PR_3)_n$

[a] dppe = $Ph_2PCH_2CH_2PPh_2$; PR_3 = PMe_2Ph or $PMePh_2$; M = Mo or W; X = Cl or Br; py = pyridine; n = 2 or 3.

We have found no evidence to substantiate the existence of a diazene ($HN{=}NH$) form of the N_2H_2 ligand either symmetrically bonded or bonded through one nitrogen atom only. The spectral evidence tentatively assigned some years ago as evidence for an asymmetrically bonded diazene ligand (*18*) seems much more likely, in the light of molecular structures since determined, to be indicative of an asymmetrically hydrogen-bonded $N—NH_2$ ligand such as occurs in $[WCl_3(NNH_2)(PMe_2Ph)_2]$ (*see* Table II) (*19*).

Some materials representing a third stage of protonation of the *bis*-dinitrogen tungsten complexes, whose first found member was formulated $[WCl_3(NHNH_2)(PMePh_2)_2]$, now have been reformulated as hydrides, for example $WCl_3H(NNH_2)(PMePh_2)_2$ (Table I). This reformulation was made because the ^{15}N NMR spectrum of the supposed

Table II. X-Ray Parameters for Hydrazido(2−) Complexes

Complex	M—N(Å)	N—N(Å)	M—N—N(°)	Reference
[MoF(NNH$_2$)(dppe)$_2$]BF$_4$[a]	1.762(12)	1.333(24)	176.4(13)	23
[Mo(NNH$_2$)(8-hq)(PMe$_2$Ph)$_3$]I[b]	1.743(4)	1.347(7)	172.3(5)	18
[WCl(NNH$_2$)(dppe)$_2$]BPh$_4$[c]	1.73(1)	1.37(2)	171(1)	24
[W(NNH$_2$)(8-hq)(PMe$_2$Ph)$_3$]I[b]	1.753(10)	1.360(17)	174.7(9)	18
[WBr(NNH$_2$)(PMe$_2$Ph)$_3$(MeC$_5$H$_4$N)Br[d]	1.75	1.34	177	25
[WCl$_3$(NNH$_2$)(PMe$_2$Ph)$_2$][e]	1.75(1)	1.30(2)	179	26

[a] F *trans* to NNH$_2$.
[b] O *trans* to NNH$_2$.
[c] Cl *trans* to NNH$_2$.
[d] Br *trans* to NNH$_2$.
[e] N—H^1 = 0.784(10)Å; N—H^2 = 1.083(12)Å; HNH = 143°; H^2—Cl of adjacent molecule = 2.27Å.

[WCl$_3$(^{15}NH—^{15}NH$_2$)(PMe$_2$Ph)$_3$] shows no splitting of the ^{15}N$_\alpha$ resonance. This type of complex does not appear to be on the direct route to ammonia. Corresponding complexes containing monophosphines and dinitrogen or the N—NH$_2$ grouping give ammonia and hydrazine in the same quantities and proportions on treatment with methanol–sulfuric acid. However the hydrido-hydrazidocomplexes representing the third stage of protonation as above, give a much greater ratio of hydrazine to ammonia (30% N$_2$H$_4$). This suggests that they are part of the side reaction on the way to the by-product hydrazine. These hydride complexes are hydrazido(2−) complexes of tungsten(VI) and so might be expected to hydrolyze by protonation of the α-nitrogen atom to give hydrazine according to Reaction 8. The product WCl$_5$H would also hydrolyze rapidly to give dihydrogen. That so much ammonia is obtained from the hydridic complexes indicates that the N—NH$_2$ group is conjugated sufficiently to the metal that the next protonation occurs mainly at the terminal nitrogen atom leading to ammonia, probably as in Reaction 9.

$$WX_3H(N—NH_2)(PR_3)_n \xrightarrow{2HX} WX_5H + N_2H_4 + nPR_3 \qquad (8)$$

$$(PR_3)_nX_3\overset{H}{\overset{/}{W}}\text{⩵}N\text{⸱⸱⸱}NH_2 \xrightarrow{H^+} (PR_3)_nX_3\overset{H}{\overset{/}{W}}\text{⩵}N—NH_3^+ \rightarrow$$

$$NH_3 + (PR_3)_nX_4W\equiv N \xrightarrow{3H^+} NH_3 + W^{(VI)} \text{ product} \quad (9)$$

It seems most probable that the third protonation of dinitrogen on the direct route to ammonia occurs at the terminal nitrogen of the M⩵N⸱⸱⸱NH$_2$ stage to produce the atomic grouping M⩵N—NH$_3^+$. No complex containing this grouping has been isolated, but it would not be expected to be stable in complexes such as these, where it can undergo degradation with oxidation of the metal as in Reaction 10.

$$\overset{\frown}{M}\text{⩵}N\overset{+}{—}NH_3 \rightarrow M\equiv N + \overset{+}{N}H_3 \qquad (10)$$

In the successive protonation reactions (Reaction sequence 11), the degradation of the strong triple bond of dinitrogen is achieved neither by splitting nor saturating it directly. Rather, as protonation of

$$M\text{⫶}N\equiv N \rightarrow M\text{⫶}N\text{⩵}NH \rightarrow M\text{⩵}N\text{⸱⸱⸱}NH_2 \rightarrow M\text{⩵}N—NH_3^+ \quad (11)$$

the terminal nitrogen atom takes place, the triple bond is transferred into the M–N position. The energy of the triple bond thus is apportioned between the new $M\equiv N$ triple bond and the three new N–H σ-bonds, degrading it to a weak N–N single bond. Only at the third protonation does the N–N bond order reach unity and since both nitrogen atoms in the grouping M⩵N—NH$_3^+$ likely carry some positive charge, it probably would be a weak and rather long N–N single bond, readily degraded by the mechanism of Reaction 10. The nitride that also is formed would hydrolyze or reduce readily to produce ammonia. Indeed, such complexes as $[MoCl_2N(PMe_2Ph)_2]$ on treatment in methanol–sulfuric acid give ammonia immediately and quantitatively (21). One would not expect to isolate complexes of nitride type from our treatment of dinitrogen complexes with acid unless it can be done by the addition of some trapping agent. The conditions needed to form the $M—N—NH_3^+$ group are just those that would lead to its rapid degradation to ammonia.

Qualitatively, we can now see the essential mechanism of the acid degradation of dinitrogen to ammonia in our tungsten complexes $[W(N_2)_2(mp)_4]$. The first step is a single protonation of the complex, possibly at the terminal nitrogen atom of one dinitrogen ligand. This causes withdrawal of electronic charge from the metal, loosening the attachment of the second dinitrogen ligand that is displaced by the HSO_4^- ion. This ion not only brings negative charge into the complex but has filled p orbitals on its oxygen-atom ligands. Their interaction with the metal will raise the energy of the metal filled d orbitals, repelling more electronic charge into the antibonding orbitals of the remaining N_2 or $N≡NH$ ligand. This in turn would cause further rapid protonation at the terminal nitrogen with effective rise in the oxidation state of the metal. This rise in oxidation state weakens the attachment of the phosphine ligands, which then probably are replaced successively by HSO_4^- or SO_4^{2-}. Effectively, it is the successive replacement of the weakly π-acceptor phosphine ligands on the metal by the π-donor HSO_4^- or SO_4^{2-} that, assisted by their overall negative charge, repels electrons from the metal through the d_π–p_π-system into the ligating $N—NH_2$. Thus, the ligand is further protonated and degraded completely to ammonia by the mechanism of Reaction 10, followed by the hydrolysis of the nitride.

Application to Nitrogenase

The production of ammonia from complexes such as $[W(N_2)_2(PMe_2Ph)_4]$ is stoichiometric. All the electrons needed to effect the reduction of one molecule of dinitrogen were stored in the tungsten(0) center before the reaction and the complex is destroyed during the reaction. However, in nitrogenase, the electrons are stored in the iron–sulfur clusters and the molybdenum center could remain in whatever oxidation state is necessary to bind the dinitrogen throughout the reaction. The electrons for the reduction would be fed from the iron–sulfur clusters through the molybdenum atom into the dinitrogen molecule, which would combine with protons in the same way as it does in the tungsten(0) complexes. However, at the end of the nitrogenase reaction when the dinitrogen molecule had been reduced and ammonia removed, the molybdenum would remain in the dinitrogen-binding oxidation state. This leads to the proposal of a catalytic cycle for the reduction of dinitrogen on the enzyme as shown in Scheme 1, which is based on our study of the tungsten complexes (7).

Scheme 1. *Proposed mechanism for the catalytic reduction of dinitrogen at a molybdenum site in nitrogenase*

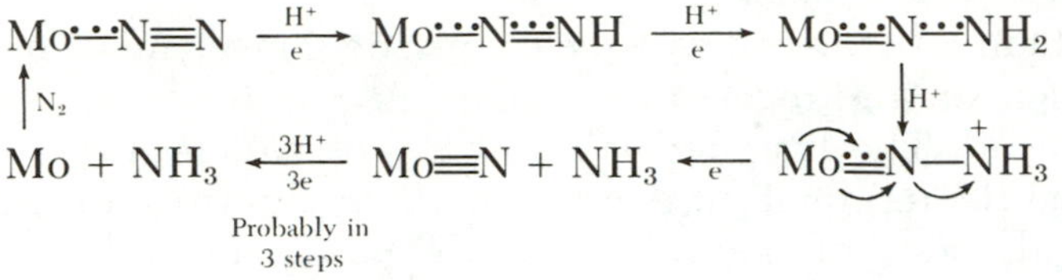

$$Mo\text{···}N\equiv N \xrightarrow[e]{H^+} Mo\text{···}N\text{≝}NH \xrightarrow[e]{H^+} Mo\text{≝}N\text{···}NH_2$$

$$\uparrow N_2 \qquad\qquad\qquad\qquad\qquad\qquad\qquad\qquad \downarrow H^+$$

$$Mo + NH_3 \xleftarrow[3e]{3H^+} Mo\equiv N + NH_3 \xleftarrow[e]{} Mo\text{≝}N\overset{+}{-}NH_3$$

Probably in
3 steps

Mo = molybdenum and attached protein and/or other ligands

The N—NH$_2$ stage represents the most stable partially reduced state of the dinitrogen molecule in our complexes and the biochemists in my group have looked for it in the functioning enzyme (22). This was done by quenching the enzymic reaction with acid or alkali and testing for hydrazine, which they found in the same quantity whether an acid or alkaline quench was used. The hydrazine undoubtedly arose from some dinitrogen hydride intermediate that was bound to the enzyme because it was found only when the enzyme was actually fixing nitrogen. When the enzyme functioned under argon or was reducing cyanide ion, no hydrazine was found. Also, the rate of ammonia production was proportional to the concentration of the hydrazine-producing intermediate. Also, it is interesting in connection with the above observation that when the N—NH$_2$ complexes of tungsten are treated with strong alkali or strong acid to give ammonia and hydrazine, the quantity of ammonia and hydrazine obtained is independent of whether acid or alkali is used.

That the mechanism of Scheme 1 is very satisfying chemically and that both the functioning enzyme and our tungsten complexes give the same quantity of hydrazine with acid as with alkali, does not prove that it is the mechanism of the reduction of dinitrogen at nitrogenase. The main virtue of the mechanism is that the initial stages of the protonation in complex compounds as models are well documented by many established examples. The structure of the $M\text{≝}N\text{···}NH_2$ grouping indicates a strong triple bond between the metal and its attached nitrogen atom, which should encourage the next stage of protonation at the terminal nitrogen. Logically, such protonation (by putting a positive charge on the terminal nitrogen and reducing still further the N–N bond order) should assist heterolytic splitting of the N–N bond to give ammonia and a ligating nitride ion. This in its turn should be removed easily by reduction to provide the second molecule of ammonia. The proposed mechanism is plausible and most of its steps have been realized as definite individual chemical reactions. How far these reactions mimic the reduction of dinitrogen at the molecular level on the active site in nitrogenase still has to be demonstrated.

Literature Cited

1. Zumft, W. G.; Mortenson, L. E. *Biochim. Biophys. Acta* **1975**, *416*, 1.
2. Orme–Johnson, W. H.; Davies, L. C. In "Enzymology of Nitrogenase in Iron Sulphur Proteins"; Lovenberg, W., Ed.; Academic: New York and London, 1977; pp. 15–59.
3. Bortels, H. *Arch. Microbiol.* **1930**, *1*, 333.
4. Smith, B. E. *J. Less-Common Met.* **1977**, *54*, 465.
5. Cramer, S. P.; Hodgson, K. O.; Gillum, W. O.; Mortenson, L. E. *J. Am. Chem. Soc.* **1978**, *100*, 3398.
6. Cramer, S. P.; Gillum, W. O.; Hodgson, K. O.; Mortenson, L. E.; Stiefel, E. I.; Chisnell, S. R.; Brill, W. J.; Shah, V. K. *J. Am. Chem. Soc.* **1978**, *100*, 3814.
7. Chatt, J.; Dilworth, J. R.; Richards, R. L. *Chem. Rev.* **1978**, *78*, 589.
8. Miura, Y.; Yamamoto, A. *Chem. Lett.* **1978**, 937.
9. Chatt, J.; Pearman, A. J.; Richards, R. L. *J. Chem. Soc., Dalton Trans.* **1977**, 1852.
10. Chatt, J.; Fakley, M. E.; Richards, R. L., unpublished data.
11. Mason, J.; Stenhouse, I. A., unpublished data.
12. Hidai, M.; Mizobe, Y.; Takahashi, T.; Uchida, Y. *Chem. Lett.* **1978**, 1187.
13. Chatt, J.; Fakley, M. E.; Richards, R. L.; Hanson, I. R.; Hughes, D. J. *J. Organomet. Chem.*, submitted for publication.
14. Shilov, A. E. In "Biological Aspects of Inorganic Chemistry"; Addison, A. E., Cullen, W. R., Dolphin, D., James, B. R., Eds.; John Wiley and Sons: New York, London, Sydney, Toronto, 1977; pp. 197–228.
15. Liebelt, W.; Dehnicke, K. *Z. Naturforsch Teil* **1979**, *34b*, 7.
16. Chatt, J.; Head, R. A.; Leigh, G. J.; Pickett, C. J. *J. Chem. Soc., Dalton Trans.* **1978**, 1638.
17. Chatt, J.; Leigh, G. J.; Neukomm, H.; Pickett, C. J. *J. Chem. Soc., Dalton Trans.*, submitted for publication.
18. Chatt, J.; Heath, G. A.; Richards, R. L. *J. Chem. Soc., Dalton Trans.* **1974**, 2074.
19. Chatt, J.; Pearman, A. J.; Richards, R. L. *J. Chem. Soc., Dalton Trans.* **1978**, 1766.
20. Chatt, J.; Pearman, A. J.; Richards, R. L. *J. Chem. Soc., Dalton Trans.* **1976**, 1520.
21. Bishop, M. W.; Chatt, J.; Dilworth, J. R.; Hursthouse, M. B.; Motevalle, M. *J. Less-Common Met.* **1977**, *54*, 487.
22. Thorneley, R. N. F.; Eady, R. R.; Lowe, D. J. *Nature* **1978**, *272*, 557.
23. Hidai, M.; Kodoma, T.; Sato, M.; Hasakawa, M.; Uchida, Y. *Inorg. Chem.* **1976**, *15*, 2694.
24. Heath, G. A.; Mason, R.; Thomas, K. M. *J. Am. Chem. Soc.* **1974**, *96*, 259.
25. Hursthouse, M. B.; Motevalli, M., unpublished data.
26. Chatt, J.; Fakley, M. E.; Hitchcock, P. B.; Richards, R. L.; Luong-Thi, N. T. *J. Organomet. Chem.*, **1979**, *172*, C55.

RECEIVED May 15, 1979.

Molybdothiol and Molybdoselenol Complex Catalysts

Acetylene Reduction and Electron Spin Resonance Characteristics

YUKIO SUGIURA, TAKANOBU KIKUCHI, and HISASHI TANAKA

Faculty of Pharmaceutical Sciences, Kyoto University, Kyoto 606, Japan

In the reduction of acetylene with molybdothiol and molybdoselenol complex catalysts, the effects of structural variation in ligands, variety of coordination-donor atom, kind of transition-metal ion, and other factors have been surveyed systematically. These factors have profound effects on the catalytic activity. The Mo complexes of cysteamine (or selenocysteamine), its N,N-dimethyl derivative, and its β,β-dimethyl derivative give ethylene, ethane, and 1,3-butadiene, respectively, as the major product. The Co(II) complexes of cysteine and cysteamine show higher catalytic activity than do the corresponding Mo complexes, and the order of the activity in the donor atom, namely S > Se ≫ O in the Co(II) complexes is consistent with that in the Mo complex systems. On the basis of electron spin resonance (ESR) features of these Mo complex catalysts, a relationship between their ESR characteristics and catalytic activities is discussed.

One of the more remarkable properties of nitrogenase is its ability to catalyze the reduction of diverse small, unsaturated molecules besides molecular nitrogen. Prominent among these substrates is acetylene, which the enzyme readily reduces to ethylene. Recently, Schrauzer and his collaborators discovered that the Mo complexes of cysteine and glutathione with $Na_2S_2O_4$ or $NaBH_4$ mimic the enzyme in this respect, catalyzing the formation of ethylene [1, 2]. The catalytic systems were based initially on the premise that nitrogenase contained Mo and sulfhydryl-containing amino acids. In contrast with the native

0-8412-0514-0/80/33-191-393$05.00/0
© 1980 American Chemical Society

enzyme, the molybdothiol model system produces 1,3-butadiene from acetylene. Therefore, the difference of the selectivity between the enzyme and the molybdothiol complex system is of special interest. In this chapter we survey the effects of various factors that affect the catalytic activity (i.e., total yield, product distribution, and reaction rate) in the acetylene reduction with the molybdothiol and molybdoselenol complex catalysts. Variations in ligand structure, the nature of the coordination-donor atom and the transition-metal ion, and other factors have been investigated. In addition, the catalytic activity of the molybdothiol and molybdoselenol complexes has been probed through the electron spin resonance (ESR) features, which reflect the structural characteristics of the complex species. The information presented here presumably opens the way for a superior catalytic system and also will define the essential factors that determine the reduction of acetylene in aqueous solution, though the molybdothiol model complex systems may not mimic the native enzyme in an important aspect.

Experimental

N- and β-Substituted derivatives of cysteamine and selenocysteamine were synthesized according to Klayman's method (3), with some modifications. Cysteamine, L-cysteine, selenocysteamine, and L-selenocystine were purchased from Sigma Company; L-selenocysteine was prepared by the reduction of L-selenocystine with sodium borohydride. Various sulfhydryl-containing peptides were synthesized according to our previous method (4, 5). The selenohydryl-containing ligands were used freshly, as they readily oxidize to diselenides. All other reagents used were of commercial reagent grade.

A typical catalytic system consisted of a 20-mL glass container fitted with a rubber serum cap containing borate buffer (pH 9.2; 3.5 mL), Na_2MoO_4 or $CoCl_2$(0.5 mL; 0.1mM aqueous solution), and the ligand (0.5 mL of 0.2mM solution; borate buffer). Water-washed acetylene (1 atm) was flushed into the solution and the reaction was initiated by the injection of 0.5 mL of $NaBH_4$(0.5 mL of 2mM solution; borate buffer). The reaction mixture was shaken at 20°C and the gas phase analyzed by gas chromatography using a Shimadzu gas chromatograph, Model GC–5A, equipped with a 0.3-cm × 2-m column of activated aluminum oxide containing 1% squalene and a flame-ionization detector. Two-component systems consisting of solutions of the metal ion and $NaBH_4$ alone exhibited no significant catalytic activity.

X-Band ESR spectra were obtained at 77 and 293 K with a JES–FE–3X spectrometer. The g-values were determined relative to Li–TCNQ(g = 2.0026) and the magnetic fields were calibrated by the splitting of Mn(II) in MgO($\Delta H_{3-4} = 86.9$ G). Magnetic circular dichroism (MCD) measurements, using a 11.7-kG magnet, were carried out on a Jasco J–20 spectropolarimeter and are expressed in terms of molecular ellipticity, $[\Theta] = 2.303$ $(4500/\pi)$ $(\epsilon_L - \epsilon_R)$, with units of (deg cm²)/d mol.

Results and Discussion

Ligand Effect. In Mo complexes containing cysteine-related ligands as potential catalysts, structural variations in the ligand have been pursued systematically. (6) The results are summarized in Table I. The presence of two methyl groups in the β-position, as in penicillamine (β,β-dimethylcysteine), clearly has compromised the reactivity of the complex. The complex of the threo form of β-methylcysteine ($R_1 = CH_3, R_2 = H$) behaves quite like that of cysteine in the reduction of acetylene, while that of the erythro form ($R_1 = H, R_2 = CH_3$) resembles penicillamine complex and has little activity. A steric factor must be responsible for the obtained results. The N,N-ethylenedicysteine complex, Mo_2O_4 (edcys)$^{2-}$, wherein an ethylene bridge links the cysteine units on each Mo atom, also was tested to clarify the requirement of sulfur donor and monomeric Mo species for the activity. Despite the

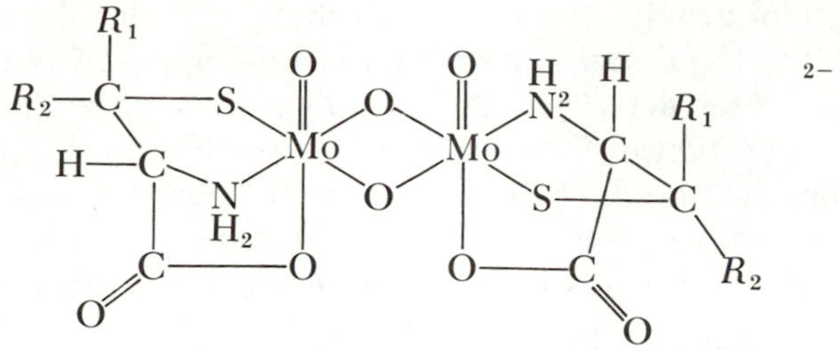

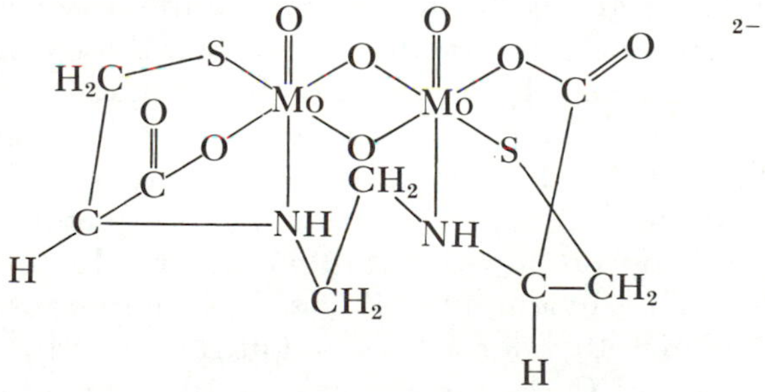

Table I. Acetylene Reduction by Mo-Complex Systems of Cysteine and Its Related Ligands[a]

Complex (ligand)	Gas Phase (μmol)			
	C_2H_4	C_2H_6	C_4H_6	C_2H_2
($R_1 = R_2 = H$)(cys)	52.5	0.6	142	362
($R_1 = R_2 = H$)(cys)[b]	130	4.8	27	450
($R_1 = R_2 = CH_3$)(pen)	4.0	1.1	3.7	668
($R_1 = CH_3, R_2 = H$)(threo-β-Me)	52.9	1.3	149	345
($R_1 = H, R_2 = CH_3$)(erythro-β-Me)	8.0	1.3	10.0	672
(edta)	4.9	0.9	0.6	672
(edcys)	0.6	0.3	0.1	668

[a] 25-min reaction times and borate buffer(pH 9.6) were used unless noted otherwise.
[b] 0.2M carbonate buffer(pH 9.6) was used in the place of borate buffer.

presence of thiol groups, the bridging ethylene group does not allow dissociation into the discrete monomeric units apparently needed for activity. The limited activity of this complex leads to the formation of a substantial proportion of C_2H_6, suggesting the ability of the reduced dinuclear complex to effect the required four-electron reduction.

Table II shows the effect of structural variations in the cysteamine-related ligands (7). In contrast with the case in cysteine

ligand, the β-substitution of cysteamine has a promotive effect on the catalytic activity. Inspection of molecular models reveals a probable explanation for the difference of activity between penicillamine and β,β-dimethylcysteamine. Penicillamine requires dissociation of the carboxyl group to exhibit its activity. However, eclipsing caused by methyl and carboxyl groups is unfavorable for the dissociation of the carboxyl group and the β,β-dimethyl groups can block partially the position *trans* to Mo—O_t. However, β,β-dimethylcysteamine, which lacks a carboxyl group, has a clearly accessible site for the metal ion and the dimethyl substitution may affect electronically the activity. On the other hand, the N,N-dimethyl substitution of cysteamine induces a drastic change of the product-distribution pattern. In this case, the major product is not ethylene but ethane. Cysteamine (or selenocysteamine), its N,N-dimethyl derivative, and its β,β-dimethyl derivative give C_2H_4, C_2H_6, and C_4H_6, respectively, as the major product. Of special interest is that seemingly small changes in ligand have remarkable effects on the product distribution and total yield.

Donor Effect. Similar catalytic activity of molybdoselenol complexes is predicted through the close resemblance of selenium to sulfur in biochemical aspects. However, the Se atom(covalent radii = 1.16 and ionic radii of Se^{2-} = 1.98 Å) is larger than the S atom (1.02 and 1.84 Å) being less electronegative and possesses somewhat more metallic character. Table II summarizes the catalytic activity in the acetylene reduction by various molybdoselenol complexes, together with those by the corresponding molybdothiol complexes. The catalytic activity of the ethanolamine–Mo complex system was negligible. The effect of coordination donor atoms on the catalytic activity clearly decreases in the order S > Se ≫ 0. In general, the replacement of S by Se gave the following effects on the catalytic activity and

Table II.   Catalytic Activity of Various Molybdothiol and Molybdoselenol Complex Catalysts in Acetylene Reduction

Donor	R_1	R_2	R_3	R_4	Product Yield $(\mu mol)^a$				Ratio			$Rate^a$ $(\mu mol/min)$
					C_2H_4	C_2H_6	C_4H_6	Total	C_2H_4 :	C_2H_6 :	C_4H_6	
S	H	H	H	H	165.5	12.0	29.8	207.3	13.8 :	1.0 :	2.5	18.5
S	H	H	CH₃	CH₃	15.3	183.1	82.2	280.6	0.08 :	1.0 :	0.4	23.8
S	CH₃	CH₃	H	H	200.0	22.3	262.8	485.1	9.0 :	1.0 :	11.8	13.7
Se	H	H	H	H	59.1	58.9	58.0	176.0	1.0 :	1.0 :	1.0	24.0
Se	H	H	CH₃	H	34.3	154.6	58.4	247.3	0.2 :	1.0 :	0.4	18.2
Se	H	H	CH₃	CH₃	14.8	256.2	30.6	301.6	0.06 :	1.0 :	0.1	21.8
Se	CH₃	H	H	H	51.9	58.8	61.5	172.2	0.9 :	1.0 :	1.0	18.5
Se	CH₃	CH₃	H	H	80.0	71.2	93.6	244.8	1.1 :	1.0 :	1.3	15.0

^{a} Yields were determined after 30-min reaction and rates for the initial 5 min.

product distribution: lower ethylene production, increased ethane production, decreased 1,3-butadiene production, and smaller difference in product-distribution ratio caused by β-substitution. The β-substitution of selenocysteamine gives little change on the catalytic activity, and their product distribution ratio is approximately $C_2H_4 : C_2H_6 : C_4H_6 = 1 : 1 : 1$. The N-substitution of selenocysteamine clearly increases the catalytic activity, as is the case for cysteamine. However, the product-distribution pattern changes considerably and in particular, the ethane production increases. The complexes with Se donor atoms show catalytic ability similar to that in complexes with S donor atoms show catalytic ability similar to that in complexes with S donor atoms in the reduction of acetylene. However, S → Se replacement has a distinguishable effect on the catalytic activity and/or product-distribution pattern. The catalytic change that occurs upon the atomic substitution probably reflects the spatial and electronic differences of the two elements, S and Se.

Metal Effect.   Schrauzer and Schlesinger surveyed the relative activity of various transition-metal ions as catalysts in the reduction of acetylene to ethylene in an aqueous solution containing 1-thioglycerol and excess $Na_2S_2O_4$ (8). Except for the remarkably specific activity of Mo, only iridium showed appreciable activity, converting acetylene to ethylene at 15% of the rate of the Mo system. In the catalytic system of cysteine and $NaBH_4$, tungsten, rhodium, rhenium, and ruthenium demonstrated the catalytic activity of approximate 7.0, 2.7, 2.0, and 1.5%, respectively, relative to the Mo system (9).

We recently studied Co(II) complexes containing cysteine- and cysteamine-related ligands that show potential as catalysts, and obtained results that are somewhat different from those mentioned above. The discrepancy presumably is attributable to the difference in pH of the reaction, concentration of the reagents, and molar ratio of the metal/ligand. No formation of finely divided metal was observed under the conditions used. Figure 1 and Table III show the yield and rate of acetylene reduction with the Co(II)–cysteine and –cysteamine ligand systems in the presence of sodium borohydride (10). These Co(II)-complex catalysts produce ethylene at a rate superior to that with the corresponding Mo-complex systems. One of the salient features is that the formation of 1,3-butadiene is negligible and the major product is ethylene. On the other hand, the major product from the acetylene reduction with the Mo(V)-cysteine catalysts in borate buffer is not ethylene but 1,3-butadiene (6). The effect of coordination donor atoms on the catalytic activity is S > Se ≫ O, consistent with that in the Mo–ligand systems. However, the Co(II) complexes of selenocysteine and selenocysteamine showed a higher ethylene–ethane product ratio than those of cysteine and cysteamine, though the total yield was lower. The activity of the Co(II)–cysteine and –cysteamine complexes

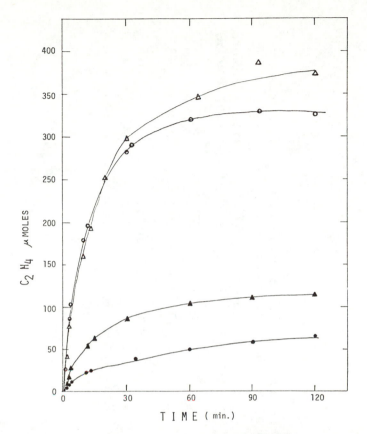

Figure 1. Catalytic activity of Co(II)–complex catalysts as compared with corresponding Mo catalysts: (△) Co–cysteine; (○) Co–cysteamine; (▲) Mo–cysteine; and (●) Mo–cysteamine

was maximum in the pH region 8.5–10.0, which is consistent with the optimum pH region (8.0–10.5) for the complexation of Co(II) complexes of these ligands. Effects with pH were parallel substantially, indicating that the complex formation plays a specific role in acetylene reduction with these Co(II)-complex systems. The results of visible and MCD spectroscopy suggest that pink-colored Co(II)–cysteine complex [500 nm(ϵ 76) and 570 nm($\theta + 0.33 \times 10^{-3}$ deg cm²/d mol)] and blue-colored Co(II)–cysteamine complex [670 nm(370) and 700 nm(-13.5×10^{-3})] have octahedral and tetrahedral geometries, respectively. The magnitude of the MCD bands associated with the *d–d* transition of the Co(II) chromophore is 50–100-fold larger when the Co(II) ion is in a tetrahedral ligand field rather than an octahedral one (*11*). The maximal amount of ethylene is formed at the Co(II)–cysteine ratio of 2:1. Probably, the low activity in the presence of excess cysteine is attributed to lack of efficient residual coordination sites of Co(II) for the substrate. These results suggest that the predominant

Table III. Yield and Rate of Ethylene and Ethane Produced from Acetylene with Co(II) Complexes of Cysteine, Cysteamine, and Their Related Ligands[a]

Ligand	C_2H_4 (μmol)	C_2H_6 (μmol)	Total Yield (μmol)	Relative Yield (%)	$\frac{C_2H_4}{C_2H_6}$	Rate ($\mu mol/min$)	Relative Rate (%)
Serine	4	0	4	0.8	—	0	0
Cysteine	428	92	519	100	4.7:1	47	100
Selenocysteine	220	14	234	45	15.7:1	25	52
Ethanolamine	3	0	3	0.6	—	0	0
Cysteamine	406	40	446	100	10.2:1	64	100
Selenocysteamine	154	5	159	36	30.8:1	38	60

[a]Yield of the products was obtained at reaction time of 30 min, and rate is represented in terms of $C_2H_4 + C_2H_6$/min at initial 5 min.

formation of ethylene by these Co(II)-complex systems is due to the monomeric Co(II) complexes as active species. Noteworthy is that the turnover number of acetylene by these Co(II)-complex catalysts is approximately 15, being higher than that (0.5) of the well known Mo–cysteine complex catalyst (*see* Table IV).

Table V summarizes the reduction activity of acetylene with the Co(II)-complex catalysts of various sulfhydryl-containing peptides. Of interest is the high activity of the sulfhydryl- and imidazole-containing peptides such as N-mercaptoacetyl-L-histidine and N-mercaptoacetyl-DL-histidyl-DL-histidine. In addition, the effect of amino-acid residues on the reduction of acetylene with these Co(II) complexes decreases in the order histidine > glycine > cysteine > tryptophan.

Other Effect. The product distribution is affected profoundly by changing the buffer ion from borate to carbonate (6). In borate buffer, the major reduction product of acetylene by the Mo–cysteine complex system is 1,3-butadiene, in contrast with ethylene in carbonate buffer. The product-distribution pattern was as follows: $C_2H_4 = 280.2$, $C_2H_6 = 21.0$, $C_4H_6 = 372.2$ μmol (in borate buffer); and $C_2H_4 = 264.2$, $C_2H_6 = 6.4$, $C_4H_6 = 82.8$ μmol (in carbonate buffer). These results are in agreement with those by Corbin et al. (*see* Table I) (6). However, further investigations are necessary to clarify the true nature of this buffer effect.

One characteristic feature of the molybdothiol complex catalysts is the stimulating effect of ATP on the reduction of acetylene (9, 12, 13). Recently, the effect of ATP as well as ADP was found to depend upon the pH of the ATP solution added to the reaction mixture. In addition, the effect of added H_2SO_4 is similar to that of ATP and ADP.

Table IV. Comparison of Some Characteristics of the Co–Cysteine and Co–Cysteamine Catalysts with Mo–Cysteine Catalyst and Nitrogenase

Characteristic	Co–Cysteine Co–Cysteamine	Mo–Cysteine	Nitrogenase
Acetylene reduction, turnover number[a]	13–16	0.5	150–200
N_2 reduction to NH_3, turnover number[a]	Unknown	1.5×10^{-6}	50
Enhancement factor of acetylene reduction by ATP	<1.5	10	>1000
Inhibition by CO	Weak	Weak	Strong
Inhibition by H_2	None	None	Competitive

[a] Turnover numbers defined as mole substrate reduced per mole metal in complex system per minute.

Table V.   Effect of Amino-Acid Residues on Acetylene Reduction

Parent Compound	R
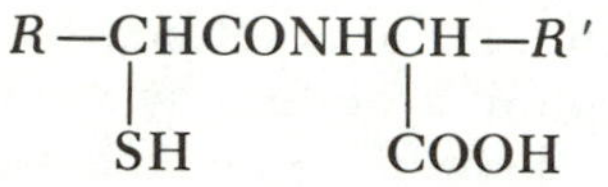	 R (phenyl ring)
	CH_3
	$(CH_3)_2CH$
$R-\underset{\underset{SH}{\|}}{CH}CONH\underset{\underset{COOH}{\|}}{CH}-R'$	CH_3
	H
	H
	CH_3
	$(CH_3)_2$
$R-\underset{\underset{SH}{\|}}{CH}CH_2CONH\underset{\underset{COOH}{\|}}{CH}-R'$	H
	H
$R-CH CONH CH CONH CHCH_2-$ (imidazole: N NH) $\quad R' \quad COOH$	H
	H

^{a} Yield of the products was obtained at reaction time of 30 min.

Therefore, this ATP-stimulation seems to be attributable to a nonspecific effect as a protic acid (14). Acid catalysis in various redox reactions is well known.

Figure 2 presents the relative effects of various charge carriers on the acetylene reduction by the Mo(V)–cysteine complex (15). The optimum enhancement of ethylene formation is observed for additives having reduction potentials around -0.9 V vs. SHE. The charge carriers such as 1,1'-trimethylene-2,2'-bipyridylium bromide and iron phthalocyanine-4,4',4'',4'''-tetrasulfonate sodium, enhance not only the charge transfer from the strong donors such as $NaBH_4$ and $Na_2S_2O_4$ to

with 2:1 Sulfhydryl-Containing Peptide–Co(II) Complex Catalysts

R'	Yield[a] $C_2H_4(\mu mol)$	$C_2H_6(\mu mol)$	Rate[b] $(\mu mol/min)$
H	154	60	19.7
H	141	35	13.7
H	59	5	9.0
 (phenyl)	159	21	23.0
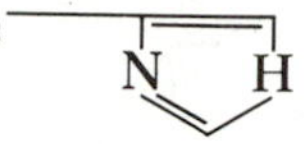 CH₂—(imidazole, N H)	256	49	27.5
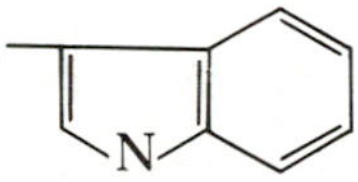 CH₂—(indole, N)	62	7	2.5
CH_2SH	83	8	6.4
CH_2SH	70	8	8.3
H	63	4	9.4
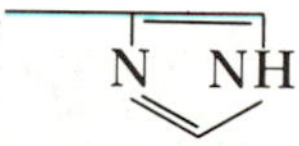 CH₂—(imidazole, N NH)	84	4	13.0
H	180	29	16.2
 CH₂—(imidazole, N NH)	291	81	47.0

[b] Rate is represented in terms of $C_2H_4 + C_2H_6/min$ at initial 5 min.

the Mo(V)–cysteine complex but also the activation of hydride in $NaBH_4$ solution to produce reactive hydrogen.

ESR Characteristics and Their Relationship to Catalytic Activity.   It is known that a diamagnetic Mo(V)–cysteine dimer dissociates slowly at pH range 6–10 to form a small amount of paramagnetic monomer. The preliminary ESR measurement of the isotopically enriched ^{95}Mo(V)–cysteine complex gave the following parameters: $g = 1.975$, $g_x = 1.931$, $g_y = 1.972$, $g_z = 2.029$, and $a = 35$ G; and the six-line hyperfine splitting follows a monomeric pattern (*16*). These results suggest that the dioxo bridge of the dimer can be broken by the

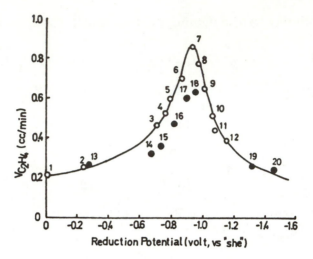

*Figure 2. The cocatalytic effects of various charge carriers on acety-
lene reduction by the Mo(V)–cysteine complex (0.42 mmol) and NaBH₄
(1.0 mmol), plotted against the reduction potentials (V vs. SHE) of the
corresponding additives.*

$V_{C_2H_4}$ *denotes the rate of ethylene formation when each charge carrier (0.17
mmol) was added to the reaction system. In the absence of the charge carriers,
the rate of ethylene formation was about 0.16 mL³/min in a solution of Na₂-
Mo₂O₄ (Cys)₂ and NaBH₄ under the similar reaction conditions. For the bi-
pyridyl diquaternary salts and phthalocyanines, the second-reduction potentials
are taken from the polarographic results: (1) Methylene Blue, (2) sodium riboflavin phosphate, (3) 1,1'-ethylene-4,4'-dimethyl-2-2'-bipyridylium bromide,
(4) 2,2'-bipyridylium chloride, (5) 1,2'-dimethyl-4,4'-bipyridylium bromide, (6)
2,2'-bipyridylium-1,1'-ethylene-5,5'-dimethyl bromide, (7) 1,1'-trimethylene-
2,2'-bipyridylium bromide, (8) iron phthalocyanine-4,4',4",4'''-tetrasulfonate
sodium, (9) copper phthalocyanine–4-(sodium sulfite), (10) metal-free phthalocyanine–4-(sodium sulfite), (11) Fe²⁺(bipy)₃SO₄, (12) sodium anthraquinone-1-
sulfonate, (13) CuSO₄, (14) NiCl₂, (15) CoCl₂, (16) CdSO₄, (17) FeSO₄, (18) RhCl₃,
(19) CrCl₃, (20) MnCl₂.*

attack of OH⁻ in aqueous solution. Recently, ESR data from well-
characterized monomeric Mo(V)-oxo complexes of *N,N'*-dimethyl-
N,N'-bis(2-mercaptoethyl)ethylenediamine(L_1) and *N,N'-bis*(2-mer-
captopropyl)ethylenediamine(L_2) have been reported by Spence et al.

$L_1: R_1{=}H, R_2{=}CH_3$
$L_2: R_1{=}CH_3, R_2{=}H$

(*17*), and could be used to obtain structural information and Mo-oxidation state by comparison. Several ESR features of our molybdothiol and molybdoselenol complexes, which were obtained in the presence of $NaBH_4$, were compared with those of the above-mentioned monomeric Mo(V)–cysteine complex, monomeric MoOClL complexes, and the Mo(V) enzyme (*see* Figure 3 and Table VI). These Mo complexes exhibit rhombic distortion in their ESR spectra, as do the monomeric Mo(V)-oxo complexes and xanthine oxidase. The anisotropic and average g-values are considerably higher than those of Mo(V) enzyme, and also the *a*-value is smaller, except for the cases of *N,N*-dimethylcysteamine and *N,N*-dimethylselenocysteamine. It might be possible to assign signals at $g = 1.97–1.98$ to Mo(V)-thiol complexes and at $g > 1.99$ to Mo(III)-thiol complexes, because a number of Mo(III)–sulfur complexes such as 2-mercaptoethanol and 3,4-dimercaptotoluene show g-values at 1.990–2.005 in solution (*18, 19*). Mitchell and Scarle also have proposed that electron delocalization is greater from Mo(III) to the sulfur ligand than from Mo(V) (*18, 19*). It is likely that sodium borohydride reduces the Mo(V)-thiol (or selenol) complexes to paramagnetic Mo(III)-thiol (or selenol) complexes. In the Mo complexes of cysteine and cysteamine, in fact, the same ESR spectra were obtained from the reaction of K_3MoCl_6 and the ligands. The close similarities of ESR parameters among the Mo complexes of cysteamine (or selenocysteamine) and its *β*-derivatives indicate strongly the formation of complexes with similar geometry and

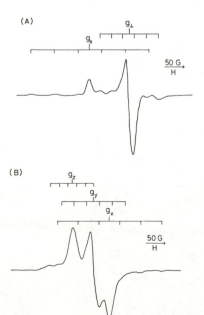

Figure 3. ESR spectra of Mo–cysteine complexes in borate buffer (pH 9.6) at 77 K: (A), cysteine + $MoCl_5$; (B), cysteine + $K_3 MoCl_6$ or (A) + $NaBH_4$

Table VI. ESR Parameters of Various Molybdothiol and Molybdoselenol Complexes

| Ligand | $g_\perp$ | | $g_\parallel$ | $g_{av}^{\ a}$ | $g_o^{\ a}$ | $a\,(G)$ | C_2H_4/C_2H_6 |
	g_x	g_y	g_z				
Cysteamine	1.965	1.991	2.025	1.994	1.994	27	13.8
N,N-Dimethylcysteamine	1.931	2.000	2.014	1.982	1.984	33	0.08
β,β-Dimethylcysteamine	1.960	1.993	2.022	1.992	1.992	26	9.0
L-Cysteine	1.967	1.992	2.024	1.994	1.993	35	13.3
Selenocysteamine	1.991	2.019	2.042	2.017	2.016	23	1.0
N-Methylselenocysteamine	1.992	1.992	2.027	2.004	2.005	—	0.2
N,N-Dimethylselenocysteamine	1.947	1.997	2.013	1.986	1.988	—	0.06
β-Methylselenocysteamine	1.991	2.014	2.036	2.014	2.013	25	0.9
β,β-Dimethylselenocysteamine	1.990	2.003	2.035	2.009	2.009	25	1.1
L-Selenocysteine	1.992	2.005	2.033	2.010	2.010	33	—
$MoOClL_1$	1.940	1.949	2.006	1.965	1.966	37.6	—
$MoOClL_2$	1.944	1.958	2.011	1.971	1.970	38.7	—
Xanthine Oxidase	1.951	1.956	2.025	1.977	—	31.7	—

^{a} The g_{av} and g_o values were obtained from $g_{av} = (g_x + g_y + g_z)/3$ and solution spectra (room temperature), respectively.

Mo oxidation state. These Mo catalytic systems show similar ethylene–ethane product ratios in the reduction of acetylene (*see* Table VI). The ESR result appears to correspond well to the recent electrochemical result (*20, 21*).

On the basis of the electrochemical reduction of a binuclear Mo(V)–cysteine complex and similar di-μ-oxo-bridged Mo(V) complexes, it has been demonstrated that these Mo(V) complexes are reduced in a single four-electron step at −1.2 to −1.3 V vs. SCE to Mo(III) products with no evidence of an intermediate oxidation state, and that a monomeric Mo(III) complex is the catalytically active species in the acetylene reduction with the molybdothiol complexes (*20, 21*). The definitive assignment for these ESR signals requires further investigations, however, because a Mo(III) oxidation state usually has a relatively short spin-lattice relaxation time and is oxidized easily to Mo(V) state. On the other hand, the *N,N*-dimethyl derivatives have largely different ESR parameters from those of the parent complexes, and these Mo complex systems reduce acetylene to form ethane as the major product, in contrast with the complex systems of cysteamine (or selenocysteamine) and its β-derivatives. In general, the ESR signals from the molybdoselenol complexes apparently have larger g-values and smaller a-values than those from the corresponding molybdothiol complexes. Thus, electron delocalization from Mo to Se donor is implicated as being more extensive than that from Mo to S donor, consistent with the lower electronegativity of Se donor atom.

In conclusion, subtle variations in the molybdothiol-complex catalysts—ligand substitution, coordination donor atom, and central metal ion—have large effects on the catalytic activity of acetylene reduction. The results presented here may offer useful information for a design of catalysts with optimal activity in such complex systems.

Literature Cited

1. Schrauzer, G. N. *J. Less-Common Met.* **1974**, *36*, 475.
2. Schrauzer, G. N. *Angew. Chem.* **1975**, *87*, 579.
3. Klayman, D. L. *J. Org. Chem.* **1965**, *30*, 2454.
4. Sugiura, Y.; Hirayama, Y. *J. Am. Chem. Soc.* **1977**, *99*, 1581.
5. Sugiura, Y. *Inorg. Chem.* **1978**, *17*, 2176.
6. Corbin, J. L.; Pariyadath, N.; Stiefel, E. I. *J. Am. Chem. Soc.* **1976**, *98*, 7862.
7. Sugiura, Y.; Kikuchi, T.; Tanaka, H. *J. Chem. Soc., Chem. Commun.* **1976**, 591.
8. Schrauzer, G. N.; Schlesinger, G. *J. Am. Chem. Soc.* **1970**, *92*, 1808.
9. Schrauzer, G. N.; Doemeny, P. A. *J. Am. Chem. Soc.* **1971**, *93*, 1608.
10. Sugiura, Y.; Kikuchi, T.; Tanaka, H. *J. Chem. Soc., Chem. Commun.* **1977**, 795.
11. Holmquist, B.; Kaden, T. A.; Vallee, B. L. *Biochemistry* **1975**, *14*, 1454.
12. Schrauzer, G. N.; Kiefer, G. W.; Doemeny, P. A.; Kisch, H. *J. Am. Chem. Soc.* **1973**, *95*, 5582.

13. Werner, D.; Russell, S. A.; Evans, H. J. *Proc. Natl. Acad. Sci. USA* **1973**, *70*, 339.
14. Khrushch, A. P.; Shilov, A. E.; Voronstsova, T. A. *J. Am. Chem. Soc.* **1974**, *96*, 4987.
15. Ichikawa, M.; Meshitsuka, S. *J. Am. Chem. Soc.* **1973**, *95*, 3411.
16. Huang, T. J.; Haight, G. P., Jr. *J. Am. Chem. Soc.* **1970**, *92*, 2336.
17. Spence, J. T.; Minelli, M.; Kroneck, P.; Scullane, M. I.; Chasteen, N. D. *J. Am. Chem. Soc.* **1978**, *100*, 8002.
18. Mitchell, P. C. H.; Scarle, R. D. *J. Less-Common Met.* **1974**, *36*, 265.
19. Mitchell, P. C. H.; Scarle, R. D. *J. Chem. Soc. Dalton* **1975**, 110.
20. Ott, V. R.; Schultz, F. A. *J. Electroanal. Chem.* **1975**, *61*, 81.
21. Ledwith, D. A.; Schultz, F. A. *J. Am. Chem. Soc.* **1975**, *97*, 6591.

RECEIVED May 15, 1979.

Metal Ion Effects in the Reactions of Phosphosulfate

W. TAGAKI[1] and T. EIKI

Gunma University, Kiryu, Gunma, Japan

The activation mechanism of phosphosulfate linkages (P—O—S) has been studied to understand the chemistry of biological sulfate-transfer reactions of phosphosulfates of adenosine (APS and PAPS). Several phosphosulfates were prepared and subjected to several nucleophilic reactions including hydrolysis. In general, phosphosulfates are stable in neutral aqueous media, but become labile under acidic conditions, resulting in selective S—O fission. This S—O fission appears to occur by unimolecular elimination of sulfur trioxide, which can react with a nucleophilic acceptor, leading to a sulfate-transfer reaction. This process can be accelerated by Mg²⁺ ion when the solvent is of low water content. Under neutral conditions, divalent metal ions also were found to catalyze nucleophilic reactions, but these occurred on phosphorus to result in exclusive P-O fission.

Inorganic sulfate is a very inert anion. It shows negligible nucleophilic reactivity and resists attack of various reagents; anhydrous sodium sulfate has been used for centuries as a safe drying agent in organic synthesis. Sulfate is also one of the most inert counteranions used to prepare reactive organic salts. Meanwhile, in the biosphere, plants and microbes utilize inorganic sulfate as the starting material to form various organic sulfur compounds including sulfur-containing amino acids and vitamins that are essential to our nutrition. The first steps in such sulfur metabolism are known to be the activa-

[1] Current address: Department of Applied Chemistry, Faculty of Engineering, Osaka City University, Sugimotocho, Sumiyoshiku, Osaka 558, Japan

0-8412-0514-0/80/33-191-409$05.00/0
© 1980 American Chemical Society

tion of sulfate by the reaction with ATP to form adenosine 5′-phosphosulfate (APS) and 3′-phosphoadenosine 5′-phosphosulfate (PAPS), catalyzed by the enzymes of ATP-sulfurylase and APS-kinase (Figure 1) (1, 2, 3). However, the mechanisms of catalysis by these enzymes are virtually unknown. For example, it may be puzzling how Mg^{2+} ion can activate ATP toward the attack of sulfate anion. Furthermore, the mechanism of subsequent reactions of phosphosulfates also have been left largely unclarified. In this chapter, we wish to discuss the mechanisms of the latter reactions based on our investigations into the reactions of some simpler phosphosulfates than APS and PAPS.

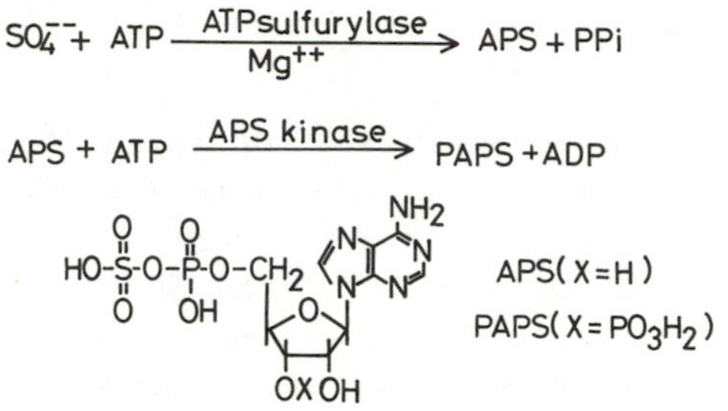

$$SO_4^{--} + ATP \xrightarrow[Mg^{++}]{ATPsulfurylase} APS + PPi$$

$$APS + ATP \xrightarrow{APS\ kinase} PAPS + ADP$$

APS (X = H)

PAPS (X = PO_3H_2)

Figure 1.   Enzyme activation of inorganic sulfate

Phosphosulfates may react with a nucleophile (Nu) in either of the two modes of P–O or S–O bond fission (Figure 2). If water is the nucleophile, both modes of fission result in the same hydrolysis products. Mechanistically, however, the enzymes that catalyze P—O fission may be regarded as phosphatases, while those that catalyze S—O fission are sulfohydrolases. In fact, many hydrolytic enzymes are assumed to be sulfohydrolases without mechanistic proof. The possibility that they might be phosphatases was suggested by Roy by taking account their metal ion dependency (4). Meanwhile, PAPS acts as the sulfate donor to numerous nucleophilic acceptors such as steroids and phenols. In such sulfate transfer reactions, S—O fission must occur. PAPS and APS also are known to act as the key intermediates in the reduction of sulfate to sulfite. Here again, the S—O fission may be the most probable mode.

Figure 2.   Two modes of P—O and S—O fission of the phosphosulfate linkage

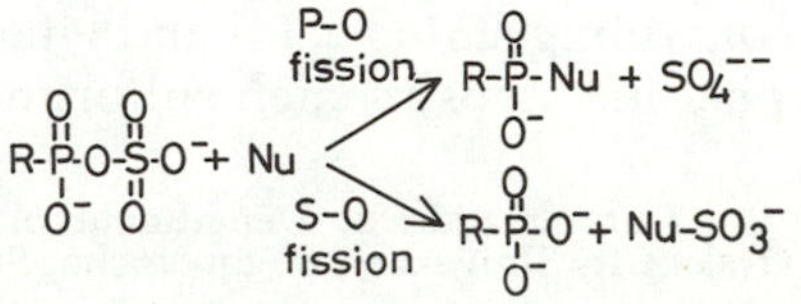

Hydrolysis of Phenylphosphosulfate

pH–Rate Profile and Predominant S—O Fission in Acid-Catalyzed Hydrolysis. Nine years ago, Benkovic and Hevey reported the hydrolysis of phenylphosphosulfate (PPS) as the first model system for the reaction of PAPS (5). Independently, we also reported similar results (6). PPS is acid labile; it is hydrolyzed to give phenylphosphate and inorganic sulfate with a half life of 3.4 min in 0.5N HCl at 30°C. By contrast, PPS is fairly stable in neutral aqueous and alkaline solutions; the half life is more than a month at pH 6–11 (30°C). Benkovic and Hevey proposed the mechanism shown in Figure 3 where the monoanion is the active species for hydrolysis at around pH 2.37 and undergoes unimolecular elimination to give sulfur trioxide. The S—O fission was confirmed by ^{18}O tracer experiments and by the formation of methylsulfate in aqueous methanol.

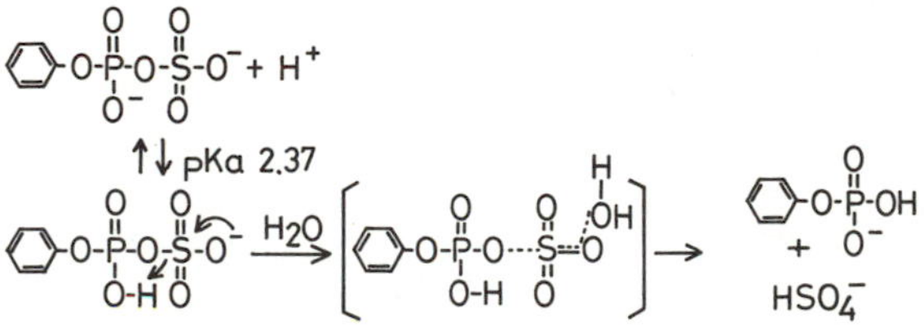

Figure 3.   Proposed mechanism for the acid-catalyzed hydrolysis of PPS (5)

METAL ION EFFECTS.   The metal ion effects on the acid-catalyzed hydrolysis of PPS also were examined by Benkovic and Hevey (5). However, they observed that in water near pH 3, the rate enhancement in the presence of an excess of metal ion was at most only threefold (Mg^{2+}, Ca^{2+}, Al^{3+}) and in some cases (Zn^{2+}, Co^{2+}, Cu^{2+}) the rate was actually retarded. We thought that the substrate PPS and Mg^{2+} ion should be hydrated heavily in water so that their complexation for rate enhancement is weak. If, however, the hydrolysis is carried out in a solvent of low water content, such complexation would not occur, and therefore, the rate enhancement might be more pronounced. This possibility appears to be supported by the fact that the active sites of many enzymes are hydrophobic. Of course, there is a possibility that the S—O fission may not require metal ion activation. In this connection, it is interesting to note that in biological phosphoryl-transfer reactions the enzymes generally require divalent metal ions for activity (7, 8, 9), but such metal ion dependency appears to be less important for sulfate-transfer enzymes. For example, many phosphatases require metal ions, but no sulfatase is known to be metal ion dependent (4). Nevertheless, it also should be noted that in a model reaction, Cu^{2+} ion enhances the rate of hydrolysis of 8-quinolylsulfate by 10^5–10^6-fold (10). The phosphosulfate linkage appears to be unique since it would serve as a measure of susceptibility of P–O and S–O bonds toward metal ion activation.

What actually is observed is shown in Figure 4a (11). Figure 4a shows a remarkable rate acceleration by Mg^{2+} ion in acetonitrile of low

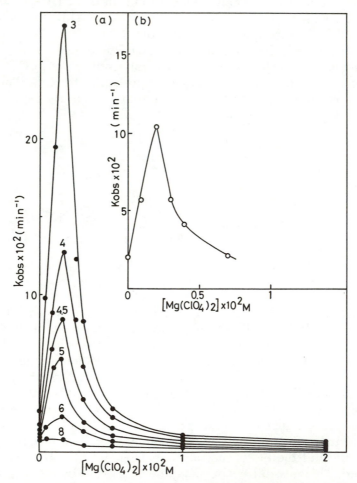

Figure 4. Mg^{2+} ion effect on the (a) hydrolysis (11) and (b) methanolysis of PPS:

(a)   [PPS] = 1.8 × 10⁻³ M,   [HClO₄] = 1.0 × 10⁻³ M,   [n-Bu₄ṄClO₄⁻] = 0.02 M, 25°C. The numbers in the plots are the molar concentrations of water in acetonitrile–water mixed solvent. (b) In acetonitrile containing methanol (1 M), water (3 M), PPS (2 × 10⁻³ M), HClO₄ (1 × 10⁻³ M), and n-Bu₄ṄClO₄⁻(0.03 M), 25°C. The k_{obs} values are for the formation of phenylphosphate. Methylsulfate was analyzed by NMR after the reaction mixture had been lyophilized (δ3.81(s) for CH₃).

water content. A greater effect can be seen for lower water content and there are rate maxima near 1:1 molar ratio of substrate and metal ion. Similar effects also were observed in the presence of methanol as shown in Figure 4b (12). In Figure 4b, the solvent is acetonitrile containing methanol (1M) and water (3M). From the reaction mixture at the rate maxima, methylsulfate was obtained in 60% yield, which again indicates an exclusive S–O bond fission under acidic conditions. This finding appears to be important since it indicates that Mg^{2+} ion enhances the rate of S—O fission.

The rate maxima of Figure 4 suggest that the substrate can be activated through the formation of a 1:1 complex with metal ion. Unfortunately, it was difficult to determine whether chelation occurs on the P—O group, the S—O group, or both. Another problem was the complicated dependency of this metal ion effect on acid concentration. Furthermore examination of this system under neutral conditions was not successful and failed to give reproducible results, because of the very slow rate even in the presence of Mg^{2+} ion.

As described above, the phosphosulfate linkage is acid labile but stable under neutral conditions. We, therefore, directed our next efforts to activating it under neutral conditions, whatever the site of bond fission. One obvious way was to use metal ions. However, it was evident that the effect of metal ion alone was small. Therefore, we examined the effects of amines that might act as nucleophiles cooperatively with metal ion.

Aminolysis of Phosphosulfates

Aminolysis of phenylphosphosulfate (PPS) by aliphatic amines occurred with exclusive P—O fission (70–85%) to give phenylphosphoramidate and inorganic sulfate as the major products (Figure 5)

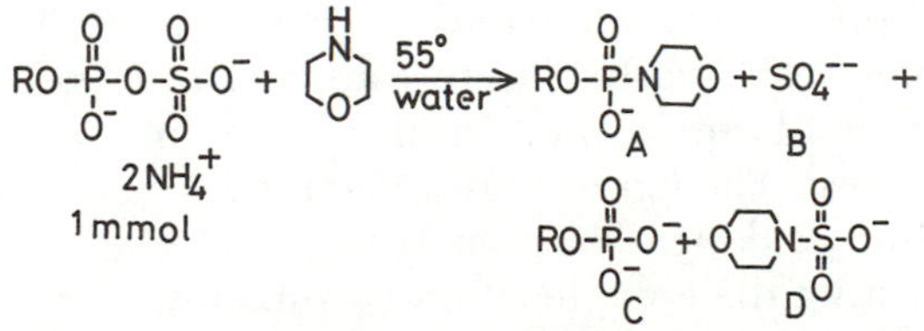

R	Mg(ClO₄)₂	A	B	C	D
⬡-	0 mmol	80	80	20	20
	10	100	100	—	—
⬡CH₂CH₂CH₂-	10	100	100	—	—

Figure 5. Aminolysis of phosphosulfates

(13). For aromatic amines (pyridines and imidazole), phosphoramidates were not detected in water. Yet the reaction appears to be essentially the same as in the former cases, because the formation of phenyl-methylphosphate, which was observed in methanol, was presumed to occur by the methanolysis of phosphoramidate.

In Figure 6 are shown the relation between the rates of formation of inorganic sulfate and pK_a of amines. The slope of the line is small (i.e., $\beta = 0.15$). Such a small β-value also is observed for other related systems *(14, 15)*. Evidence for nucleophilic reaction also is obtained from the observation that N-ethylmorpholine and 2,6-lutidine are virtually unreactive toward PPS.

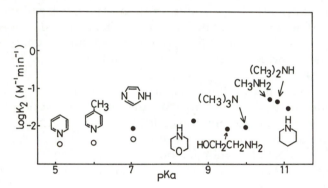

Figure 6.   Brönsted plots for the aminolysis of PPS (55°C in water).

The k_2 values were obtained as the slopes of linear plots of k_{obs} for the formation of inorganic sulfate vs. free amine concentration. Points of open circles were obtained in 0.2M N-ethylmorpholine buffer. N-Ethylmorpholine and 2,6-lutidine showed no detectable reaction.

Metal Ion effect.   Divalent metal ions (Mg^{2+}, Co^{2+}, Zn^{2+}, Ni^{2+}) catalyzed the above nucleophilic reaction on the phosphorus atom. The yields of phosphoramidate and inorganic sulfate become 100% in the presence of Mg^{2+} ion. In Figure 7 are shown the effect of Mg^{2+} ion on the reaction of imidazole. In this case, imidazole acts as nucleophilic catalyst. The figure shows that the k_{obs} values tend to deviate from the first-order dependency on Mg^{2+} ion at high metal ion concentration, suggesting the complexation of substrate with metal ion.

Meanwhile, PPS liberated phenol in a Tris buffer solution *(16)*. Ethanolamine works similarly and the reaction proceeded first with the expulsion of sulfate by nucleophilic attack of amine followed by liberation of phenol, presumably through the formation of a five-membered cyclic intermediate (Figure 8). Both steps of SO_4^{2-} and phenolate liberation were found to be catalyzed by Mg^{2+} ion. In Figures 9 and 10 the data are shown for the liberation of phenol. Figure 9 indicates clear rate saturation attributable to complexation in acetonit-

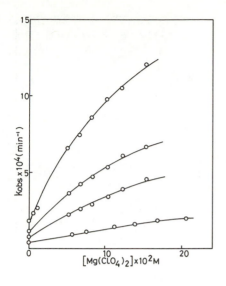

Figure 7. Mg²⁺ ion effect on the imidazole-catalyzed hydrolysis of PPS: [PPS] = 2 × 10⁻²M, N-ethyl-morpholine buffer (0.2M, pH 8), and 55°C. Imidazole concentrations are, from the bottom: 0.0, 0.01, 0.02, and 0.04M.

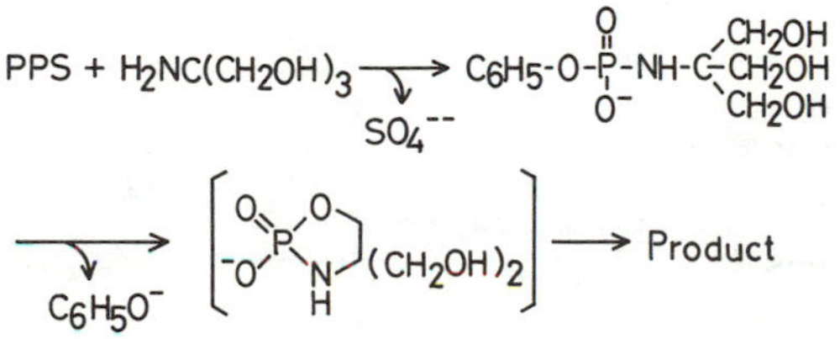

Figure 8. Proposed mechanism for the reaction of PPS with Tris

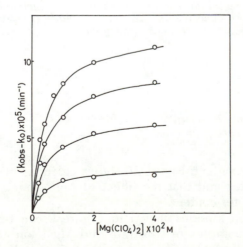

Figure 9. Mg²⁺ ion effect on the Tris-promoted liberation of phenol from PPS in water(7M)–acetonitrile, 55°C: [PPS] = 2 × 10⁻³M. Tris concentrations are, from the bottom: 0.01, 0.021, 0.031, and 0.04M.

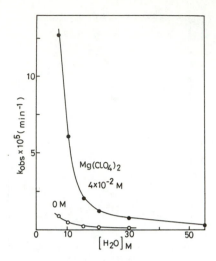

Figure 10.   Effect of water concen-
tration on the reaction rate of PPS
and Tris (see Figure 9)

$$S + Mg^{++} \underset{\longleftarrow}{\overset{K_m}{\rightleftharpoons}} Sm$$

$$Sm + B \xrightarrow{k_B^m} P$$

$$S + B \xrightarrow{k_B} P$$

$$k_{obs} = \frac{k_B[B]_T + k_B^m K_m[B]_T[Mg^{++}]_T}{1 + K_m[Mg^{++}]_T} \qquad \left([Mg^{++}]_T \gg [S]_T\right)$$

$$\frac{1}{(k_{obs} - k_B[B]_T)} = \frac{1}{(k_B^m - k_B)[B]_T} + \frac{1}{(k_B^m - k_B)K_m[B]_T[Mg^{++}]_T}$$

Figure 11.   Proposed scheme and rate equations for the reaction of phosphosulfate–metal-ion complex with base or nucleophile. Several variations may be conceivable such as to include a substrate–metal-ion–reagent ternary complex.

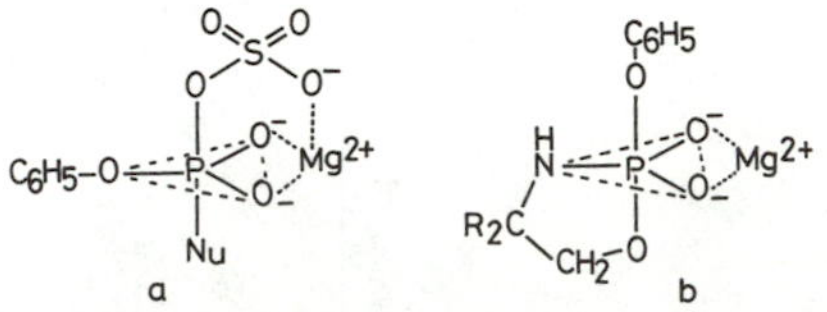

Figure   12.   Proposed   metal-ion-stabilized   pentacovalent   interme-diates

rile of low water content. Figure 10 shows again, as in Figure 4, that the rates are faster and that the effect of Mg^{2+} ion is larger in the solvent of low water content.

The saturation kinetics shown in Figure 9 can be analyzed by assuming a 1:1 complex of substrate and Mg^{2+} ion as (Sm) shown in the reaction scheme of Figure 11, and by using the derived rate equations. The following values were obtained (554°C, 7M water in acetonitrile): $k_B = 2.04 \times 10^{-4}$ ($M^{-1}min^{-1}$), $k_B^m = 3.24 \times 10^{-3}$

$(M^{-1}\text{min}^{-1})$, and $K_m = 225\ (M^{-1})$. These results indicate a fairly large association constant (K_m) for the Sm complex and a rate acceleration of sixteenfold (k_B^m/k_B) by Mg^{2+} ion. These values may become larger as the water content of solvent becomes lower.

A plausible rationale for the above Mg^{2+} ion catalysis may be electrostatic facilitation of nucleophilic attack to form a pentacovalent intermediate as depicted in the structures of Figure 12. This rationale already has been proposed by Steffens et al. for the metal ion catalysis in the intramolecular displacement reaction of phenolate ion by carboxylate (17).

Catalysis of Zn²⁺–Pyridine-2-Carboxaldoxime Complex

In the above aminolysis reactions, Mg^{2+} ion always activates P—O fission. Other divalent metal ions behave similarly. Change of nucleophile might result in the change of P—O to S—O fission. To pursue the possibility of metal ion catalyzed S—O fission under neutral conditions, we examined the catalysis of the Zn^{2+}–pyridine-2-carboxaldoxime ($Zn^{2+}PCA$) complex, since catalytic activity of this complex has been well known in some acyl and phosphoryl transfer reactions (18, 19, 20). However, the results described below again indicate that the complex promotes an exclusive P–O bond fission (21).

Predominant P—O Fission.   In the absence of Zn^{2+} ion, the reactions of PPS and PCA were very slow. Therefore, Zn^{2+} ion is essential for faster reaction. The kinetics described later indicate that the reaction proceeds through the formation of ternary complex (A) as illustrated in Figure 13. The oximate anion in A may either attack phosphorus (Path a) or sulfur (Path b). Inorganic sulfate was obtained quantitatively. This itself is not proof of Path a, because C (prepared separately) was found to be hydrolyzed readily to give sulfate under the same reaction conditions. However, the other isolated major product was B instead of the oxime catalyst that would be regenerated from C. The product B gave methylphenylphosphate when solvolyzed in methanol in the presence of Zn^{2+} ion. Methylphenylphosphate also was obtained directly from A in the reaction in methanol, whereas the formation of methylsulfate was not detected. Thus, these results all indicate that the $Zn^{2+}PCA$ complex promotes predominant P—O fission.

Kinetics.   The pH–rate profile shown in Figure 14a indicates that the PCA anion is the active nucleophile. On the other hand, the saturation kinetics of Figure 14b demand a complexation mechanism such as that postulated in Figure 11, or the ternary complex A in Figure 13. Cooperman and Lloyd already have discovered that Zn^{2+} catalyzes phosphoryl transfer from phosphoryl imidazole to the oxygen atom of PCA, just as in Path a of Figure 13 (cf. Figure 15) (22). They observed a

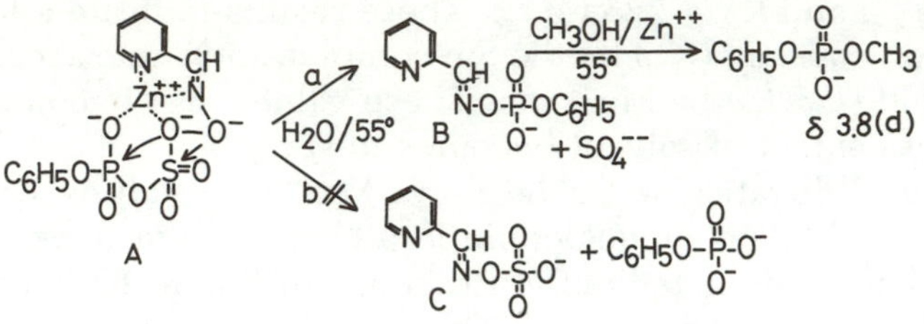

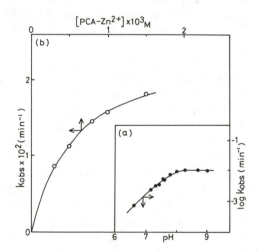

Figure 13. *Reaction of PPS with Zn²⁺ PCA complex.*

PPS (10 mmol) and Zn²⁺PCA (10 mmol) were reacted for 8 hr, 55°C, in a buffer solution of N-ethylmorpholine (pH 8–9). Inorganic sulfate was isolated as BaSO₄ (>90% yield). B was characterized by UV and NMR spectra and was estimated to be formed quantitatively.

Figure 14. *Reaction kinetics of PPS and Zn²⁺PCA complex, 55°C. (a) pH–rate profile: [PPS] = 1 × 10⁻²M, [Zn²⁺PCA] = 5 × 10⁻⁴M, N-ethylmorpholine/HClO₄ buffer (0.1M). The rate was determined by following consumption of PCA (310 nm) spectrophotometrically. (b) Effect of complex concentration: [PPS] = 1 × 10⁻²M, pH 8.7.*

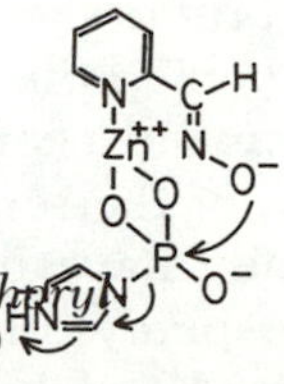

Figure 15. *Proposed phosphoryl transfer mechanism (19)*

bell-shaped pH–rate profile with a maximum at pH 6, which appears to differ from that in Figure 14a. Such difference likely is caused by the difference in pK_a of the leaving group (sulfate dianion vs. neutral imidazole).

To compare the nucleophilic reactivities of the systems described above, the saturation curves of Figures 7 and 14b were analyzed based on the scheme in Figure 11; the calculated data are shown in Table 1.

Table I.   Nucleophilic Reactivities for the P—O Fission of Phenylphosphosulfate (PPS) in Aqueous Solution (55°C)

Reaction	$k_2(M^{-1}min^{-1})$	k_{rel}
PPS + PCA	—[a]	<1
PPS + OH⁻	6.72×10^{-3}	1
PPS + imidazole	4.72×10^{-3} [b]	0.7
PPS + Mg²⁺ + imidazole	1.94×10^{-1} [c]	28.9
PPS + Zn²⁺ + PCA	5.01^c	746

[a] PPS = PCA = $0.01M$, pH 8; less than 1% reaction after 100 hr.
[b] k_B, *see* Figure 11.
[c] k_B^m, *see* Figure 11.

These data indicate a very large metal ion effect. Thus, the nucleophilic reactivities of imidazole and oxime anion toward the phosphoryl group are enhanced by Mg²⁺ to 41- and more than 746-fold, respectively, in water. The combined effect of metal ion and nucleophile may amount to some 10^4–10^6-fold rate enhancement when compared with the water rate ($\sim 10^{-5}$ min⁻¹). Such effects would be even larger in the solvents of low water content and they may be very good models of phosphatase activity. However, they do not answer our question for the activation mechanism of S—O fission under neutral conditions.

Further Search for Selective S—O Fission

Reactions of 2-Pyridylmethylphosphosulfate (23).   As illustrated in Figure 16, two modes of S—O fission are conceivable for the reaction of 2-pyridylmethylphosphosulfate (PMPS): (a) A divalent metal ion may bridge pyridyl and phosphoryl groups, leaving the sulfate group free. Such chelation would lower the pK_a of the leaving phosphoryl group and assist S—O fission by either unimolecular or

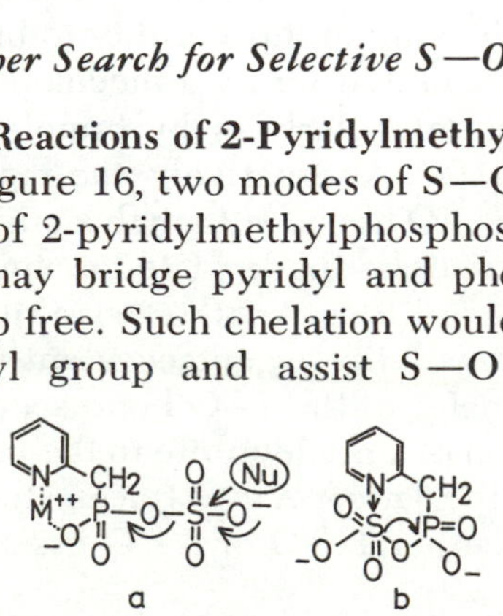

Figure 16.   Two possible modes of S—O fission of PMPS

bimolecular nucleophilic process. (b) In the absence of metal ion, the pyridyl nitrogen may attack at sulfur, resulting in intramolecular S—O fission, although S_N2 type linear transition state of N $\cdots$ S $\cdots$ O would produce unfavorable strain energy.

pH–Rate Profile.   In Figure 17 are shown the pH–rate profiles for the hydrolyses of PMPS, benzylphosphosulfate (BPS), and PPS observed in the absence of metal ion. In the acid pH region, the reactivities of these three esters do not differ greatly. This trend is similar to our previous observation that the substituent effect on the phenyl ring is small in the acid hydrolysis of PPS (Hammett $\rho = +0.22$) (6). Although the difference in reactivity is small, the order of PPS > PMPS > BPS may be significant, since the order appears to be that of the leaving group's ability to bridge the phosphoryl moiety when S—O fission occurs. In accordance with this reasoning, methylsulfate was observed to be the major product for all three esters when the acid hydrolysis was carried out in aqueous methanol.

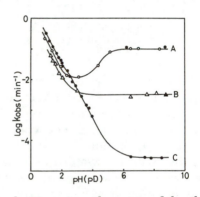

Figure 17.   log k_{obs}–pH profiles for the hydrolyses (formation of inorganic sulfate) of: A (○), PMPS; B (△), BPS; C (●), PPS; 55°C, $\mu = 1.0$ (KCl). Buffers (0.1 M): HCl–KCl (pH 1–2), ClCH$_2$CO$_2$H–ClCH$_2$CO$_2$Na (pH 2–3), HCO$_2$H–HCO$_2$Na (pH 3–4), CH$_3$CO$_2$H–CH$_3$CO$_2$Na (pH 4–6), N-ethylmorpholine (pH 7). Filled points on A and B plots were obtained in D$_2$O.

On the other hand, in the neutral pH region, the rates of the three esters are widely different, spanning more than a 10^3-fold range. The reactivity order is also changed to PMPS > BPS > PPS. The order of BPS > PPS, which is contrary to the leaving group's ability to bridge the phosphoryl moiety, may not be accounted for by a mechanism of S—O fission, whether it happens unimolecularly or by bimolecular nucleophilic attack on sulfur (Figure 16a). Alternatively, the reaction may occur on phosphorus to result in P—O fission, although a one-step S_N2 type expulsion of sulfate dianion may be inadequate to explain a 100-fold rate difference between BPS and PPS. Another possibility is addition–elimination on phosphorus involving a pentacovalent intermediate. Since the double-bond character of the P–O bond seems to be larger for BPS than for PPS, addition of a nucleophile to the former P–O bond would be easier than to the latter. A ^{18}O tracer study is necessary to settle the question of whether P—O or S—O fission occurs.

Another interesting feature of Figure 17 is a sigmoid pH–rate profile for PMPS, with an inflection at around pH 5 that corresponds to the ionization of the pyridinium cation. Thus, as illustrated in Figure 18, the neutral pyridyl group appears to act either as an intramolecular general base or as a nucleophilic catalyst. Actually, it was observed that the solvent isotope effect is close to unity ($k_{H_2O}/k_{D_2O} = 0.90$), which suggests nucleophilic catalysis. In such nucleophilic catalysis, the pyridyl nitrogen may attack either sulfur or phosphorus. In the former case, the intermediate pyridinium sulfate would act as a sulfate donor to various sulfate acceptors. If this occurs, it would provide an interesting model of sulfate-transfer enzymes. However, our attempts to trap sulfate in the presence of a large excess of acceptor such as morpholine, phenylphosphate, or methanol have all been unsuccessful. On the contrary, in aqueous methanol, the major products were found to be methylpyridylmethylphosphonate and inorganic sulfate, which indicates an exclusive P—O fission (Figure 19). The solvent isotope

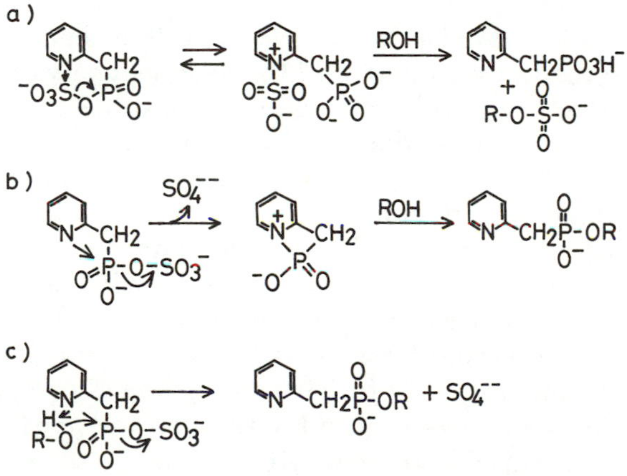

Figure 18.   *Possible intramolecular catalyses by pyridyl group*

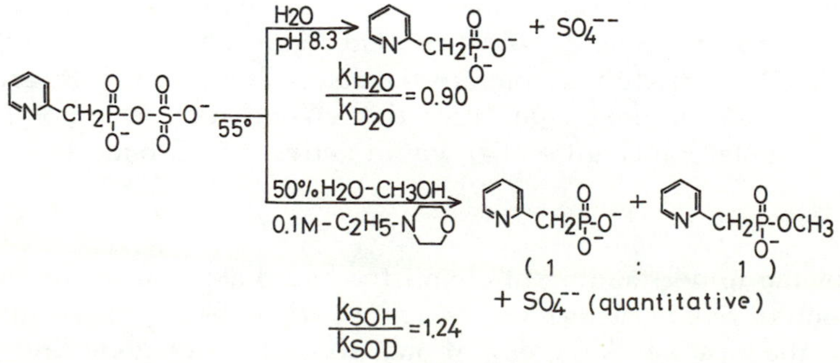

Figure 19.   *Products and solvent deuterium-isotope effect* (see Figure 17)

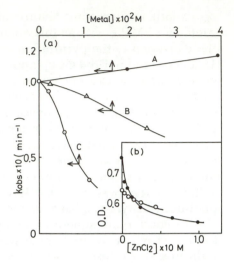

Figure 20. Metal ion effect on the hydrolysis of PMPS. (a) Effect on the rates, pH 5.33, 55°C: (A) Mg-$(ClO_4)_2$; (B) $Zn(ClO_4)_2$; (C) Cu-$(ClO_4)_2$. (b) Change of optical density (O.D.): (○) [PMPS] = 1 × 10^{-4}M (264 nm); (●) [PMP] = 1 × 10^{-4}M (268 nm); μ = 1.0 (KCl), 25°C.

effect again was found to be near unity. Therefore, a mechanism in accordance with the isotope effect and the product analysis seems to be nucleophilic attack of pyridyl nitrogen on phosphorus to form a four-membered cyclic intermediate. However, catalysis involving a four-membered ring may be unlikely because of unfavorable ring strain. Unfortunately, at the moment we do not have further evidence to prove or disprove such a cyclic intermediate.

Metal Ion Effect. Complexation of PMPS with Zn^{2+} ion could be observed by the decrease of absorption intensity of the pyridine ring (Figure 20b). A similar decrease was observed for the parent pyridyl-methylphosphonate (Figure 20b). Therefore, the 1:1 complex in which a metal ion is bridged between pyridyl nitrogen and phosphoryl oxyanion likely is involved (Figure 16a). Meanwhile, no detectable change of spectrum was observed in the case of Mg^{2+} ion, in accordance with very weak chelation of Mg^{2+} ion with a pyridine base.

The effects of these metal ions on the rates of hydrolysis are shown in Figure 20a. It is clear that Zn^{2+} and Cu^{2+} ions inhibit the hydrolysis, while Mg^{2+} ion slightly accelerate the rate. Thus, the results are the opposite of our original expectation that chelation of phosphoryl oxygen with metal ion (Figure 16a) would activate S–O bond fission.

Final Remarks

For the understanding of chemistry of sulfate transfer or sulfate reduction in biological systems, we must know the expulsion mechanism of the terminal SO_3 group of phosphosulfates of adenosine (APS and PAPS). Model studies have disclosed that it can occur by acid catalysis with a mechanism of unimolecular elimination of sulfur trioxide (Figure 3). This acid catalysis can be enhanced by divalent metal ions when the reaction occurs in a solvent of low water content. One may consider such a metal ion effect to be electrophilic catalysis, as illustrated in Figure 21a. However, this metal ion effect is highly acid dependent; under neutral conditions such a metal-ion assisted S—O fission was not detected with divalent metal ions (Mg^{2+}, Zn^{2+}, Ni^{2+}, Co^{2+}). Therefore, the results tentatively may be accounted for by a mechanism of metal-ion assisted, proton-transfer catalysis as illustrated in Figure 21b. This catalysis of proton transfer to the leaving phosphoryl group is essentially the same as that considered for the acid catalysis by a neighboring carboxyl or imidazolium group in the hydrolysis of aryl sulfates (*24*, *25*). In this connection, it is interesting to note that electrophilic catalysis also was proposed in Cu^{2+}-ion-catalyzed hydrolysis of 8-quinolyl sulfate (*10*) as shown in Figure 22.

Acid-catalyzed S—O fission may occur under neutral conditions under the influence of enzymic catalysis. Perhaps a functional carboxyl group obtains at the active site of the enzyme, a group which occurs in undissociated acid form and works as a powerful acid catalyst because of its location at a special hydrophobic pocket. However, we have examined other possibilities of metal-ion-catalyzed S—O fission.

What we found is that all metal ions catalyze P—O fission. Selective P—O fission by amines was increased from 80% to 100% in the presence of Mg^{2+} ion, which also enhanced the rate. Exclusive P—O fission also occurred in the attack by the oxyanion of PCA in the presence of Zn^{2+} ion. A plausible rationale is that such a path, which involves metal ion assistance in a pentacovalent intermediate as illustrated in Figure 12a, is energetically much more favorable than that of S_N2 displacement on sulfur. Conversely, if an enzyme that catalyzes the reaction of phosphosulfate is metal ion dependent, the reaction probably involves P—O fission, as suggested by Roy (*4*).

Inhibition, rather than acceleration, of PMPS hydrolysis by Zn^{2+} and Cu^{2+} ions (Figure 20a) seems to be important, since it suggests the unimportance of electrophilic catalysis for S–O bond fission, as expected in Figure 16a. Related observations already have been reported by Cooperman in the hydrolysis of some polyphosphates. He

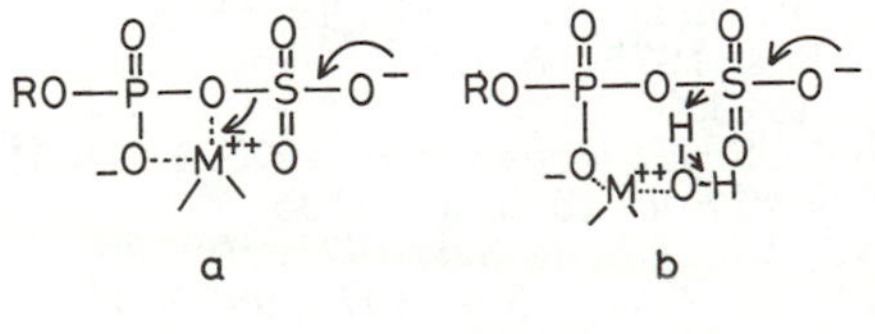

Figure 21. Metal ion effect on the acid hydrolysis of phosphosulfate: (a) electrophilic catalysis and (b) metal-ion-assisted, proton-transfer catalysis

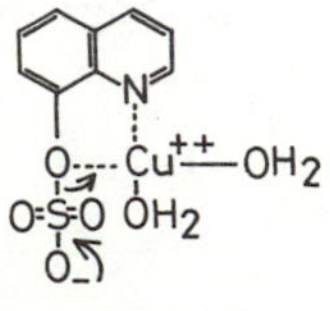

Figure 22. Proposed Cu^{2+} ion effect on the hydrolysis of 8-quinolylsulfate (10)

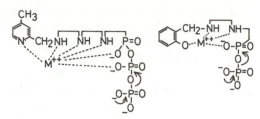

Figure 23. Hypothetical metal ion effect on the hydrolysis of poly-phosphate (26).

attempted and failed to demonstrate metal-ion-catalyzed elimination of metaphosphate anion from the terminus of a polyphosphate chain, as illustrated in Figure 23 (*26*).

Finally, we must emphasize again that many things are left to be clarified with respect to the role of divalent metal ions.

Acknowledgment

The authors thank Y. Asai, K. Tomuro, S. Aoshima, M. Suda, and S. Kawada for their collaboration as students in carrying out this research.

Literature Cited

1. Lipmann, F. *Science* **1958,** *128,* 575.
2. Roy, A. B.; Trudinger, P. A. "The Biochemistry of Inorganic Compounds of Sulphur"; Cambridge Univ. Press: Cambridge, 1970.
3. Schiff, J. A.; Hodson, R. C. "The Metabolism of Sulfate,"*Annu. Rev. Plant Physiol.* **1973,** *24,* 381.
4. Roy, A. B. *Enzyme,* **1971,** 8, 1.
5. Benkovic, S. J.; Hevey, R. C. *J. Am. Chem. Soc.* **1970,** 92, 4971.
6. Tagaki, W.; Eiki, T.; Tanaka, I. *Bull. Chem. Soc. Jpn.* **1971,** *44,* 1139.
7. Benkovic, S. J.; Schray, K. J. *Enzyme* **1973,** 8, 201.
8. Cooperman, B. S. "Metal Ions in Biological Systems"; Sigel, H., Ed.; Marcel Dekker: New York, 1976; Vol. V.
9. Kluger, R. "Bioorganic Chemistry"; Tamelen, V., Ed.; Academic Press: New York, 1978; Vol. IV.
10. Hay, R. W.; Clark, C. R.; Edmons, J. A. G. *J. Chem. Soc. Dalton Trans.* **1974,** 9.
11. Tagaki, W.; Asai, Y.; Eiki, T. *J. Am. Chem. Soc.* **1973,** 95, 3037.
12. Eiki, T.; Tagaki, W., unpublished work.
13. Tagaki, W.; Eiki, T.; Tomuro, K., presented at the 4th annual meeting of *Enzyme Simulated Organic Reaction,* Tokyo, 1976.
14. Benkovic, S. J.; Benkovic, P. A. *J. Am. Chem. Soc.* **1966,** 88, 5505.
15. Williams, A.; Douglas, K. T. *Chem. Rev.* **1975,** 75, 627.
16. Asai, Y., Thesis, Gunma University, 1973.
17. Steffens, J. J.; Siewers, I. J.; Benkovic, S. J. *Biochemistry* **1975,** *14,* 2431.
18. Breslow, R.; Chipman, D. *J. Am. Chem. Soc.* **1965,** 87, 4195.
19. Breslow, R. "Bioorganic Chemistry," *Adv. Chem. Ser.* **1971,** *100,* 20.
20. Hsu, Chih-Min; Cooperman, B. S. *J. Am. Chem. Soc.* **1976,** 98, 5657.
21. Tomuro, K. Thesis, Gunma University, 1976.
22. Lloyd, G. L.; Cooperman, B. S. *J. Am. Chem. Soc.* **1971,** 93, 4883.
23. Eiki, T.; Kawada, S.; Tagaki, W., unpublished work.
24. Benkovic, S. J. *J. Am. Chem. Soc.* **1966,** 88, 5511.
25. Benkovic, S. J.; Dunikowski, L. K. *Biochemistry* **1970,** 9, 1930.
26. Cooperman, B. S. *Biochemistry,* **1969,** 8, 5005.

RECEIVED May 15, 1979.

INDEX

INDEX

Jacket design by Carol Conway.
Editing and production by Robin Allison.

The book was composed by Bi-Comp, Inc., York, PA,
printed and bound by The Maple Press Co., York, PA.